Organometallics of the f-Elements

NATO ADVANCED STUDY INSTITUTES SERIES

Proceedings of the Advanced Study Institute Programme, which aims
at the dissemination of advanced knowledge and
the formation of contacts among scientists from different countries

The series is published by an international board of publishers in conjunction
with NATO Scientific Affairs Division

A	Life Sciences	Plenum Publishing Corporation
B	Physics	London and New York
C	Mathematical and Physical Sciences	D. Reidel Publishing Company Dordrecht, Boston and London
D	Behavioral and Social Sciences	Sijthoff International Publishing Company Leiden
E	Applied Sciences	Noordhoff International Publishing Leiden

Series C – Mathematical and Physical Sciences

Volume 44 – Organometallics of the f-Elements

Organometallics of the f-Elements

Proceedings of the NATO Advanced Study Institute
held at Sogesta, Urbino, Italy, September 11-22, 1978

edited by

TOBIN J. MARKS
Department of Chemistry, Northwestern University,
Evanston, Illinois, U.S.A.

and

R. DIETER FISCHER
Institut für Anorganische und Angewandte Chemie,
Universität Hamburg, Hamburg, West Germany

D. Reidel Publishing Company

Dordrecht : Holland / Boston : U.S.A. / London : England

Published in cooperation with NATO Scientific Affairs Division

Library of Congress Cataloging in Publication Data

Nato Advanced Study Institute, Sogesta, Italy, 1978.

    Organometallics of the f-elements.

    (NATO advanced study institutes series: Series C, Mathematical and physical sciences; v. 44)
    Includes index.
    1. Organometallic compounds—Congresses. 2. Rare earth metal compounds—Congresses. 3. Actinide elements—Congresses. I. Marks, Tobin, J. II. Fischer, Rainer Dieter, 1919–   III. Title. IV. Series.
QD410.N37 1978   547′.05′4   79-11871
ISBN-13: 978-94-009-9456-0   e-ISBN-13: 978-94-009-9454-6
DOI: 10.1007/978-94-009-9454-6

Published by D. Reidel Publishing Company
P.O. Box 17, Dordrecht, Holland

Sold and distributed in the U.S.A., Canada, and Mexico
by D. Reidel Publishing Company, Inc.
Lincoln Building, 160 Old Derby Street, Hingham, Mass. 02043, U.S.A.

TABLE OF CONTENTS

While the organometallic chemistry of the d-block transition elements has been a flourishing field for the past 25 years, it has only been in the last several years that dramatic activity and progress has occurred in the area of lanthanide and actinide organometallic chemistry. The f-element organometallic research effort has been truly multinational and multidisciplinary. In a large number of countries, scientists have become increasingly interested in the synthesis, reactivity, spectroscopy, and the molecular and electronic structures of f-element organometallic compounds. The backgrounds of these scientists range from organic, inorganic, nuclear, and catalytic chemistry to chemical and nuclear physics. The motivations for the study of f-element organometallics have been equally varied. In the area of basic research, there has been a growing realization that the lanthanides and actinides represent two unique and, to a great extent, neglected families of elements in which many fascinating aspects of chemistry and bonding remain to be explored. On a more practical level, an increasing number of these elements play important roles in nuclear energy production and in industrial catalytic processes. It has become apparent that efficiency and safety in both areas could greatly benefit from increased knowledge.

In the past there has been no suitable international forum available for bringing together researchers in the diverse areas of f-element organometallic science mentioned above. This state of affairs has had a deleterious effect on the balanced growth and full development of this rapidly expanding field. It was the belief of many of us working in this

field that increased communication and interdisciplinary interaction were
needed, and in the fall of 1976 we decided to organize a NATO Advanced
Study Institute on the topic "Organometallics of the f-Elements." The
most distinctive goal of the Institute was to develop a more complete and
coherent understanding of the chemical and physical properties of organo-
f-element compounds both among the participants, and ultimately,
within the scientific community as a whole. This Institute took place
at the ultramodern SOGESTA Conference Center near Urbino, Italy from
September 11 to September 22, 1978. It assembled highly qualified
scientific specialists and students from a large number of countries
and sought to develop via a course of lectures, seminars, and discussions,
a comprehensive and meaningful picture of the current state of knowledge about f-element organometallic compounds. It was the enthusiastic
consensus of the participants that this goal was achieved.

The chapters assembled in this volume constitute the principal
lectures in the "Organometallics of the f-Elements" Advanced Study
Institute. They are arranged in approximately the order in which
they were presented, and endeavor to provide as clear and as complete
a discussion as space and the current state of our knowledge will allow.
Topics which build upon each other have been arranged so that the
subject matter progresses from the general to the specific, and from the
theory to its application. The authors were given a great degree of
freedom in the preparation of their chapters, so that vital personal
viewpoints would be preserved and much heretofore unpublished

information could be included. Besides the principal lectures, the program of the Institute included introductory lectures on f-elements and organometallic compounds, as well as a number of shorter, contributed seminars. The titles and authors of these contributions are contained in the Appendix.

Both the Institute and this volume would not have been possible without the generous help and support of a number of people and organizations. We are greatly indebted to the NATO Division of Scientific Affairs not only for major financial support, but also through the offices of Drs. Tilo Kester and Mario di Lullo, for the advice and interest necessary to bring this project to a successful conclusion. In the area of travel grants for the participants, we thank the DAAD, Bonn-Bad Godesberg (West Germany), the CNR, Rome (Italy), and the NSF, Washington (United States) for generous financial assistance. We also greatly appreciate the efforts of those who obtained financial assistance from their universities, companies, etc. Our Italian co-organizers and colleagues, Professor Alessandro Mazzei and Dr. Gabriele Lugli, were invaluable in helping us to organize the ASI in Italy. Dr. John Taylor and the SOGESTA staff are thanked for their efficient and courteous service both before and during the conference. We also thank the chapter authors for their enthusiasm and diligence in this project. The Reidel Publishing Company, Dordrecht and Boston, has generously provided the advice and service necessary to facilitate the rapid editing and production of this volume. Also, we will

be forever grateful to Ms. R. Bach and Frau C. Heymann, without
whose valiant secretarial efforts on both sides of the Atlantic,
"Organometallics of the f-Elements" would never have come about.

Last but not least, we wish to express our gratitude to all of the
participants of the Advanced Study Institute, who attended quantitatively
and without repression or rebellion, the entire course of lectures.
These unsung heroes contributed a great deal through numerous discussions and get-togethers during the twelve days (and nights) of the
Institute. Only the future will show whether the 1978 Nato Advanced
Study Institute, "Organometallics of the f-Elements" and this monograph have sufficient impact on the scientific community to justify
a similar endeavor in the years to come.

Tobin J. Marks   R. Dieter Fischer
(director)   (co-director)

Evanston and Hamburg
November, 1978

LIST OF PARTICIPANTS

Akarkan, A.   University Hacettepe – Ankara
Faculty of Pharmacy, Dept. of Chemistry

Amberger, H. D.   Institut für Anorganische und Angewandte
Chemie Der Universität Hamburg,
2000 Hamburg 13, Martin-Luther-King Platz 6,
W. Germany

Ammon, R. von   Kernforschungszentrum, Institut für Heisse
Chemie, D 75 Karlsruhe, Postfach 3640
W. Germany

Ascenso, J.   Centro de Quimica Estentual, Complexo
Interdisciplinar, I.S.T. Lisbon I Portugal

Bagnall, K. W.   Chemistry Dept., Univ. of Manchester,
Manchester M13 9PL, England

Basolo, F.   Dept. of Chemistry, Northwestern University,
Evanston, IL U.S.A. 60201

Bau, R.   Chemistry Dept. USC, Los Angeles, CA 90007
U.S.A.

Bielang, G.   Institut für Anorganische und Angewandte
Chemie, Martin-Luther-King Platz 6,
2000 Hamburg 13, W. Germany

Bombieri, G.   Laboratorio di Chimica e Tecnologia dei
Radioelementi del C.N.R., Corso Stati Uniti,
35100, Padova, Italy

Bruncks, N.   Institut f. Anorganische und Analytische
Chemie der TU Berlin, Strasse der 17 Juni 135
1000 Berlin 12, W. Germany

Brunelli, M.   SNAM Progetti, S.p.A. Diris, 2097 S. Donato
Milanese, Italy

Campos, A. de
Pires de Matos   Laboratorio de Fisica e Engenharia Nucleares,
Radiochemistry Dept., Est. Nacional No. 10,
Sacavem, Portugal

Carnall, W. T.   Chemistry Division, Argonne National Lab.,
Argonne, IL 60439, U.S.A.

Cetincelik, M.   Turkish Nuclear Energy Institution,
P.O. Box 37 Bakanliklar, Ankara, Turkey

Chassapsis, D.   Univ. of Athens, Dept. of Chemistry,
Inorganic Chemistry Lab., 13 A Navarinou Str.
Athens, Greece

Ciliberto, E.   Istituto Dipartimentale di Chimica e Chimica
                          Industriale, Viale A. Doria 6, Catania, Italy
Day, V. W.   Department of Chemistry, University of
                          Nebraska, Lincoln, NE 68588, U.S.A.
Dell'Amico, D. B.   Istituto di Chimica Generale,
                          Via Risorgimento 35, 56100 Pisa, Italy
Dornberger, E.   Kernforschungszentrum Karlsruhe, Institut
                          für Heisse Chemie, 75 Karlsruhe, W. Germany
Edelstein, N.   Materials and Molecular Research Division
                          Lawrence Berkeley Laboratory, Bldg. 70 A,
                          Rm. 1149, Berkeley, CA 94720, U.S.A.
Egdell, R. G.   Inorganic Chemistry Dept., South Parks Road,
                          Oxford OXI 3QR, England
Eichberger, K.   Institut für Radiochemie der Technischen
                          Universität München, 8046 Garching,
                          W. Germany
Eigenbrot, C. W. Jr.   Dept. of Chemistry, University of California
                          Berkeley, CA 94720, U.S.A.
Eller, P. G.   CNC-4 MS 346, Los Alamos Scientific Laboratory
                          Los Alamos NM 87544, U.S.A.
Evans, W. J.   Dept. of Chemistry, University of Chicago,
                          5735 S. Ellis Avenue, Chicago, IL 60637,
                          U.S.A.
Fagan, P. J.   Chemistry Dept., Northwestern University,
                          Evanston, IL 60201, U.S.A.
Fischer, R. D.   Universität Hamburg, Institut für Anorgani-
                          sche und Angewandte Chemie, D-2000 Hamburg 13
                          Martin-Luther-King Platz 6, W. Germany
Fragalà, I.   Dipartimento di Chimica V. le Doria 6,
                          95125 Catania, Italy
Frisch, G.-M.   Institut f. Anorganische und Analytische
                          Chemie der TU Berlin, Strasse des 17 Juni 135
                          1000 Berlin 12, W. Germany
Gansow, O. A.   Dept. of Chemistry, University of Michigan,
                          E. Lansing, MI, U.S.A.
Genthe, W.   Institute für Anorganische Chemie der
                          Technischen Universität Berlin, Strasse des
                          17 Juni 135, I Berlin 12, W. Germany
Goffart, J.   Laboratory of Radiochemistry, Sart Tilman
                          B-4000, Liège, Belgium
Görlich, E.   Anorg. Chem. Institut der Techn. Universität
                          Clausthal, Paul Ernst Str. 4,
                          3392 Clausthal - Zellerfeld
Granozzi, G.   Istituto di Chimica Generale dell'
                          Universitàdi Podavo, Via Loredan,
                          35100 Padova, Italy
Grape, W.   Anorg. Chem. Institut der Techn. Universität
                          Paul Ernst Str. 4, 3392 Clausthal-
                          Zellerfeld, W. Germany

Green, J.	Inorganic Chemistry Laboratory, South Parks Road, Oxford OXI 3QR, England
Havemann, H.	c/o Professor R.D. Fischer, Institut für Anorganische Chemie,Universität Hamburg, Martin-Luther King Platz 6, W. Germany
Jagur, J.	Plastics Research Dept., Weizmann Institute of Science, Rehovot, Israel
Jonas, K.	Max-Planck-Institut für Kohlenforschung, Kaiser-Wilhelm-Platz 7, 4330 Mulheim/Ruhr W. Germany
Kanellakopulos, B.	Kernforschungszentrum Karlsruhe, Institut für Heisse Chemie, Postfach 3640, 7500 Karlsruhe, W. Germany
Karraker, D.	E.I. duPont de Nemors and Co., Savannah River Lab., Aiken, SC 29801, U.S.A.
Klähne, E.	Institut für Anorganische und Angewandte Chemie, Martin-Luther-King Platz 6, 2000 Hamburg 13, W. Germany
Klenze, R.	Kernforschungszentrum Karlsruhe, Institut für Heisse Chemie, 7500 Karlsruhe, W. Germany
Lappert, M. F.	School of Molecular Sciences, University of Sussex, Brighton BN1 9QJ, England
Levin, G.	c/o Levy, M., State University of N.Y., C.E.S.F., Chemistry Dept. Syracuse N.Y. 13210, U.S.A. Plastics Dept., The Weizmann Institute of Science, Rehovot, Israel
Lugli, G.	Assoreni, San Donato Milanese 20097, Milano, Italy
Lux, F.	Institut für Radiochemie der Technischen Universität München D-8046 Garching, W. Germany
Maas, E. T. Jr.	Corporate Research Laboratories, Exxon Research and Engineering Co., P.O. Box 45, Linden N.J. 07036, U.S.A.
Manriquez, J.	Dept. of Chemistry, Northwestern University, Evanston, IL 60201, U.S.A.
Marks, T. J.	Chem. Dept., Northwestern University, Evanston, IL 60201, U.S.A.
Marquet-Ellis, H.	Centre d'Etudes Nucleaires Saclay DGI/SEP CP Boîte Postale 2 Gif-sur-Yvette, 91190 France
Masino, A. P.	Department of Chemistry, University of Alberta, Edmonton, Alberta, Canada T6G 2G2
Mastoroudi, E.	Kernforschungszentrum Karlsruhe, Institut für Heisse Chemie, 7500 Karlsruhe, W. Germany

Mazzei, A.   SNAM Progetti, S.p.A., Direzione Ricerca e
                          Sviluppo, 20097 San Donato Milanese, Milano
                          Italy

McGarvey, B. R.   Dept. of Chemistry, University of Windsor,
                          Windsor, Ontario, N9B 3P4, Canada

McLaren, A. B.   Australian Atomic Energy Commission,
                          Private Mail Bag, Sutherland 2233,
                          Australia

Meijer Deliefde,   Laboratorium voor Anorganische Scheikunde,
  H. J.   Rijksuniversiteit, Nijenborgh 16,
                          9747 AG Groningen, The Netherlands

Miller, S. S.   Dept. of Chemistry, Northwestern University,
                          Evanston, IL 60201, U.S.A.

Miyake, C.   Dept. of Nuclear Engineering, Faculty of
                          Engineering, Osaka University, Suita,
                          Osaka, Japan

duMont, W.-W.   Institut für Anorg. u. Analyt. Chemie der
                          TU Berlin, Strasse des 17 Juni 135 Berlin,
                          W. Germany

Moody, D. C.   CNC-4, Los Alamos Scientific Laboratory,
                          Los Alamos NM 87545 U.S.A.

Müller, J.   Institut f. Anorganische und Analytische
                          Chemie der T 6 Berlin, Strasse des 17 Juni
                          135 Berlin, W. Germany

Nevald, R.   Laboratory of Electrophysics, Techn. Univ.
                          of Denmark, DK-2800 Lyngby, Denmark

Pagni, R.   Dept. of Chemistry, University of Tennessee,
                          Knoxville, TN 37916, U.S.A.

Paoli, G. de   Laboratorio di Chimica e Tecnologia dei
                          Radioelementi del C.N.R., Area della Ricerca,
                          Corso Stati Uniti, 35100 Padova, Italy

Pedretti, U.   Assoreni, Direzione Ricerca e Sviluppo,
                          20097 San Donato Milanese, Italy

Puschmann, K.   Anorg. Chem. Institut der Technischen
                          Universität Clausthal, Paul Ernst Str. 4,
                          3392 Clausthal, W. Germany

Raymond, K. N.   Chemistry Department, University of
                          California, Berkeley, NE 94720, U.S.A.

Schumann, H.   Institut für Anorganische und Analytische
                          Chemie der Technischen Universität Berlin,
                          Anorganische Chemie III, Strasse des 17
                          Juni 135, D-1000 Berlin 12, W. Germany

Seyam, A. M.   Department of Chemistry, University of
                          Jordan, Amman, Jordan

Sienel, G. R.   D 7891 Weilheim, Hs.-Nr. 95, W. Germany

Slater, J. L.   Dept. of Chemistry, College of Steubenville,
                          Steubenville, OH 43952, U.S.A.

Soulie, E.   c/o Dr. Norman Edelstein,
                          Materials and Molecular Research Division,

	Lawrence Berkeley Laboratory, Bldg. 70A, Rm 1149, Berkeley, CA 94720 U.S.A.
Stollenwerk, A.	Kernforschungszentrum Karlsruhe, Institut für Heisse Chemie, Postfach 3640, 7500 Karlsruhe, W. Germany
Streitwieser, A.	Department of Chemistry, University of California, Berkeley, CA 94720, U.S.A.
Takats, J.	Department of Chemistry, University of Alberta, Edmonton, Alberta, T6G 2G2, Canada
Texeira dos Santos Domingos, A. M. A.	Laboratorio de Fisica e Engenharia nucleares Radiochemistry Department, Estrada Nacional No. 10, Sacavem, Portugal
Vasamiliette, J.-L.	Kernforschungszentrum Karlsruhe, Institut für Heisse Chemie, 7500 Karlsruhe, W. Germany
Woodwark, D. R.	Inorganic Chemistry Laboratory, South Parks Road, Oxford OXI 3QR, England
Yara, B.	Dept. of Chemistry, Middle East Technical University, Ortadogu - Ankara, Turkey
Yarrow, P.	School of Molecular Sciences, University of Sussex, Brighton BN 1 9QJ, England
Zanella, P.	Laboratorio di Chimica e Tecnologia dei Radioelementi del C.N.R., Corso Stati Uniti, 35100 Padova, Italy

CYCLOPENTADIENYL COMPOUNDS OF THE ACTINIDE ELEMENTS

Basil Kanellakopulos

Institut für Heiße Chemie, Kernforschungszentrum
Karlsruhe GmbH, Postfach 3640, 7500 Karlsruhe,
Germany

1. INTRODUCTION

Among compounds of the 5f-elements, organometallic complexes
have been studied during the two last decades. The discovery of
ferrocene in 1951 [1,2] and the synthesis of the first organo-
uranium compound, triscyclopentadienyluranium chloride,
$(C_5H_5)_3UCl$, in 1956 [3], opened a new and interesting field of
coördination chemistry of the 5f-transition elements.

The rapid expansion of new knowledge has resulted from the
successful synthesis and extended investigation of a great number
of organometallic complexes of lanthanides and actinides, by
using modern physical and physical chemical methods like NMR,
ESR, magnetic susceptibility, Mössbauer spectroscopy, Raman
spectroscopy and so on.

A great deal of attention has been paid to the theoretically
interesting question of the role of the f-orbitals in the bonding
of these compounds. A critical comparison of the organometallics
of the trivalent actinides with homologous complexes of the
lanthanides is now possible.

The organometallic chemistry of the actinides is of
special interest because of the varied oxidation states of the
f-elements. On the other hand, because of the spatial expansion
of the 5f-orbitals compared to those of the 4f-orbitals, it can
be expected that organoactinides show larger covalency than the
corresponding organolanthanides.

1

T. J. Marks and R. D. Fischer (eds.), Organometallics of the f-Elements, 1–35.

2.   CLASSIFICATION

Organoactinides can be classified according to the number of electrons of the organic ligand as follows:

Number of the ligand electrons	Class	Ligand
3	enyl	π-allyl
5	dienyl	$\left\{\begin{array}{l}\pi\text{-cyclopentadienyl}\\ \pi\text{-indenyl}\end{array}\right.$
8	tetraenyl	π-cyclooctatetraenyl

3.   CYCLOPENTADIENYL COMPOUNDS

Reynolds and Wilkinson [3] reported the synthesis of the first cyclopentadienyl compound of the actinides, the triscyclo-pentadienyl uranium chloride, $(C_5H_5)_3UCl$, in 1956. The existence of the first organometallic neptunium compound was established by Baumgärtner, Fischer and Laubereau [4] in 1965. A great number of cyclopentadienyl compounds of the actinides are now known [5-15].

3.1 PREPARATION METHODS

The acidity of cyclopentadiene $(P_{Ka}\ 17)$ [16] leads quickly to formation of ionic compounds with alkali metals, which can be used as starting materials for the reaction with actinide halides.

3.1.1 Preparation in Solution

The general method for preparation of cyclopentadienyl complexes of the actinides is the reaction of an anhydrous actinide (=An) halide with ionic cyclopentadienides, like potassium or sodium cyclopentadienide, in an organic solvent such as benzene, tetrahydrofuran (THF), ethylether, etc:·

a) $An^{IV}Hal_4\ +\ 4\ K(C_5H_5)\ \xrightarrow{org.solv.}\ An^{IV}(C_5H_5)_4$

b) $An^{IV}Hal_4\ +\ 3\ Na(C_5H_5)\ \longrightarrow\ (C_5H_5)_3An^{IV}Hal$

c) $An^{III}Hal_3\ +\ 3\ K(C_5H_5)\ \longrightarrow\ An^{III}(C_5H_5)_3$

d) $An^{III}Hal_3\ +\ 2\ K(C_5H_5)\ \longrightarrow\ (C_5H_5)_2\ An^{III}Hal$

The actinide compounds are usually isolated and purified by extraction with a convenient organic solvent or by sublimation under reduced pressure if they are of sufficient volatility.

3.1.2 Preparation in the Melt

For the synthesis of compounds of highly alpha-radioactive elements such as transuranic elements, it is essential to adopt a preparation procedure which minimizes the fire and explosion risk in a glove box. The radiolysis problem of the reactants and of the solvent during the long reaction time (required for a good yield) must also be avoided. For these reasons, the reaction of actinide halides with a melt of beryllium-[17] or magnesium-[18] cyclopentadienide has the advantages of the fire security , short reaction time, and compact apparatus. However, the (α,n) reaction of the beryllium

$$\ce{^{9}_{4}Be} + \alpha \longrightarrow \ce{^{12}_{6}C} + n$$

and the resulting neutron flux, must be taken into account when working with shortlived transuranic elements.

The $Be(C_5H_5)_2$ and $Mg(C_5H_5)_2$ have relatively low melt points at 59-60 and 177-178°C respectively:

$$An^{IV}Hal_4 + x\ Be(C_5H_5)_2 \xrightarrow[65°C]{melt} An^{IV}(C_5H_5)_4 \text{ or } (C_5H_5)_3An^{IV}Hal$$

$$An^{III}Hal_3 + y\ Be(C_5H_5)_2 \xrightarrow{180-185°C} An^{III}(C_5H_5)_3$$

While the synthesis in an organic solvent requires long reaction times in order to obtain high yields, the preparation in the melt is normally quick, in most cases within only a few hours. After removal of the alkaline earth cyclopentadienide excess, by sublimation under vacuum at low temperatures, the organo-actinides can be isolated by extraction with a solvent or by sublimation under vacuum at elevated temperature.

3.1.3 Special Preparation Methods

Many modified preparation methods have been used for the synthesis of cyclopentadienyl compounds of the actinides: Triscyclopentadienyluranium(IV)fluoride, $(C_5H_5)_3UF$, can be prepared in good yields (>90%) by heating the homologous bromide or chloride compound with sodium fluoride in the solid state in a sealed glass tube at 150-180°C [19]:

$$(C_5H_5)_3UBr + NaF \longrightarrow (C_5H_5)_3UF + NaBr$$

$(C_5H_5)_3UF$ reacts with NaOH in the solid phase to yield the base triscyclopentadienyluranium hydroxide, $(C_5H_5)_3UOH$ [20].
Pure $(C_5H_5)_3ThCl$ can be prepared by sublimation of thermally stable $Th(C_5H_5)_4$ through a layer of thoriumtetrachloride [14].
By passing well dried hydrochloric acid into a solution of tetrakis(cyclopentadienyl)uranium(IV) in benzene, the tris-cyclopentadienyluranium(IV)chloride is formed quantitatively at room temperature:

$$(C_5H_5)_4 U + HCl_{(g)} \xrightarrow[RT]{C_6H_6} (C_5H_5)_3UCl + C_5H_6$$

Gaseous H_2S under the same conditions yields $(C_5H_5)_3USH$ [20].
The reaction of freshly prepared liquid HCN with $(C_5H_5)_4U$ or $(C_5H_5)_3U$ leads to the substitution of a cyclopentadienyl ring by a CN^- group [21]:

$$(C_5H_5)_4U^{IV} + HCN_{(liq)} \xrightarrow[RT]{C_6H_6} (C_5H_5)_3U^{IV}CN + C_5H_6$$

$$(C_5H_5)_3U^{III} + HCN_{(liq)} \xrightarrow[RT]{THF} (C_5H_5)_2U^{III}CN + C_5H_6$$

Borontrifluoride reacts with $(C_5H_5)_3UF$ in solution to the tetrafluoroborate compound [20]:

$$(C_5H_5)_3UF + BF_{3(g)} \longrightarrow (C_5H_5)_3UBF_4$$

A special preparation method for substitution reaction of the $C_5H_5^-$ ligand was developed in our laboratory.
Ammonium salts react with pure cyclopentadienyl compounds of the actinides to yield the monosubstituted compounds [22]:

$$An^{IV}(C_5H_5)_4 + NH_4X \longrightarrow (C_5H_5)_3An^{IV}X + NH_3\uparrow + C_5H_6$$

$$An^{III}(C_5H_5)_3 + NH_4X \longrightarrow (C_5H_5)_2An^{III}X + NH_3\uparrow + C_5H_6$$

The method can also be used for lanthanides and transition metals and is readily applicable for microscale preparation, when only small quantities of less common metals or radioactive elements are available.

3.1.4 Preparation in Aqueous Solutions

A great number of $(C_5H_5)_3UX$ compounds has been prepared in aqueous solutions of $(C_5H_5)_3UCl$. G.W. Sienel's thesis [23] gives an excellent review of this subject and demonstrates a new field of metallorganic chemistry of the tetravalent uranium. Similar to the behaviour of $(C_5H_5)_3UCl$ in oxygen-free water [3], most tris(cyclopentadienyl)uranium(IV) compounds are soluble in water saturated with an inert gas. The green colour of the aqueous solution is due to the hydrated $[\pi-(C_5H_5)_3U]^+$ cation which reacts with inorganic salts [23,43]:

$$(C_5H_5)_3UX \; + \; n \; H_2O \; \rightleftharpoons \; [(C_5H_5)_3U(H_2O)_n]^+ \; + \; X^-$$

$$[(C_5H_5)_3U(H_2O)_n]^+ \; + \; KY \; \rightleftharpoons \; (C_5H_5)_3UY$$

3.2 CYCLOPENTADIENYL COMPOUNDS OF THE TETRAVALENT ACTINIDES

Cyclopentadienyl compounds of the tetravalent actinides can be classified in special types:
$An(C_5H_5)_4$, $(C_5H_5)_3AnX$, $(C_5H_5)_2AnX_2$, $(C_5H_5)AnX_3$.

Bridged compounds of the type $(C_5H_5)_3An-Y-An(C_5H_5)_3$ (where Y is a dianion) and several adducts are also known.

3.2.1 Tetrakis(cyclopentadienyl)compounds, $An(C_5H_5)_4$

The pure cyclopentadienyl compounds of thorium [24], protactinium [25], uranium [26,27], neptunium [28] are listed in Table 3.1. Attempts to prepare the plutonium compound by reacting Cs_2PuCl_6 with magnesium cyclopentadienide [29], $[PuCl_6]^{2-}$ with potassium cyclopentadienide [30] or PuF_4 with $Be(C_5H_5)_2$ [19] have been unsuccessful. The product was $Pu(C_5H_5)_3$.

Table 3.1.   Tetrakis(cyclopentadienyl)actinides, $An^{IV}(C_5H_5)_4$

An^{IV}	Colour	Subl.Temp. (^{o}C)	yield	Ref
Th	colorless	250-290	40	24
Pa	orange	220 dec.	54	25
			43	31
U	red	200-220 dec.	6	26
			99	19
Np	reddish-brown	200-220 dec.	72/86	28/48

The compounds seem to have the same arrangement of the
ligands around the central ion. They show similar ir spectra,
which suggest that all compounds possess similar molecular
structures. Powder X-ray diffraction investigations have shown
that the Th, U and Np complexes are isostructural [18,39].
The four ligands are bonded centrosymmetrically to the central
ion. The vanishingly small dipole moment of $U(C_5H_5)_4$ [26,27]
and the proton magnetic resonance spectra [33] showing only a
single resonance indicate an almost tetrahedral displacement of
the cyclopentadienyl rings about the central ion.

The crystal structure of the $U(C_5H_5)_4$ has been investigated
by Burns [34,35]. The individual molecules of $U(C_5H_5)_4$ have the
point symmetry S_4. The site symmetry D_{2d} was proposed with a
placement of the rings at the apices of a tetrahedron.
For detailed discussion of the crystal structure see ref. 35
and 36.

The tetrakis(cyclopentadienyl)actinides can be reduced to
the actinide metal by metallic potassium in an organic solvent.
After extraction of the potassium cyclopentadienide formed, a
very pyrophoric and extremely reactive metal powder can be isolated. $U(C_5H_5)_4$ can be reduced to $U(C_5H_5)_3$ by using uranium
metal as reducing reagent [37]. The compounds are sensitive to
air and moisture. They are stable until 200-250°C. Above 200°C
they can be sublimed under vacuum. The thorium compound is stable
up to 400°C, while the Pa, U and Np complexes partially decompose
at temperatures higher than 200-220°C. Their solubility in
organic solvents is very small (<1mg/ml).

3.2.2 Tris(cyclopentadienyl)actinide(IV) Compounds, $(C_5H_5)_3AnX$

A great number of compounds with a variation of the ligand
X are known. (Table 3.2)

Table 3.2. Some Actinide(IV)-cyclopentadienyl compounds of the
type $(C_5H_5)_3AnX$

An(IV)	Ligand (X$^-$)	Colour	Subl. Temp. ($^\circ$C)	Prep. yield %	m.p	Ref.
Th(5f^0)	F	pale yellow	200	–	–	38
	Cl	colourless	200	44(80)	–	39(19)
	Br	pale yellow	180	–(70)	–	38(19)
	I	pale yellow	190	–	–	38
	OCH$_3$	colourless	–	–	–	39
	OC$_2$H$_5$	colourless	–	–	–	39
	n-OC$_4$H$_9$	colourless	135	38	148–150	39
	t-OC$_4$H$_9$	colourless	110	92	–	19
	BH$_4$	pale yellow	220	93	–	19
Pa(5f^1)	Cl	red-brown	–	90	–	31,22
U(5f^2)	F	green	170	80	–	40
	Cl	brown	120–130	40–90	–	3,17,27,42
	Br	dark-brown	160	80	–	40
	I	brown	170	80	–	40,42
	BH$_4$	red	170	95	–	27,41
	BF$_4$	green	–	90	–	20
	NCS	green	–	95	dec.>180	20
	OCN	green	180	56	–	20
	OH	green	–	85	–	20
	SH	brown	–	>90	–	20
	CN	green	250	–	–	21
	C(CN)$_3$	green	–	>95	–	43
	NO$_3$	brown	–	100	–	22
	ClO$_4$	deep-green	–	100	–	22
	ReO$_4$	yellow-brown	–	96	–	22
	B(C$_6$H$_5$)$_4$	beige	–	86	–	22
	N$_3$	green	–	74	–	19
	OCH$_3$	green	120	299–302	74	39,44
	OC$_2$H$_5$	green	120	210–213	54	27,44
	(i)OC$_3$H$_7$	green	130	200–201	90	44
	(n)OC$_4$H$_9$	green	120	149–151	15/83	39,44
	(t)OC$_4$H$_9$	green	120	–	78	44
	OC$_6$H$_{13}$	green	–	76	96	45
	OC$_6$H$_{11}$	green	120	247–248	>80	45
	(n)OC$_8$H$_{17}$	green	–	38–40	25	20
	OC$_{27}$H$_{45}$	green	–	dec.245	>80	45
	OCCH$_3$ $\\ \parallel \\ O$	green	150	–	56	19

Table 3.2 (continued)

An(IV)	Ligand	Colour	Subl. Temp. ($^\circ$C)	Prep. yield %	m.p ($^\circ$C)	Ref.
$U(5f^2)$	$SCCH_3$ ‖ S	brown	–	–	–	19
	SC_2H_5	–	–	–	–	51
	NR_2	–	–	–	–	51
	$t-SC_4H_9$	–	–	5	–	51
$Np(5f^3)$	F	green-brown	170	–	(36)	46,47,(48)
	Cl	deep-brown	100	dec.350	45/99	4,46,48
	Br	brown	–	–	97	48
	I	pale-brown	–	–	92	48

3.2.3 Bis(cyclopentadienyl)actinide(IV) Compounds, $(C_5H_5)_2AnX_2$

The bis(cyclopentadienyl)uranium(IV)chloride $[(C_5H_5)_2UCl_2]$ has been previously described [49]. The green-brown compound was obtained by the reaction of UCl_4 with TlC_5H_5 in dimethoxyethane (DME). New extended studies of the synthesis reaction have shown that the reaction product has the composition $[(C_5H_5)_3U]UCl_6\cdot2DME$ [50]. The DME-free compound was also prepared [50].

Proton NMR and vibrational spectra of "$(C_5H_5)_2UCl_2$" are interpreted [12] in terms of a mixture of $(C_5H_5)_3UCl$ and $(C_5H_5)UCl_3\cdot DME$. The problem of the exact nature of "$(C_5H_5)_2UCl_2$" and its existence still remains open. "$(C_5H_5)_2UCl_2\cdot(tppo)_2$" and "$(C_5H_5)_2UCl_2\cdot L$" (L = dma, dmpva) appear to be more stable [55]. Recently bis(pentamethylcyclopentadienyl)dialkyls of thorium and uranium have been synthesized and investigated [63].

The reaction of tetrakis(diethylamido)uran(IV) with cyclopentadiene results in bis(cyclopentadienyl)bis(diethylamido)-uranium(IV) [51]:

$$UCl_4 \ + \ 4 \ LiN(C_2H_5)_2 \ \xrightarrow{THF} \ U[N(C_2H_5)_2]_4 \ + \ 4 \ LiCl$$

$$U[N(C_2H_5)_2]_4 \ + \ 2 \ C_5H_6 \ \xrightarrow{Pentane} \ (C_5H_5)_2U[N(C_2H_5)_2]_2 \ + \ 2HN(C_2H_5)_2$$

The compound is a useful starting reagent for substitution reactions. It reacts with toluene-3,4-dithiol (tdt), o-mercaptophenol (omp), catechol (cat), 1,2-ethanedithiol (edt) to produce bis(cyclopentadienyl)uranium(IV) compounds. Mass spectroscopic and NMR investigations on the reaction products with carbon disulfide show the formation of a dithiocarbamate, $(C_5H_5)_2U[S_2CN(C_2H_5)_2]_2$. The synthesis of the red-brown bis(cyclopentadienyl)uranium(IV)bis(tetrahydroborate) has been reported [52]. The compound was prepared from $UCl_2(BH_4)_2$ with TlC_5H_5 at room temperature in DME or THF. Ir investigations favour a tridentate ligation by the tetrahydroborate group.

3.2.4 Mono(cyclopentadienyl)actinide(IV) Compounds, $(C_5H_5)UX_3$

Only a few organoactinides containing only one C_5H_5 -ring are known. The reaction of uranium tetrachloride with the stoichiometric amount of cyclopentadienylthallium in DME at room temperature leads to the formation of cyclopentadienyluranium(IV)trichloride, $(C_5H_5)UCl_3 \cdot DME$ [53]. The green compound is soluble in tetrahydrofuran but insoluble in hydrocarbons. The THF adduct, $(C_5H_5)UCl_3 \cdot 2THF$ was prepared [54] by the same method described for $(C_5H_5)UCl_3 \cdot DME$ when THF was used as solvent. The compound is a bright-green, air-sensitive solid. It decomposes in acetone and benzene and upon heating above 120°C.

$(C_5H_5)UCl_3 \cdot 2THF$ reacts in THF with hydrotris(1-pyrazolylborate) $(HBpz_3)^-$ to yield cyclopentadienyldichlorouranium-hydrotris(1-pyrazolylborate),$(C_5H_5)UCl_2[HBpz_3]$, a bright-green, airsensitive solid [54]. The compound is soluble in acetone and tetrahydrofuran, but it is insoluble in dichloromethane. The ir and NMR indicate that all three pyrazolyl rings are equivalent at room temperature. Oxygen-donor complexes of $(C_5H_5)UCl_3$ with the general formula $(C_5H_5)UCl_3 \cdot L_2$ (L = $CH_3CON(CH_3)_2$ (dma), $(CH_3)_3CCON(CH_3)_2$ (dmpva), $(C_6H_5)_3PO$ (tppo) have also been prepared [55].

3.2.5 Bridged Organoactinides

Fluorine-bridged adducts of the tris(cyclopentadienyl)-uranium(IV) fluoride with triscyclopentadienyls of uranium(III) and ytterbium were synthesized [56]. $(C_5H_5)_3UF$ behaves as a Lewis acid towards bases and reacts with Lewis acids like $(C_5H_5)_3M^{III}$ (M = lanthanide or actinide) to form 1:1 adducts.

The green compounds $[(C_5H_5)_3U]_2ox$ and $[(C_5H_5)_3U]_2SO_4$ can be prepared from $U(C_5H_5)_4$ and the corresponding ammonium salts in THF [22]. The bright brown neptunium compound $[(C_5H_5)_3Np]_2SO_4$ was also synthesized by the same method [48].

The oxygen and sulfur bridged uranium complexes $[(C_5H_5)_3U]_2O$ and $[(C_5H_5)_3U]_2S$ are known [19]. The first compound can be prepared by the reaction of $(C_5H_5)_3UCl$ with Ag_2O, or by careful oxidation of $(C_5H_5)_3U\cdot THF$ with oxygen in benzene. Another preparation method is the rapid heating of the base $(C_5H_5)_3UOH$ at 400°C. The oxygen bridged compound sublimes with small yields. Heating $(C_5H_5)_3USH$ leads to the sulfur bridged compound. However, the best method of preparation is the reaction of $(C_5H_5)_3UBr$ with freshly prepared K_2S in an organic solvent [20].

3.2.6 Miscellaneous Compounds

A large number of σ-bonded organometallic compounds of the type $(C_5H_5)_3UR$ have been described in the literature. [9-14,57-61] They are of special interest for studies of reaction kinetics, pyrolysis, catalytic, spectroscopic, and magnetochemical properties. Organometallic uranium complexes with more than one metal-carbon σ-bond have also been described [62,63].

3.3 CYCLOPENTADIENYL COMPOUNDS OF THE TRIVALENT ACTINIDES

The known cyclopentadienyl actinide(III) compounds of thorium to californium are listed in Table 3.3.

Table 3.3   Tris(cyclopentadienyl)compounds of the Actinide Elements

	Compound	Colour	Subl.point (°C)	Prep. yield (%)	Ref.
$(5f^1)$	$Th(C_5H_5)_3$	violet	dec	69	64
	$Th(C_5H_5)_3$	green	–	92	65
	$(C_5H_5)_2ThCl$	brown	–	96	22
$(5f^3)$	$U(C_5H_5)_3$	bronze	–dec	10-40	37
	$(C_5H_5)_3U\cdot THF$	brown	–	95	37
	$(C_5H_5)_3U\cdot nic$	brown	–	75	37
	$(C_5H_5)_3U\cdot CNC_6H_{11}$	pale-brown	–	75	37
$(5f^4)$	$(C_5H_5)_3Np\cdot 3THF$	–	–	–	32
	$(C_5H_5)_3Np\cdot THF$	–	–	>90	19
	$(C_5H_5)_3Np\cdot CNC_6H_{11}$	–	–	>90	67,68

Table 3.3 (continued)

	Compound	Colour	Subl.point ($^\circ$C)	Prep. yield (%)	Ref.
($5f^5$)	$Pu(C_5H_5)_3$	moss green	140–165	60–80	17,29,66
	$(C_5H_5)_3Pu \cdot THF$	green	–	90	19
	$(C_5H_5)_3Pu \cdot CNC_6H_{11}$	green	–	>90	67,68
	$(C_5H_5)_3Pu \cdot nic$	green	–	70	19
	$(C_5H_5)_2PuCl$	green-blue	–	60	19
($5f^6$)	$Am(C_5H_5)_3$	rose	160–200	50–62	29,67,69
	$(C_5H_5)_3Am \cdot THF$	rose	–	–	19
	$(C_5H_5)_3Am \cdot CNC_6H_{11}$	rose	–	85	67,68
($5f^7$)	$Cm(C_5H_5)_3$	colourless	180	–	70,71
($5f^8$)	$Bk(C_5H_5)_3$	amber	–	–	72
	$(C_5H_5)_2BkCl$	–	–	–	73
($5f^9$)	$Cf(C_5H_5)_3$	ruby-red	–	–	72

$Th(C_5H_5)_3$ was prepared by reduction of $Th(C_5H_5)_3Cl$ with metallic sodium and naphthalene in tetrahydrofuran [64]. The violet compound shows a very low magnetic moment (0.403 μ_B at room temperature). Recently a dark green $Th(C_5H_5)_3$ was synthesized by photolysis of $(C_5H_5)_3Th(i-C_3H_7)$ [65]. In a similar manner, the methylcyclopentadienyl compound can be prepared. The magnetic moment of the green $(C_5H_5)_3Th$ at 298 K is 2.10 μ_B. [65]

The tris(cyclopentadienyl)actinide(III) compounds are analogous to the corresponding 4f-complexes. Both compound series show similar ir-spectra and have a Lewis-acid character.(The Cp_2-uranium compound is dimeric [19]).They form 1:1 adducts with bases like tetrahydrofuran, 1-nicotine, NH_3, PH_3, cyclohexyl-isocyanide, etc. HCl reacts with $An(C_5H_5)_3$ to yield the $(C_5H_5)_2AnCl$ (An = U, Pu) species like the lanthanide compounds. The covalency of the actinide triscyclopentadienides is very small (2.8 % in $Am(C_5H_5)_3$ and 2.5 % in $Cm(C_5H_5)_3$) and is comparable with the magnitude of the covalency in the homologous lanthanide compounds [74]. In spite of this similarity it seems that the nature of the chemical bonding, especially in the complexes of the higher actinides, is somewhat different from that in the lanthanide compounds. For example the C≡N absorption bands in the ir spectra of the $(C_5H_5)_3An \cdot CNC_6H_{11}$ adducts are shifted towards higher frequencies relative to   (free ligand at 2136 cm^{-1}) the corresponding frequencies of the lanthanide compounds. (Table 3.4)

Table 3.4   C≡N-Frequencies of $(C_5H_5)_3M \cdot C \equiv NC_6H_{11}$-compounds

f^q	M(Ln)	$\nu_{C \equiv N}(cm^{-1})$	M(An)	$\nu_{C \equiv N}(cm^{-1})$
f^0	Y	2208		
f^1	Ce	2197	Th	2140
f^2	Pr	2203	Pa	-
f^3	Nd	2207	U	2160
f^4	Pm	-	Np	2168
f^5	Sm	2202	Pu	2190
f^6	Eu	2200	Am	2202
f^7	Gd	2196	Cm	-
f^8	Tb	2205	Bk	-
f^9	Dy	2204	Cf	-
f^{10}	Ho	2205	Es	-
f^{11}	Er	2206	Fm	-
f^{12}	Tm	2204	Md	-
f^{13}	Yb	2203	No	-
f^{14}	Lu	2208	Lr	-

The C≡N frequency increases rapidly in the actinide series with the atomic number; (Fig. 3.1) this phenomenon is consistent with the increasing thermostability of the triscyclopentadienides of the transuranic elements. Increasing of the frequency means increasing of the electron density in the C≡N bond.
$Pu(C_5H_5)_3$ and $Am(C_5H_5)_3$ are thermally stable compounds like the lanthanide complexes. The bonding stabilization energies in the Ce^{3+}, Pr^{3+}, Tb^{3+} and Er^{3+} is connected with the ligand field influence. This phenomenon has been measured in octahedrally coordinated compounds of the 3d transition metals, in the CF-splitting energy (or 10 Dq). In the case of the lanthanide compounds the form of the C≡N frequency curve is due to the non-radially symmetric shielding of the nuclear charge, when the spherical symmetry of the 4f-orbitals is distorted by the CF.

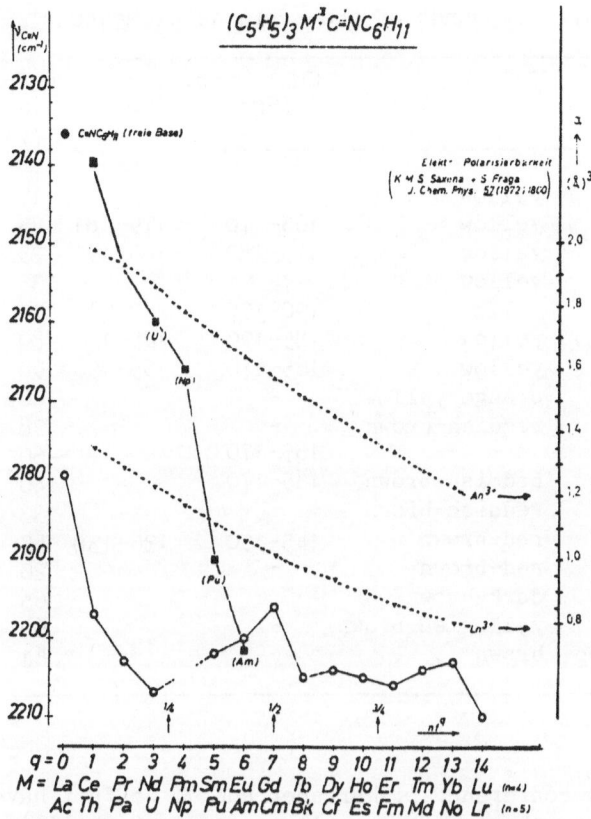

Fig. 3.1.   C≡N frequencies of the $(C_5H_5)_3(Ln,An)\cdot C\equiv NC_6H_{11}$ compounds [68]

4.   INDENYL COMPOUNDS

During the past 10 years many actinide organometallics have been synthesized since the preparation of the first uranium compound with the indenyl anion as ligand, namely trisindenyl-uranium(IV)chloride, $(C_9H_7)_3UCl$ in 1971 [75].

4.1 Preparation Methods

Indenyl compounds of the actinide elements can be prepared by using methods similar to those used for the synthesis of cyclopentadienyl complexes. The general method is the reaction of actinide halides with potassium- or sodiumindenyl in an organic solvent. The indenyl compounds are tabulated in Table 4.1.

Table 4.1   Indenyl compounds of the actinide elements

Compound	Colour	Subl.Temp. ($^{\circ}$C)	m.p ($^{\circ}$C)	Prep. yield (%)	Ref.
$(C_9H_7)_3ThCl$	yellow	–	–	17	75
	yellow	165–170	179–181	55	76
$(C_9H_7)_3ThBr$	yellow	175–180	189–191	35	76
$(C_9H_7)_3ThI$	yellow	–	–	35	77
$(C_9H_7)_3ThCH_3$	yellow	190–195	213–215	70	78
$(C_9H_7)_3Th(n-C_4H_9)$	yellow	185–190	125–126	80	78
$(C_9H_7)_3ThOCH_3$	yellow	165–170	155–156	90	78
$(C_9H_7)_4Th$	orange-yellow	–	–	60	79
$(C_9H_7)_3UCl$	reddish-brown	–	–	28	75
		165–170	–	60	76
$(C_9H_7)_3UBr$	reddish-brown	165–170	–	40	76
$(C_9H_7)_3UI$	reddish-brown	–	–	33	77
$(C_9H_7)_3UOCH_3$	red-brown	145–150	136–138	57	78
$(C_9H_7)_3UOC_2H_5$	red-brown	–	–	20	78
$(C_9H_7)_4U$	dark-brown	–	–	65	79
$(C_9H_7)_2U(CH_3)_2$	yellowish-brown	–	–	60	80
$(C_9H_7)_2U(t-C_4H_9)_2$	brown	–	–	61	80

During this conference D. Karraker and J. Goffart have also reported on the synthesis of indenyl compounds of Th(III), U(III), Np(III) and Np(IV).
It should be noted that the indenyl ligand is a more powerful reductant than the cyclopentadienyl anion.

5.   MAGNETIC INVESTIGATIONS

Magnetic susceptibilities of several organometallic compounds of the higher actinides (thorium to americium) have been measured.
Contrary to the lanthanide organometallics which at room temperature have a magnetic moment comparable to that calculated for the free ion, the magnetic moment of the organometallic compounds of the actinides is much smaller than the theoretical value.

The predominant influence on the temperature function of the individual effective magnetic moment is due to the ligand field potential, however, the "true" wave functions of populated states contain mixtures of other states through following interactions:

a) Intermediate coupling which mixes multiplets with the same total angular momentum through the spin orbit coupling parameters. Intermediate coupling increases in 5f-systems with atomic number and plays naturally an important role.

b) J-mixing depends on the crystal field splitting relative to the spacing of free ion terms. J-mixing is dominant especially in the first members of the trivalent lanthanides and highly ionized actinides.

c) Mixtures of configurations with the same parity as the ground state, which although individually quite small, may account for am accumulative contribution to the total wave function.

Due to the low symmetry of most organoctinides a theoretical treatment of the magnetic susceptibility data has been made for only a few compounds.

In the following some experimental results are presented only in a qualitative manner.

$5f^1$-Configuration ($^2F_{5/2}$)

Th(C$_5$H$_5$)$_3$. The magnetic moment of Th(C$_5$H$_5$)$_3$ (Fig. 5.1) is very

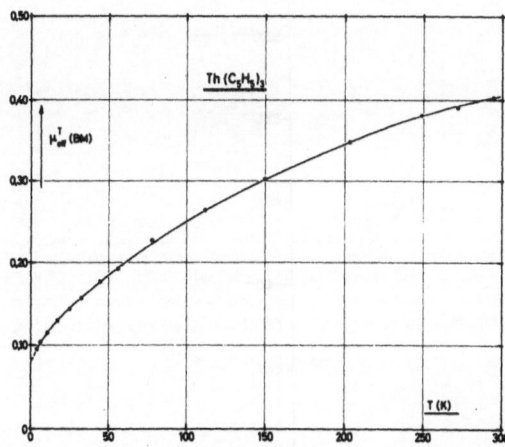

Fig. 5.1.

μ_{eff}-value of Th(C$_5$H$_5$)$_3$ as a function of temperature.

low and can not be explained by either a $5f^1$- nor by a $6d^1$ configuration. Although the energy separation $5f^q \rightarrow 5f^{q-1}6d$ is small (9.193 x 10^3 cm^{-1}) [81] the assumption of a $6d^1$ configuration for the Th^{3+} ion can not explain the magnitude of the moment ((μ_{eff}^2)^{RT} for Ti(C$_5$H$_5$)$_3$ is 2.704 μ_B^2 [82] and the corresponding value of TiCl$_3$·THF is 2.956 μ_B^2 [83]) or the shape

of the paramagnetism. Even in the case of an orbit reduction
factor of 0.60 to 0.70 the expected magnetic moment of
$\mu_{eff}^2 = 0.267 \ \mu_B^2$ for a doublet (3/2 ± 3/2) as ground state is
much higher than the experimentally observed. The magnetic
moment of tris(indenyl)Th·THF reported by J. Goffart during this
conference is 0.8 μ_B at room temperature. On the other hand the
moment reported for the "green" Th(C$_5$H$_5$)$_3$ (μ_{eff}^{298} = 2·10 μ_B) [65]
appears a little high, so the temperature dependence of the
paramagnetism could provide useful information.

Pa(C$_5$H$_5$)$_4$

    Pa(C$_5$H$_5$)$_4$ shows also a very low magnetic moment (Fig. 5.2).

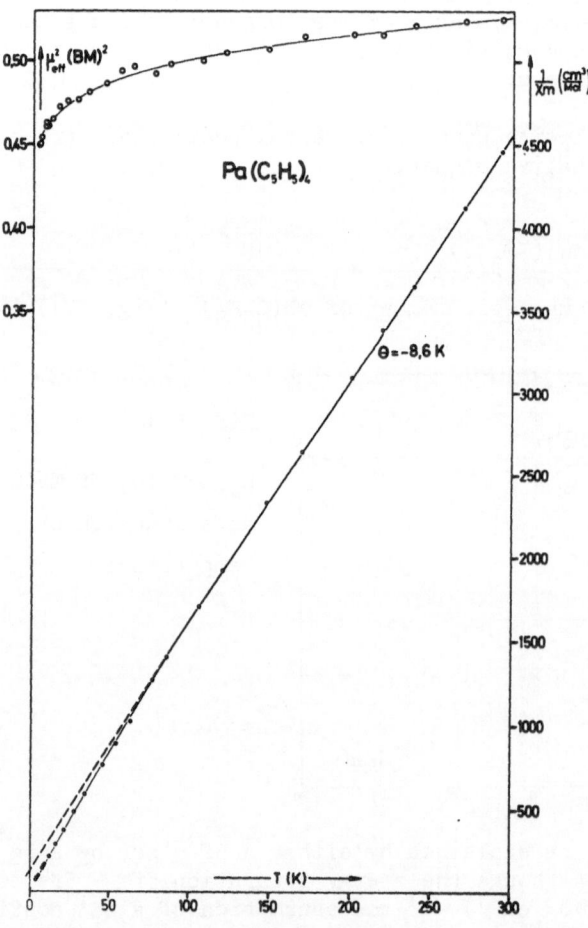

Fig. 5.2. The magnetic susceptibility of Pa(C$_5$H$_5$)$_4$.

Above 90 K the magnetic susceptibility shows a Curie-Weiss behaviour with $\Theta = -8.6$ K. The value of μ_{eff} at room temperature (0.525 μ_B^2) indicates a strong splitting of the $^2F_{5/2}$ terms by the crystal field. From the three possible crystal field levels of the $^2F_{5/2}$ term of the Pa^{4+} ion, the two lower levels are energetically separated by 15-30 cm^{-1}. The third level is separated by some 600 cm^{-1}. The low symmetry of the compound or a J-J-mixing can not explain the magnitude of the magnetic moment.

$5f^2$-Configuration ($^3H_4$)

Uranium(IV) compounds, $U(C_5H_5)_4$; $(C_5H_5)_3UX$

There are two classes of compounds, those with an almost T_d-symmetry like $U(C_5H_5)_4$, $(C_5H_5)_3UBH_4$, $(C_5H_5)_3UBF_4$ and $(C_5H_5)_3UOR$, and those with a monoclinic structure like most $(C_5H_5)_3UX$ compounds.

Compounds of the first class show a small electric dipole moment and a magnetic moment of 2.2 to 2.8 μ_B ($\mu = 2.84\sqrt{\chi(T-\Theta)}$). Compounds of the second class have large dipole moments ($(C_5H_5)_3UCl = 3.88$, $(C_5H_5)_3UJ = 4.89$ Debye units) and an effective magnetic moment of about 3.3 μ_B). Fig. 5.3 shows a schematic crystal field splitting of the $^3H_4$ groundstate of both classes.

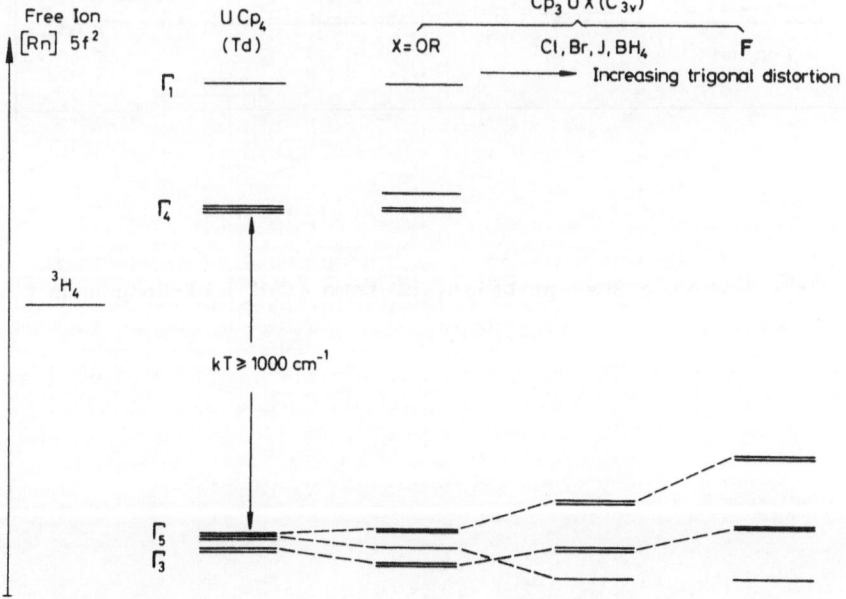

Fig. 5.3. CF-splitting of the ionic $^3H_4$ ground state of U^{4+}. 84

With increasing trigonal distortion the temperature range of the
temperature independent paramagnetism of the compounds $(C_5H_5)_3UX$
also increases, as shown in Fig. 5.4 to 5.6.

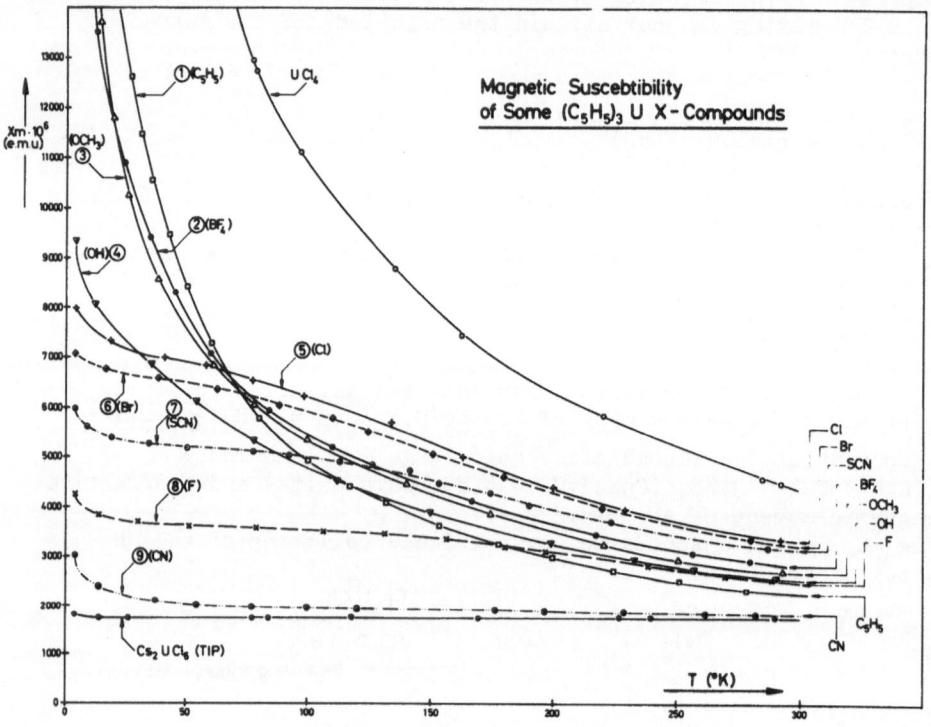

Fig. 5.4. Magnetic Susceptibility of some $(C_5H_5)_3UX$-Compounds

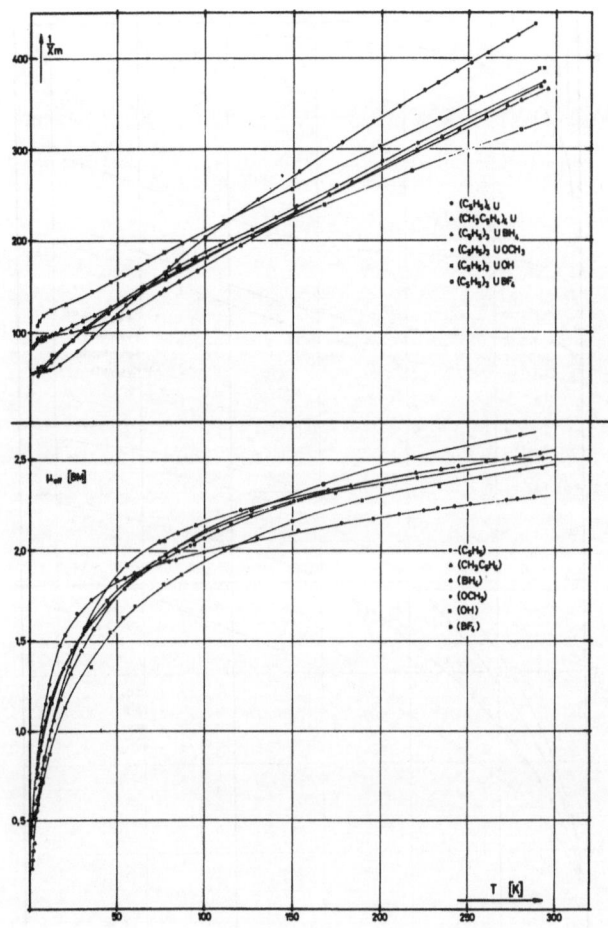

Fig. 5.5 $1/\chi_m$ and μ_{eff} values of U^{4+} compound with almost
Td symmetry.

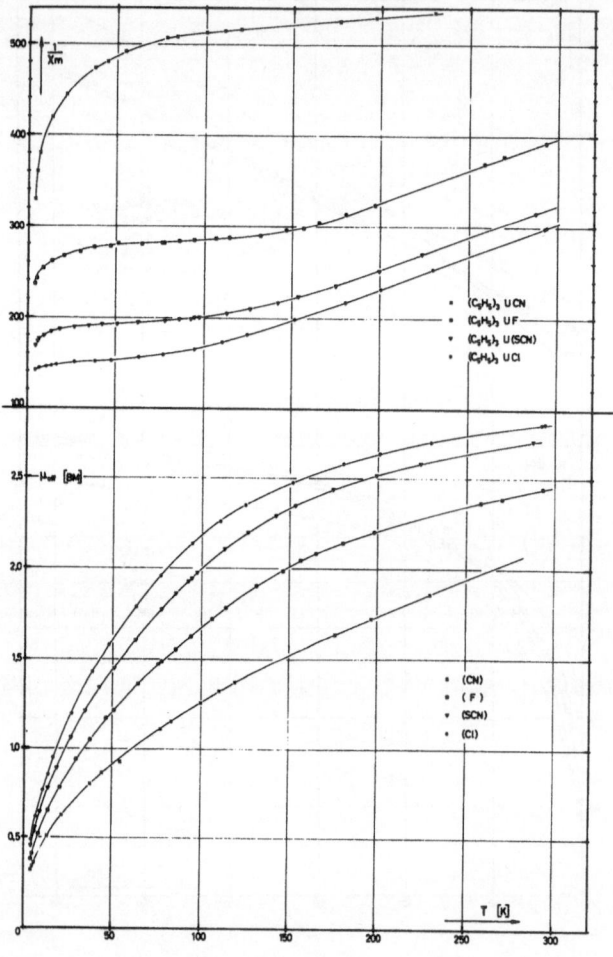

Fig. 5.6 $1/\chi_m$ and μ_{eff} values of U^{4+} compound with trigonal distortion.

The magnetic susceptibility of $U(C_5H_5)_4$ has been discussed [85] by using different semi-empirical approaches with the assumption of rather weakly perturbed tetrahedral crystal fields. Above 200 K the paramagnetic susceptibility obeys a Curie-Weiss law with $\Theta \cong - 91$ K. (Fig. 5.7).

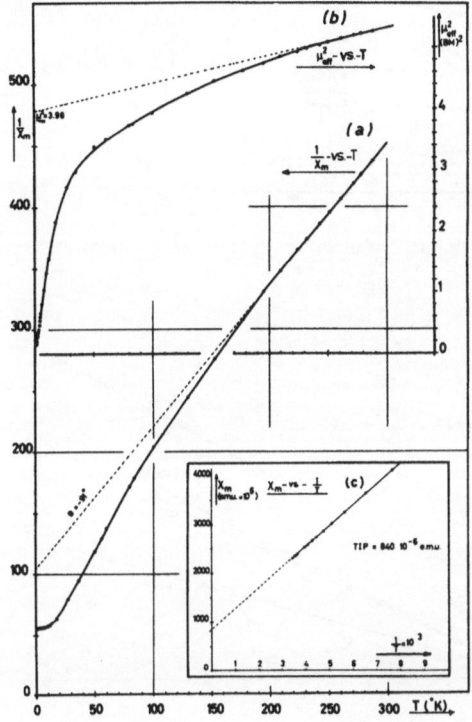

Fig. 5.7

Magnetic suscepti-

bility data of

$U(C_5H_5)_4$.

The susceptibilities of the complexes $(C_5H_5)_3UX$ (X = F, Cl, Br, J) have been analyzed and compared to the susceptibilities of the uranium tetrahalides [86]. The compounds do not obey a Curie-Weiss law. The magnetic moments at room temperature show maximally 70% of the moment calculated for the free U^{4+}-ion. (Fig. 5.8/5.9).
The influence of the ligand X on the ground state of the U^{4+}-ion can be seen in the μ_{eff}^2-curves at low temperatures (Fig. 5.10). Spectroscopic and magnetic susceptibility results are consistent. Magnetic measurements have also been used to characterize the adducts shown in Fig. 5.11 [20].

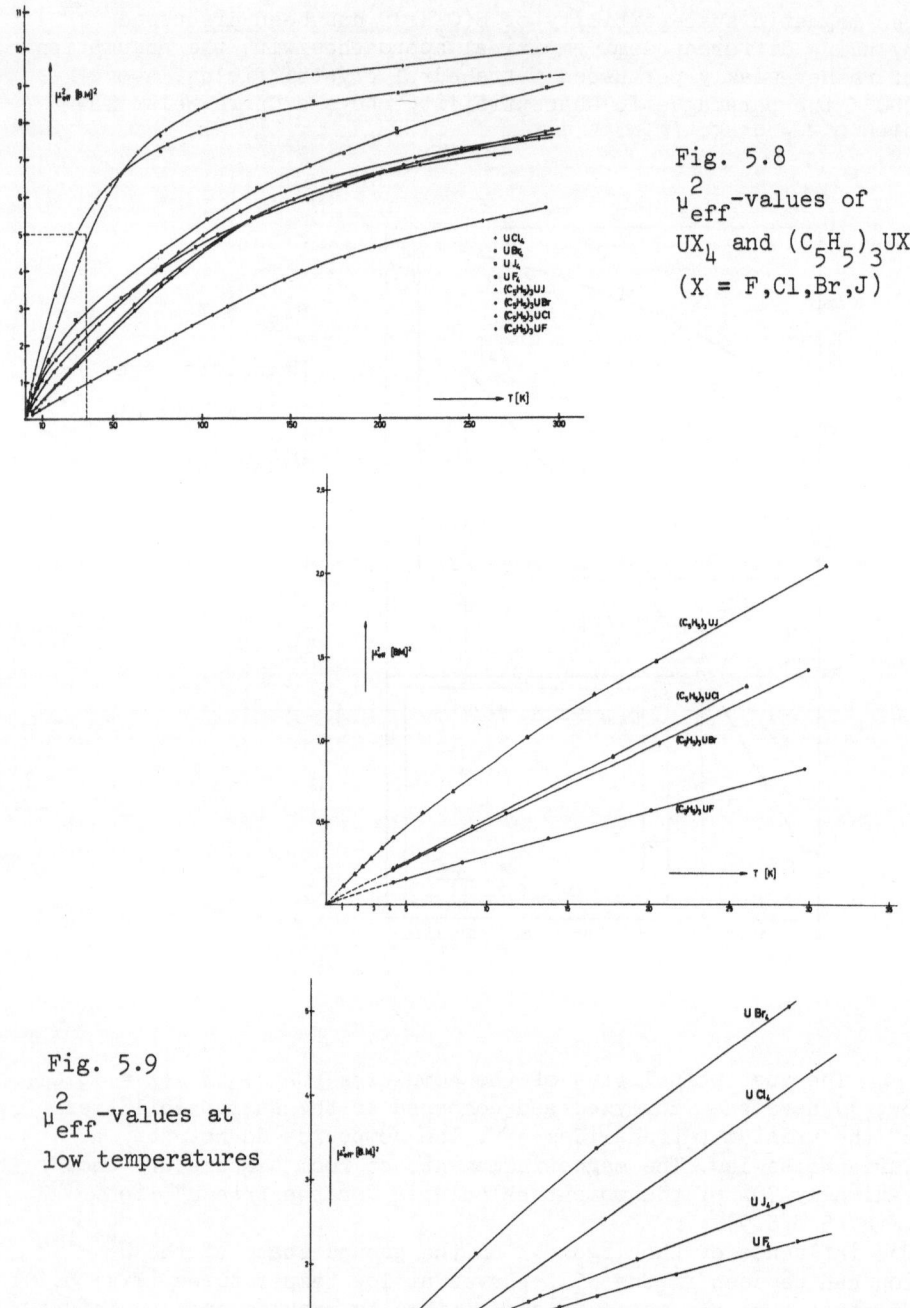

Fig. 5.8
μ_{eff}^2-values of
UX_4 and $(C_5H_5)_3UX$
$(X = F, Cl, Br, J)$

Fig. 5.9
μ_{eff}^2-values at
low temperatures

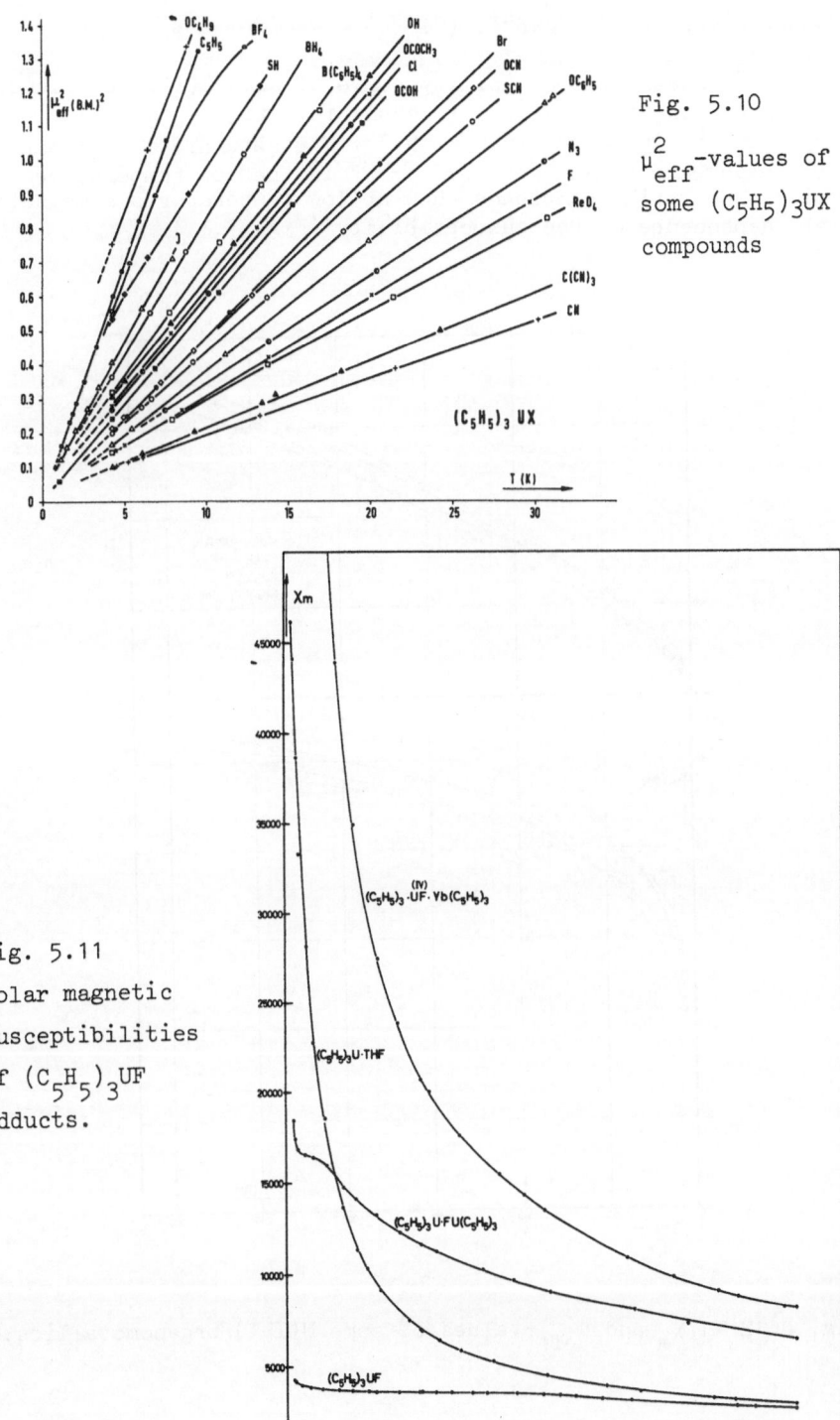

Fig. 5.10

μ_{eff}^2-values of some $(C_5H_5)_3UX$ compounds

Fig. 5.11 molar magnetic susceptibilities of $(C_5H_5)_3UF$ adducts.

f^3-configuration (U^{3+}, Np^{4+}) ($^4I_{9/2}$)

The magnetic susceptibility curves of some U(III) compounds are plotted in Fig. 5.12 and recently investigated Np(IV) compounds [87] are shown in Fig. 5.13. The data are in agreement with measurements made earlier [88,89] in the low temperature region. The Np(IV) compounds show at low temperatures a magnetic field dependence of the susceptibility (Fig. 5.14)

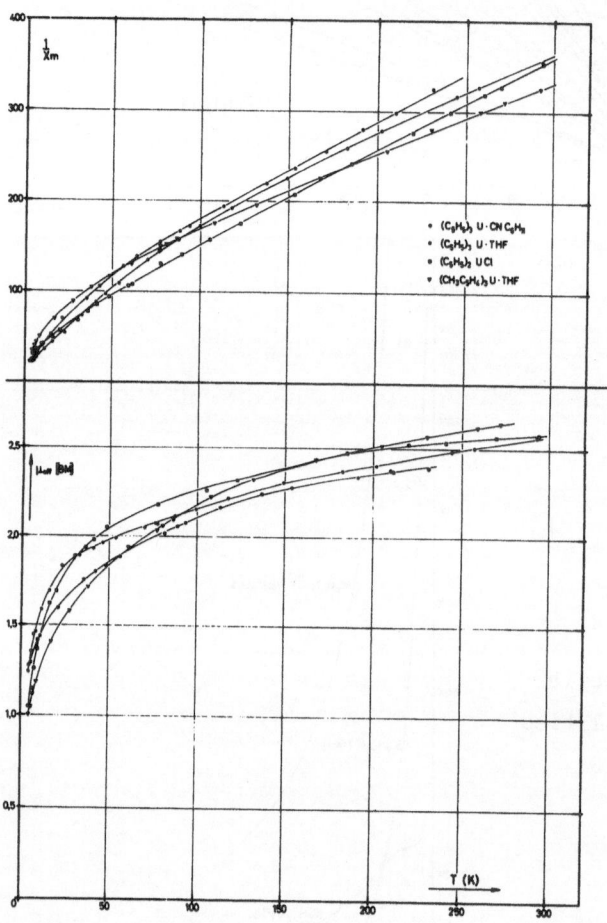

Fig. 5.12 $1/\chi_m$ and μ_{eff}-values of some U(III) organometallics.

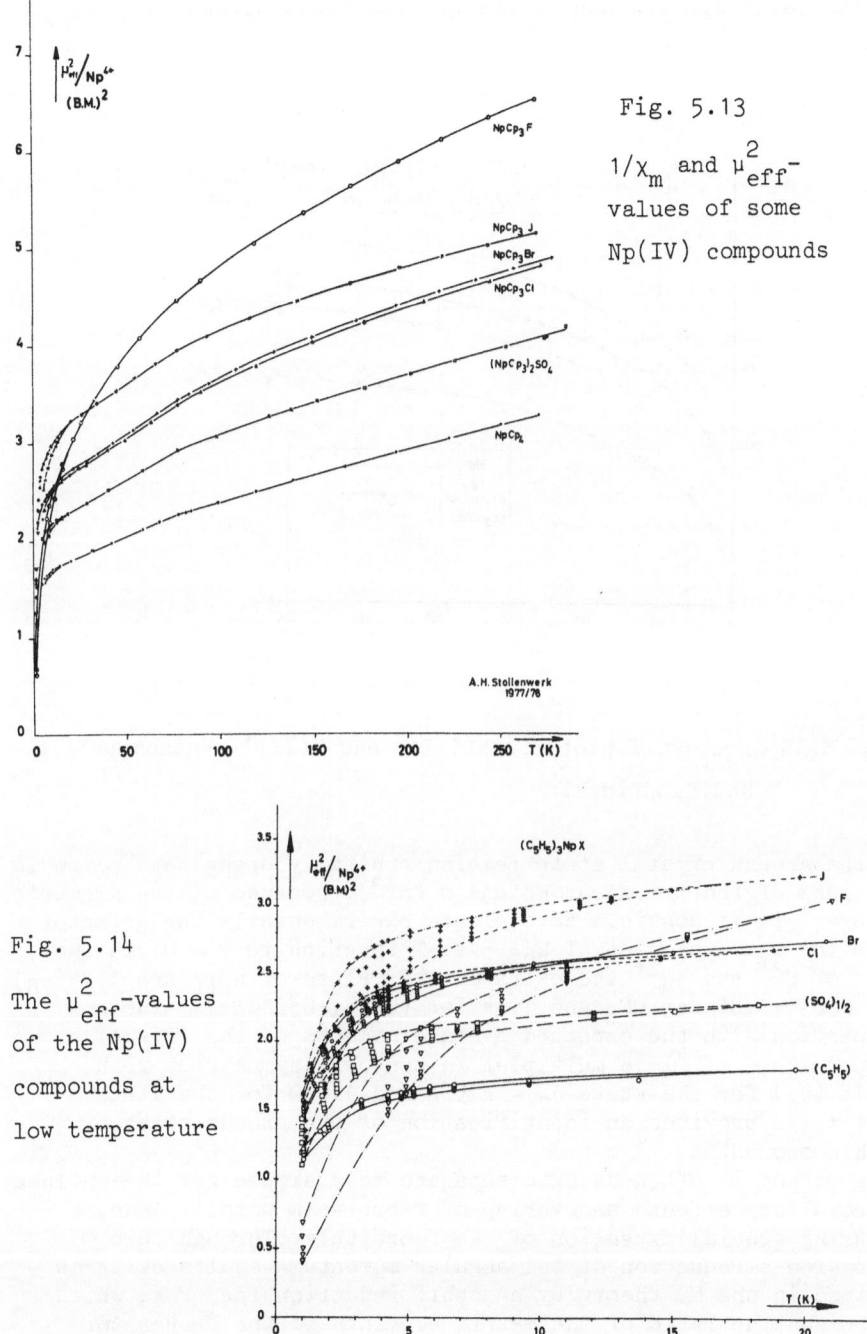

Fig. 5.13

$1/\chi_m$ and μ_{eff}^2-
values of some

Np(IV) compounds

Fig. 5.14

The μ_{eff}^2-values
of the Np(IV)
compounds at
low temperature

The temperature dependence of μ_{eff} in Fig. 5.15 gives information on the level spacing and on the crystal field strength.

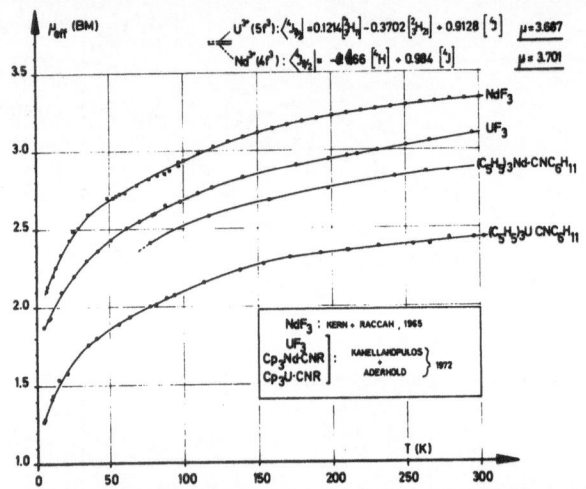

Fig. 5.15 μ_{eff}-vs.T plot of Nd(III)- and U(III)-organometallics and fluorides.

If the ground crystal state remains orbitally degenerate, like in the case of the trivalent Nd and U (nf^3), because of the Kramer's theorem, it is possible to evaluate experimentally the g-factor from the extrapolation of μ^2_{eff}-vs.-T function to T = 0. In the case of U^{3+} and Nd^{3+} the extrapolated values of μ_{eff} are 1.95(μ_B) for NdF_3, 1.80 for UF_3 and 1.14 for $(C_5H_5)_3U\cdot CNC_6H_{11}$. The comparison with the expected average moments of the crystal ground state of $^4I_{9/2}$ multiplet in noncubic symmetries with 1.818 (μ_B) for the state $J_z = \pm 5/2$ and 1.090 for the state $J_z = \pm 3/2$ provides an identification of the ground state of each compound.

This method is often used to separate in a simple way the influence of covalency effects as overlap of f-orbitals with ligands, assuming special extension of the f-orbitals. Through this extension a reduction of the angular momentum results as it is defined in the MO theory by an orbit reduction factor k, which expresses the ratio of the matrix elements of the $\hat{L}$ operator between the MO of the complex and those of the unpertubated f-orbitals.

In NdF_3 three crystal field states are populated as indicated
by the significant changes of the slope of the μ_{eff}-curve . This
is due mainly to second order Zeeman effects. On the other hand
the slowly monotonous increase of the μ_{eff}-value in UF_3 until
150 K suggests that not only the ± 5/2 but also the ± 3/2 state
is populated. This is also supported by the anisotropic ESR-
spectrum 90.

Strong interactions between ligand field and central ion
in the organometallics is indicated by the μ_{eff}-values. All
investigated ionic Nd^{3+}-compounds show μ_{eff}-values greater than
the UF_3 values, whereas the values of organometallics lie below
those of UF_3.

f^4-Configuration ($^5I_4$)

The magnetic susceptibility of $(C_5H_5)_3Np\cdot3THF$ has been
measured by Karraker from 2.5 to 100 K. The Np^{3+} ion in this
compound is in a site descended from O_h symmetry 88.
Fig. 5.16 shows the μ^2_{eff}-vs.-T diagram of $(C_5H_5)_3Np\cdot CNC_6H_{11}$
and the calculated curve for a pure $(C_5H_5)_3Np$ 67.

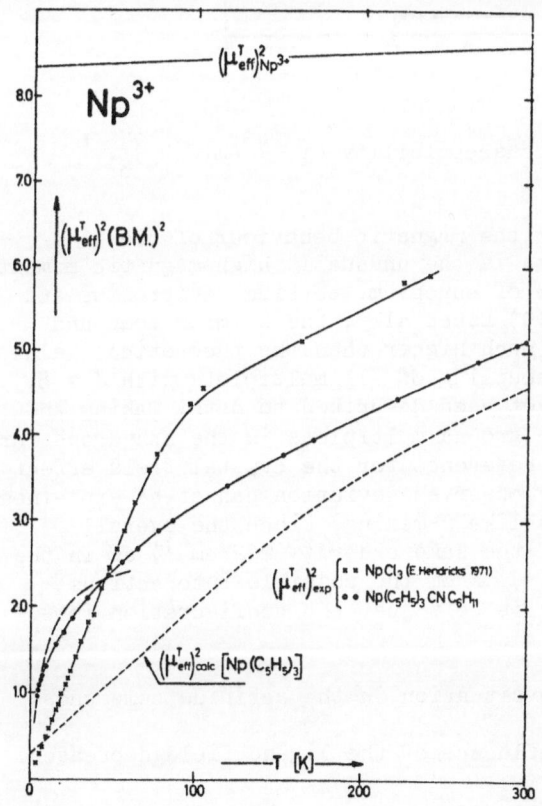

Fig. 5.16

μ^2_{eff}-values of
Np(III) compounds.

$5f^5$-Configuration ($^6H_{5/2}$)

Pu(C_5H_5)$_3$ and some adducts with Lewis bases have been measured [67]. The data have been theoretically treated by using Van Vleck's theory.

$5f^6$-Configuration (^7F_o)

Am(C_5H_5)$_3$ shows a temperature independent paramagnetism of $(716 \pm 14) \cdot 10^{-6}$ emn (Fig. 5.17) as it is expected for the ^7F_o ground state.

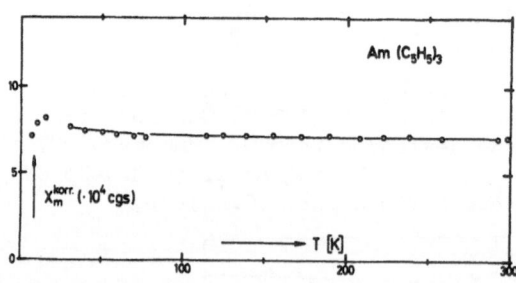

Fig. 5.17 The magnetic susceptibility of ^{243}Am(C_5H_5)$_3$. [69]

A striking point in the magnetic behaviour of the (C_5H_5)$_3$Ln·CNC$_6$H$_{11}$ adducts is the unusually high magnetic moment, particularly in the case of europium, terbium, dysprosium and holmium. (Fig. 5.18, 5.19) Especially, the holmium compound shows a moment which is much higher than the theoretical value including intermediate coupling of all multiplets with J = 8. Such an effect has not yet been described to date. Taking the overall splitting of the ground multiplets in the anhydrous rare earth trichlorides as a reference for the crystal field effect at room temperature, the observed deviation cannot be explained by crystal field effects like J-mixing, since the overall splitting of J = 8 is of the same order ($\sim$ 212 cm^{-1}) as in the case of PrCl$_3$ ($\sim$ 200 cm^{-1}). From the possible interactions influencing the g factor is left just the configuration interaction.

Fig. 5.20 shows the situation in the actinide compounds (U, Np, Pu, Am).
With increasing Z the influence of the ligand field decreases.

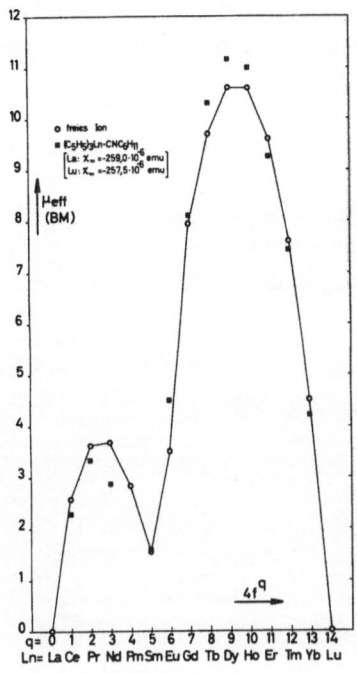

Fig. 5.18

μ_{eff}-values of $(C_5H_5)_3Ln \cdot CNC_6H_{11}$ at room temperature

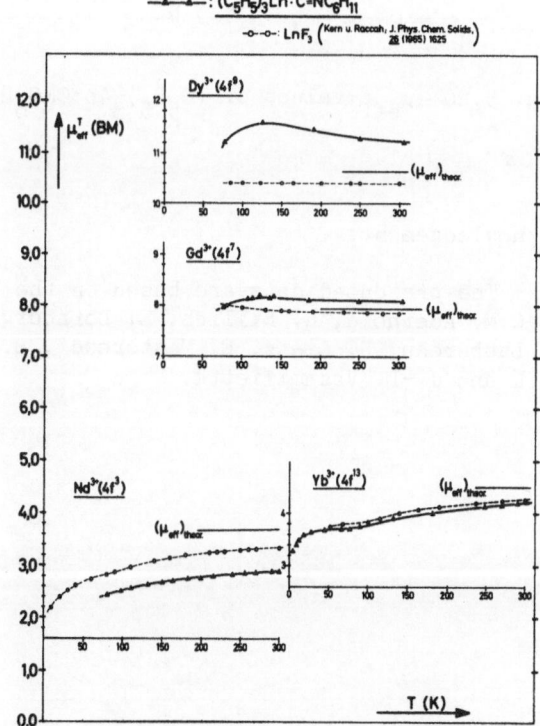

Fig. 5.19

magnetic moments of the Nd, Gd, Dy and Yb compounds

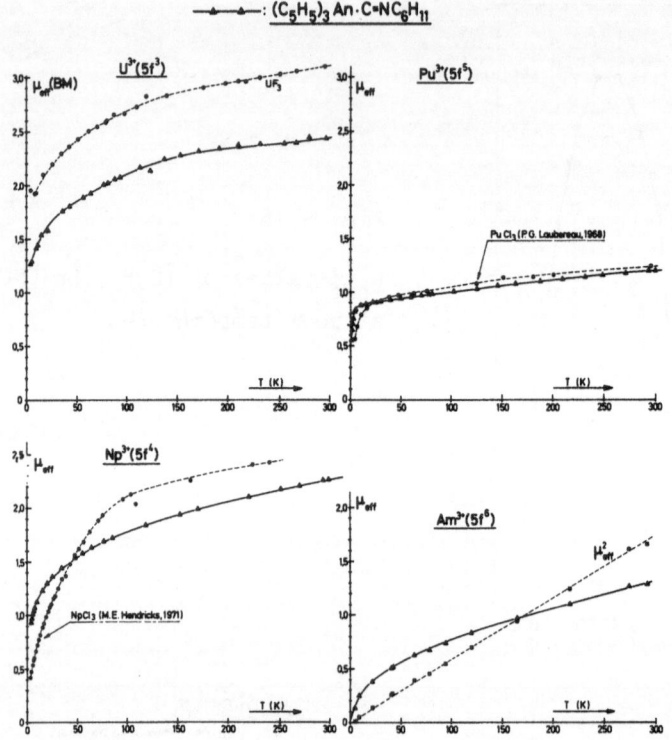

Fig. 5.20 μ_{eff}-values of $(C_5H_5)_3An \cdot CNC_6H_{11}$ compounds.

Acknowledgements

The presented data are based on the experimental work of C.M. Aderhold, H. Billich, E. Dornberger, R. Klenze, P. Laubereau, H. Lietz, E. Mastoroudi, B. Powietzka, A. Stollenwerk and J.-L. Vasamiliette.

References

1. T.J. Kealy, P.L. Pauson, Nature, 168, 1039 (1951).

2. S.A. Miller, J.A. Tebboth, J.F. Tremaine, J. Chem. Soc., 632 (1952).

3. L.T. Reynolds, G. Wilkinson, J. Inorg. Nucl. Chem., 2, 246 (1956).

4. F. Baumgärtner, E.O. Fischer, P. Laubereau, Naturwiss., 52, 560 (1965).

5. H. Gysling, M. Tsutsui, Adv. Organomet. Chem., 9, 361 (1970).

6. R.G. Hayes, J.L. Thomas, Organomet. Chem. Rev., A7, 1 (1971).

7. B. Kanellakopulos, K.W. Bagnall, MPT Intern. Rev. Sci., Series One, 7, 217 (1972).

8. F. Baumgärtner, B. Kanellakopulos, Gmelin's Handbuch, Vol.4, C 19, 271 (1972).

9. E. Cernia, A. Mazzei, Inorg. Chim. Acta, 10, 239 (1974).

10. F. Calderazzo, J. Organomet. Chem., 79, 175 (1974).

11. T.J. Marks, J. Organomet. Chem., 79, 181 (1974).

12. T.J. Marks, J. Organomet. Chem., 95, 301 (1975).

13. T.J. Marks, J. Organomet. Chem., 119, 229 (1976).

14. T.J. Marks, J. Organomet. Chem., 138, 157 (1977).

15. B. Kanellakopulos, F. Baumgärtner, Gmelin's Handbuch der anorg. Chemie, Vol.2, 96 (1977).

16. A. Streitwieser, Jr., Tetrahedron Letters, No.6, 23 (1960).

17. P. Laubereau, Dissertation, Technische Universität, München (1966).

18. A.F. Reid, P.C. Wailes, Inorg. Chem., 7, 1213 (1965).

19. F. Baumgärtner, E. Dornberger, B. Kanellakopulos, unpublished results

20. B. Kanellakopulos, Habilitationsschrift, Universität Heidelberg (1972).

21. B. Kanellakopulos, E. Dornberger, H. Billich, J. Organomet.
    Chem., 76, C 42 (1974).

22. E. Dornberger, R. Klenze, B. Kanellakopulos, Inorg. Nucl.
    Chem. Letters, 14, 319 (1978)

23. G.R. Sienel, Dissertation, Universität Erlangen-Nürnberg
    (1976).

24. E.O. Fischer, A. Treiber, Z. Naturforschung, 17^b, 276 (1962).

25. F. Baumgärtner, E.O. Fischer, B. Kanellakopulos, P. Laubereau,
    Angew. Chem., 81, 182 (1969); Int. Ed., 8, 202 (1969).

26. E.O. Fischer, Y. Hristidu, Z. Naturforschung, 17^b, 275 (1962).

27. Y. Hristidu, Dissertation, Ludwig-Maximilian Universität,
    München (1962).

28. F. Baumgärtner, E.O. Fischer, B. Kanellakopulos, P. Laubereau,
    Angew. Chem., 80, 661 (1968); Int. Ed., 7, 634 (1968).

29. L.R. Grisler, W.G. Eggerman, J. Inorg. Nucl. Chem., 36,
    1424 (1974).

30. D.G. Karraker, private communications (1971).

31. H. Billich, Diplomarbeit, Universität Heidelberg (1975).

32. D.G. Karraker, J.A. Stone, Inorg. Chem., 11, 1742 (1972).

33. R. von Ammon, B. Kanellakopulos, R.D. Fischer, Chem. Phys.
    Letters, 2, 513 (1968).

34. J.H. Burns, J. Amer. Chem. Soc., 95, 3815 (1973).

35. J.H. Burns, J. Organomet. Chem., 69, 225 (1974).

36. E.C. Baker, G.W. Halstead, K.N. Raymond, Structure and
    Bonding, Vol. 25, 23-68 (1976); Springer-Verlag, Berlin,
    Heidelberg, New York.

37. B. Kanellakopulos, E.O. Fischer, E. Dornberger, F. Baum-
    gärtner, J. Organomet. Chem., 24, 507 (1970).

38. P. Laubereau, private communications (1968).

39. G.L. Ter Haar, M. Dubeck, TID 18749 (1963); Inorg. Chem.,
    3, 1848 (1964).

40. R.D. Fischer, R. von Ammon, B. Kanellakopulos, J. Organomet.
    Chem., 25, 123 (1970).

41. M.L. Anderson, L.R. Crisler, CDRL-940327-2 (1968); J. Organo-
    met. Chem., 17, 345 (1969).

42. H. Marquet-Ellis, G. Folcher, CEA-R-4668; J. Organomet. Chem.,
    257 (1977).

43. a) R.D. Fischer, G.R. Sienel, Z. Anorg. Allg. Chem., 419,
       126 (1976).

    b) R.D. Fischer, G.R. Sienel, G. Landgraf, H. Wagner,
       VIIth Intern. Conf. On Organomet. Chem., Venice,
       Sept. 1-5 (1975) Abstr.1.

44. R. von Ammon, B. Kanellakopulos, R.D. Fischer, Radiochim.
    Acta, 11, 162 (1969).

45. R. von Ammon, R.D. Fischer, B. Kanellakopulos, Chem. Rev.,
    105, 45-62 (1972).

46. E.O. Fischer, P. Laubereau, F. Baumgärtner, B. Kanellakopulos,
    J. Organomet. Chem., 5, 583 (1966).

47. W.T. Carnall, P.R. Fields, R.G. Pappalardo, Progr. in Coord.
    Chem., Haifa-Jerusalem, Sept. 1968, page 411 (Ed. by M. Cais)
    Elsebrier.

48. A. Stollenwerk, E. Dornberger, B. Kanellakopulos, (1977/78)
    unpublished data.

49. P. Zanella, S. Faleschini, L. Doretti, G. Faraglia,
    J. Organomet. Chem., 26, 353 (1971).

50. B. Kanellakopulos, C. Aderhold, E. Dornberger, J. Organomet.
    Chem., 66, 447 (1974).

51. J.D. Jamerson, J. Takats, J. Organomet. Chem., 78, C23 - C24
    (1974).

52. P. Zanella, G. de Paoli, G. Bombieri, G. Zanotti, R. Rossi,
    J. Organomet. Chem., 142, C 21 (1977).

53. L. Doretti, P. Zanella, G. Faraglia, S. Faleschini, J. Organo-
    met. Chem., 43, 339 (1972).

54. K.W. Bagnall, J. Edwards, J. Organomet. Chem., 80, C 14 (1974)

55. K.W. Bagnall, J. Edwards, A.C. Tempest, J. Chem. Soc., (Dalton) 295 (1978).

56. B. Kanellakopulos, E. Dornberger, R. von Ammon, R.D. Fischer, Angew. Chem., 82, 956 (1970); Int. Ed., 9, 957 (1970).

57. T. Marks, A.M. Seyam, J. Amer. Chem. Soc., 94, 6545 (1972).

58. A.E. Gebala, M. Tsutsui, J. Amer. Chem. Soc., 95, 91 (1973).

59. T. Marks, A.M. Seyam, J.R. Kolb, J. Amer. Chem. Soc., 95, 5529 (1973).

60. W. Wagner, Dissertation, Univ. of Heidelberg, 1974.

61. V.K. Vasilev, V.N. Sokolov, G.P. Kondratenkov, J. Organomet. Chem., 142, C 7 (1977).

62. G. Lugli, M. Brunelli, A. Mazzei, 7th Intern.Conf. on Organomet. Chem., Venice-Italy, Sept. 1-5, 1975.

63. J.M. Manriquez, P.J. Fagan, T.J. Marks, J. Amer. Chem. Soc., 100, 3939 (1978).

64. B. Kanellakopulos, E. Dornberger, F. Baumgärtner, Inorg. Nucl. Chem. Letters, 10, 155 (1974).

65. D.G. Kalina, T.J. Marks, W.A. Wachter, J. Amer. Chem. Soc., 99, 3877 (1977).

66. F. Baumgärtner, E.O. Fischer, B. Kanellakopulos, P. Laubereau, Angew. Chem., 77, 866 (1965); Int. Ed., 4, 878 (1965).

67. C.M. Aderhold, Dissertation, Universität Heidelberg, 1975.

68. B. Kanellakopulos, C. Aderhold, XXIV. IUPAC Conf. on Coord. Chem., Hamburg 1973, Sept. 2-8.

69. B. Kanellakopulos, C. Aderhold, E. Dornberger, W. Müller, R. Baybarz, Radiochim. Acta (in press).

70. F. Baumgärtner, E.O. Fischer, H. Billich, E. Dornberger, B. Kanellakopulos, W. Roth, L. Stieglitz, J. Organomet. Chem., 22, C 17 (1970).

71. P.G. Laubereau, J.H. Burns, Inorg. Nucl. Chem. Letters, 6, 59 (1970).

72. P.G. Laubereau, J.H. Burns, Inorg. Chem., 9, 1091 (1970).

73. P.G. Laubereau, Inorg. Nucl. Chem. Letters, 6, 611 (1970).

74. L.J. Nugent, P.G. Laubereau, G.K. Werner, K.L. Van der Sluis, J. Organomet. Chem., 27, 365 (1971).

75. P.G. Laubereau, L. Ganguly, J.H. Burns, B.U. Benjamin, J.L. Atwood, J. Selbin, Inorg. Chem., 10, 2274 (1971).

76. J. Goffart, J. Fuger, B. Gilbert, L. Hocks, D. Duyckaerts, Inorg. Nucl. Chem. Letters, 11, 169 (1975).

77. J. Goffart, G. Duyckaerts, Inorg. Nucl. Chem. Letters, 14, 15 (1978).

78. J. Goffart, B. Gilbert, G. Duyckaerts, Inorg. Nucl. Chem. Letters, 13, 189 (1977).

79. B. Kanellakopulos, E. Dornberger, unpublished results.

80. A.M. Seyam, G. Ala' Eddein, Inorg. Nucl. Chem. Letters, 13, 115 (1977).

81. L. Brewer, J. Opt. Soc. Am., 61, 1666 (1971).

82. R.L. Martin, G. Winter, J. Chem. Soc., 4709 (1965).

83. R.J. Clark, J. Lewis, D.J. Machlin, R.S. Nyholm, J. Chem. Soc., 379 (1963).

84. R. von Ammon, B. Kanellakopulos, R.D. Fischer, XIIIth Intern. Conf. on Coord. Chem., Krakov-Zakopane, Poland, Sept. 14-22, 1970.

85. H.-D. Amberger, R.D. Fischer, B. Kanellakopulos, Z. Naturforsch., 316, 12 (1976).

86. C. Aderhold, F. Baumgärtner, E. Dornberger, B. Kanellakopulos, Z. Naturforsch. in press.

87. A. Stollenwerk, Thesis in progress.

88. D. Karraker, J. Stone, Inorg. Chem., 11, 1742 (1972).

89. R.D. Fischer, P. Laubereau, B. Kanellakopulos, Z. Naturforsch., 24, 616 (1969).

90. J. M. Baker and R.S. Rubius, Proc. Phys. Soc. (London) 78, 1313 (1961).

ELECTRONIC STRUCTURE OF f-BLOCK COMPOUNDS

Norman M. Edelstein

Materials & Molecular Research Division, Lawrence
Berkeley Laboratory, Berkeley, California 94720 USA

Introduction

The lanthanide and actinide series differ from the more
widely studied d-transition metal series in that the 4f and 5f
shells are inner electrons shielded by the $5s^2 5p^6$ (or $6s^2 6p^6$)
closed shells. The result of this shielding is that the f shell
interacts much less strongly with its environment than the d-
transition series. Figures 1 and 2 illustrate the radial charge
density for Pr^{3+} and U^{4+}. The electronic structure of an f^n ion
is dominated by different interactions than for the more familiar
d-transition ions. In this paper we will review the methods and
nomenclature used to describe the electronic structure of f^n compounds.

Brief Review of Atomic Theory [1-4].

For an N electron atom with a nuclear charge Ze where e is
the charge of the electron and Z is the atomic number, the non-
relativistic Hamiltonian may be written (assuming the nuclear
mass is infinite)

$$\mathcal{H} = \sum_{i=1}^{N} \frac{p_i^2}{2m} - \sum_{i=1}^{N} \frac{Ze^2}{r_i} + \sum_{i<j}^{N} \frac{e^2}{r_{ij}} \quad . \tag{1}$$

The first term in this equation represents the kinetic energy of
all the electrons, the second term the potential energy of all
the electrons in the electric field of the nucleus, and the third
term the repulsive Coulomb potential between pairs of electrons.

T. J. Marks and R. D. Fischer (eds.), Organometallics of the f-Elements, 37–79.

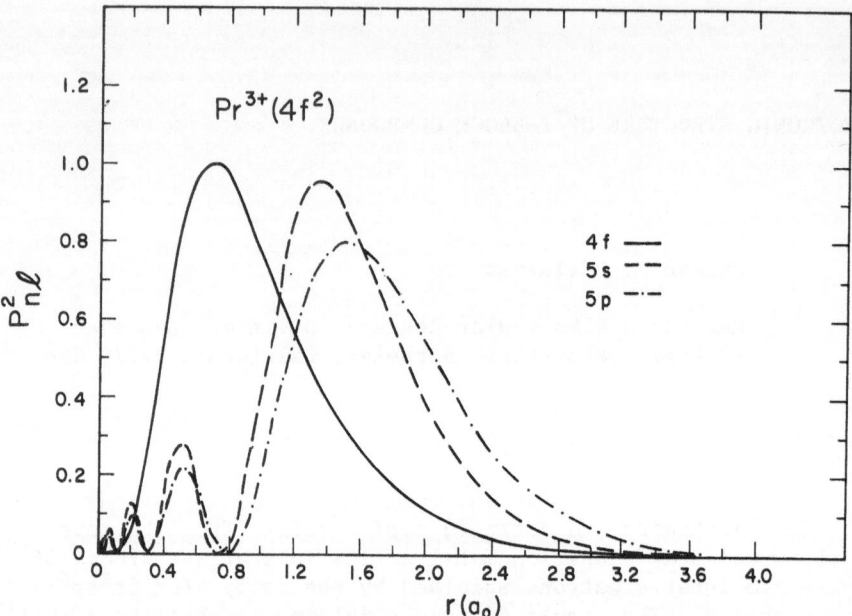

Figure 1. Radial charge density for Pr^{3+}.

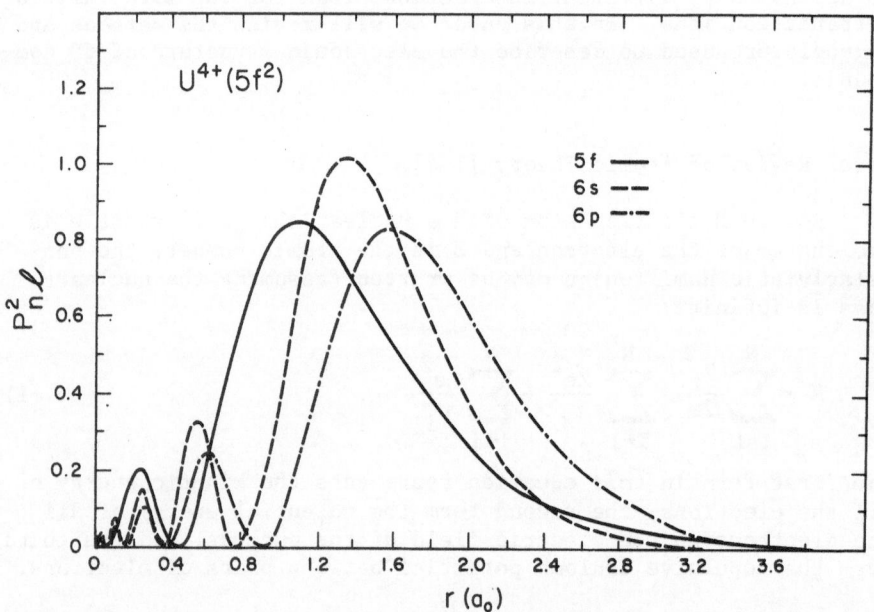

Figure 2. Radial charge density for U^{4+}.

In order to solve this equation we make use of the central field approximation for which the following assumptions are made:
1) Each electron is assumed to move independently.
2) There is a central field made up of the spherically averaged potential fields of each of the other electrons and the nucleus; that is each electron is said to be moving in a spherically symmetric field (potential), $-\dfrac{U(r_i)}{e}$.

Then we may write the central field Hamiltonian

$$\mathcal{H}_{CF} = \sum_{i=1}^{N} \left[\frac{p_i^2}{2m} + U(r_i) \right] \quad . \tag{2}$$

This central field Hamiltonian results in a Schrodinger equation which may be readily solved in polar coordinates with wavefunctions of the form

$$\Psi = r^{-1} R_{n\ell}(r) \, Y_{\ell\ell_z} (\Theta,\phi) \quad . \tag{3}$$

These wavefunctions are products of the radial functions $R_{n\ell}(r)$ times the spherical harmonics $Y_{\ell\ell_z}(\Theta,\phi)$ and the energy levels are highly degenerate. The energy levels are labeled by the principal quantum number n and the orbital quantum number ℓ. This degeneracy is removed by considering a number of perturbing effects. An electronic configuration is described by a particular set of quantum numbers n and ℓ. For example, the electronic configuration of the U^{4+} ion is

$$1s^2 2s^2 2p^6 3s^2 3p^6 3d^{10} 4s^2 4p^6 4d^{10} 4f^{14} 5s^2 5p^6 5d^{10} 6s^2 6p^6 5f^2$$

or as is commonly written [Rn]$5f^2$. Since the electrons in filled subshells ($n\ell^{4\ell+2}$) do not contribute to the electronic structure of the low-lying levels we consider in this paper only the properties of the electrons in the unfilled shell.

For f electrons the most important perturbation is the term obtained by subtracting equation 2 from equation 1:

$$\mathcal{H} - \mathcal{H}_{CF} = \sum_{i=1}^{N} \left[-\frac{Ze^2}{r_i} - U(r_i) \right] + \sum_{i<j}^{N} \frac{e^2}{r_{ij}} \quad . \tag{4}$$

The first summation shifts all the levels in a given configuration equally so we will not consider it. The second term

$$\mathcal{H}_1 = \sum_{i<j}^{N} \left(\frac{e^2}{r_{ij}}\right) \tag{5}$$

represents the electrostatic Coulomb repulsion between pairs of electrons.

It is convenient at this point to introduce the operators (note: throughout this paper all operators are typed in script;)

$$L = \sum_i \ell_i \quad \text{and} \quad S = \sum_i \delta_i$$

where ℓ_i and δ_i are the orbital and spin angular momentum operators of the ith electron. $\mathcal{H}_1$ is diagonal in L and S which means we can label the eigenstates with particular eigenvalues of L and S in the form ^{2S+1}L. This type of coupling is called Russell-Saunders coupling or L-S coupling.

To allow for relativistic corrections in the Hamiltonian we introduce $\mathcal{H}_2$, the spin-orbit interaction as

$$\mathcal{H}_2 = \sum_i \xi(r_i)\, \delta_i \cdot \ell_i \tag{6}$$

or $\quad \mathcal{H}_2 = \zeta_{n\ell}\, S \cdot L$

where $\quad \xi(r) = \dfrac{\hbar^2}{2m^2 c^2 r} \dfrac{dU}{dr}$

and $\quad \zeta_{n\ell} = \displaystyle\int_0^\infty R_{n\ell}^2 \xi(r)dr$.

The term $\mathcal{H}_2$ becomes progressively more important as Z, the atomic number increases. This spin-orbit interaction is diagonal in J where $J = L + S$. The ^{2S+1}L multiplet is split into levels labeled by their J eigenvalues; $J = |L+S|$, $|L+S-1|, \ldots |L-S|$ where each J level has a degeneracy of 2J+1. The interaction $\mathcal{H}_2$ will couple ^{2S+1}L states whose value of S and L differ by not more than one. The spin-orbit interaction is especially important for actinide ions because of their high values of Z. Then the L-S coupling scheme is no longer a valid approximation and we speak of inter-mediate coupling.

In a complex a transition metal ion is surrounded by a number of ligands. If we assume the ion of interest has purely electrostatic interactions with its surroundings we can write

$$\mathcal{H}_3 = - e\sum_i V(r_i, \Theta_i, \phi_i) \qquad (7)$$

where the sum extends over all the electrons of the central ion. $V(r,\Theta,\phi)$ is the potential at the central ion from the surrounding ligands. This potential may be expanded in a series of spherical harmonics and we obtain

$$\mathcal{H}_3 = \sum_{i,k,q} D_k^q \langle r_i^k \rangle y_k^q (\Theta_i, \phi_i) \qquad . \qquad (8)$$

The number of terms that need to be considered in this series depends on the symmetry of the problem. This crystal-field or ligand field perturbation breaks the 2J+1 degeneracy of a particular J level and the eigenvalues of the operator J_z are needed to label the states.

The last perturbation to be discussed will be the Zeeman operator

$$\mathcal{H}_4 = \beta(L+2S) \cdot H \qquad (9)$$

To first order this operator can be replaced by

$$\mathcal{H}_4 = g_J \beta H \cdot J \qquad (10)$$

where g_J is the Landé g value which depends on L, S, and J for a particular free ion multiplet, H is the external magnetic field, and β is the Bohr magneton. This term is important for the discussion of magnetic susceptibility and electron paramagnetic resonance (EPR) results.

We can picture the results of this discussion in Figure 3. Here we show the energy levels of the f^2 configuration and schematically show the results of successively applying the perturbation Hamiltonians $\mathcal{H}_1$, $\mathcal{H}_2$, $\mathcal{H}_3$ to this configuration.

Ligand Field and Spin Orbit Hamiltonians for f^1.

In this section we will discuss in some detail the application of $\mathcal{H}_2$ and $\mathcal{H}_3$ to a configuration consisting of one f electron outside a closed shell. We shall use as an example a ligand field of axial symmetry such as one might find in uranocene-type molecules.

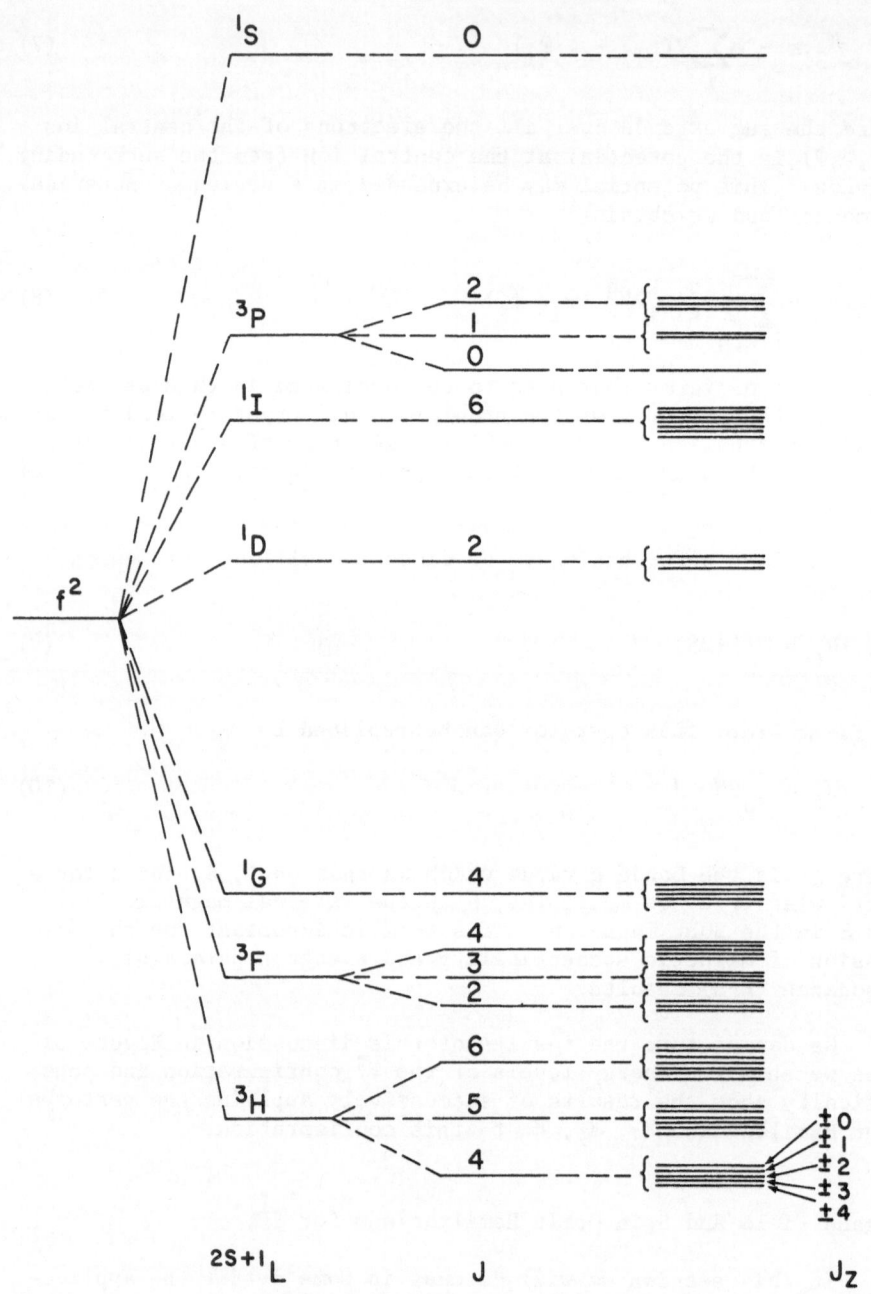

Figure 3. Energy levels of the f^2 configuration with the perturbations $\mathcal{H}_1$, $\mathcal{H}_2$, and $\mathcal{H}_3$ applied successively.

The ligand-field Hamiltonian was written as (equation 8)

$$\mathcal{H}_3 = \sum_{i,k,q} D_k^q \langle r_i^i \rangle Y_k^q(\Theta,\phi)$$

for an f electron. The value of k is restricted to be $k \leqslant 2\ell$ where ℓ is the value of the orbital angular momentum for the electrons. For f electrons $k \leqslant 6$. Furthermore, from a consideration of the parity of the matrix elements involving the crystal field potential, k must be even. This further restricts values of k to be 2, 4, or 6 for f electrons. Values allowed for q are restricted by the rule $|q| \leqslant k$. Further restrictions on q are determined by the symmetry of the ligand field. If the highest symmetry axis (the quantization axis) contains an n-fold rotation axis then $|q| = \lambda n$, $\lambda = 0,1,2,\ldots$. If q is non-zero, then only J_z eigenstates with eigenvalues that differ by $\pm q$ can be mixed by the crystal-field interaction. For the C_8 axis found in uranocene type molecules only $q = 0$ is allowed. Therefore, the non-zero crystal field parameters are D_2^0, D_4^0, and D_6^0. We shall later assume that D_2^0 is the dominant term, i.e., $D_2^0 \gg D_4^0$ or D_6^0, and determine the energy levels for an f^1 system.

A straightforward method for evaluating crystal field matrix elements was developed by Stevens [5], the operator-equivalent method. The basis for this method is derived from the Wigner-Eckart theorem from which it can be shown that within a particular J (or L) manifold all operators of the same rank have matrix elements which are proportional to one another. The matrix elements of these operators have been tabulated along with the proportionality constants for the ground terms of the f^n ions. A convenient source for these tables is the appendices of Abragam and Bleaney[6].

The usual way of writing the matrix element is

$$\langle J,J_z | \mathcal{H}_3 | J,J_z \rangle = \sum_{k,q} D_k^q \langle r^k \rangle \langle J,J_z | Y_k^q | J,J_z \rangle$$

$$= \sum_{k,q} K_k A_k^q \langle r^k \rangle \langle J,J_z | O_k^q | J,J_z \rangle$$

The O_k^q's are the equivalent operators which are proportional to the spherical harmonic tensors of Eq.(8) with the proportionality constants K_k being α, β, and γ; the second, fourth, sixth order operator equivalent factors, respectively. The A_k^q's are usually treated as crystal field parameters.

We will now evaluate the second order crystal field matrix elements for an f^1 ion. The equivalent operator is $O_2^0 = 3J_z^2 - J^2$.

Table 1

Matrix elements of the type $\langle \ell, \ell_z | O_k^0 | \ell, \ell_z \rangle$ for $L = 3$ ($\ell = 3$ for a single f electron) from Abragam and Bleaney. Each matrix element must be multiplied by $(-2/45)A_2^0 \langle r^2 \rangle$.

ℓ_z	0	±1	±2	±3
0	-12	0	0	0
±1	0	-9	0	0
±2	0	0	0	0
±3	0	0	0	15

From Table 17 of Appendix B in Abragam and Bleaney we find $J = 3$ (or L in our case) and the L_z states 0, ±1, ±2, ±3. The matrix is shown in Table 1. From Table 18 (Abragam and Bleaney) the second rank operator equivalent for an f electron is -2/45.

The energy levels may then be listed as

$$E_{\pm 3} = -\frac{2}{45} \times 15 \times A_2^0 \langle r^2 \rangle = -\frac{2}{3} A_2^0 \langle r^2 \rangle$$

$$E_{\pm 2} = -\frac{2}{45} \times 0 \times A_2^0 \langle r^2 \rangle = 0$$

$$E_{\pm 1} = -\frac{2}{45} \times (-9) \times A_2^0 \langle r^2 \rangle = +\frac{2}{5} A_2^0 \langle r^2 \rangle$$

$$E_0 = -\frac{2}{45} \times (-12) \times A_2^0 \langle r^2 \rangle = +\frac{8}{15} A_2^0 \langle r^2 \rangle \tag{12}$$

The factor $\langle r^2 \rangle$ is the expectation value of the radial wavefunction. Since A_2^0 and $\langle r^2 \rangle$ are functions of the radial wavefunctions, the usual practice is to evaluate these parameters empirically. The matrix elements and factors we have found are dependent only on the angular parts of the wavefunction and may be evaluated exactly. The energy levels in terms of the second, fourth, and sixth order crystal field parameters are given later in Appendix C.

The other interaction which is important for the f^1 case is the spin-orbit interaction $\mathcal{H}_2$. We may use the ℓ, ℓ_z basis set described above and evaluate exactly the angular part of this interaction. This is described in detail in Appendix A. We may

also use the J, J_z basis set which is diagonal in J to calculate
the energies of the states. This calculation is also shown in
Appendix A. Finally, we may draw a correlation diagram which goes
from the limit of strong spin-orbit interaction to the limit of
strong crystal-field interaction [7]. This diagram is shown in
Figure 4. Although there is no data for organoactinide or

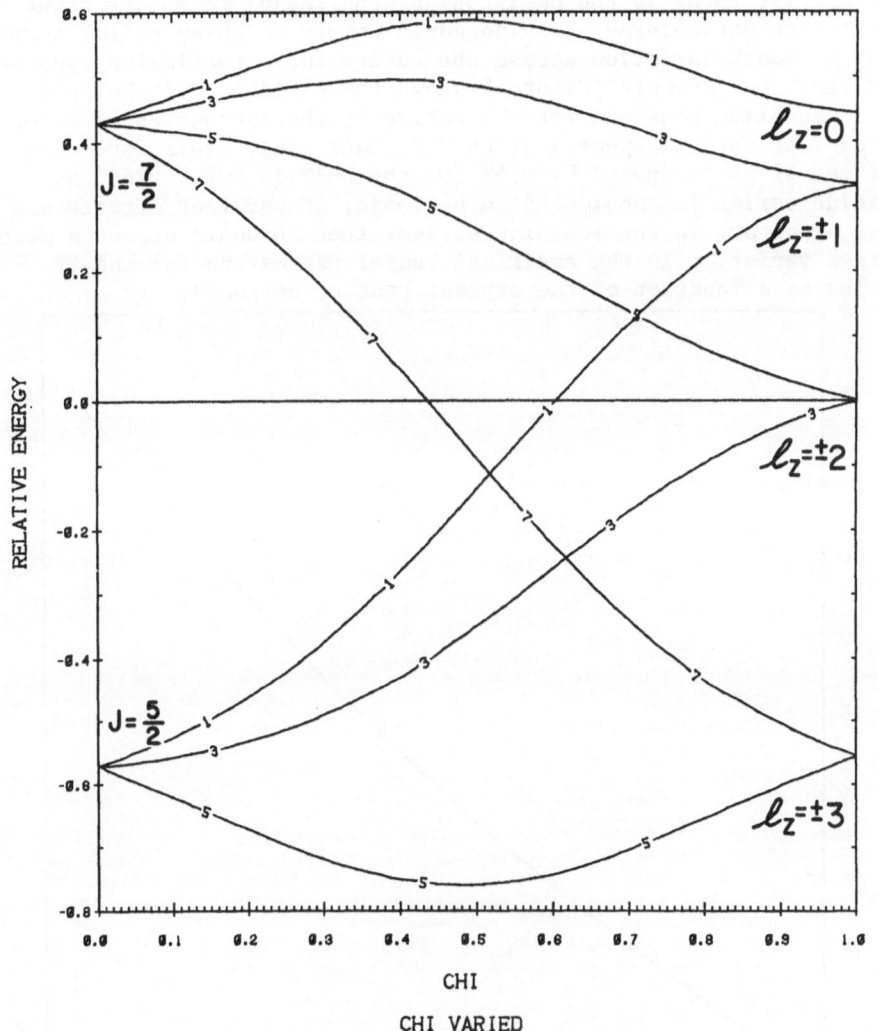

CHI VARIED

Figure 4. Energy levels of the f^1 configuration as a function of
the relative strengths of the spin-orbit and crystal field interactions. For chi = 0, only the spin-orbit interaction is considered; for chi = 1, only the crystal field interaction is
considered. The energy levels are numbered by $2 \times J_z$.

lanthanide compounds on the relative strengths of the two interactions we may extrapolate from inorganic complexes and estimate
that chi is approximately .4 for the actinides and possibly .3 or
.2 for the lanthanides. If covalent bonding is very important
for the actinides we would expect chi to be larger.

It should be emphasized that we are evaluating exactly only
the angular terms in the Hamiltonian, the radial terms are being
treated as parameters. For inorganic complexes these radial terms
show a smooth variation across the series for a particular type of
complex. For example, Figure 5 shows the variation of the spin-
orbit coupling constant for the entire lanthanide series as determined from optical spectra of Ln^{3+} in $LaCl_3$ [8]. This curve is
shifted slightly upward by ∿.5% for the Ln^{3+} in LaF_3. The lanthanide series is considered to be ionic; if covalent effects are
more important in the actinide series, then we would expect a much
larger variation in the empirical radial parameters for the 5f
series as a function of the crystal host or compound.

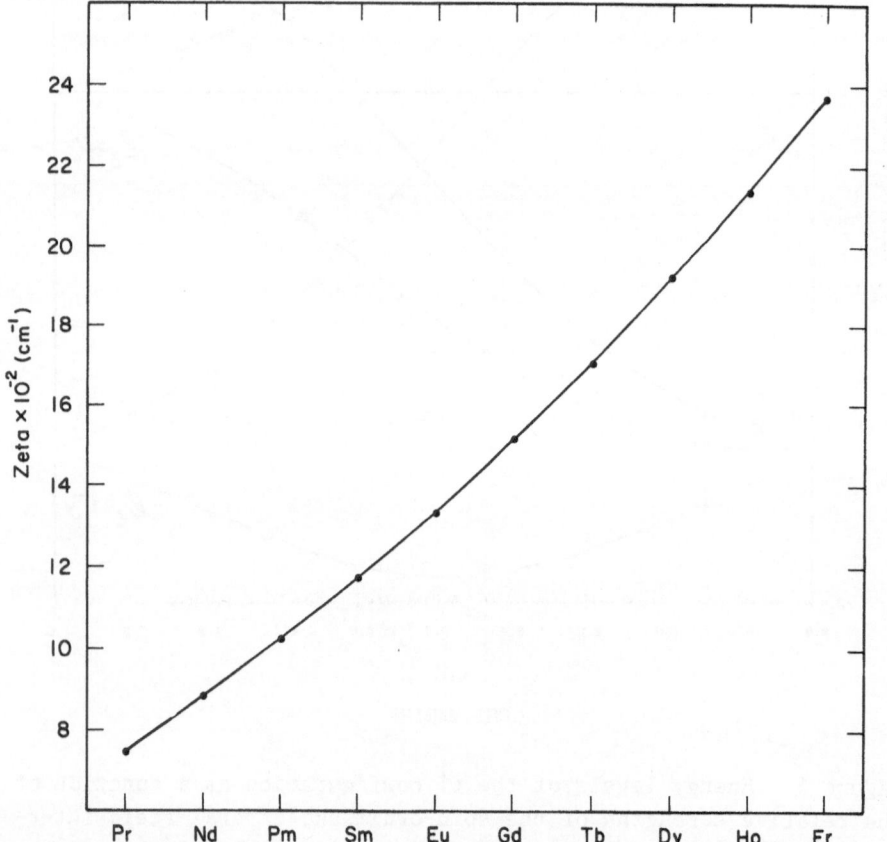

Figure 5. Variation of the empirically determined spin-orbit
coupling constant for Ln^{3+} ions in $LaCl_3$ [8].

The Electrostatic and Spin-Orbit Hamiltonians for f^2.

For an f^n configuration the energy levels of the electrostatic interaction

$$\mathcal{H}_1 = \frac{e^2}{r_{ij}}$$

are written in terms of the Slater integrals

$$F^{(k)} = e^2 \int_0^\infty \int_0^\infty \frac{r_<^k}{r_>^{k+1}} \left[R_{nf}(r_i) R_{nf}(r_j) \right]^2 dr_i \, dr_j \tag{13}$$

where $r_<$ is the lesser and $r_>$ is the greater of r_i and r_j. The limitations on k are obtained from the properties of Legendre polynomials and are: k must be even, and $k \leqslant 2\ell$, which for f electrons means k is restricted to k = 0, 2, 4, 6. For a much more detailed discussion see Condon and Shortley or Judd [2,3].

It is convenient to define a related set of parameters which avoids the occurrence of fractional coefficients for $F^{(k)}$'s:

$$F_0 = F^{(0)}$$

$$F_2 = \frac{F^{(2)}}{225}$$

$$F_4 = \frac{F^{(4)}}{1089}$$

$$F_6 = \frac{25F^{(6)}}{184041} \quad . \tag{14}$$

The electrostatic interaction is diagonal in the L-S representation. This is the representation usually used in calculations involving f electrons. We show in Appendix B the calculation of the electrostatic energies in this representation using the diagonal sum method.

The effect of the spin-orbit interaction, $\mathcal{H}_2$, may be readily evaluated within a particular L-S multiplet. This is equivalent to assuming $\mathcal{H}_1 \gg \mathcal{H}_2$. We may write

$$\langle SLJ | \zeta(SL) | SLJ \rangle = \zeta(SL) \left\{ \frac{J(J+1) - L(L+1) - S(S+1)}{2} \right\} \tag{15}$$

From this expression it can easily be shown that the energy interval between two states of the same multiplet which differ by 1 in their J value is

$$E_{L,S,J} - E_{L,S,J-1} = J\zeta(SL) \quad . \tag{16}$$

This is called the Landé interval rule.

For the lanthanide series as mentioned previously, Russell–Saunders coupling is a reasonable approximation for the ground terms except for the ions Sm^{3+} and Eu^{3+}. In the actinide series this approximation is worse because the effects of spin-orbit coupling increase as Z increases. The relatively simple methods we have discussed earlier are inadequate to determine the energies of the configuration f^n with $n > 2$. In the early 1940's Racah [9] developed more powerful methods which have since been applied to atomic spectroscopy. We shall simply mention the results of these methods here.

Results of Tensor Operator Methods

Racah defined a new set of radial parameters (called Racah parameters) which are related to the Slater integrals by

$$E^0 = F_0 - 10F_2 - 33F_4 - 286F_6$$
$$E^1 = (70F_2 + 231F_4 + 200F_6) \times \frac{1}{9}$$
$$E^2 = (F_2 - 3F_4 + 7F_6) \times \frac{1}{9}$$
$$E^3 = (5F_2 + 6F_4 - 91F_6) \times \frac{1}{3} \tag{17}$$

Expressions for the angular parts of the electrostatic and spin-orbit matrix elements may be obtained in terms of fractional parentage coefficients and can be relatively easily evaluated by a digital computer. These matrix elements have been tabulated for all f^n configurations and published in book form by Nielson and Koster [10].

The spin-orbit matrix elements may be evaluated by the expression

$$\langle \ell^N \alpha SLJ | \zeta_{n\ell} \sum (\delta_i \cdot \ell_i) | \ell^N \alpha' S' L' J \rangle$$

$$= \zeta_{n\ell} (-1)^{J+L+S'} \begin{Bmatrix} L & L' & 1 \\ S' & S & J \end{Bmatrix} \times \left[\ell(\ell+1)(2\ell+1) \right]^{1/2}$$

$$\times \langle \ell^N \alpha SL \| V^{(11)} \| \ell^N \alpha' S' L' \rangle \qquad (18)$$

In this expression the part in the curly brackets $\{ \}$ is a 6-j symbol and the $\langle \alpha SL \| V^{(11)} \| \ell^N \alpha' S' L \rangle$ is the $V^{(11)}$ reduced matrix element. The 6-j symbols are tabulated in the book by Rotenberg, Bivins, Metropolis, and Wooten [11] while the $V^{(11)}$ reduced matrix elements are given in Nielson and Koster for the entire f^n series. The symbol α in these matrix elements represent additional quantum numbers which are necessary to specify a particular state. Note that these formulas allow us to calculate off-diagonal elements also.

Nielson and Koster also tabulate another useful series of reduced matrix elements, the $U^{(k)}$ reduced matrix elements. These are written as

$$\langle f^N \alpha SL \| U^{(k)} \| f^N \alpha' S' L' \rangle \qquad (19)$$

where $k \leqslant 6$ for f electrons.

In Figure 6 the energy levels of the f^2 configuration are plotted as a function of eta, ξ, the relative magnitude of the electrostatic and spin-orbit interactions [12]. At the left hand side of the figure $\xi = 0$, and the energy levels represent the limit for pure Russell-Saunders coupling, while on the right hand side $\xi = 1$, which represents the energy levels in the limit of j-j coupling. For the lanthanide ion Pr^{3+} ξ is $\sim$.1 while for U^{4+} ξ is $\sim$.3. As can be seen the ordering of the energy levels are very sensitive to the relative strengths of these interactions and can change from one compound to another.

The discussion up to this point has been concerned with the free ion model. We have defined the electrostatic and spin-orbit radial parameters which can be used as variables in order to fit spectra. Ideally these parameters should be independent of the compound in which they are measured. In fact it is found that, as mentioned earlier for the spin-orbit coupling constant, the values of the Slater parameters for Ln^{3+} in $LaCl_3$ are slightly different than those found in LaF_3. For example, the data for $F^{(2)}$ are plotted in Figure 7. In the actinide series these effects are more important. For the octahedral UX_6^{2-} series (X = F, Cl, Br, I) [13] the value of $F^{(2)}$ is found to vary by 20% as X goes from

F to I. This point will be discussed later.

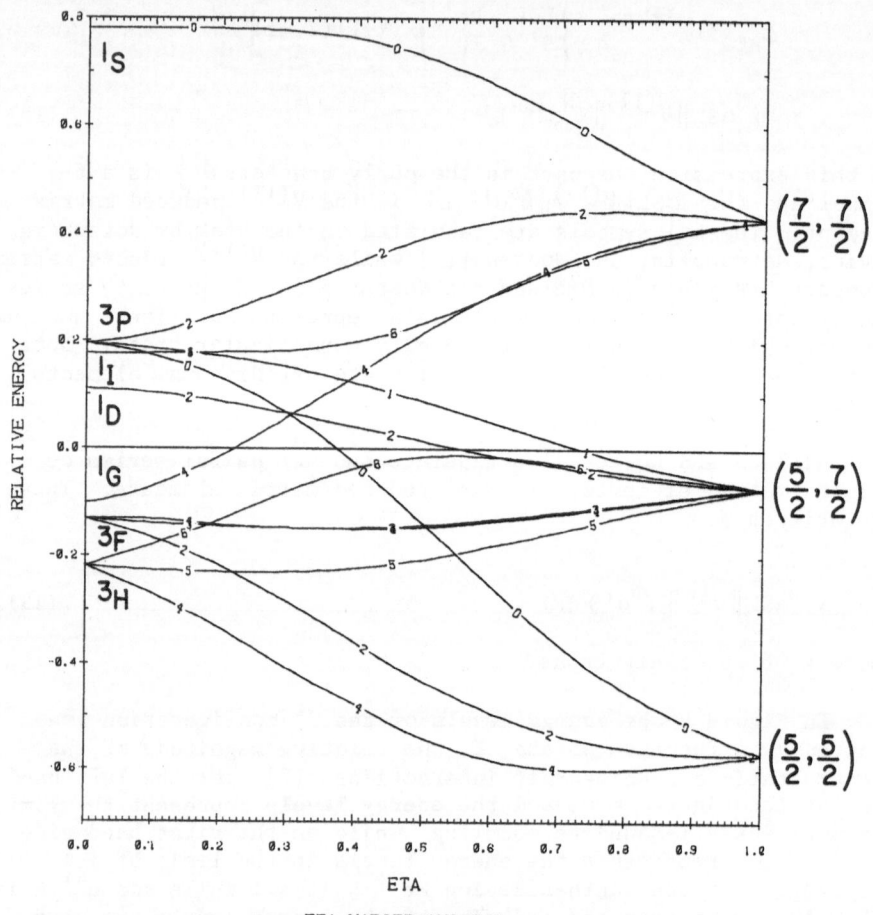

ETA VARIED, CHI=0.0 .

Figure 6. Energy levels of the f^2 configuration for relative
values of the spin-orbit and electrostatic interactions. For
eta = 0, only the electrostatic interaction is considered; for
eta = 1, only the spin-orbit interaction is considered. The
energy levels are numbered by the J values.

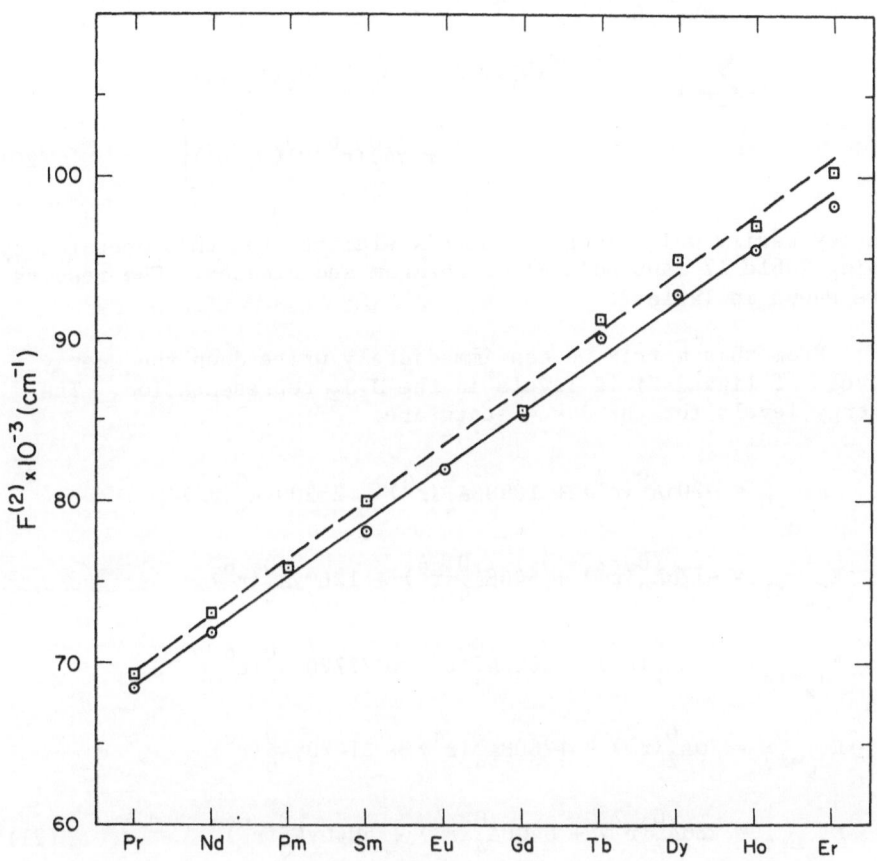

Figure 7. Variation of $F^{(2)}$ for the Ln^{3+} ions diluted in LaF_3 (dashed line) and $LaCl_3$ (solid line) [8].

Ligand Field Effects for the f^2 Configuration - Uranocene.*

We may apply the operator equivalent method discussed earlier for the f^2 case on the lowest term for the U^{4+} ion, $^3H_4$. In this example we will use all the terms in the ligand field Hamiltonian allowed by the symmetry of the molecules. Since uranocene has a C_8 axis there will be no off-diagonal terms allowed (q = 0) and

---

*For energy levels in the strong ligand field limit, see Appendix E. For a discussion of uranocene chemistry see Chapt. 5.

$$\mathcal{H}_3 = \sum_{i,k=2,4,6} K_k A_k^0 \langle r^k \rangle O_k^0(\Theta_i, \phi_i)$$

$$= \sum_i \left[\alpha A_2^0 \langle r^2 \rangle O_2^0(\Theta_i, \phi_i) + \beta A_4^0 \langle r^4 \rangle O_4^0(\Theta_i, \phi_i) \right.$$

$$\left. + \gamma A_6^0 \langle r^6 \rangle O_6^0(\Theta_i, \phi_i) \right] \quad . \quad (20)$$

We may easily write down the matrix elements for this operator by using Table 17 (Appendix B) of Abragam and Bleaney. The results are shown in Table 2.

From this matrix we can immediately write down the energy levels of ligand field levels in the J, J_z representation. The energy levels for the $J = 4$ state are:

$$E_{J_z=0} = -20\alpha A_2^0 \langle r^2 \rangle + 1080\beta A_4^0 \langle r^4 \rangle - 25200\gamma A_6^0 \langle r^6 \rangle$$

$$E_{J_z=\pm1} = -17\alpha A_2^0 \langle r^2 \rangle + 540\beta A_4^0 \langle r^4 \rangle + 1260\gamma A_6^0 \langle r^6 \rangle$$

$$E_{J_z=\pm2} = -8\alpha A_2^0 \langle r^2 \rangle - 660\beta A_4^0 \langle r^4 \rangle + 27720\gamma A_6^0 \langle r^6 \rangle$$

$$E_{J_z=\pm3} = 7\alpha A_2^0 \langle r^2 \rangle - 1260\beta A_4^0 \langle r^4 \rangle - 21420\gamma A_6^0 \langle r^6 \rangle$$

$$E_{J_z=\pm4} = 28\alpha A_2^0 \langle r^2 \rangle + 840\beta A_4^0 \langle r^4 \rangle + 5040\gamma A_6^0 \langle r^6 \rangle \quad . \quad (21)$$

Table 2

Ligand field matrix for a $J = 4$ state for a molecule with a C_8 symmetry axis. In this matrix $B_2 = \alpha A_2^0 \langle r^2 \rangle$, $B_4 = \beta 60 A_4^0 \langle r^4 \rangle$, and $B_6 = \gamma 1260 A_6^0 \langle r^6 \rangle$.

J_z	±0	±1	±2	±3	±4
0	$-20B_2+18B_4-20B_6$	0	0	0	0
±1	0	$-17B_2+9B_4+B_6$	0	0	0
±2	0	0	$-8B_2-11B_4+22B_6$	0	0
±3	0	0	0	$7B_2-21B_4-17B_6$	0
±4	0	0	0	0	$28B_2+14B_4+4B_6$

Now let us assume as was done earlier, that the second order term is the most important, i.e., $\alpha A_2^0 \langle r^2 \rangle \gg \beta A_4^0 \langle r^4 \rangle$ or $\gamma A_6^0 \langle r^6 \rangle$. Then if αA_2^0 is positive the $J_z = \pm 4$ ligand field state is lowest in energy.

It is interesting at this point to determine the energy levels of the f^2 configuration as a function of the strength of the crystal field. Figures 8a and 8b show the energy levels of this configuration drawn for two values of eta, the ratio of the spin-orbit interaction to the electrostatic interaction as defined earlier.

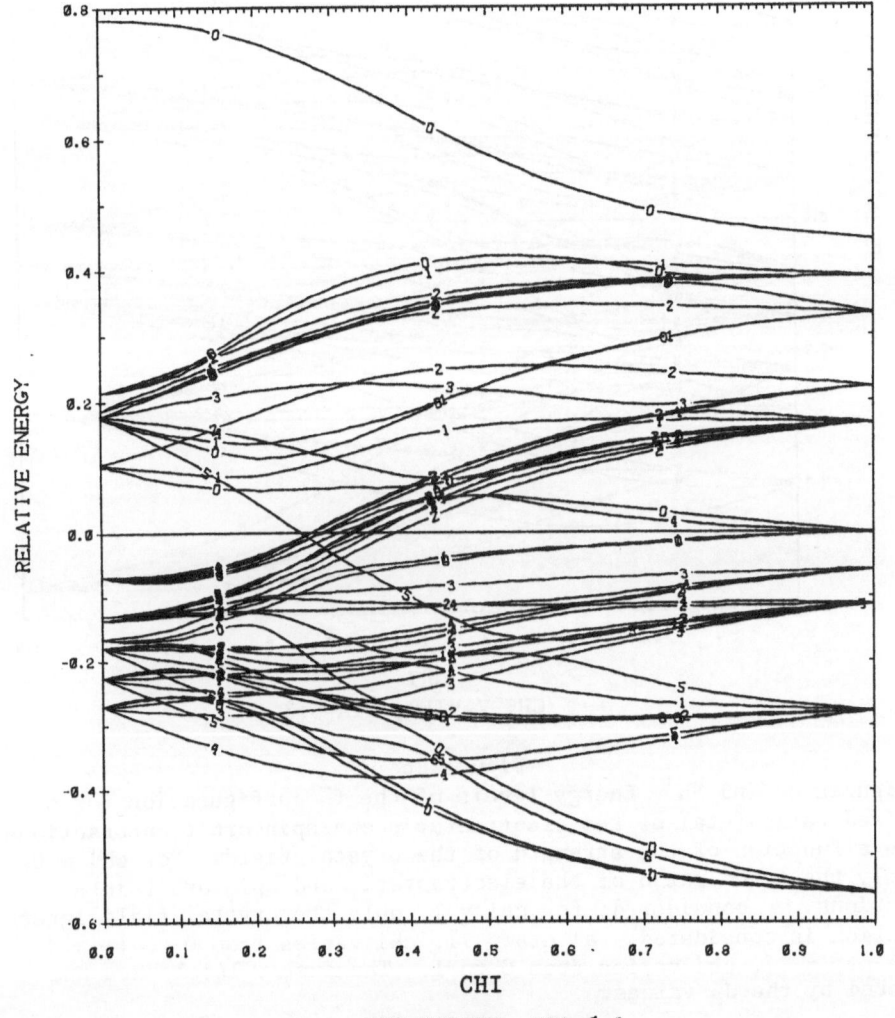

CHI VARIED, ETA=0.1

Figure 8a

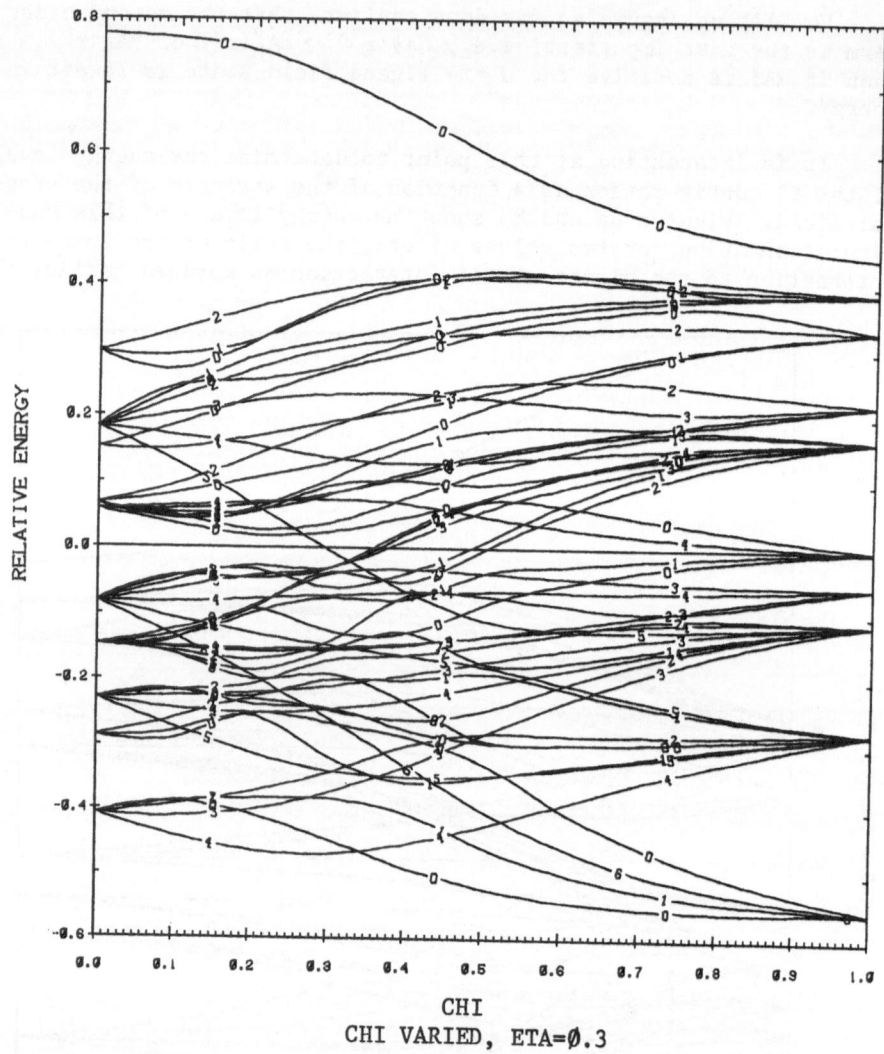

CHI

CHI VARIED, ETA=∅.3

Figure 8b

Figures 8a and 8b. Energy levels of the f^2 configuration for a fixed ratio (eta) of the electrostatic and spin-orbit interactions as a function of the strength of the crystal field. For chi = 0, only the fixed ratio of the electrostatic and spin-orbit inter-actions is considered; for chi = 1, only the crystal field inter-action is considered. a) eta = .1, chi varies from 0 to 1; b) eta = .3, chi varies from 0 to 1. The energy levels are num-bered by the J_z values.

At the right hand side of the figures, chi = 1, the energy levels are drawn considering only the crystal field interaction. At the left hand side of the figures, chi = 0, the energy levels are drawn for the particular value of eta and no crystal field interaction as obtained from Figure 6. The values of eta = .1 and .3 are reasonable estimates for the lanthanide and actinide series respectively. The value of chi, the ratio of the crystal field interaction to the electrostatic and spin-orbit interaction is probably about $\sim$.2-.3 for the two series. These graphs represent Tanabe-Sugano diagrams for the f^2 configuration [14].

The magnetic susceptibility of uranocene has been measured from 4.2°K to room temperature [15-18]. On the basis of the above much over-simplified model we may calculate the magnetic susceptibility of this compound. We assume only the ground crystal field state contributes to the susceptibility and second order effects can be neglected.

The magnetic susceptibility X is defined for our purposes as

$$X = \frac{N \sum \left(\frac{W_1^2}{kT} - 2W_2 \right) e^{-W_0/kT}}{\sum e^{-W_0/kT}} \qquad (22)$$

where N is Avogadro's number and the summation is over all energy levels. We assume the energy level E may be expanded in a power series of H, the magnetic field, as

$$E = W_0 + W_1 H + W_2 H^2 + \ldots \qquad (23)$$

Since we are neglecting second order effects $W_2 = 0$.

To evaluate W_1 for our example we need to calculate the matrix element

$$\langle LSJJ_z | L + 2S | LSJJ_z' \rangle = g_J \langle J_z | J | J_z' \rangle$$

$$= g_J \langle J_z | J_x + J_y + J_z | J_z' \rangle \qquad (24)$$

If $J_z = J_z'$ then

$$\langle J_z | J_z | J_z \rangle = J_z \quad . \qquad (25)$$

If $J_z \neq J_z'$ this component is zero.

$$J_x + iJ_y = J_+$$

$$J_x - iJ_y = J_-$$

and

$$\langle JJ_z | J_\pm | JJ_z \pm 1 \rangle = \left[(J(J+1) - J_z(J_z \pm 1) \right]^{1/2} \tag{26}$$

If $\Delta J_z > 1$ the off-diagonal matrix elements are zero.

In our simplified model with $\alpha A_2^0 > 0$ we have found the degenerate $J_z = \pm 4$ state is lowest. In a magnetic field of magnitude H parallel to the symmetry axis z the energies of the two levels are

$$E_{J_z=+4} = g_J\beta H \langle J_z=4 | J_z | J_z=4 \rangle = 4g_J\beta H$$

$$E_{J_z=-4} = g_J\beta H \langle J_z=-4 | J_z | J_z=-4 \rangle = -4g_J\beta H \tag{27}$$

Since $\Delta J_z > 1$ for the off-diagonal components the susceptibility is zero in the perpendicular direction. Then

$$\chi_\parallel = \frac{(16g_J^2\beta^2) \times 2)}{2kT} = \frac{16g_J^2\beta^2}{kT}$$

$$\chi_\perp = 0$$

$$\chi_{Ave} = \frac{1}{3} (\chi_\parallel + 2\chi_\perp)$$

$$= \frac{1}{3} \frac{(16g_J^2\beta^2)}{kT} \tag{28}$$

$g_J = \frac{4}{5}$ for U^{4+} and then

$$\chi_{Ave} = \frac{\mu_{eff}^2}{3kT}$$

$$\mu_{eff} = 3.2\beta \quad .$$

The experimental value for the magnetic susceptibility for uranocene in the low temperature range ($4.2°K < T < \sim80°K$) is $\mu_{eff} = 2.4\beta$. In the original calculation [15] the orbital reduction factor k was introduced by replacing the Zeeman operator $L + 2S$ with

$kL + 2S$. The deviation of the value of the orbital reduction factor from unity is a measure of the covalency of the molecule. The experimental value of the magnetic susceptibility was fit with $k = .8$.

This same type of calculation was performed for the higher Z compounds, neptunocene and plutonocene. For $Pu(COT)_2$ the ground term is a $^5I_4$. Since $J = 4$ for this term the crystal field Hamiltonian is of exactly the same form as shown earlier for $U(COT)_2$ except the value of α, β, and γ are different. If we assume again $B_2^0 \gg B_4^0$ and B_6^0, then the only difference between the f^2 and f^4 ion will be the value of α. (We have also assumed $A_2^0 \langle r^2 \rangle$ does not change much in the two compounds.) From Table 20 (Appendix B) in Abragam and Bleaney we find

$$\langle f^{23}H_4 \| \alpha \| f^{23}H_4 \rangle = \frac{-2^2 \times 13}{3^2 \times 5^2 \times 11}$$

$$\langle f^{45}I_4 \| \alpha \| f^{45}I_4 \rangle = \frac{2 \times 7}{3 \times 5 \times 11^2} \tag{29}$$

Since the signs of these factors are different, if the $J_z = \pm 4$ state is lowest for $U(COT)_2$ then the $J_z = 0$ state must be lowest for $Pu(COT)_2$. The experimental magnetic susceptibility data for $Pu(COT)_2$ show this compound is diamagnetic which is consistent with the above calculations.

Another approach to evaluating the energy levels of uranocene-type molecules is the effective crystal field model. In this model, first suggested for actinide COT complexes by Hayes and Edelstein [19], the ligand field splittings were calculated for an f^1 system. Hayes and Edelstein used the Wolfsberg-Helmholz approximation to determine the one electron orbitals derived from the metal f orbitals. This calculation showed the filled ring orbitals were quite a bit lower in energy than the metal orbitals. The level scheme for the metal orbitals is shown in Figure 9. These splittings were used to evaluate the crystal field parameters ($A_k^q \langle r^k \rangle$) given for the f^1 case in axial symmetry in Appendix C. The crystal field parameters evaluated in this fashion were then used with the values of the electrostatic and spin-orbit parameters derived for U^{4+} in UCl_4 and the energy splittings and magnetic moment of the ground crystal field state were obtained.

The results of these calculations should be treated with caution. Even in well-characterized systems of f-transition ions it appears the empirical fitting of radial parameters to optical and magnetic data is open to some question. As mentioned previously, in the octahedral UX_6^{2-} series (X = F, Cl, Br, I) the value of $F^{(2)}$ varied by $\sim 20\%$ as X changed from F to I [13]. These systems

$$\ell_z$$

$$\pm 2 \text{——} 3313 \text{ cm}^{-1} = 0B_2^0 - \frac{56}{33} B_4^0 + \frac{160}{11 \times 13} B_6^0$$

$$0 \text{——} 2511 \text{ cm}^{-1} = \frac{8}{15} B_2^0 + \frac{16}{11} B_4^0 + \frac{1600}{33 \times 13} B_6^0$$

$$\pm 1 \text{——}$$
$$\pm 3 \text{——}$$

$$166 \text{ cm}^{-1} = \frac{2}{5} B_2^0 + \frac{8}{33} B_4^0 - \frac{400}{11 \times 13} B_6^0$$

$$0 \text{ cm}^{-1} = -\frac{2}{3} B_2^0 + \frac{8}{11} B_4^0 - \frac{80}{33 \times 13} B_6^0$$

Figure 9. Results of the Wolfsberg–Helmholz calculation [19] for the metal–like orbitals of uranocene. $B_k^0 = A_k^0 \langle r^k \rangle$.

were characterized by fitting the electrostatic, spin–orbit, and crystal field parameters to optical data. It was suggested that the value of the Slater parameter $F^{(2)}$ is affected by the type of ligand in the complex and may have adsorbed some of the effects of the ligand field. This appeared to be true to a lesser extent for the spin–orbit coupling constant. If the value of $F^{(2)}$ and ζ were affected by the ligands, then the values found for the ligand field parameters would also not be the "correct" value.

General Method for Calculating Crystal Field Matrix Elements for f^n Configuration.

The operator equivalent method is useful for determining crystal field splittings for the lowest J state of a lanthanide or actinide ion since the necessary α, β, γ's are tabulated. However, if we include higher lying J states, the effects of intermediate coupling and the mixing of various J states by the relatively strong crystal field are important, especially for actinide ions. Then it is much simpler to calculate the necessary matrix elements by the tensor operator technique. Unfortunately, the definition used for the crystal field parameters B_q^k's in the tensor operator method is different than that of the operator equivalent method. Note also that we are now using B_q^k for the parameters where k is the superscript and q the subscript; this is the opposite to the earlier B_k^q's which are defined differently. Table 3, taken from Kassman [20], shows the relationship between the tensor operator notation B_q^k and the operator equivalent

Table 3

Relationship of the tensor operator parameters B_q^k to the operator equivalent parameters A_k^q (from Kassman).

$$B_0^2 = 2A_2^0 \langle r^2 \rangle$$

$$B_0^6 = 16A_6^0 \langle r^6 \rangle$$

$$B_2^2 = (1/3)(6)^{1/2} A_2^2 \langle r^2 \rangle$$

$$B_2^6 = (16/105)(105)^{1/2} A_6^2 \langle r^6 \rangle$$

$$B_0^4 = 8A_4^0 \langle r^4 \rangle$$

$$B_3^6 = -(8/105)(105)^{1/2} A_6^3 \langle r^6 \rangle$$

$$B_2^4 = (2/5)(10)^{1/2} A_4^2 \langle r^4 \rangle$$

$$B_4^6 = (8/21)(14)^{1/2} A_6^4 \langle r^6 \rangle$$

$$B_3^4 = -(2/35)(35)^{1/2} A_4^3 \langle r^4 \rangle$$

$$B_6^6 = (16/231)(231)^{1/2} A_6^6 \langle r^6 \rangle$$

$$B_4^4 = (4/35)(70)^{1/2} A_4^4 \langle r^4 \rangle$$

notation $A_k^q \langle r^k \rangle$. (Note that Table 6.1 of Wybourne [4] contains a number of errors. The phase factor in Equation 6-5 of Wybourne is also incorrect.)

A general formula for crystal field matrix elements is given in Appendix D along with an example of its application.

Electron Paramagnetic Resonance

We have previously discussed the Zeeman operator $\mathcal{H}_4$ which was written as

$$\mathcal{H}_4 = \beta H \cdot (L + 2S)$$

which for an isolated J level, which we will consider here, may be rewritten

$$\mathcal{H}_4 = g_J (H \cdot J) = g_J \beta (H_z J_z + H_x J_x + H_y J_y)$$

where g_J is the free ion g value for a particular J level, and is the Bohr magneton, and H the magnetic field. Now EPR experiments are usually described in terms of a phenomenological Hamiltonian called a spin-Hamiltonian [6]

$$\mathcal{H} = \beta H \cdot g \cdot S' = (g_{\parallel} H_z S_z' + g_x H_x S_x' + g_y H_y S_y') \tag{30}$$

We see by comparison of the two Hamiltonians that if the wave-function of the ground crystal field state is known we can calculate the g values which will be found in the EPR spectrum, i.e.,

$$g_\| = 2g_J \langle SLJJ_z | J_z | SLJJ_z \rangle$$

$$g_x = 2g_J \langle SLJJ_z | J_x | SLJJ_z \pm 1 \rangle$$

$$g_y = 2g_J \langle SLJJ_z | J_y | SLJJ_z \pm 1 \rangle \tag{31}$$

If $g_x = g_y$, then the g value in the xy plane is called $g_\perp$. If $\Delta J_z \neq 0, \pm 1$ the matrix elements will be zero and no EPR spectrum will be observed. In the f transition series, the orbital angular momentum is not quenched as in the d transition series and the measured g values may vary from 0 to 18. The magnitude of the g value is a very sensitive test of the crystal field wavefunction.

In the above Hamiltonian we have not included the hyperfine interaction term. This term is written as

$$\mathcal{H}_{hf} = AI \cdot J \tag{32}$$

for an isolated J state is proportional to free ion g_J value and the free hyperfine coupling constant value a_J:

$$\frac{A}{g} = \frac{a_J}{g_J} = \frac{A_\|}{g_\|} = \frac{A_\perp}{g_\perp} \tag{33}$$

If these proportionalities do not hold, it is an indication that crystal field mixing of different J states is important.

Acknowledgements

    I would like to thank Mr. T. Hayhurst for carefully reading and correcting this paper and Mrs. K. Janes for her skill and patience in typing and preparing the manuscript.

    This work was supported by the Division of Nuclear Sciences, Office of Basic Energy Sciences, U. S. Department of Energy.

References and Footnotes

[1] For a more complete discussion, see references 2 through 4.
[2] E. U. Condon and G. H. Shortley, "The Theory of Atomic Spectra", Cambridge University Press, New York, 1935.
[3] B. R. Judd, "Operator Techniques in Atomic Spectroscopy",
McGraw-Hill Book Company, New York, 1963.
[4] B. G. Wybourne, "Spectroscopic Properties of Rare Earths",
John Wiley and Sons, New York, 1965.
[5] K.W.H. Stevens, Proc. Phys. Soc. $\underline{A65}$, 209 (1952).
[6] A. Abragam and B. Bleaney, "Electron Paramagnetic Resonance
of Transition Ions", Clarendon Press, Oxford, 1970.
[7] This diagram is drawn for B_0^2 positive. The x axis is labeled

by the parameter chi where $\dfrac{\text{chi}}{(1-\text{chi})} = \dfrac{3.5B_0^2}{.6\ \text{zeta}}$. The y axis or rela-

tive energy axis is equal to $E/[(.6\ \text{zeta})^2 + (3.5B_0^2)^2]^{1/2}$.
[8] W. T. Carnall, H. Crosswhite, and H. M. Crosswhite, "Energy
Level Structure and Transition Probabilities in the Spectra of the
Trivalent Lanthanides in LaF_3", Argonne National Laboratory Report,
1976.
[9] G. Racah, Phys. Rev. $\underline{61}$, 186 (1942); Phys. Rev. $\underline{62}$, 438
(1942); Phys. Rev. $\underline{63}$, 367 (1943); Phys. Rev. $\underline{76}$, 1352 (1949).
[10] C. W. Nielson and G. F. Koster, "Spectroscopic Coefficients
for the p^n, d^n, and f^n Configuration", The M.I.T. Press, Cambridge,
MA, 1963.
[11] M. Rotenberg, R. Bivins, N. Metropolis, and J. K. Wooten, Jr.,
"The 3-j and 6-j Symbols", The M.I.T. Press, Cambridge, MA, 1959.
[12] In this figure the x axis is labeled by the parameter eta

where $\dfrac{\text{eta}}{(1-\text{eta})} = \dfrac{7\ \text{zeta}}{148.1949\ F_2}$. The y axis or relative energy axis

is equal to $E/[(F_2 \times 148.1949)^2 + (7 \times \text{zeta})^2]^{1/2}$. The electrostatic parameters F_4 and F_6 are set at their respective hydrogenic
ratios of F_2.
[13] W. Wagner, N. Edelstein, B. Whittaker, and D. Brown, Inorg.
Chem. $\underline{16}$, 1021 (1977).
[14] This diagram is drawn for B_0^2 positive. The x axis is labeled

by the parameter chi where $\dfrac{\text{chi}}{1-\text{chi}} = \dfrac{B_0^2}{[\text{zeta}^2 + F_2^2]^{\frac{1}{2}}}$. The values for

zeta and F_2 are obtained from the parameter eta as described
earlier [12]. The y axis or relative energy axis is equal to
$E/[(1.2 \times B_0^2)^2 + (F_2 \times 148.1949)^2 + (7 \times \text{zeta})^2]^{1/2}$. The electrostatic parameters F_4 and F_6 are set at their respective hydrogenic
ratios of F_2.
[15] D. G. Karraker, J. A. Stone, E. R. Jones, Jr., and N. Edelstein, J. Amer. Chem. Soc. $\underline{92}$, 4841 (1970). This paper on the
magnetic susceptibility of uranocene reported Curie-Weiss behavior
in the temperature range studied; subsequently it was shown that

the magnetic susceptibility of uranocene becomes temperature
independent below 10°K. See references [16-18] for further discussion of this point.

[16] D. Karraker, Inorg. Chem. 12, 1105 (1973).
[17] H. D. Amberger, R.D. Fischer, and B. Kanellakopulos, Theor.
Chim. Acta 37, 105 (1975).
[18] N. Edelstein, A. Streitwieser, Jr., D. G. Morrell, and R.
Walker, Inorg. Chem. 15, 1397 (1976).
[19] R. G. Hayes and N. Edelstein, J. Amer. Chem. Soc. 94, 8688
(1972).
[20] A. J. Kassman, J. Chem. Phys. 53, 4118 (1970).

Appendix A

To evaluate the spin-orbit coupling in this basis set we need the following matrix elements

$$\zeta \ell \cdot s = \zeta \left[\frac{1}{2}(\ell_+ s_- + \ell_- s_+) + \ell_z s_z \right]$$

$$\langle \ell \ell_z s s_z | \ell_+ s_- | \ell', \ell_z - 1, s, s_z + 1 \rangle$$

$$= \left[(\ell + \ell_z)(\ell - \ell_z + 1)(s - s_z)(s + s_z + 1) \right]^{1/2}$$

$$\langle \ell \ell_z s s_z | \ell_- s_+ | \ell, \ell_z + 1, s, s_z - 1 \rangle$$

$$= \left[(\ell - \ell_z)(\ell + \ell_z + 1)(s + s_z)(s - s_z + 1) \right]^{1/2}$$

$$\langle \ell \ell s s_z | \ell_z s_z | \ell \ell_z s s_z \rangle = \ell_z s_z$$

Now for $\ell = 3$, $s = \frac{1}{2}$

$$\langle 3, \ell_z, \frac{1}{2}, s_z | \ell_+ s_- | 3, \ell_z - 1, \frac{1}{2}, s_z + 1 \rangle$$

$$= \left[(3 + \ell_z)(4 - \ell_z)(\frac{1}{2} - s_z)(\frac{3}{2} + s_z) \right]^{1/2}$$

$$\langle 3, \ell_z, \frac{1}{2}, s_z | \ell_- s_+ | 3, \ell_z + 1, s, s_z - 1 \rangle$$

$$= \left[(3 - \ell_z)(4 + s_z)(\frac{1}{2} + s_z)(\frac{3}{2} - s_z) \right]^{1/2}$$

$$\langle 3, \frac{1}{2} | \ell_+ s_- + \ell_- s_+ | 2, -\frac{1}{2} \rangle = \langle 3, \frac{1}{2} | \ell_+ s_- | 2, -\frac{1}{2} \rangle \ ^*$$

$$+ \ \langle 3, \frac{1}{2} | \ell_- s_+ | 2, -\frac{1}{2} \rangle = 0 + 0$$

*We have changed the notation here and are using only the $\ell_z s_z$ values since ℓ and s are fixed at 3 and ½ respectively, i.e., $\langle \ell_z, s_z | \quad | \ell_z, s_z \rangle$.

$$\langle 2,+\tfrac{1}{2}\,|\,\ell_+ s_- + \ell_- s_+\,|\,3,-\tfrac{1}{2}$$

$$= \left[(3-2)(4+2)(\tfrac{1}{2}+\tfrac{1}{2})(\tfrac{3}{2}-\tfrac{1}{2})\right]^{1/2} = (6)^{1/2}$$

$$\langle 2,+\tfrac{1}{2}\,|\,\ell_+ s_- + \ell_- s_+\,|\,1,-\tfrac{1}{2}\rangle = 0$$

$$\langle 3,-\tfrac{1}{2}\,|\,\ell_+ s_- + \ell_- s_+\,|\,2,\tfrac{1}{2}\rangle$$

$$= \left[(6)(1)(\tfrac{1}{2}-(-\tfrac{1}{2}))(\tfrac{3}{2}+-\tfrac{1}{2})\right]^{1/2} = (6)^{1/2}$$

$$\langle 1,-\tfrac{1}{2}\,|\,\ell_+ s_- + \ell_- s_+\,|\,0,\tfrac{1}{2}\rangle$$

$$= \left[(4)(3)(\tfrac{1}{2}--\tfrac{1}{2})(1)\right]^{1/2} = 2(3)^{1/2}$$

$$\langle 0,\tfrac{1}{2}|\,\ell_+ s_- + \ell_- s_+\,|\,1,-\tfrac{1}{2}\rangle = 2(3)^{1/2}$$

$$\langle -1,-\tfrac{1}{2}\,|\,\ell_+ s_- + \ell_- s_+\,|\,-2,\tfrac{1}{2}\rangle$$

$$= \left[(3-1)(4-(-1))(1)(1)\right]^{1/2} = (10)^{1/2}$$

$$\langle -2,\tfrac{1}{2}|\,\ell_+ s_- + \ell_- s_+\,|\,-1,-\tfrac{1}{2}\rangle$$

$$= \left[(3-(-2))(4+(-2))(1)(1)\right]^{1/2} = (10)^{1/2}$$

$$\langle -3,-\tfrac{1}{2}\,|\,\ell_+ s_- + \ell_- s_+\,|\,-2,\tfrac{1}{2}\rangle = 0$$

Using these results (and those of Eq. (12)) we may write down the matrices of the spin-orbit interaction in the ℓ_z, s_z representation as follows:

ℓ_z, s_z	$3, -\frac{1}{2}$	$2, +\frac{1}{2}$
$3, -\frac{1}{2}$	$-\frac{3}{2}\zeta + \epsilon_3$	$\frac{1}{2}(6)^{\frac{1}{2}}\zeta$
$2, +\frac{1}{2}$	$\frac{1}{2}(6)^{\frac{1}{2}}\zeta$	$+\zeta + \epsilon_2$

	$1, -\frac{1}{2}$	$0, \frac{1}{2}$
$1, -\frac{1}{2}$	$-\frac{1}{2}\zeta + \epsilon_1$	$(3)^{\frac{1}{2}}\zeta$
$0, \frac{1}{2}$	$(3)^{\frac{1}{2}}\zeta$	$+\epsilon_0$

	$-1, -\frac{1}{2}$	$-2, \frac{1}{2}$
$-1, -\frac{1}{2}$	$\zeta/2 + \epsilon_1$	$\frac{1}{2}(10)^{\frac{1}{2}}\zeta$
$-2, \frac{1}{2}$	$\frac{1}{2}(10)^{\frac{1}{2}}\zeta$	$-\zeta + \epsilon_2$

	$-3, -\frac{1}{2}$
$-3, -\frac{1}{2}$	$\frac{3}{2}\zeta + \epsilon_3$

Similarly,

$$\langle 3, \tfrac{1}{2} | \ell_+ s_- + \ell_- s_+ | 2, -\tfrac{1}{2} \rangle = 0$$

$$\langle 2, -\tfrac{1}{2} | \ell_+ s_- + \ell_- s_+ | 1, \tfrac{1}{2} \rangle = \left[(3+2)(4-2)(1)(1) \right]^{1/2} = (10)^{1/2}$$

$$\langle 1, \tfrac{1}{2} | \ell_+ s_- + \ell_- s_+ | 2, -\tfrac{1}{2} \rangle = \left[(2)(5)(1)(1) \right]^{1/2} = (10)^{1/2}$$

$$\langle 1, \tfrac{1}{2} | \ell_+ s_- + \ell_- s_+ | 0, -\tfrac{1}{2} \rangle = 0$$

$$\langle 0, -\tfrac{1}{2} | \ell_+ s_- + \ell_- s_+ | -1, \tfrac{1}{2} \rangle = \left[(3)(4)(1)(1) \right]^{1/2} = 2(3)^{1/2}$$

$$\langle -1, \tfrac{1}{2} | \ell_+ s_- + \ell_- s_+ | -2, -\tfrac{1}{2} \rangle = 0$$

$$\langle -2,-\tfrac{1}{2} \,| \ell_+\delta_- + \ell_-\delta_+ |-3,\tfrac{1}{2}\rangle = \left[(1)(6)^{1/2}(1)(1)\right]^{1/2} = (6)^{1/2}$$

and the spin-orbit matrices are:

	$3,\tfrac{1}{2}$
$3,\tfrac{1}{2}$	$\tfrac{3}{2}\zeta+\varepsilon_3$

	$2,-\tfrac{1}{2}$	$1,\tfrac{1}{2}$
$2,-\tfrac{1}{2}$	$-\zeta+\varepsilon_2$	$\tfrac{1}{2}(10)^{1/2}\zeta$
$1,\tfrac{1}{2}$	$\tfrac{1}{2}(10)^{1/2}\zeta$	$\tfrac{1}{2}\zeta+\varepsilon_1$

	$0,-\tfrac{1}{2}$	$-1,\tfrac{1}{2}$
$0,-\tfrac{1}{2}$	$+\varepsilon_0$	$(3)^{1/2}\zeta$
$-1,\tfrac{1}{2}$	$(3)^{1/2}\zeta$	$-\tfrac{1}{2}\zeta+\varepsilon_1$

	$-2,-\tfrac{1}{2}$	$-3,\tfrac{1}{2}$
$-2,-\tfrac{1}{2}$	$+\zeta+\varepsilon_2$	$\tfrac{1}{2}(6)^{1/2}\zeta$
$-3,\tfrac{1}{2}$	$\tfrac{1}{2}(6)^{1/2}\zeta$	$-\tfrac{3}{2}\zeta+\varepsilon_3$

Let us now consider another basis set, the $LSJJ_z$ quantum numbers. For f^1 $L = 3$, $S = 1/2$, $J = 5/2$ or $7/2$, that is $^2F_{5/2}$ or $^2F_{7/2}$. We may find the magnitude of the spin-orbit splitting from the equation for the diagonal elements

$$E_{so}(J=\tfrac{7}{2}) = \zeta/2[J(J+1)-L(L+1)-S(S+1)]$$

$$= \zeta/2[\tfrac{7}{2}(\tfrac{9}{2})-3(4)-\tfrac{1}{2}(\tfrac{3}{2})]$$

$$= \zeta/2[\tfrac{63}{4} - \tfrac{48}{4} - \tfrac{3}{4})] = \tfrac{3}{2}\zeta$$

$$E_{so}(J=\tfrac{5}{2}) = \zeta/2[\tfrac{5}{2}\times\tfrac{7}{2} - \tfrac{48}{4} - \tfrac{3}{4}] = -2\zeta$$

We now follow the procedures for operator equivalents as we did previously. This time we are interested in the J = 5/2 and J = 7/2 states. From Table 17 of Abragam and Bleaney we can find the operator equivalent factors for the J = 5/2 and the J = 7/2 states. The energy matrices obtained including spin-orbit coupling are

$J = \dfrac{5}{2}$

J_z	$\pm\dfrac{1}{2}$	$\pm\dfrac{3}{2}$	$\pm\dfrac{5}{2}$
$\pm\dfrac{1}{2}$	$-2\zeta-8\alpha A_2^0\langle r^2\rangle$		0
$\pm\dfrac{3}{2}$	0	$-2\zeta-2\alpha A_2^0\langle r^2\rangle$	0
$\pm\dfrac{5}{2}$	0		$-2\zeta-10\alpha A_2^0\langle r^2\rangle$

$J = \dfrac{7}{2}$

J_z	$\pm\dfrac{1}{2}$	$\pm\dfrac{3}{2}$	$\pm\dfrac{5}{2}$	$\pm\dfrac{7}{2}$
$\pm\dfrac{1}{2}$	$\tfrac{3}{2}\zeta-15\alpha'A_2^0\langle r^2\rangle$	0	0	0
$\pm\dfrac{3}{2}$	0	$\tfrac{3}{2}\zeta-9\alpha'A_2^0\langle r^2\rangle$	0	0
$\pm\dfrac{5}{2}$	0	0	$\tfrac{3}{2}\zeta+3\alpha'A_2^0\langle r^2\rangle$	0
$\pm\dfrac{7}{2}$	0	0	0	$\tfrac{3}{2}\zeta+21\alpha'A_2^0\langle r^2\rangle$

The values for the second order operator equivalent factors may be obtained from Table 20 (Appendix B) of Abragam and Pryce. The factors are

$$\alpha = \langle f^1\,{}^2F_{5/2}|\alpha|f^1\,{}^2F_{5/2}\rangle = \frac{-2}{5\times7}$$

$$\langle f^1\,{}^2F_{5/2}|\alpha|f^1\,{}^2F_{7/2}\rangle = \frac{2^2}{3\times5\times7}$$

$$\alpha' = \langle f^1 \, {}^2F_{7/2} | \alpha | f^1 \, {}^2F_{7/2} \rangle = -\langle f^{13} \, {}^2F_{7/2} | \alpha | f^{13} \, {}^2F_{7/2} \rangle =$$

$$\frac{-2}{3^2 \times 7}$$

Now the operator O_2^0 will mix states of different J but the same J_z value, i.e., matrix elements of the type

$$\langle J = \frac{5}{2}, J_z | O_2^0 | J = \frac{7}{2}, J_z \rangle \neq 0$$

so that 2×2 matrices need to be diagonalized in order to solve the problem exactly. We can check the energy of the $J_z = \pm 7/2$ state and compare it to that of the ℓ_z state of 3,1/2 and 3,-1/2 calculated previously.

$$E_{J=\frac{7}{2}, J_z=\frac{7}{2}} = \frac{3}{2}\zeta + 21\alpha' A_2^0 \langle r^2 \rangle$$

$$= \frac{3}{2}\zeta + 21 \times \frac{-2}{3 \times 3 \times 7} A_2^0 \langle r^2 \rangle$$

$$= \frac{3}{2}\zeta - \frac{2}{3} A_2^0 \langle r^2 \rangle$$

$$E_{\ell=3, \ell_z=3} = \frac{3}{2}\zeta + \epsilon_3 = \frac{3}{2}\zeta - \frac{2}{3} A_2^0 \langle r^2 \rangle$$

Thus, as must be true, the energies do not depend on the basis functions chosen to do the calculations.

Appendix B

We start out the calculation of the electrostatic energies by constructing a table of the complete sets of electrons which belong to the configuration using all allowed ℓ_z and s_z values consistent with the Pauli principle. The resulting table classified by the L_z and S_z values is shown (Table 4). We have given only a little more than half the table since the negative L_z values may be obtained by interchanging the signs of the ℓ_z values for the positive L_z values. From this table we can determine which L-S states are allowed. The largest value of the orbital angular momentum is L = 6 and the spins are paired. This term must come from an ^{1}I state which also must have singlet L_z = 6,5,4,3,...., 0,....-5,-6 substates. There is an S_z = 1, L_z = 5 state which must come from a ^{3}H term and must have triplet and singlets L_z = 5,4,3,....,-3,-4,-5 substrates. By this type of elimination we arrive at the following terms allowed for an f^2 configuration

$$^1I, \ ^3H, \ ^1G, \ ^3F, \ ^1D, \ ^3P, \ ^1S \ .$$

Now we can calculate the electrostatic energy of the various terms by use of the diagonal sum rule. This rule states that the sum of the roots of the secular equation is equal to the sum of the diagonal matrix elements. All levels of a particular term have the same energy. There are no matrix elements connecting states with different L_z or S_z values, thus the secular equation for the configuration factors into a series of secular equations. Thus writing $E(^1I)$ for the energy of this state and (3^+3^-) for its diagonal matrix element we find

Table 4

f^2	$S_z = 1$	$S_z = 0$	$S_z = -1$
$L_z = 6$		(3^+3^-)	
$L_z = 5$	(3^+2^+)	$(3^+2^-)(3^-2^+)$	(3^-2^-)
$L_z = 4$	(3^+1^+)	$(3^+1^-)(3^-1^+)(2^+2^-)$	(3^-1^-)
$L_z = 3$	$(3^+0^+)(2^+1^+)$	$(3^+0^-)(3^-0^+)(2^+1^-)(2^-1^+)$	$(3^-0^-)(2^-1^-)$
$L_z = 2$	$(3^+-1^+)(2^+0^+)$	$(3^+-1^-)(2^+0^-)(1^+1^-)(3^--1^+)(2^-0^+)$	$(3^--1^-)(2^-0^-)$
$L_z = 1$	$(3^+-2^+)(1^+0^+)(2^+-1^+)$	$(3^+-2^-)(3^--2^+)(1^+0^-)(1^-0^+)(2^+-1^-)(2^--1^+)$	$(3^--2^-)(1^-0^-)(2^--1^-)$
$L_z = 0$	$(3^+-3^+)(2^+-2^+)(1^+-1^+)$	$(3^+-3^-)(2^+-2^-)(1^+-1^-)(0^+-0^-)(3^--3^+)(2^--2^+)(1^--1^+)$	$(3^--3^-)(2^--2^-)(1^--1^-)$

L_z

6    $E(^1I) = (3^+3^-)$

5    $E(^3H) = (3^+2^+)$

4    $E(^3H)+E(^1I)+E(^1G) = (3^+1^-)+(3^-1^+)+(2^+2^-)$

3    $E(^3F)+E(^3H) = (3^+0^+)+(2^+1^+)$

2    $E(^3F)+E(^3H)+E(^1I)+E(^1G)+E(^1D) = (3^+-1^-)+(2^+0^-)+(1^+1^-)$

$$+(3^--1^+)+(2^-0^+)$$

1    $E(^3F)+E(^3H)+E(^3P) = (3^+-2^+)+(1^+0^+)+(2^+-1^+)$

0    $E(^3F)+E(^3H)+E(^1I)+E(^1G)+E(^1D)+E(^3P)+E(^1S) =$

$$(3^+-3^-)+(2^+-2^-)+(1^+-1^-)+(0^+0^-)+(3^--3^+)+(2^--2^+)+(1^--1^+)$$

In order to use this method we need to define two types of
integrals: the direct integral

$$(ab|q|ab) = J(a,b) = \sum_{k=0}^{\infty} a^k(\ell^a\ell_z^a, \ell^b\ell_z^b)F^k(n^a\ell^a, n^b\ell^b)$$

and the exchange integral

$$(ab|q|ba) = K(a,b) = \delta(s_z^a, s_z^b)\sum_{k=0}^{\infty} b^k(\ell^a\ell_z^a, \ell^b\ell_z^b)G^k(n^a\ell^a, n^b\ell^b)$$

and $E(A) = \sum_{a>b=1}^{N} \left[J(a,b)-K(a,b) \right]$

where $q = \dfrac{e^2}{r_{12}}$ and the coefficients $a^k(\ell^a\ell_z^a, \ell^b\ell_z^b)$ and $b^k(\ell^a\ell_z^a, \ell^b\ell_z^b)$
are tabulated in Condon and Shortley (Table 2^6, page 180 and Table
1^6, page 178) for s,p,d, and f electrons. For equivalent electrons
as we are dealing with here $F^{(k)}(n^a\ell^a, n^a\ell^a) = G^{(k)}(n^a\ell^a, n^b\ell^b)$.
Note also that because of the δ function for s_z^a, s_z^b the exchange
integral is zero for singlet states.

$$E(^1I) = F_0+25F_2+9F_4+F_6$$

$$E(^3H) = F_0+0F_2-21F_4-6F_6-25F_2-30F_4-7F_6$$

$$= F_0-25F_2-51F_4-13F_6$$

$$E(^1G) = (3^+1^-)+(3^-1^+)+(2^+2^-)-E(^3H)-E(^1I)$$

$$= F_0-15F_2+3F_4+15F_6$$

$$F_0-15F_2+3F_4+15F_6$$

$$F_0+0F_2+49F_4+36F_4$$

$$-F_0+25F_2+51F_4+13F_6$$

$$-F_0-25F_2-9F_4-F_6$$

$$= F_0-30F_2+97F_4+78F_6$$

$$E(^3F) = (3^+0^+)+(2^+1^+)-E(^3H)$$

$$= F_0-20F_2+18F_4-20F_6$$

$$+0F_2-63F_4-84F_6$$

$$+ F_0+0F_2-7F_4-90F_6$$

$$-15F_2-32F_4-105F_6$$

$$-F_0+25F_2+51F_4+13F_6$$

$$= F_0-10F_2-33F_4-286F_6$$

and so on.

Appendix C

The energy levels for $\ell = 3$ with the complete crystal field Hamiltonian of Equation 20 are (using Table 17, Appendix B of Abragam and Bleaney) and the notation $A_k^0 \langle r^k \rangle = B_k^0$

$$E_{\ell_z = 0} = -12\alpha B_2^0 + 360\beta B_4^0 - 3600\gamma B_6^0$$

$$E_{\ell_z = \pm 1} = -9\alpha B_2^0 + 60\beta B_4^0 + 2700\gamma B_6^0$$

$$E_{\ell_z = \pm 2} = 0\alpha B_2^0 - 420\beta B_4^0 + 1080\gamma B_6^0$$

$$E_{\ell_z = \pm 3} = 15\alpha B_2^0 + 180\beta B_4^0 + 180\gamma B_6^0$$

Now from Table 18, Appendix B

for $\ell = 3$

$$\langle 3 \| \alpha \| 3 \rangle = \frac{-2}{45}$$

$$\langle 3 \| \beta \| 3 \rangle = \frac{2}{11 \times 45}$$

$$\langle 3 \| \gamma \| 3 \rangle = \frac{-4}{11 \times 13 \times 27}$$

Then we obtain

$$E_{\ell_z = 0} = +\frac{8}{15} A_2^0 \langle r^2 \rangle + \frac{16}{11} A_4^0 \langle r^4 \rangle + \frac{1600}{3 \times 11 \times 13} A_6^0 \langle r^6 \rangle$$

$$E_{\ell_z = \pm 1} = \frac{2}{5} A_2^0 \langle r^2 \rangle + \frac{8}{33} A_4^0 \langle r^4 \rangle - \frac{400}{11 \times 13} A_6^0 \langle r^6 \rangle$$

$$E_{\ell_z = \pm 2} = 0 - \frac{56}{33} A_4^0 \langle r^4 \rangle + \frac{160}{11 \times 13} A_6^0 \langle r^6 \rangle$$

$$E_{\ell_z = \pm 3} = -\frac{2}{3} A_2^0 \langle r^2 \rangle + \frac{8}{11} A_4^0 \langle r^4 \rangle - \frac{80}{3 \times 11 \times 13} A_6^0 \langle r^6 \rangle$$

Appendix D

A general equation for crystal field matrix elements may be obtained in terms of tensor operators as follows:

$$\langle f^n\alpha SLJJ_z | \mathcal{H}_3 | f^n\alpha' SL'J'J'_z \rangle$$

$$= \sum_q B_q^k \langle f^n\alpha SLJJ_z | U_q^{(k)} | f^n\alpha SL'J'J'_z \rangle \langle f \| C^{(k)} \| f \rangle \quad .$$

Now for f electrons

$$\langle f \| C^{(k)} \| f \rangle = \langle 3 \| C^{(k)} \| 3 \rangle = (-1)^3 \left[(7)(7) \right]^{1/2} \begin{pmatrix} 3 & k & 3 \\ 0 & 0 & 0 \end{pmatrix}$$

$$\langle f^n\alpha SLJJ_z | U_q^{(k)} | f^n\alpha' SL'J'J'_z \rangle$$

$$= (-1)^{J-J_z} \begin{pmatrix} J & k & J' \\ -J_z & q & J'_z \end{pmatrix} \langle f^n\alpha SLJ \| U^{(k)} \| f^n\alpha' SL'J' \rangle$$

and

$$\langle f^n\alpha SLJ \| U^{(k)} \| f^n\alpha SL'J' \rangle$$

$$= (-1)^{S+L'+J+k} \left[(2J+1)(2J'+1) \right]^{1/2} \begin{Bmatrix} J & J' & k \\ L' & L & S \end{Bmatrix} \langle f^n\alpha SL \| U^{(k)} \| f^n\alpha' S'L' \rangle$$

We may gather all the terms together and write for f electrons

$$\langle f^n\alpha SLJJ_z | V | f^n\alpha' SL'J'J'_z \rangle$$

$$= \sum_q B_q^{(k)} (-1)^{3-J_z+S+L'+2J+k} (7) \begin{pmatrix} 3 & k & 3 \\ 0 & 0 & 0 \end{pmatrix} \begin{pmatrix} J & k & J' \\ -J_z & q & J'_z \end{pmatrix} \begin{Bmatrix} J & J' & k \\ L' & L & S \end{Bmatrix}$$

$$\times \left[(2J+1)(2J'+1) \right]^{1/2} \langle f^n\alpha SL \| U^{(k)} \| f^n\alpha' SL \rangle$$

In the above equation the () are 3-j symbols, $\{\}$ is a 6-j symbol, and $\langle \| U^{(k)} \| \rangle$ is a reduced matrix element which is tabulated for all f^n configurations by Nielson and Koster. Note that S = S', if this is not true, the matrix element is zero. The above general equation for crystal field matrix elements may be readily evaluated by computer techniques.

To illustrate these calculations, we will evaluate the $B_0^{(6)}$ matrix element for the $J = 4$, $J_z = 4$ state of the $^3H_4$ term of f^2.

$$\langle f^n \alpha SLJJ_z | \mathcal{H}_3 | f^n \alpha' SL'J'J_z' \rangle$$

$$= B_0^{(6)} \langle f^2\, {}^3H44 | U_0^{(6)} | f^2\, {}^3H44 \rangle\ \langle 3 \| C^{(6)} \| 3 \rangle$$

$$= B_0^{(6)} (-1)^{3-4+1+5+8+6} \times (7) \begin{pmatrix} 3 & 6 & 3 \\ 0 & 0 & 0 \end{pmatrix} \begin{pmatrix} 4 & 6 & 4 \\ -4 & 0 & 4 \end{pmatrix} \begin{Bmatrix} 4 & 4 & 6 \\ 5 & 5 & 1 \end{Bmatrix}$$

$$\times\ (2 \times 4 + 1)\ \langle f^2\, {}^3H \| U^{(6)} \| f^2\, {}^3H \rangle$$

Now from Rotenberg, Bivens, et al.

$$\begin{pmatrix} 3 & 6 & 3 \\ 0 & 0 & 0 \end{pmatrix} = \begin{pmatrix} 6 & 3 & 3 \\ 0 & 0 & 0 \end{pmatrix} = \left(\frac{2^2 \times 5^2}{3 \times 7 \times 11 \times 13} \right)^{1/2}$$

$$\begin{pmatrix} 4 & 6 & 4 \\ -4 & 0 & 4 \end{pmatrix} = \begin{pmatrix} 6 & 4 & 4 \\ 0 & 4 & -4 \end{pmatrix} = \begin{pmatrix} 6 & 4 & 4 \\ 0 & -4 & 4 \end{pmatrix} = \left(\frac{2^2}{3^2 \times 5 \times 11 \times 13} \right)^{1/2}$$

$$\begin{Bmatrix} 4 & 4 & 6 \\ 5 & 5 & 1 \end{Bmatrix} = \begin{Bmatrix} 6 & 4 & 4 \\ 1 & 5 & 5 \end{Bmatrix} = \begin{Bmatrix} 6 & 5 & 5 \\ 1 & 4 & 4 \end{Bmatrix} = \left(\frac{2^2 \times 17}{3^3 \times 5^2 \times 11^2} \right)^{1/2}$$

From Nielson and Koster

$$\langle f^2\, {}^3H \| U^{(6)} \| f^2\, {}^3H \rangle = -\left(\frac{5 \times 17}{3^2 \times 7} \right)^{1/2}$$

Substituting we obtain

$$\langle f^2\, {}^3H44 | V_0^6 | f^2\, {}^3H44 \rangle$$

$$= B_0^{(6)} (-1)^{19} \times 7 \times \left(\frac{2^2 \times 5^2}{3 \times 7 \times 11 \times 13} \right)^{1/2} \times \left(\frac{2^2}{3^2 \times 5 \times 11 \times 13} \right)^{1/2}$$

$$\times \left(\frac{2^4 \times 17}{3^3 \times 5^2 \times 11^2} \right)^{1/2} \times 9 \times (-1) \times \left(\frac{5 \times 17}{3^2 \times 7} \right)^{1/2}$$

$$= B_6^0 \left(\frac{2^8 \times 17^2}{3^4 \times 11^4 \times 13^2} \right)^{1/2}$$

$$= B_6^0 \left(\frac{2^4 \times 17}{3^2 \times 11^2 \times 13} \right)$$

We can check this matrix element by comparing it with the value calculated by the operator equivalent method.

Now from Table 3 in operator equivalent notation this matrix element is

$$\langle f^2 \; {}^3H44 | V_0^6 | f^2 \; {}^3H44 \rangle = 4 \times 1260 \times \gamma \; A_6^0 \langle r^6 \rangle$$

Now from Table 20, Appendix B, Abragam and Bleaney

$$\langle {}^3H_4 | \gamma | {}^3H_4 \rangle = \frac{2^4 \times 17}{3^4 \times 5 \times 7 \times 11^2 \times 13}$$

so

$$4 \times 1260 \times \frac{2^4 \times 17}{3^4 \times 5 \times 7 \times 11^2 \times 13} \; A_6^0 \langle r^6 \rangle = B_0^6 \left(\frac{2^4 \times 17}{3^2 \times 11^2 \times 13} \right)$$

$$16 \; A_6^0 \langle r^6 \rangle = B_0^6$$

This value agrees with the entry in Table 3.

Appendix E

We have shown in Fig. 4 that the energy levels for an f^1 system in a strong axial ligand field (chi = 1 and considering only a B_0^2 crystal field term in the Hamiltonian) are:

$$\ell_z = 0 \quad \text{or } a_2 \quad , \quad E = .444 \quad ;$$

$$\ell_z = \pm 1 \quad \text{or } e_1 \quad , \quad E = .333 \quad ;$$

$$\ell_z = \pm 2 \quad \text{or } e_2 \quad , \quad E = .000 \quad ;$$

$$\ell_z = \pm 3 \quad \text{or } e_3 \quad , \quad E = -.556 \quad ;$$

Now let us consider the case of two f electrons in the strong crystal or ligand field limit.

There are 10 different ways we may combine the above set of orbitals by pairs:

$$a_2 a_2 \quad e_1 e_1 \quad e_2 e_2 \quad e_3 e_3$$

$$a_2 e_1 \quad e_1 e_2 \quad e_2 e_3$$

$$a_2 e_2 \quad e_1 e_3$$

$$a_2 e_3$$

The crystal field operator is a one electron operator so if we consider the two orthogonal orbitals ϕ_1 and ϕ_2 we may write

$$\langle \phi_1 \phi_2 | V_c | \phi_1 \phi_2 \rangle = \langle \phi_1 \phi_2 | V_1 + V_2 | \phi_1 \phi_2 \rangle = \langle \phi_1 | V_1 | \phi_1 \rangle$$

$$+ \langle \phi_2 | V_2 | \phi_2 \rangle$$

For example, the energy of two electrons in the $a_2 e_1$ orbital is written as

$$\langle a_2 e_1 | V_c | a_2 e_1 \rangle = \langle a_2 | V_1 | a_2 \rangle + \langle e_1 | V_2 | e_1 \rangle \quad .$$

Now as noted previously $\langle a_2 | V_1 | a_2 \rangle = .444$,

$$\langle e_1 | V_2 | e_1 \rangle = .333 \quad ,$$

so $\langle a_2 e_1 | V_c | a_2 e_1 \rangle = .444 + .333 = .777.$

In order to obtain the normalization used in Fig. 8 the above energy .777 must be divided by 2. The following energies are obtained:

Orbital	Normalized energy
$a_2 a_2$	.444
$e_1 a_2$	.389
$e_1 e_1$	.333
$e_2 a_2$	.222
$e_2 e_1$	.167
$e_2 e_2$	.000
$e_3 a_2$	-.056
$e_3 e_1$	-.111
$e_3 e_2$	-.278
$e_3 e_3$	-.556

We may determine the degeneracies of each of the two electron orbitals from a probability argument. An electron may be placed in an a_2 orbital with spin up or down, that is, in two different ways. An electron may be placed in an e_i orbital in four different ways. If the second e_1 orbital is equivalent to the first orbital there are limitations in the number of ways an electron may be placed in the second orbital due to the Pauli principle. For example, let us consider the $e_2 e_2$ orbitals. We may place the first electron in one of four ways in the first e_2 orbital, one of three ways in the second e_2 orbital so there are 4×3 or 12 ways of placing the electrons in this orbital pair. However, we must divide this number by two because the electrons are indistinguishable and we are counting only distinct pairs; that is there are $4 \times 3/2$ or 6 different ways of putting two electrons in the e_2e_2 orbital pair. The degeneracies of the orbitals are:

Orbital	Degeneracy
a_2a_2	$2 \times 1/2 = 1$
e_1a_2	$4 \times 2 = 8$
e_2a_2	$4 \times 2 = 8$
e_3a_2	$4 \times 2 = 8$
e_1e_1	$4 \times 3/2 = 6$
e_2e_1	$4 \times 4 = 16$
e_3e_1	$4 \times 4 = 16$
e_2e_2	$4 \times 3/2 = 6$
e_2e_3	$4 \times 4 = 16$
e_3e_3	$4 \times 3/2 = 6$
	TOTAL = 91

We can easily calculate for the purpose of comparison the degeneracies in the LSJ coupling scheme (Fig. 3), which will turn out to be the same as in the strong ligand field case. For each J level, there is a (2J+1) degeneracy.

Level	Degeneracy
$^3H_4$	$2 \times 4 + 1 = 9$
$^3H_5$	$2 \times 5 + 1 = 11$
$^3H_6$	$2 \times 6 + 1 = 13$
$^3F_2$	$2 \times 2 + 1 = 5$

Level	Degeneracy
$^3F_3$	$2 \times 3 + 1 = 7$
$^3F_4$	$2 \times 4 + 1 = 9$
$^1G_4$	$2 \times 4 + 1 = 9$
$^1D_2$	$2 \times 2 + 1 = 5$
$^3P_0$	$2 \times 0 + 1 = 1$
$^3P_1$	$2 \times 1 + 1 = 3$
$^3P_2$	$2 \times 2 + 1 = 5$
$^1S_0$	$2 \times 0 + 1 = 1$
	TOTAL = 91

ORGANOMETALLIC COMPOUNDS WITH LANTHANIDE - CARBON SIGMA BONDS,
ONE OF THE LAST PROBLEMS IN ORGANOMETALLIC CHEMISTRY

Herbert Schumann

Institut für Anorganische und Analytische Chemie der
Technischen Universität Berlin, Berlin, W-Germany

1. INTRODUCTION AND HISTORY

Cadet's Fuming Liquid, containing the first organoelemental
compound, was described in 1760 (1). About 100 years later, Cahours
and Rieche established the correct formula for this first compound
having a covalent σ-bond between the carbon atom and the metalloid
as shown by $(CH_3)_2As-As(CH_3)_2$ (2). The first genuine organometallic
compound containing a metal - carbon - σ - bond, C_2H_5ZnI was prepared in 1849 by Frankland (3). This work ignited the fascinating
evolution of organometallic chemistry during the past century.
But only σ-bonds between carbon and metal atoms having completely
filled or empty d-orbitals, i.e. the main group elements and the
metals of the Zn-group could be made in this early period. Not
until 1952, when the first compound containing a covalent Ti-C-
bond was isolated, was the characterization of the first organometallic compound containing a carbon - transition metal - σ -
bond accomplished (4).

In contrast to the σ - bonded organometallic compounds of
the main group elements, all organometallic compounds of the
transition metals containing metal - carbon - σ - bonds are in
general extremely sensitive and unstable substances. This feature
can be explained by the high reactivity of these compounds due to
the incompletely filled d-orbitals. Whereas in the case of the
main group elements stable metal - carbon σ - bonds are built up
if the element has eight electrons available for bonds, the transition metal needs eighteen.

The situation will become much more complicated, when going
from the transition metals to the inner transition metals, which

81

T. J. Marks and R. D. Fischer (eds.), Organometallics of the f-Elements, 81–112.

involve f-orbitals in their outer valence shells. A. von Grosse postulated in 1925 the nonexistence of alkyl or aryl derivatives of the lanthanides (5).

But in 1935, the possible formation of methyl lanthanum compounds was mentioned by Rice and Rice. Using the Paneth Technique, they found that free radicals reacted with a variety of metals including lanthanum. No mention was made of the isolation or identification of any of these alkyl lanthanum species (6).

In 1938, the first organometallic compounds of the lanthanides appeared in the literature (7). Triethylscandium and triethylyttrium etherates were prepared from $ScCl_3$ and YCl_3, respectively, by their reaction with C_2H_5MgBr. The scandium compound, a yellowish transparent liquid (boiling point 170 - 172°C), was inflamed in air. On contact with water it forms $(C_2H_5)_2ScOH$, and by the reaction with HBr white needles, probably $(C_2H_5)_2ScBr$. Analogous reactions have been found in the case of the colorless yttrium compound, which showed a boiling point 222 - 225°C !!! But all attempts to repeat these results were unsuccessful (8).

In 1945, Gilman and Jones described the next attempts to prepare organometallic compounds of lanthanum (9). But biphenyl was the only product isolated from the reaction of lanthanum trichloride with phenyl lithium in ether. The same result was obtained from the reaction of metallic lanthanum with diphenyl mercury at 135°C in a sealed tube for 100 days. Attempts at stoichiometric coupling of phenylmagnesium bromide, phenyl lithium, and pentafluorophenylmagnesium bromide and -lithium, as well as of some alkyl magnesium and lithium derivatives in the presence of different lanthanide trichlorides, have been studied in the early 1960's. The formation of biphenyl and related hydrocarbons in all these reactions suggest the possibility of a transitory bond formation between the organic groups and the metal. Aside from some spontaneously igniting powders, no compound with satisfactory elemental analyses could be isolated (10).

Organometallic chemistry of the transition metals, especially after the discovery of the π-complexes, and of the main group elements, came to its most rapid growth during the last two decades. There was, however, one important problem open in this field of chemistry: the synthesis of organometallic compounds containing covalent σ-bonds between carbon and the metals of the rare earths.

This review also includes compounds of scandium and yttrium. Although they also belong to group IIIa of the periodic table, they are not lanthanides. The article is mainly concerned with $\sigma(\eta^1)$-bonded organometallics and includes synthetic aspects of sandwich complexes, allyl derivatives and related compounds, as well.

A number of excellent review articles covering this area of organometallic chemistry appeared in recent years (11). Since 1964 there is an annual survey available, covering the literature in the field of organolanthanide and organoactinide chemistry (12).

2. CYCLOPENTADIENYL DERIVATIVES

Wilkinson and Birmingham (13) reported the synthesis of the first organometallic compounds of the lanthanides, the tricyclo-pentadienyl complexes of Sc, Y, La, Ce, Pr, Nd, Sm and Gd in 1954. The compounds were synthesized by the reaction of the correspon-ding lanthanide trichloride with sodium cyclopentadienide in tetrahydrofuran (14):

$$MCl_3 + 3\ NaC_5H_5 \longrightarrow M(C_5H_5)_3 + 3\ NaCl$$

M = Sc, Y, La, Ce, Pr, Nd, Sm, Gd, Dy, Er, Yb

The synthesis of these complexes was the starting point for an exhaustive research in this field of organometallic chemistry.

2.1. Tricyclopentadienyl compounds.

The general method for the preparation of tricyclopentadienyl lanthanides is the reaction of anhydrous metal halide with cyclo-pentadienyl potassium or -sodium in tetrahydrofuran. The products are usually isolated by evaporating the solvent and subliming them from the residue under vacuum, or by extracting the residue with a suitable organic solvent, such as tetrahydrofuran, from which the complex can be recovered by evaporation of the solvent. The particular thermal instability of the europium compound pre-vented its preparation by this method. It could only be isolated as an adduct with tetrahydrofuran (15). Tetrahydrofuran could, however, be removed by warming the complex up to 70°C under vacuum (16). The Tb, Ho, Tm and Lu derivatives were synthesized by a modification of the normal procedure. Benzene or diethyl ether were used as solvents and potassium cyclopentadienide as the alkali compound (17).

The unsolvated metal cyclopentadienides of Sc, Ce, Nd and Sm could also be prepared by a method without using a solvent. Molten $Mg(C_5H_5)_2$ reacts with the anhydrous trifluorides of Sc, Ce, Nd or Sm in a sealed tube at 200 - 260°C, forming the cor-responding tricyclopentadienyl complexes in high yields (18):

$$2\ MF_3 + 3\ Mg(C_5H_5)_2 \xrightarrow[\text{200-260°C}]{\text{melt}} 2\ M(C_5H_5)_3 + 3\ MgF_2$$

M = Sc, Ce, Nd, Sm

The remaining compound of this series, $Pm(C_5H_5)_3$, has been synthesized by two different methods. The initial preparation took into account the radioactivity of promethium, and was therefore carried out by radiochemical techniques starting with $Nd(C_5H_5)_3$. This was subjected to neutron bombardment, yielding some $Pm(C_5H_5)_3$ in a matrix of the starting material (19). Larger quantities of the pure compound were prepared in analogy to the method mentioned above, but using $Be(C_5H_5)_2$, because of its lower melting point (20):

$$2\ PmCl_3 + 3\ Be(C_5H_5)_2 \xrightarrow{\text{melt}} 2\ Pm(C_5H_5)_3 + 3\ BeCl_2$$

$Yb(C_5H_5)_3$ can be prepared as a by-product, along with $Yb(C_5H_5)_2$ from the reaction of ytterbium metal with cyclopentadiene in liquid ammonia (21).

All of these compounds are air- and moisture-sensitive, but surprisingly stable to heat, and, with the exception of the europium derivative, sublime at elevated temperatures. A summary of some physical properties is given in Table 1. Many of the other physical properties including vibrational spectra, NMR spectra, optical spectra, mass spectra, magnetic susceptibiliy, thermodynamic and other data as well as X-ray investigations, have been the subject of several papers which are not to be discussed in this work. The metal-to-ring bonding in these compounds is assumed to be predominantly electrostatic (24). The original assignments were based primarily on the physical and magnetochemical data. The f-electron covalency decreases with increasing atomic number in the lanthanide series due to the lanthanide contractions weakening overlap between metal f and ring-carbon orbitals.

Attention should be drawn only to some structural aspects. $Sm(C_5H_5)_3$ shows a curious disordered structure in which the molecules form zig-zag chains with a complicated bridging system between neighboring Sm atoms (30). $Nd(C_5H_4CH_3)_3$ has an interesting tetrameric structure in which each Nd atom is bound to three pentahapto- and one monohapto-cyclopentadienyl groups. The monohapto ligand is pentahapto for another Nd and vice versa (27). Tricyclopentadienylscandium is also a dimer with two pentahapto-C_5H_5 ligands bound to each Sc and with two bridging monohapto-pentahapto-C_5H_5-groups inbetween (31).

2.2. Compounds of the type $M(C_5H_5)_2$

Sm, Eu and Yb form compounds of the type $M(C_5H_5)_2$, which can be prepared by different methods. Because of the solubility of metallic Eu and Yb in liquid ammonia, their derivatives can

Table 1. Physical Properties of Tricyclopentadienyl Lanthanides

Compound	Color	m.p. (°C)	μ_{eff} (B.M.)	Ref.
$Sc(C_5H_5)_3$	straw	240	diamagn.	13,14,18
$Y(C_5H_5)_3$	pale yellow	295	diamagn.	13,14
$La(C_5H_5)_3$	colorless	395	diamagn.	13,14
$La(C_5H_4CH_3)_3$	colorless	155	diamagn.	25
$Ce(C_5H_5)_3$	orange-yellow	435	2.46	13,14,18
$Pr(C_5H_5)_3$	pale green	420	3.61	13,14,20
$Nd(C_5H_5)_3$	pale blue	380	3.63	13,14,18,19
$Nd(C_5H_4CH_3)_3$	blue-violett	163		26,27
$Pm(C_5H_5)_3$	orange	250 (dec.)		19,20,22
$Sm(C_5H_5)_3$	orange	365	1.54	13,14,18,20
$Sm(C_5H_4CH_3)_3$				28
$Eu(C_5H_5)_3$	brown		3.74	15,16,23
$Gd(C_5H_5)_3$	yellow	350	7.98	13,14,20
$Gd(C_5H_4CH_3)_3$				28
$Tb(C_5H_5)_3$	colorless	316	8.9	17,20
$Dy(C_5H_5)_3$	yellow	302	10.0	14
$Dy(C_5H_4CH_3)_3$				28
$Ho(C_5H_5)_3$	yellow	295	10.2	17,23
$Ho(C_5H_4CH_3)_3$				28
$Er(C_5H_5)_3$	pink	285	9.45	14
$Er(C_5H_4CH_3)_3$				28
$Tm(C_5H_5)_3$	yellow-green	278	7.1	17
$Tm(C_5H_4CH_3)_3$				28
$Yb(C_5H_5)_3$	dark green	273	4.00	14,17,24
$Lu(C_5H_5)_3$	colorless	264	diamagn.	17

be synthesized by the reaction of the metals with cyclopentadiene in this solvent (17,21,32). The dicyclopentadienide of ammonia-insoluble samarium was prepared starting with $Sm(C_5H_5)_3$ (33), and the ytterbium derivative, as well, can be made by the reduction of Yb(III) by Na or Yb in THF (24):

$$M + 3\ C_5H_6 \xrightarrow{\text{liq.NH}_3} M(C_5H_5)_2 + C_5H_8$$

M = Eu, Yb; colour of "$Yb(C_5H_5)_2$": red (17,32)

$$Sm(C_5H_5)_3 + KC_{10}H_8 \xrightarrow{\text{THF}} Sm(C_5H_5)_2 \cdot \text{THF} + KC_5H_5 + C_{10}H_8$$

$C_{10}H_8$ = naphthalene

$$Yb(C_5H_5)_3 + 2\ Na \xrightarrow{\text{THF}} Yb(C_5H_5)_2 + NaC_5H_5$$
colour of "$Yb(C_5H_5)_2$": green (24)

$$(C_5H_5)_2YbCl + Na \longrightarrow Yb(C_5H_5)_2 + NaCl$$

$$3 \ (C_5H_5)_2YbCl + Yb \longrightarrow 3 \ Yb(C_5H_5)_2 + YbCl_3$$

Magnetic moments, infrared spectra and the ^{151}Eu-Mössbauer spectrum of the europium compound indicate the presence of predominantly ionically bonded, symmetrical five-membered rings, in these compounds (34).

2.3. Compounds of the type $M(C_5H_5)_4$

Ce$(C_5H_5)_4$ is described to be formed by the interaction of dipyridinium cerium hexachloride, with sodium cyclopentadienide in THF as a red-orange crystalline compound, which is quite stable, both thermally and chemically, but melts with decomposition at 225 - 226°C!!! The compound is said to be unaffected by water and dilute acids, but is decomposed by concentrated mineral acids and by dilute hot alkalies. The IR spectrum contains a band at 2980 cm^{-1}, in addition to the usual C-H stretching vibrations of the cyclopentadienyl group, which indicates that the C_5H_5 ring may be more covalently bonded to the Ce atom (35):

$$(C_5H_6N)_2CeCl_6 + 4 \ NaC_5H_5 \longrightarrow Ce(C_5H_5)_4 + 4 \ NaCl + 2 \ C_5H_6NCl$$

2.4. Compounds of the types $(C_5H_5)_2MX$ and $C_5H_5MX_2$

Dicyclopentadienyllanthanide chlorides of scandium and of the heavier lanthanides have been prepared by the stoichiometric reaction of the corresponding anhydrous chloride with Mg$(C_5H_5)_2$ (36), TlC_5H_5 (37) or NaC_5H_5 (38,39), or by the reaction of the trichlorides with the corresponding tricyclopentadienyl derivatives:

$$2 \ ScCl_3 + 3 \ Mg(C_5H_5)_2 \longrightarrow 2 \ (C_5H_5)_2ScCl + 3 \ MgCl_2$$

$$ScCl_3 + 2 \ TlC_5H_5 \longrightarrow (C_5H_5)_2ScCl + 2 \ TlCl$$

$$MCl_3 + 2 \ NaC_5H_5 \longrightarrow (C_5H_5)_2MCl + 2 \ NaCl$$

M = Sm, Gd, Dy, Ho, Er, Yb, Lu

$$MCl_3 + 2 \ M(C_5H_5)_3 \longrightarrow 3 \ (C_5H_5)_2MCl$$

M = Sm, Gd, Dy, Ho, Er, Yb

The terbium (40) and thulium derivative (29) may be prepared in the same way, as well as the di(methylcyclopentadienyl)lanthanide chlorides of Gd, Er, and Yb (38). Similar compounds of

Y, La, Ce, Pr, Nd, Pm, and Eu could not be prepared until now. Attempts to prepare some of them by several methods failed, and have been ascribed to the lanthanide contraction (38). Molecular weight determinations show a dimeric character for the compounds in benzene, while monomers are present in tetrahydrofuran, due to solvation of the complexes by this base.

The molecular structures of $(C_5H_5)_2ScCl$ (41) and $(CH_3C_5H_4)_2YbCl$ (42) have been determined, showing a chlorine-bridged dimeric arrangement. The bridge unit is symmetric in both cases, and the carbon atoms in the cyclopentadienyl rings are equidistant from the Sc or Yb atoms, with average M–C-distances of 2.46 (Sc) or 2.58 Å (Yb).

Magnetic susceptibility measurements indicate that ionic type forces bond the cyclopentadienyl rings to the metals, which are formally in a tripositive state.

Dicyclopentadienyl erbium iodide was prepared by the reaction of $Er(C_5H_5)_3$ with iodine in tetrahydrofuran (38); and cyanide derivatives have been made in benzene solution starting from $M(C_5H_5)_3$ and HCN (43):

$$(C_5H_5)_3Er + I_2 \longrightarrow (C_5H_5)_2ErI + C_5H_5I$$

$$(C_5H_5)_3M + HCN \longrightarrow (C_5H_5)_2M\text{-}CN + C_5H_6$$

M = Nd, Yb

Methoxides, phenoxides, and acylate derivatives have been prepared by displacement reactions of the dicyclopentadienyl lanthanide chlorides of Gd, Dy, Er and Yb with sodium derivatives NaX (38). These compounds show a marked increase in the oxidative stability over the analogous chlorides. But they remain quite hydrolytically unstable. In the same way some amides (23,38) and di-tert-butylphosphine derivatives (40) have been synthesized. These, too, are still air sensitive:

$$(C_5H_5)_2MCl + NaX \longrightarrow (C_5H_5)_2MX + NaCl$$

M = Sc, X = O_2CCH_3, acac (36);

M = Gd, X = O_2CCH_3; M = Dy, X = OCH_3; M = Er, X = OCH_3, O_2CH, O_2CCH_3; M = Yb, X = OCH_3, OC_6H_5, O_2CH, O_2CCH_3, $O_2CC_6H_5$; and $(CH_3C_5H_4)_2MO_2CCH_3$ with M = Gd, Er (38)

$$(C_5H_5)_2ErCl + NaNH_2 \longrightarrow (C_5H_5)_2ErNH_2 + NaCl$$

$$Yb(C_5H_5)_3 \cdot NH_3 \longrightarrow (C_5H_5)_2YbNH_2 + C_5H_6$$

$$(C_5H_5)_2MCl + LiP(^tC_4H_9)_2 \longrightarrow (C_5H_5)_2M-P(^tC_4H_9)_2 + LiCl$$

M = Ho, Tb

$$(C_5H_5)_2ErCl + (CH_3)_3SiP(^tC_4H_9)_2 \longrightarrow$$

$$(C_5H_5)_2Er-P(^tC_4H_9)_2 + (CH_3)_3SiCl$$

The THF complexes of $(C_5H_5)_2MBH_4$ (M = Sm, Er, Yb) could be prepared by the reaction between the dicyclopentadienyl metal chlorides and $NaBH_4$ in tetrahydrofuran. The Sm compound has a tridentate BH_4^- ligation, whereas the Yb derivative is proposed to have bidentate ligation. THF can be removed in the cases of M = Er and Yb, yielding compounds of the composition $(C_5H_5)_2MBH_4$, for which a polymeric structure with bridging BH_4-groups, as well as appreciable ionic character in the bonding of the BH_4-ligand to the trivalent lanthanides, is suggested (44):

$$(C_5H_5)_2MCl + NaBH_4 \xrightarrow{THF} (C_5H_5)_2M-BH_4 \cdot THF + NaCl$$

M = Sm, Er, Yb

$$(C_5H_5)_2M-BH_4 \cdot THF \longrightarrow (C_5H_5)_2M-BH_4 + THF$$

M = Er, Yb

Monocyclopentadienyllanthanide dichlorides are only known as the tristetrahydrofuranates of the metals Sm, Gd, Eu, Dy, Ho, Er, Yb, and Lu. They can be prepared by three different methods:

$$MCl_3 + NaC_5H_5 \xrightarrow{THF} C_5H_5MCl_2 \cdot 3THF + NaCl$$

M = Sm, Eu, Gd, Dy, Ho, Er, Lu

$$(C_5H_5)_3Er + 2 ErCl_3 \xrightarrow{THF} 3 C_5H_5ErCl_2 \cdot 3THF$$

$$(C_5H_5)_2YbCl + HCl \xrightarrow{THF} C_5H_5YbCl_2 \cdot 3THF + C_5H_6$$

They are extremely sensitive to moisture and oxygen, insoluble in nonpolar solvents, and their bonding situation should show a ionic metal-ring interaction (45).

Special types of cyclopentadienyl compounds of cerium are in the literature. The reaction of dipyridinium cerium hexachloride with tetracyclopentadienyl cerium or with sodiumcyclopentadienide is described yielding tricyclopentadienylcerium(IV) chloride (46), which was the starting material for a considerable number of other derivatives of the type $(C_5H_5)_3CeX$, which are sometimes chemically quite inert:

$$(C_5H_5)_3CeCl + KX \longrightarrow (C_5H_5)_3CeX + KCl$$

X = CN, NCO, NCS, N_3 (47)

X = phenol, 1- or 2-naphthol, resorcinol, pyrogallol, phloroglucinol (48)

X = OCH_3, OC_2H_5, OC_3H_7, $O^iC_3H_7$, OC_4H_9, $O^iC_4H_9$, $O^iC_5H_{11}$ (49)

X = O_2CH, O_2CCH_3, $O_2CC_2H_5$, $O_2CC_3H_7$, $O_2CC_6H_5$ (50)

X = SCH_3, SC_2H_5, SC_3H_7, $S^iC_3H_7$, SC_4H_9, $S^iC_4H_9$, $S^iC_5H_{11}$ (51)

X = H, NH_2 (52)

X = BH_4 (53).

The reaction between $Ce(O^iC_3H_7)_4$ and $Mg(C_5H_5)_2$ gave $(C_5H_5)_3CeO^iC_3H_7$ in the form of black microcrystals (54):

$$2\ Ce(O^iC_3H_7)_4 + 3\ Mg(C_5H_5)_2 \longrightarrow$$
$$2\ (C_5H_5)_3CeO^iC_3H_7 + 3\ Mg(O^iC_3H_7)_2$$

2.5. Donor complexes of tricyclopentadienyl lanthanides

Tricyclopentadienyl lanthanides are Lewis acids. They form adducts with the common solvents used for their preparation, such as THF and other oxygen and nitrogen donor molecules. Complexes with NH_3 have been isolated for the Pr and Sm compounds by dissolution of $M(C_5H_5)_3$ in liquid ammonia and removal of the solvent at room temperature under vacuum (14). The green complex $(C_5H_5)_3Yb \cdot NH_3$ can be sublimed at 150°C (23,24), whereas complexes of the lower lanthanides lose the coordinated base on heating, reverting to the tricyclopentadienides. Other complexes are described like $(C_5H_5)_3Yb \cdot py$ (55,56), $(C_5H_5)_3Nd \cdot methylpyrrolidone$ (55), $(C_5H_5)_3M \cdot (-)nicotine$ (M = Pr, Nd, Tm) (55), and $(C_5H_5)_3Yb \cdot P(C_6H_5)_3$ (23).

$(C_5H_5)_3M$ reacts with cyclohexylisonitrile in benzene with the formation of very stable 1:1 complexes, which can be sublimed without decomposition at about 150°C (23,57,58,59):

$$(C_5H_5)_3M + C_6H_{11}NC \longrightarrow (C_5H_5)_3M \cdot CNC_6H_{11}$$

M = Y, La, Ce, Pr, Nd, Sm, Eu, Gd, Tb, Dy, Ho, Er, Tm, Yb, Lu.

These adducts have been the first examples of organometallic compounds having a monohapto metal to carbon sigma bond with a lanthanide element. The shift of the νCN observed in these complexes by about 75 cm^{-1} to higher wavenumbers relative to the free ligand, supports the formation of a metal-carbon sigma donor bond with the isonitrile (23). The crystal structure of $(C_5H_5)_3PrCNC_6H_{11}$ shows three pentahapto cyclopentadienyl rings forming the base of a trigonal pyramid and the isonitrile carbon at the apex. The short Pr-C distance of 2.65 Å, the triple bond in the isonitrile with a C-N bond lenght of 1.11 Å and a C-N-C angle of 177.8° further confirm the hypothesis of a sigma-donor bond between Pr and C (60).

$(C_5H_5)_3Yb$ reacts with pyrazine in benzene with formation of a dinuclear complex in which two $(C_5H_5)_3Yb$ units are linked together nearly linearly by a pyrazine ring through its nitrogens. An X-ray structural investigation showed a Yb-N distance of 2.61 Å (61).

Recently it has been shown that tricyclopentadienyl lanthanide compounds form adducts with a variety of metal carbonyl and metal nitrosyl compounds. Examples are $(MeCp)_3Sm \cdot CrCp(NO)_2Cl$, $Cp_3Er \cdot CrCp(NO)_2Cl$, $Cp_3Yb \cdot CrCp(NO)_2Cl$, $Cp_2ClYb \cdot CrCp(NO)_2Cl$, $(MeCp)_2ClYb \cdot CrCp(NO)_2Cl$, $(MeCp)_3Sm \cdot Mn(MeCp)(CO)_3$, $Cp_3Er \cdot Mn(MeCp)(CO)_3$, $Cp_3Yb \cdot Mn(MeCp)(CO)_3$, $Cp_2ClYb \cdot Mn(MeCp)(CO)_3$, $Cp_3Sm \cdot FeCp(CO)_2$, $(MeCp)_3Sm \cdot FeCp(CO)_2$, $(MeCp)_3Sm \cdot 2Co(CO)_4$ (28,62).

3. CYCLOOCTATETRAENYL AND OTHER SANDWICH DERIVATIVES

The first cyclooctatetraenyl derivatives of the lanthanides, EuC_8H_8 and YbC_8H_8, were prepared in 1969 by the direct reaction of Eu or Yb with cyclooctatetraene in liquid ammonia (63):

$$M + C_8H_8 \xrightarrow{\text{liq.NH}_3} MC_8H_8$$

M = Eu, Yb

Only one year later complexes of the types $K[M(C_8H_8)_2]$ (64), and two years later compounds $[(C_8H_8)MCl \cdot 2THF]_2$ (65), were described for most of the lanthanide metals. The scandium (66) and yttrium derivatives (67) have been recently prepared:

$$MCl_3 + 2 K_2C_8H_8 \xrightarrow{THF} K[M(C_8H_8)_2]$$

M = Y, La, Ce, Pr, Nd, Sm, Gd, Tb

$$MCl_3 + K_2C_8H_8 \xrightarrow{THF} [(C_8H_8)MCl \cdot 2THF]_2$$

M = Ce, Pr, Nd, Sm

$$ScCl_3 \cdot 3THF + K_2C_8H_8 \xrightarrow{THF} C_8H_8ScCl \cdot THF$$

$$C_8H_8ScCl \cdot THF + K_2C_8H_8 \xrightarrow{THF} K[Sc(C_8H_8)_2]$$

Chemical and spectral data strongly suggest that these compounds are highly ionic, in contrast to the analogous actinide derivatives (68). The crystal structures of $[K(CH_3OCH_2CH_2)_2O]$ $[Ce(C_8H_8)_2]$ (69) and of the complex $[Ce(C_8H_8)Cl \cdot 2THF]_2$ (70) have been determined. The first compound contains discrete $[Ce(C_8H_8)_2]^-$ anions. The two cyclooctatetraene rings adopt a staggered configuration with equal Ce-C distances of 2.742 Å. The latter complex is a dimer with asymmetric chlorine bridges. The planar cyclooctatetraene dianion is symmetrically bound to Ce with an average Ce-C-distance of 2.710 Å.

Two other cerium complexes can be made by the reduction of $Ce(O^iC_3H_7)_4$ by $Al(C_2H_5)_3$ in the presence of cyclooctatetraene. The first compound, $Ce(C_8H_8)_2$, is isomorphous with uranocene as shown by X-ray investigations. The second compound, $Ce_2(C_8H_8)_3$, is easily converted into the first by an excess of cyclooctatetraene (54):

$$Ce(O^iC_3H_7)_4 \cdot {}^iC_3H_7OH + 5 Al(C_2H_5)_3 \xrightarrow{C_8H_8, \text{ excess}}$$

$$Ce(C_8H_8)_2 + 5 (C_2H_5)_2AlO^iC_3H_7 + C_2H_6 + 4 C_2H_5^\bullet$$

$$2 Ce(O^iC_3H_7)_4 \cdot {}^iC_3H_7OH + 10 Al(C_2H_5)_3 + 3 C_8H_8 \longrightarrow$$

$$Ce_2(C_8H_8)_3 + 10 (C_2H_5)_2AlO^iC_3H_7 + 2 C_2H_6 + 8 C_2H_5^\bullet$$

Finally tetrafluorenyl cerium(IV) (71) and dicycloheptatrienylcerium(IV) dichloride (72) are in the literature. But their synthesis could not be repeated.

$$(C_5H_6N)_2CeCl_6 + 4 C_{13}H_9Na \longrightarrow Ce(C_{13}H_9)_4 + 4 NaCl + 2 C_5H_6NCl$$

$$(C_5H_6N)_2CeCl_6 + 2\ C_7H_8 \longrightarrow (C_7H_7)_2CeCl_2 + 2\ HCl + 2\ C_5H_6NCl$$

Recently the synthesis of mixed sandwich compounds $(C_5H_5)M(C_8H_8)$ was described using two methods (66):

$$C_5H_5MCl_2 \cdot 3THF + K_2C_8H_8 \xrightarrow[-30^\circ]{THF} (C_5H_5)M(C_8H_8) \cdot THF + 2\ KCl$$

M = Y, Sm, Ho, Er, Yb, Lu

$$[(C_8H_8)MCl \cdot 2THF]_2 + 2\ NaC_5H_5 \xrightarrow[-30^\circ]{THF} (C_5H_5)M(C_8H_8) \cdot THF + 2\ NaCl$$

M = Y, Pr, Nd, Sm

Both reactions also produce $M_2(C_8H_8)_3$ as by-product, separable on the basis of solubility. The slightly soluble $M_2(C_8H_8)_3$ was removed during first crystallization of the mixed sandwich complex, and the mixed compound purified by recrystallization from tetrahydrofuran/hexane.

$$(C_5H_5)M(C_8H_8) \cdot THF \xrightarrow[50^\circ,\ 1-2\ h]{vacuum} (C_5H_5)M(C_8H_8) + THF$$

M = Y, Sm, Nd, Ho, Er, Lu

IR evidence indicates electrostatic interaction between the lanthanide metal and the organic ligands. The Lewis acidity of the complexes was ascertained by the isolation of complexes $(C_5H_5)M(C_8H_8) \cdot L$, where L = NH_3, pyridine, THF or cyclohexylisonitrile (67). The corresponding scandium compounds were prepared as well (66).

A novel asymmetrical neodymium cyclooctatetraene compound, $[Nd(C_8H_8)(THF)_2][Nd(C_8H_8)_2]$, was prepared by codeposition of Nd atoms, prepared by high temperature vacuum evaporation, with cyclooctatetraene at -196°C. The molecular structure consists of an anion-cation pair: $[Nd(C_8H_8)_2]^-$ and $[Nd(C_8H_8)(THF)_2]^+$, in which the anion has a sandwich configuration. The coordination sphere about the cationic Nd consists of a C_8H_8 ring, average Nd-C distance of 2.673 Å, together with the two THF oxygen atoms and a part of one C_8H_8 ring of the anion, showing Nd-C distances of 2.700, 2.896 and 2.896 Å respectively (73).

In a recent communication, the synthesis of a new class of organolanthanide compounds is described (74). Cocondensation of 1,3-butadiene with Er metal at -196°C in a Timms-reactor produces a matrix of brown solids. The major product is $Er(C_4H_6)_3$. In addition to this compound, some other complexes of the metals La, Nd, Sm and Er with the ligands butadiene or dimethylbutadiene have been characterized. Examples include $La[C_4H_4(CH_3)_2]_2$, $Nd(C_4H_6)_3$, $Sm(C_4H_6)_3$ and $Er[C_4H_4(CH_3)_2]_2$.

4. INDENYL COMPOUNDS

A special attention should be drawn to the indenyl derivatives of the lanthanides, because of some evidence to their being the first covalent sigma bonded organolanthanide compounds. The THF adducts of the trisindenylderivatives of La, Sm, Gd, Dy, Tb, and Yb have been prepared by the reaction of the anhydrous metal trichlorides with sodium indeneide in THF (75,76):

$$MCl_3 + 3 NaC_9H_7 \xrightarrow{THF} M(C_9H_7)_3 \cdot THF + 3 NaCl$$

M = La, Sm, Gd, Dy, Tb, Yb

The NMR spectrum of the La derivative showed the A_2X pattern for the protons of the five membered ring of the indenyl group. This would be expected for a π-bonded indenyl compound, whereas the NMR spectrum of the Sm compound had an ABX pattern for the analogous protons, suggesting a covalently sigma bonded compound.

In contrast, X-ray structural analysis of the solvent free $Sm(C_9H_7)_3$, prepared from $SmCl_3$ and $Mg(C_9H_7)_2$ in benzene, gave a trigonal configuration of the five-membered rings around the Sm atom. There is no evidence for preferential bonding of the Sm atom to the electron rich C-1 position of the indenyl groups, because of no significant difference in the distances between Sm and all carbon atoms of the five membered rings (77).

Cerium forms a compound of the type $Ce(C_9H_7)_4$, which is prepared by the reaction between $(C_5H_6N)_2CeCl_6$ and NaC_9H_7 in tetrahydrofuran. The IR spectra and other properties indicate that the indenyl group is covalently bonded to Ce (35,36), but some structural information would be very helpful in formulating a final statement about the bonding situation. In addition to this Ce derivative, $(C_9H_7)_2CeCl_2$ (46), compounds of the type $(C_9H_7)_2CeX_2$ with X = H (52), BH_4 (53), NH_2 (52), CN, NCO, NCS, N_3 (47), SR (51), O_2CR (50), OR (48), are known.

5. ALLYL- AND ALKYNYL - DERIVATIVES

The first organolanthanide compounds containing the allyl group have been prepared by the reaction of the appropriate dicyclopentadienyl lanthanide chloride with allylmagnesium bromide in a THF-ether solution at -78°C (78); the Sc derivative was prepared in an analogous way using the allylmagnesium chloride (36):

$$(C_5H_5)_2ScCl + C_3H_5MgCl \longrightarrow (C_5H_5)_2ScC_3H_5 + MgCl_2$$

$$(C_5H_5)_2MCl + C_3H_5MgBr \longrightarrow (C_5H_5)_2MC_3H_5 + MgBrCl$$

M = Sm, Er, Ho

These compounds are highly sensitive to air and moisture, but in contrast, quite thermally stable. The IR spectra show a strong band at 1533 cm^{-1} attributed to the delocalized C-C stretching vibration of a π-allyl group. This indicates a preference of an η^3-allyl-lanthanide bond over a monohapto M-C-σ-bond, which implies the expansion of the coordination number of the metal from 7 for the monohapto bonding situation to 8 (78).

On the borderline between an ionic bond and a covalent lanthanide - carbon sigma bond are the phenylacetylide derivatives, which are sensitive toward oxygen and water (36,79,80):

$$(C_5H_5)_2ScCl + NaC\equiv CC_6H_5 \xrightarrow{THF} (C_5H_5)_2ScC\equiv CC_6H_5 + NaCl$$

$$(C_5H_5)_2MCl + LiC\equiv CC_6H_5 \xrightarrow[-20°]{THF} (C_5H_5)_2MC\equiv CC_6H_5 + LiCl$$

M = Gd, Ho, Er, Yb

$$C_5H_5HoCl_2 \cdot 3THF + 2 LiC\equiv CC_6H_5 \longrightarrow C_5H_5Ho(C\equiv CC_6H_5)_2 + 2 LiCl$$

Magnetic measurements confirm that the metals are in the trivalent state. The visible spectra of these compounds, with the exception of the ytterbium complex, show a charge-transfer band which is absent in the $(C_5H_5)_2MCl$ compounds, and which is shifted to lower energy in the spectrum of $C_5H_5Ho(C\equiv CC_6H_5)_2$ in comparison with $(C_5H_5)_2HoC\equiv CC_6H_5$. This indicates that the charge-transfer involves ligand to metal interactions, and that the lanthanide carbon sigma bond is not purely ionic. In addition the visible spectrum of the Er compound has bands with rather high values, possibly indicating some covalent character in the metal carbon bond.

$ScCl_3$ reacts with lithium phenylacetylide in THF-hexane with formation of scandium phenylacetylide, a pyrophoric material, stable to 250°C in vacuo and giving phenylpropiolic acid upon carbonation (81,82):

$$ScCl_3 + 3 LiC\equiv CC_6H_5 \longrightarrow Sc(C\equiv CC_6H_5)_3 + 3 LiCl$$

Solutions of Eu and Yb in liquid ammonia react at -78°C with propyne to form the methylacetylides. These pyrophoric materials are hydrolyzed by water to yield the hydroxide of the divalent metal and propyne. The Eu compound forms the carbide EuC_2 on heating to 90°C in vacuo (83). Reactions of bis(phenylethynyl)mercury with Yb, and of bis(pentafluorophenyl)ytterbium with phenylacetylene, yield bis(phenylethynyl)ytterbium, which has been isolated without supporting ligands and is exceptionally air sensitive (84):

$$M + 2\ CH_3C \equiv CH \xrightarrow{\ \text{liq. } NH_3\ } M(C \equiv CCH_3)_2 + H_2$$

M = Eu, Yb

$$Yb + Hg(C \equiv CC_6H_5)_2 \longrightarrow Yb(C \equiv CC_6H_5)_2 + Hg$$

$$Yb(C_6F_5)_2 + 2\ C_6H_5C \equiv CH \longrightarrow Yb(C \equiv CC_6H_5)_2 + 2\ C_6F_5H$$

Since a coordination number of 2 is highly improbable in an organolanthanide, an associated structure is likely for this compounds. Molecular weights indicative of trimeric - tetrameric species were obtained by boiling point elevation in THF. This association may involve either sigma-bridging through the α-carbon atom, or, more likely, π-bonding between the triple bond and Yb, as observed in phenylethynylcopper.

6. COMPOUNDS WITH METAL TO CARBON SIGMA BONDS.

The question of covalency in organometallic compounds of the lanthanides could not be resolved in terms of a metal to carbon "monohapto" sigma bond by the synthesis of the cyclopentadienyl, indenyl, cyclooctatetraenyl, allyl or even alkynyl complexes. Spectroscopic investigations as well as some crystallographic results and the magnetic properties of the new compounds have shown that there is no evidence for an exclusively sigma bonded organometallic compound.

This last question in organometallic chemistry could be solved in this decade, starting with the paper of Hart and coworkers in 1968, which reported the preparation of triphenylscandium (81), and in 1970, which described the synthesis of the first phenyl derivatives of Y, La and Pr (82).

Starting with this pioneering work, research in this area of chemistry grew slowly for the first 5 years, but now, I belive, is rapidly accelerating.

6.1. Compounds of the types $(C_5H_5)_nMR$ and $(C_9H_7)_nMR_2$

Dicyclopentadienyl alkyl and aryl derivatives of Gd, Er and Yb have been synthesized by the reaction of methyllithium or phenyllithium with the appropriate dicyclopentadienyl metal chloride in THF (80,85):

$$(C_5H_5)_2MCl + RLi \xrightarrow{\text{THF}} (C_5H_5)_2MR + LiCl$$

$$M = Gd, R = C_6H_5; M = Er, R = CH_3, C_6H_5; M = Yb, R = CH_3, C_6H_5$$

The infrared spectra of these compounds support the presence of sigma bonded alkyl or aryl groups. The electronic spectra of the Er derivatives show no hypersensitivity, but variable-temperature magnetic susceptibility measurements show that the magnetic moments of the lanthanide ions in the new compounds are distinctly temperature-dependent.

In 1976, Lappert and coworkers discovered that the methyl derivatives are dimeric, having electron-deficient three-center bonding between a lanthanide metal and a methyl bridge (86). The compounds were prepared in 70% yield from di-μ-methyl-(dimethyl-aluminium)biscyclopentadienyl-lanthanides, by reaction with pyridine in toluene. They are air-sensitive, crystalline materials, soluble in CH_2Cl_2 or toluene, but insoluble in saturated hydrocarbons:

$$2 \ (C_5H_5)_2M \underset{CH_3}{\overset{CH_3}{\diagdown\diagup}} Al(CH_3)_2 + 2 \ C_5H_5N \longrightarrow$$

$$2 \ (CH_3)_3Al \cdot C_5H_5N + (C_5H_5)_2M \underset{CH_3}{\overset{CH_3}{\diagdown\diagup}} M(C_5H_5)_2$$

M = Y, Dy, Ho, Er, Yb

The symmetrical $M(CH_3)_2M$ double-methyl bridge is established for the Y compound in solution from NMR spectra, and for the solid Yb analog, from a single crystal X-ray analysis.

In contrast, the pale yellow Sc derivative is monomeric and forms a complex with pyridine. This different behaviour of the Sc complex with pyridine is related to its non-fluxionality at room temperature in contrast to the other compounds (77):

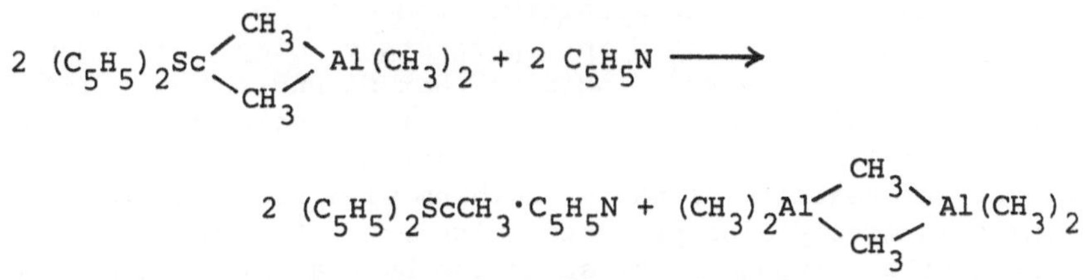

$$2 \ (C_5H_5)_2Sc \overset{CH_3}{\underset{CH_3}{<\quad>}} Al(CH_3)_2 + 2 \ C_5H_5N \longrightarrow$$

$$2 \ (C_5H_5)_2ScCH_3 \cdot C_5H_5N + (CH_3)_2Al \overset{CH_3}{\underset{CH_3}{<\quad>}} Al(CH_3)_2$$

Analogous derivatives with the bulky ligands $CH_2Si(CH_3)_3$ and $t\text{-}C_4H_9$ could be prepared by the reactions of dicyclopentadienyl lanthanide chlorides or t-butylates with the appropriate alkyl lithium compounds in THF or pentane, at low temperatures (88):

$$(C_5H_5)_2MCl + LiCH_2Si(CH_3)_3 \xrightarrow{\text{THF}} (C_5H_5)_2MCH_2Si(CH_3)_3 + LiCl$$

M = Sm, Er, Yb

$$(C_5H_5)_2Mo^tC_4H_9 + Li^tC_4H_9 \xrightarrow{\text{pentane}} (C_5H_5)_2M^tC_4H_9 + LiO^tC_4H_9$$

M = Er, Yb

The solid compounds are air sensitive. They contain different amounts of the solvent THF or diethylether coordinated to the metal. The steric hindrance of the bulky alkyl groups increases the stability, but an exact determination of the structure has not been possible until now.

A number of alkyl and aryl derivatives of cerium(IV) cyclopentadienyls and indenyls were prepared following the Grignard or alkyl lithium route (89):

$$(C_5H_5)_3CeCl + RX \xrightarrow{\text{THF}} (C_5H_5)_3CeR + XCl$$

$$(C_9H_7)_2CeCl_2 + 2 \ RX \xrightarrow{\text{THF}} (C_9H_7)_2CeR_2 + 2 \ XCl$$

X = Li or MgBr; R = CH_3, C_2H_5, C_6H_5, $CH_2C_6H_5$, $C(O)C_6H_5$

The IR spectra suggest that the linkage between the cyclopentadienyl, as well as indenyl rings, and the Ce atom retains the character of delocalized π-bonds while the alkyl and aryl groups are probably attached to the metal by σ-bonds.

6.2. Compounds MR_3, Li_nMR_{3+n}, and related systems

The reaction of $ScCl_3$, YCl_3, $LaCl_3$ or $PrCl_3$ with phenyl- or

methyllithium gives air-sensitive products. Only the phenyl derivatives could be isolated in a pure state. Simple reactions, infrared spectra and analysis indicate that Sc and Y form solid tri-
phenyl derivatives which are pyrophoric, insoluble in benzene, but
soluble in THF. In vacuo, or under nitrogen, they are stable up to
215°C. The reaction of phenyl lithium with lanthanum or praseodymium chlorides yields products which are not simple phenyls. The
air-sensitive products are suggested by analysis to be $LiM(C_6H_5)_4$.
They are soluble in benzene when first obtained from THF solution,
but, after complete drying in vacuo, they are insoluble in benzene.
The original product should be a THF complex, which polymerizes
upon removal of THF (81,82):

$$MCl_3 + 3 \ LiC_6H_5 \xrightarrow{\text{THF}} M(C_6H_5)_3 + 3 \ LiCl$$

M = Sc, Y

$$MCl_3 + 4 \ LiC_6H_5 \xrightarrow{\text{THF}} LiM(C_6H_5)_4 + 3 \ LiCl$$

M = La, Pr

The polymeric nature of these compounds led to the use of
some other ligands which should stabilize organolanthanide compounds with coordination number 3 or 4, like bulky alkyl or aryl
groups. The first compound of this type, tetrakis(tetrahydrofuran)-
lithium tetrakis(2,6-dimethylphenyl)lutetiate has been prepared
by the action of 2,6-dimethylphenyllithium with anhydrous lutetium
chloride in THF at -78°C, followed by repeated concentration, filtration, and redissolution in THF. This procedure ultimately gave,
on addition of benzene, colourless needles. X-ray diffraction
shows the lanthanide ion to be bound to four monohapto-aryl groups
in an approximately tetrahedral array; Lu-C distances range from
2.359 to 2.385 Å and C-Lu-C angles from 99.9° to 118.4°. The analogous ytterbium compound has been similarly prepared and is isostructural. The Yb compound smoulders instantly on exposure to air
while the lutetium compound appears less readily affected (90):

$$MCl_3 + 4 \ LiC_6H_3(CH_3)_2 \xrightarrow{\text{THF}} [Li(THF)_4] \ [M\{C_6H_3(CH_3)_2\}_4] + 3 \ LiCl$$

M = Yb, Lu

A considerable stabilization of three coordinated organolanthanide compounds can be achieved by the use of chelated organic
ligands. The reaction of $LiCH_2C_6H_4$-o-$N(CH_3)_2$ with $ScCl_3$ led to
the isolation of thermally stable, air-sensitive, pale-yellow
$Sc[CH_2C_6H_4N(CH_3)_2]_3$ (91):

$$ScCl_3 + 3 \ LiCH_2C_6H_4N(CH_3)_2 \longrightarrow Sc\left\{\begin{array}{c} CH_2 \\ \\ N \\ (CH_3)_2 \end{array}\right\}_3 + 3 \ LiCl$$

Other types of alkyl groups, bulky, and with no possibility for ß-H elemination reactions, have been used by Lappert and co-workers for the preparation of interesting new organolanthanide complexes. Sc and Y compounds with the groups $CH_2C(CH_3)_3$ (92), $CH_2Si(CH_3)_3$ (92), $CH[Si(CH_3)_3]_2$ (93) and $CH_2Si(CH_3)_2C_6H_4$-o-OCH_3 (92) have been isolated from the reaction of the appropriate lithium reagent with $ScCl_3$ ode YCl_3. The complexes were obtained as analytically pure, air-sensitive, colourless, crystals from pentane, containing 2 THF molecules coordinated to the metal. The NMR data are consistent with a trigonal bipyramidal structure, the THF molecules occupying the axial sites. The coordinated THF could not be removed in vacuo and only the compounds $Y\{CH[Si(CH_3)_3]_2\}_3$ and $Sc[CH_2Si(CH_3)_2C_6H_4OCH_3]_3$, with extremely bulky ligands, could be obtained solvent free. The Tb, Er, and Yb compounds of the type $M[CH_2Si(CH_3)_3]_3 \cdot 2THF$ were obtained by a similar procedure and also some ionic derivatives $LiMR_4$, $LiMR_3Cl$ and $LiMR_2Cl_2$ were reported (94). Their structures could be established by infrared and NMR spectra as well as by a single crystal X-ray analysis for $[Li(THF)_4][Yb\{CH(Si(CH_3)_3)_2\}_3Cl]$:

$$MCl_3 + 3 \ LiCH_2C(CH_3)_3 \xrightarrow{THF} M[CH_2C(CH_3)_3]_3 \cdot 2THF + 3 \ LiCl$$

M = Sc, Y

$$ScCl_3 + 3 \ LiCH_2Si(CH_3)_2C_6H_4OCH_3 \xrightarrow{THF} Sc[CH_2Si(CH_3)_2C_6H_4OCH_3]_3$$
$$+ 3 \ LiCl$$

$$MCl_3 + 3 \ LiCH_2Si(CH_3)_3 \xrightarrow{THF} M[CH_2Si(CH_3)_3]_3 \cdot 2THF + 3 \ LiCl$$

M = Sc, Y, Tb, Er, Yb

$$MCl_3 + 3 \ LiCH[Si(CH_3)_3]_2 \xrightarrow{THF} M\{CH[Si(CH_3)_3]_2\}_3 \cdot 2THF + 3 \ LiCl$$

M = Sc, Y

$$YCl_3 + 4 \ LiCH_2Si(CH_3)_3 \xrightarrow{THF} [Li(THF)_4][Y\{CH_2Si(CH_3)_3\}_4]$$
$$+ 3 \ LiCl$$

$$MCl_3 + 4 \ LiCH_2Si(CH_3)_3 \xrightarrow{tmed} [Li(tmed)_2][M\{CH_2Si(CH_3)_3\}_4]$$

M = Y, Er, Yb $+ 3 \ LiCl$

$$MCl_3 + 4\ LiCH[Si(CH_3)_3]_2 \xrightarrow{THF} [Li(THF)_4][MCl\{CH((Si(CH_3)_3)_2\}_3]$$

M = Er, Yb   + 3 LiCl

$$YCl_3 + 2\ LiC_5H_4Si(CH_3)_3 \xrightarrow{L} [LiL_2][YCl_2\{\eta C_5H_4Si(CH_3)_3\}_2]$$

                                          + 2 LiCl

L = THF, 1/2 tmed

$$tmed = (CH_3)_2NCH_2CH_2N(CH_3)_2$$

Independently, we also studied the reactions of several lanthanide trichlorides with lithiotrimethylsilylmethane (95,96).
However, in contrast to Lappert's results, we obtained two species
of such complexes in which either two or three molecules of THF
are bound to the metal. With $ErCl_3$ and $TmCl_3$, complexes with three
THF were obtained as pink or white air-sensitive crystals from
pentane at low temperatures. Above -35 to -25°C one molecule of
THF is irreversibly lost and the di-solvated complex precipitates
on recrystallization:

$$MCl_3 + 3\ LiCH_2Si(CH_3)_3 \xrightarrow[THF]{Et_2O} M[CH_2Si(CH_3)_3]_3 \cdot 3THF$$

$$M[CH_2Si(CH_3)_3]_3 \cdot 3THF \longrightarrow M[CH_2Si(CH_3)_3]_3 \cdot 2THF + THF$$

M = Er, Tm

With $YbCl_3$ and $LuCl_3$ we observed the same results as Lappert:
the corresponding complexes with two molecules of THF. The ratio
of the ligands and the structure of the compounds could be established by infrared, NMR, and chemical reactions, e.g. quantitative
hydrolysis. As for the Lu compound, the NMR data are consistent
with a trigonal bipyramidal structure, with the THF ligands occupying axial sites.

On keeping a pentane solution of $M[CH_2Si(CH_3)_3]_3 \cdot 2THF$ at
room temperature for several days, the complexes loose THF and
tetramethylsilane, and an extremely pyrophoric material precipitates. Quantitative experiments showed one mole of $Si(CH_3)_4$ per
mole of complex to be split off. Chemical analysis, infrared
spectra and deuterolysis experiments indicate polymeric compounds
of the formula $\{M[CH_2Si(CH_3)_3][CHSi(CH_3)_3]\}_n$. The polymeric nature
of the compounds is also evident from the unusually high decomposition temperature of 380 to 390°C.

$$M[CH_2Si(CH_3)_3]_3 \cdot 2THF \longrightarrow \{M[CH_2Si(CH_3)_3][CHSi(CH_3)_3]\}$$
$$+ \ 2 \ THF + Si(CH_3)_4$$

M = Er, Tm, Lu

We consider it resonable to assume an α-elimination route for these complexes, which should have a structure involving $CH_2Si(CH_3)_3$ and bridging $CHSi(CH_3)_3$ groups:

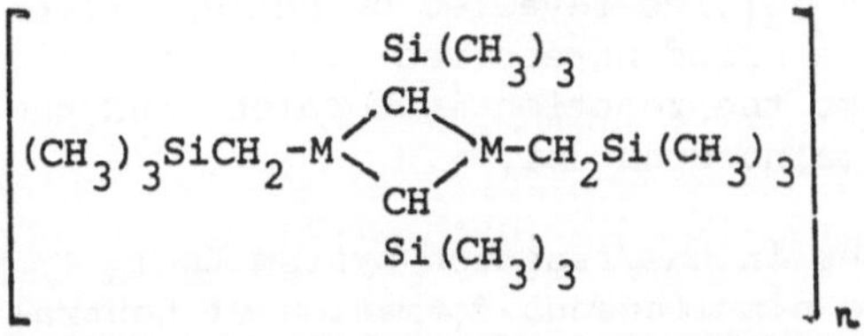

$$\left[(CH_3)_3SiCH_2-M \underset{CH}{\overset{CH}{\diagdown\diagup}} M-CH_2Si(CH_3)_3 \right]_n$$

(with $Si(CH_3)_3$ substituents on the bridging CH groups)

The "ate" complexes $Li(tmed)_2M[CH_2Si(CH_3)_3]_4$, reported by Lappert, could be verified for the Lu compound. IR and NMR data suggest an approximately tetrahedral array of alkyl ligands. The white Lu compound decomposes at 138 to 142°C and smoulders on exposure to air.

However, the situation becomes rather complicated when replacing tmed by ethers such as diethylether or THF. The benzene soluble complex $[Li(Et_2O)_4][Lu\{CH_2Si(CH_3)_3\}_4]$ shows broadening for the CH_2 resonance in the NMR spectrum, which indicates a relatively slow dissociation with respect to the NMR time scale.

$$[Li(Et_2O)_4][Lu\{CH_2Si(CH_3)_3\}_4] \longrightarrow Lu[CH_2Si(CH_3)_3]_3 \cdot 2Et_2O$$
$$+ \ LiCH_2Si(CH_3)_3$$

In a following reaction in benzene solution the complex looses one molecule of $Si(CH_3)_4$ by an α-elimination route during a period of a week. A more quantitative evidence was obtained by an NMR experiment in a sealed tube. At -10°C a green-white compound was isolated from pentane which by analytical data, deuteriolysis and spectroscopic means, was characterized as $Li\{Lu[CH_2Si(CH_3)_3]_2[CHSi(CH_3)_3]\}$.

$$[Li(Et_2O)_4][Lu\{CH_2Si(CH_3)_3\}_4] \longrightarrow 4 \ Et_2O + Si(CH_3)_4$$
$$+ \ Li\{Lu[CH_2Si(CH_3)_3]_2[CHSi(CH_3)_3]\}$$

Addition of bases such as tmed, THF, or DME leads to stabilized, but benzene insoluble compounds. The NMR spectrum of $Li(tmed)Lu[CH_2Si(CH_3)_3]_2[CHSi(CH_3)_3]$ in THF-d8 shows only some of the expected signals. At δ = 0.02 and -0.05 ppm (vs. TMS) two peaks appear for the $(CH_3)_3Si$-groups, the one at -0.05 having a half width of 5 Hz. The CH_2 resonance at -1.06 ppm is also broadened to 5 Hz; the CH signal cannot be located with certainty due to its low intensity. Cooling to -35°C leads to two sharp signals for the $(CH_3)_3Si$-protons with an approximate integrated ratio of 2:1, suggesting a definite kinetically rather stable compound.

Further splitting off of one molecule of $Si(CH_3)_4$ occurs for $LiLu[CH_2Si(CH_3)_3]_2[CHSi(CH_3)_3]$, as revealed by NMR experiments. The spectrum collapses to a broad unresolved signal centered around 0.40 ppm. After three weeks the reaction is complete and the $Si(CH_3)_4$ concentration remains constant.

Careful investigations in the reaction system $LnCl_3$ + $LiCH_2Si(CH_3)_3$ revealed the simultaneous formation of neutral and ionic species, even using an excess of $LnCl_3$. The ionic derivative decomposes by the outlined α-elemination mechanism, finally resulting in an extremely pyrophoric polymeric compound with the unusual stoichiometry Li:Ln = 1:2. All analytical figures are best rationalized by the formulation:

$$\left[\{Li(THF)_2\}\ \{Ln_2[CH_2Si(CH_3)_3]_2[CHSi(CH_3)_3][CSi(CH_3)_3]\}\right]_n$$

The mechanism of formation is at present only speculative. The unusual stoichiometry suggests the reaction of a decomposition product of the ionic "ate" complex with neutral $Ln[CH_2Si(CH_3)_3]_3 \cdot xTHF$ and subsequent loss of $Si(CH_3)_4$. The resemblance to the polymeric $\{Ln[CH_2Si(CH_3)_3][CHSi(CH_3)_3]\}_n$ is quite striking. Thus the overall reaction in the system $LnCl_3$ + $LiCH_2Si(CH_3)_3$ (Ln = Er, Tm, Yb) should be expressed by the equation:

$$LnCl_3 + 3\ LiCH_2Si(CH_3)_3 \longrightarrow 3\ LiCl + 3x\ Si(CH_3)_4 +$$

$$y\ Ln[CH_2Si(CH_3)_3]_3 \cdot 3THF +$$

$$x\{Li(THF)_2Ln_2[CH_2Si(CH_3)_3]_2[CHSi(CH_3)_3][CSi(CH_3)_3]\}_n$$

y varies from 37% (Er) to 41% (Yb)

x+y: % of total reaction = 92& for Er, 95 % for Yb.

Comparable results which also can only be interpreted with some difficulties are obtained in the system $LnCl_3$ and $Li^tC_4H_9$ (96). $TbCl_3$ and $ErCl_3$ react with t-butyl lithium in THF or in diethylether at -78 to -10°C with formation of pyrophoric compounds. These yield i-butane upon hydrolysis. Analytical data are consistent with 1:1 complexes of LnR_3 and the corresponding ether, but it was not possible to get the products absolutely free of lithium.

Using $Er(O^tBu)_3$ instead of the trichloride in the reaction with t-butyl lithium, it was possible to isolate a complex containing lithium, erbium, t-butyl- and t-butoxy groups, and, running this reaction in tmed instead of pentane, we obtained a complex of ErR_3 with two moles of tmed. It is not entirely free of LiO^tBu:

$$Er(O^tC_4H_9)_3 + 3\ Li^tC_4H_9 \longrightarrow Er(^tC_4H_9)_3 \cdot 3LiO^tC_4H_9$$

$$Er(O^tC_4H_9)_3 + 3\ Li^tC_4H_9 \xrightarrow{tmed} Er(^tC_4H_9)_3 \cdot 2tmed + 3\ LiO^tC_4H_9$$

MCl_3 and $M(O^tBu)_3$ react with an excess of t-butyl lithium in ether or tmed, with formation of "ate" complexes, which can be characterized by elemental analysis, infrared and NMR spectroscopy:

$$MCl_3 + 4\ Li^tC_4H_9 \xrightarrow{Et_2O} [Li(Et_2O)_4][M(^tC_4H_9)_4] + 3\ LiCl$$

M = Tb, Er, Lu

$$M(O^tC_4H_9)_3 + 4\ Li^tC_4H_9 \xrightarrow{tmed} [Li(tmed)_2][M(^tC_4H_9)_4] +$$
$$3\ LiO^tC_4H_9$$

M = Er, Lu

There is a need for further investigations in this area to establish the exact structure of these interesting compounds.

$ErBr_3$ reacts with methyl lithium in THF/ether, with the formation of extremely pyrophoric materials, which probably contain complexes such as $Er(CH_3)_3 \cdot xTHF$ and $[Li(THF)_4][Er(CH_3)_4]$. This formulation is derived from analytical and spectroscopic evidence and because methane is formed on hydrolysis of these yellow compounds (96). On the other hand, using tmed as a stabilizing base, it was possible to synthesize the first peralkylated complexes of the lanthanides containing only methyl ligands (97). Dropwise addition of an etheral solution of methyl lithium to a suspension of $ErCl_3$, $YbCl_3$ or $LuCl_3$ in diethyl ether with a stoichiometric amount of the amine, affords good yields of the reaction products, as analytically pure, pink or white crystals which were freed of excess methyl lithium by recrystallization:

$$MCl_3 + 6\ LiCH_3 + 3\ tmed \longrightarrow [Li(tmed)_3][M(CH_3)_6] + 3\ LiCl$$

M = Er, Yb, Lu

The new compounds are extremely sensitive to air and moisture; however, they exhibit surprising thermal stability and decompose only slowly at their melting point of 138 - 139°C (Er), 141 - 142°C (Yb), and 141 - 142°C (Lu). Hydrolysis and alkoholysis expectedly liberate 6 moles of methane per mole of lanthanide. The NMR spectrum of the diamagnetic Lu compound shows a singlet for the methyl groups bound to Lu at δ = -1.18 ppm (vs. TMS) in addition to the peaks of tmed.

An X-ray single crystal structural analysis is in progress (98). First results show a hexagonal unit cell with a = 16.61 Å, c = 13.30 Å and a space group R3m for the Er derivative. The Er-C distance is found to be 2.6 Å. But further refinement is nescessary.

6.3. Ylide complexes

The addition of trimethylmethylenphosphorane to a suspension of lanthanide trichlorides in pentane or hexane results in the formation of pyrophoric phosphonium salts in quantitative yields. While no dehydrochlorination of these salts was observed by the reaction of them with an excess of the ylide with formation of $[(CH_3)_4P]Cl$, the compounds react with butyl lithium in ether-hexane, with formation of LiCl, butane, and a new class of uncharged homoleptic lanthanide alkyls (99):

$$MCl_3 + 3 \ (CH_3)_3P=CH_2 \longrightarrow M[CH_2\overset{\oplus}{P}(CH_3)_3]_3 \ 3Cl^{\ominus}$$

$$M[CH_2\overset{\oplus}{P}(CH_3)_3]_3 \ 3Cl^{\ominus} + 3 \ C_4H_9Li \longrightarrow 3 \ LiCl + 3 \ C_4H_{10}$$

$$+ \ M\left\{\begin{array}{c} CH_2 \\ \diagup \\ CH_2 \end{array} P(CH_3)_2 \right\}_3$$

M = La, Pr, Nd, Sm, Gd, Ho, Er, Lu

These highly air-sensitive complexes apparently owe their high thermal stability (decomposition points between 140 and 210°C!) to the incorporation of chelating phosphorus ylide ligands. The neutral complexes are monomeric in benzene solution and the La derivative shows an NMR spectrum of two doublet signals for the CH_2 and CH_3 protons, having a J(HP) of 10 and 12 Hz respectively. The other compounds give only very broad signals owing to the paramagnetism of the lanthanide ions.

6.4. Compounds of the types MR_2, RMX, and related systems

Eu and Yb metals react with alkyl and aryl iodides in THF with formation of brown solutions. Magnetic susceptibility measurements by NMR methods showed that 83 - 93% of the Yb is present in the bivalent state. Similar values were obtained by hydrolysis of the solutions with dilute acetic acid and immediate titration of the Yb^{2+} ion by iodometry. Sm reacts less readily and the percentage of the bivalent state is considerably lower (100,101):

$$M + RI \xrightarrow{\text{THF}} RMI$$

$$M = Sm, \ R = C_6H_5; \ M = Eu, \ R = C_6H_5; \ M = Yb, \ R = CH_3, \ C_5H_5,$$

$$C_6H_5, \ 2\text{-}CH_3C_6H_4, \ 2,6\text{-}(CH_3)_2C_6H_3, \ 2,4,6\text{-}(CH_3)_3C_6H_2$$

These compounds react with water, trimethylchlorosilane and benzophenone in typical Grignard-like reactions.

Bis(pentafluorophenyl)ytterbium has been isolated as a highly air-sensitive complex with four THF, after the transmetallation reaction between ytterbium metal and the corresponding mercury compound (102,103):

$$Yb + Hg(C_6F_5)_2 \xrightarrow{\text{THF}} Yb(C_6F_5)_2 + Hg$$

The orange crystals are stable at room temperature for only short periods of time. They yield pentafluorobenzene on acidolysis. A THF complex of bis(2,3,5,6-tetrafluorophenyl)Ytterbium has been prepared in a similar manner and was identified spectroscopically. Reaction was also observed between Yb and bis(2,3,4,5-tetrafluorophenyl)mercury, but the product decomposed too rapidly for isolation.

The synthesis of another derivative of bivalent Yb, bis(phenylethynyl)ytterbium, was mentioned earlier (see 5.).

6.5. Other sigma bonded derivatives

Reaction of dicyclopentadienyl yttrium chloride with $LiAl(CH_3)_4$ in toluene at $0°C$ gives the colorless, crystalline, doubly methyl bridged compound $[(C_5H_5)_2Y(CH_3)_2Al(CH_3)_2]$ in high yield (104):

$$(C_5H_5)_2YCl + LiAl(CH_3)_4 \longrightarrow (C_5H_5)_2Y\underset{CH_3}{\overset{CH_3}{<}}Al(CH_3)_2 + LiCl$$

The compound shows remarkable thermal stability, subliming without decomposition at about $120°C$ and 0.05 mm Hg. The ^{1}H-NMR spectrum at $-45°C$ showed resonances at $\tau = 10.32$ for the bridging methyl groups with $J(^{89}YH) = 5$ Hz and $\tau = 10.98$ for the terminal aluminium methyls in CD_2Cl_2, which collapse at 40°C to a singlet, indicating rapid exchange of the bridge and terminal methyl groups.

Lappert and coworkers (86,87) published the synthesis and structural information on the corresponding derivatives of Sc, Gd, Dy, Ho, Er, Tm, and Yb, as well as the ethyl bridged derivatives of Sc and Y:

$$(C_5H_5)_2MCl + LiAlR_4 \longrightarrow (C_5H_5)_2M\overset{R}{\underset{R}{<\quad>}}AlR_2 + LiCl$$

M = Sc, Gd, Dy, Ho, Er, Tm, Yb, R = CH_3; M = Sc, Y, R = C_2H_5

^{1}H- and ^{13}C-NMR studies reveal that the diamagnetic Y complex to be fluxional at room temperature, undergoing rapid bridge-terminal R group exchange. The Sc complex, on the other hand, is rigid in this temperature range. The molecular geometries of the Y and Yb complexes were determined by X-ray diffraction. The structure contains the dimethyl bridge; the average bridging bond distances being 2.58 Å for the Y-C and 2.10 Å for the Al-C, and the Y-Al separation is 3.056 Å (105).

7. COMPOUNDS WITH METAL TO SILICON, GERMANIUM, AND TIN SIGMA BONDS

Because of similarities in the bonding situation, and because of the generally high stability of compounds of the group IV elements silicon, germanium, and tin, organometallic derivatives of these elements with lanthanide-to-element IV sigma bonds should be included in this review. The first examples of such compounds have been prepared recently (106), opening a new and interesting field in organometallic chemistry.

$PrCl_3$, $NdCl_3$, $GdCl_3$ and $ErCl_3$ react with a suspension of lithio triphenylgermane or lithio triphenyltin in THF with formation of air sensitive compounds with lanthanide-to-germanium or tin sigma bonds. They contain different amounts of THF coordinated to the metals and are isolable in a pure state only with some difficulties (107):

$$MCl_3 + 3 LiGe(C_6H_5)_3 \longrightarrow M[Ge(C_6H_5)_3]_3 + 3 LiCl$$

M = Nd, Gd, Er

$$MCl_3 + 3 LiSn(C_6H_5)_3 \longrightarrow M[Sn(C_6H_5)_3]_3 + 3 LiCl$$

M = Pr, Nd, Gd

Analogous compounds are formed by the reaction of tris[bis-(trimethylsilyl)-amido]praseodymium and -neodymium with the organohydrides of germanium and tin in dimethoxyethane. The coloured crystalline solids are formed in yields up to 80% and give the expected reactions with water, bromine, 1,2-dibromoethane, and benzoyl peroxide (108):

$$3 \ R_3EH + M\{N[Si(CH_3)_3]_2\}_3 \xrightarrow{\text{dme}} M(ER_3)_3 + 3 \ [(CH_3)_3Si]_2NH$$

$$M = Pr, \ Nd; \ E = Ge, \ Sn; \ R = C_6H_5, \ CH_2Si(CH_3)_3$$

$$Pr(ER_3)_3 + 3 \ H_2O \longrightarrow 3 \ R_3EH + Pr(OH)_3$$

$$Nd[Ge(C_6H_5)_3]_3 + 3 \ BrCH_2CH_2Br \longrightarrow 3 \ (C_6H_5)_3GeBr + PrBr_3$$
$$+ 3 \ CH_2{=}CH_2$$

$$Nd[Ge(C_6H_5)_3]_3 + 3 \ Br_2 \longrightarrow 3 \ (C_6H_5)_3GeBr + NdBr_3$$

$$Pr(ER_3)_3 + 3 \ (C_6H_5COO)_2 \longrightarrow 3 \ R_3EOC(O)C_6H_5 + Pr(OC(O)CC_6H_5)_3$$

$$E = Ge, \ Sn; \ R = C_6H_5, \ CH_2Si(CH_3)_3$$

Pentafluorophenyl groups stabilize metal-metal bonds in bimetallic organometallic compounds. Therefore attempts to prepare organometallic compounds with Ge-Pr bonds have been succesful by a transmetallation reaction between bis[tris(pentafluorophenyl)-germyl]mercury and metallic praseodymium in dimethoxyethan. A complex of the starting material with $Pr[Ge(C_6F_5)_3]_3$ is formed, which reacts with HCl or $(C_6F_5)_3GeBr$, with formation of two other new compounds containing Ge-Pr bonds (109):

$$5 \ [(C_6F_5)_3Ge]_2Hg + 2 \ Pr \longrightarrow 2 \ [(C_6F_5)_3Ge]_3Pr{\cdot}Hg[Ge(C_6F_5)_3]_2$$
$$+ 3 \ Hg$$

$$[(C_6F_5)_3Ge]_3Pr{\cdot}Hg[Ge(C_6F_5)_3]_2 + HCl \longrightarrow [(C_6F_5)_3Ge]_2Hg$$
$$+ (C_6F_5)_3GeH + [(C_6F_5)_3Ge]_2PrCl$$

$$[(C_6F_5)_3Ge]_3Pr{\cdot}Hg[Ge(C_6F_5)_3]_2 + (C_6F_5)_3GeBr \longrightarrow$$

$$[(C_6F_5)_3Ge]_2Hg + [(C_6F_5)_3Ge]_2 + (C_6F_5)_3GeH + [(C_6F_5)_3Ge]_2PrBr$$

Finally, dicyclopentadienyl derivatives of lanthanide germanium or -tin compounds are formed in the reactions of dicyclopentadienyl lanthanide chlorides with lithio triphenylgermane or -tin in THF at $-78^\circ C$ (107):

$$(C_5H_5)_2MCl + LiGe(C_6H_5)_3 \longrightarrow (C_5H_5)_2M\text{-}Ge(C_6H_5)_3 + LiCl$$

M = Pr, Nd, Gd, Ho, Er, Yb

$$(C_5H_5)_2MCl + LiSn(C_6H_5)_3 \longrightarrow (C_5H_5)_2M\text{-}Sn(C_6H_5)_3 + LiCl$$

M = Pr, Nd, Gd, Tb, Er, Yb

Both pink Er derivatives could be isolated solvent-free. They melt without decomposition at 98 to $100^\circ C$ (Ge) and 200 to $205^\circ C$ (Sn), whereas the yellow ytterbium tin compound is coordinated with two molecules THF, showing a melting point of 185 to $190^\circ C$ without decomposition (107).

8. FURTHER PROBLEMS

It has been shown within the last 8 years that organometallic compounds of the lanthanides having metal-to-carbon sigma bonds can exist. But despite the intensive search of some research groups around the world for preparative routes to such compounds, the number of clearly characterized derivatives is not very great. New methods must be developed to knit lanthanide metals with organic groups, to stabilize organometallic compounds of lanthanides which are formed as intermediates in some reactions, and to analyze and characterize these new intermediates by new and better spectroscopic methods. On the other hand, research will go on, and after the establishment of compounds having lanthanide-to-carbon sigma bonds, compounds with covalent bonds between lanthanide elements and other nonmetals and metals will recieve ever greater interest.

First examples in this area are organometallic compounds with lanthanide metal-to-transition metal bonds, prepared by salt-elimination methods (28) or by the reaction of finely divided lanthanide metals with organometallic compounds of transition metals (110):

$$(C_5H_5)_2MCl + NaW(C_5H_5)(CO)_3 \xrightarrow{\text{THF}} (C_5H_5)_2M\text{-}W(C_5H_5)(CO)_3 + NaCl$$

M = Dy, Ho, Er, Yb

$$ErCl_3 + 3\ NaMo(C_5H_5)(CO)_3 \xrightarrow{\text{H}_2\text{O}} Er[Mo(C_5H_5)(CO)_3]_3 \cdot 7H_2O$$
$$+ 3\ NaCl$$

The bonding situation in these compounds is not completely known. The tungsten compounds have IR spectra which imply the presence of isocarbonyl links and a polymeric structure has been suggested (28). The reaction of Yb and other lanthanide metals with $(CO)_5MnBr$ in THF produce air- and water-sensitive red solutions containing Grignard-like species $[(CO)_5Mn]_xMBr_y$. The IR spectra of these solutions are speaking against isocarbonyl linkages (110). Further work in this area is nescessary like in the whole field of this part of organometallic chemistry.

References

1. L.D.Cadet de Gassicourt, Mémoires de Mathématique et de Physique 3 (1760) 623.
2. A.Cahours, A.Rieche, Ann. 92 (1854) 361.
3. E.Frankland, J.Chem.Soc. 2 (1849) 263.
4. D.F.Herman, W.K.Nelson, J.Amer.Chem.Soc. 74 (1952) 2693.
5. A.von Grosse, Z.Anorg.Allgem.Chem. 152 (1925) 133.
6. Rice, Rice, The Aliphatic Free Radicals, The Johns Hopkins Press, Baltimore 1935 pp.58.
7. W.M.Pletz, Dokl.Akad.Nauk SSSR 20 (1938) 27.
8. B.N.Afanasev, P.A.Tsyganova, Zh.Obshch.Khim. 18 (1948) 306; F.A.Cotton, Chem.Rev. 1955, 544.
9. H.Gilman, R.G.Jones, J.Org.Chem. 10 (1945) 505.
10. H.Schumann, unpublished work 1963 - 1965; M.Tsutsui, unpublished work 1965.
11. H.Gysling, M.Tsutsui, Advan.Organometal.Chem. 9 (1970) 361; R.G.Hayes, J.L.Thomas, Organometal.Chem.Rev. A 7 (1971) 1; B.Kanellakopolus, K.W.Bagnall, M.T.P. International Review of Science, Inorg.Chem. Ser.One, 7 (1972) 217; M.Tsutsui, N.Ely, R.Dubois, Acc.Chem.Res. 9 (1976) 217; S.A.Cotton, J.Organometal.Chem.Libr. 3 (1977) 189.
12. D.Seyferth, R.B.King, Ann.Survey Organometal.Chem. 1 (1965) 165; 2 (1966) 203; 3 (1967) 281; F.Calderazzo, Organometal.Chem.Rev. B 4 (1968) 10; 5 (1969) 545; 6 (1970) 997; 9 (1972) 131; F.Calderazzo, J.Organometal.Chem. 53 (1973) 173; 79 (1974) 175; T.J.Marks, J.Organometal.Chem. 79 (1974) 181; 95 (1975) 301; 119 (1976) 229; 138 (1977) 157.
13. G.Wilkinson, J.M.Birmingham, J.Amer.Chem.Soc. 76 (1954) 6210.
14. J.M.Birmingham, G.Wilkinson, J.Amer.Chem.Soc. 78 (1956) 42.
15. S.Manastyrskyi, M.Dubeck, Inorg.Chem. 3 (1964) 1647.
16. M.Tsutsui, T.Takino, D.Lorenz, Z.Naturforsch. 21b (1966) 1.
17. E.O.Fischer, H.Fischer, J.Organometal.Chem. 3 (1965) 181.
18. A.F.Reid, P.D.Wailes, Inorg.Chem. 5 (1966) 1213.
19. F.Baumgärtner, E.O.Fischer, P.Lauberau, Radiochim.Acta 7 (1967) 188.
20. P.G.Lauberau, J.H.Burns, Inorg.Chem. 9 (1970) 1091.

21. R.G.Hayes, J.L.Thomas, Inorg.Chem. 8 (1969) 2521.

22. R.Kopunec, F.Macasek, V.Mikulaj, P.Drienovsky, Radiochim.
    Radioanal.Letters 1 (1969) 117.

23. E.O.Fischer, H.Fischer, J.Organometal.Chem. 6 (1966) 141.

24. F.Calderazzo, R.Pappalardo, S.Losi, J.Inorg.Nucl.Chem. 28
    (1966) 987.

25. W.Strohmeier, H.Landsfeld, F.Gernert, W.Langhäuser, Z.Anorg.
    Allgem.Chem. 307 (1960) 120.

26. L.T.Reynolds, G.J.Wilkinson, J.Inorg.Nucl.Chem. 9 (1958) 86.

27. J.H.Burns, W.H.Baldwin, F.H.Fink, Inorg.Chem. 13 (1974) 1916.

28. A.E.Crease, P.Legzdins, J.Chem.Soc., Dalton Trans. 1973, 1501.

29. R.Pappalardo, J.Mol.Spectry. 29 (1969) 13.

30. C.Wong, T.Lee, Y.Lee, Acta Cryst. B 25 (1969) 2580.

31. J.L.Atwood, K.D.Smith, J.Amer.Chem.Soc. 95 (1973) 1488.

32. E.O.Fischer, H.Fischer, Angew.Chem. 76 (1964) 52.

33. G.W.Watt, E.W.Gillow, J.Amer.Chem.Soc. 91 (1969) 775.

34. P.Brix, S.Hüfner, P.Kienle, D.Quitmann, Phys.Letters 13
    (1964) 140.

35. B.L.Kalsotra, S.P.Anand, R.K.Multani, B.D.Jain, J.Organome-
    tal.Chem. 28 (1971) 87.

36. R.S.P.Coutts, P.C.Wailes, J.Organometal.Chem. 25 (1970) 117.

37. L.E.Manzer, J.Organometal.Chem. 110 (1976) 291.

38. R.E.Magin, S.Manastyrskyi, M.Dubeck, J.Amer.Chem.Soc. 85
    (1963) 672.

39. F.Gomez-Beltran, L.A.Oro, F.Ibanez, J.Inorg.Nucl.Chem. 37
    (1975) 1541.

40. H.Schumann, H.Jarosch, Z.Anorg.Allgem.Chem. 426 (1976) 127.

41. J.L.Atwood, K.D.Smith, J.Chem.Soc., Chem.Commun. 1972, 593;
    J.Chem.Soc., Dalton Trans. 1973, 2487.

42. E.C.Baker, L.D.Brown, K.N.Raymond, Inorg.Chem. 14 (1975) 1376.

43. B.Kanellakopulos, E.Dornberger, H.Billich, J.Organometal.
    Chem. 76 (1974) C42.

44. T.J.Marks, G.W.Grynkewich, Inorg.Chem. 15 (1976) 1302.

45. S.Manastyrskyi, R.E.Magin, M.Dubeck, Inorg.Chem. 5 (1963) 904.

46. B.L.Kalsotra, R.K.Multani, B.D.Jain, Israel J.Chem. 9 (1971)
    569.

47. B.L.Kalsotra, R.K.Multani, B.D.Jain, J.Inorg.Nucl.Chem. 34
    (1972) 2265.

48. S.Kapur, B.L.Kalsotra, R.K.Multani, J.Chin.Chem.Soc. (Taipei)
    19 (1972) 197.

49. S.Kapur, B.L.Kalsotra, R.K.Multani, J.Chin.Chem.Soc. (Taipei)
    20 (1973) 171.

50. B.L.Kalsotra, R.K.Multani, B.D.Jain, J.Chin.Chem.Soc. (Taipei)
    18 (1971) 189.

51. S.Kapur, B.L.Kalsotra, R.K.Multani, J.Inorg.Nucl.Chem. 35
    (1973) 3966.

52. S.Kapur, B.L.Kalsotra, B.K.Multani, J.Inorg.Nucl.Chem. 36
    (1974) 932.

53. S.Kapur, B.L.Kalsotra, R.K.Multani, B.D.Jain, J.Inorg.Nucl.
    Chem. 35 (1973) 1689.

54. A.Greco, S.Cesca, G.Bertolini, J.Organometal.Chem. 113 (1976)
    321.
55. R.von Ammon, B.Kanellakopulos, R.D.Fischer, P.Lauberau, Inorg.
    Nucl.Chem.Letters 5 (1969) 315.
56. R.D.Fischer, H.Fischer, J.Organometal.Chem. 4 (1965) 412.
57. E.O.Fischer, H.Fischer, Angew.Chem. 77 (1965) 261.
58. R.von Ammon, B.Kanellakopulos, Ber.Bunsenges.Phys.Chem. 76
    (1972) 995.
59. R.von Ammon, R.D.Fischer, B.Kanellakopulos, Chem.Ber. 104
    (1971) 1072.
60. J.H.Burns, W.H.Baldwin, J.Organometal.Chem. 120 (1976) 361.
61. E.C.Baker, K.N.Raymond, Inorg.Chem. 16 (1977) 2710.
62. A.E.Crease, P.Legzdins, Chem.Commun. 1972, 268.
63. R.G.Hayes, J.L.Thomas, J.Amer.Chem.Soc. 91 (1969) 6876.
64. F.Mares, K.Hodgson, A.Streitwieser, J.Organometal.Chem. 24
    (1970) C68.
65. F.Mares, K.O.Hodgson, A.Streitwieser, J.Organometal.Chem. 28
    (1971) C24.
66. A.Westerhof, H.J.de Liefde Meijer, J.Organometal.Chem. 116
    (1976) 319.
67. J.D.Jamerson, A.P.Masino, J.Takats, J.Organometal.Chem. 65
    (1974) C33; A.P.Masion, J.Takats, pers.communication 1978.
68. K.O.Hodgson, F.Mares, D.F.Starks, A.Streitwieser, J.Amer.
    Chem.Soc. 95 (1973) 8650.
69. K.O.Hodgson, K.N.Raymond, Inorg.Chem. 11 (1972) 3030.
70. K.O.Hodgson, K.N.Raymond, Inorg.Chem. 11 (1972) 171.
71. B.L.Kalsotra, R.K.Multani, B.D.Jain, J.Inorg.Nucl.Chem. 34
    (1972) 2679.
72. B.L.Kalsotra, R.K.Multani, B.D.Jain, J.Organometal.Chem. 31
    (1971) 67.
73. S.R.Ely, T.E.Hopkins, C.W.Dekock, J.Amer.Chem.Soc. 98 (1976)
    1624.
74. W.J.Evans, S.C.Engerer, A.C.Neville, J.Amer.Chem.Soc. 100
    (1978) 331.
75. M.Tsutsui, H.J.Gysling, J.Amer.Chem.Soc. 90 (1968) 6880.
76. M.Tsutsui, H.J.Gysling, J.Amer.Chem.Soc. 91 (1969) 3175.
77. J.L.Atwood, J.H.Burns, P.G.Lauberau, J.Amer.Chem.Soc. 95
    (1973) 1830.
78. M.Tsutsui, N.Ely, J.Amer.Chem.Soc. 97 (1975) 3551.
79. M.Tsutsui, N.Ely, J.Amer.Chem.Soc. 96 (1974) 4042.
80. N.M.Ely, M.Tsutsui, Inorg.Chem. 14 (1975) 2680.
81. F.A.Hart, M.S.Saran, Chem.Commun. 1968, 1614.
82. F.A.Hart, A.G.Massey, M.S.Saran, J.Organometal.Chem. 21
    (1970) 147.
83. E.Murphy, G.E.Toogood, Inorg.Nucl.Chem.Letters 7 (1971) 755.
84. G.B.Deacon, A.J.Koplick, J.Organometal.Chem. 146 (1978) C43.
85. M.Tsutsui, N.M.Ely, J.Amer.Chem.Soc. 97 (1975) 1280.
86. J.Holten, M.F.Lappert, D.G.H.Ballard, R.Pearce, J.L.Atwood,
    W.E.Hunter, J.Chem.Soc., Chem.Commun. 1976, 480.

87. J.Holten, M.F.Lappert, G.R.Scollary, D.G.H.Ballard, R.Pearce,
    J.Chem.Soc., Chem.Commun. 1976, 425.
88. H.Schumann, W.Genthe, unpublished results.
89. B.L.Kalsotra, R.K.Multani, B.D.Jain, J.Inorg.Nucl.Chem. 35
    (1973) 311.
90. S.A.Cotton, F.A.Hart, M.B.Hursthouse, J.A.Welch, Chem.
    Commun. 1972, 1225.
91. L.E.Manzer, J.Organometal.Chem. 135 (1977) C6.
92. M.F.Lappert, R.Pearce, J.Chem.Soc., Chem.Commun 1973, 126.
93. G.K.Barker, M.F.Lappert, J.Organometal.Chem. 76 (1974) C45.
94. J.L.Atwood, W.E.Hunter, R.D.Rogers, J.Holten, J.Mc Meeking,
    R.Pearce, M.F.Lappert, J.Chem.Soc., Chem.Commun. 1978, 140.
95. H.Schumann, J.Müller, J.Organometal.Chem. 146 (1978) C5.
96. H.Schumann, J.Müller, unpublished results.
97. H.Schumann, J.Müller, Angew.Chem. 90 (1978) 307.
98. H.Schumann, N.Bruncks, unpublished results.
99. H.Schumann, S.Hohmann, Chemiker-Ztg. 100 (1976) 336.
100. D.F.Evans, G.V.Fazakerley, R.F.Phillips, Chem.Commun. 1970,
     244.
101. D.F.Evans, G.V.Fazakerley, R.F.Phillips, J.Chem.Soc. A 1971,
     1931.
102. G.B.Deacon, D.G.Vince, J.Organometal.Chem. 112 (1976) C1.
103. G.B.Deacon, W.D.Raverty, D.G.Vince, J.Organometal.Chem. 135
     (1977) 103. .
104. D.G.H.Ballard, R.Pearce, J.Chem.Soc., Chem.Commun. 1975, 621.
105. G.R.Scollary, Aust.J.Chem. 31 (1978) 411.
106. H.Schumann, M.Cygon, Proc. XVIII intern. Conference on Coor-
     dination Chemistry, Sao Paulo, Brazil, 1977, pp. 185.
107. H.Schumann, M.Cygon, J.Organometal.Chem. 144 (1978) C43.
108. G.A.Razuvaev, G.S.Kalinina, E.A.Fedorova, L.N.Bochkarev,
     S.P.Markelov, Proc. VIII.intern.Conf.Organometal.Chem.,
     Kyoto, Japan, 1977, pp.64.
109. G.A.Razuvaev, L.N.Bochkarev, G.S.Kalinina, M.N.Bochkarev,
     Inorg.Chim.Acta 24 (1977) L40.
110. A.E.Crease, P.Legzdins, J.Chem.Soc., Chem.Commun. 1973, 775.

PROPERTIES OF ACTINIDE-TO-CARBON SIGMA BONDS

Paul J. Fagan, Juan M. Manriquez,* and Tobin J. Marks

Department of Chemistry, Northwestern University,
Evanston, Illinois 60201

I. INTRODUCTION

Two-electron sigma bonds between metal ions and carbon
atoms of organic molecules represent one of the fundamental
building blocks of organometallic chemistry. Furthermore, the
creation and transformations of such bonds are crucial steps in
many of the catalytic processes by which main group, d-element,
and f-element (see Chapter 12) metal ion catalysts can be used to
transform organic and inorganic feedstocks into useful materials.
Until quite recently, very little was known about the nature of
actinide-to-carbon sigma bonds, and there was even some question
as to whether such species would be isolable. We now know that
this is not the case, and that a rich and varied field of 5f
element organometallic chemistry involving sigma bonds lies on
the horizon. In this chapter we summarize and analyze what is
currently known about the properties of actinide-to-carbon sigma
bonds. We begin by discussing routes for the synthesis of the
various classes of compounds and survey some of the more distinctive chemical and physicochemical characteristics of each class.
We next focus more sharply on the chemical behavior of actinide-
to-carbon sigma bonds. Those features which stabilize this linkage are discussed, and then important reaction patterns involving
electrophilic reagents, hydrogen, and carbon monoxide are summarized.

II. TRIS(CYCLOPENTADIENYL) ALKYLS AND ARYLS

The first isolable sigma-bonded actinide organometallics
were prepared independently by three different research groups

113

T. J. Marks and R. D. Fischer (eds.), Organometallics of the f-Elements, 113–148.
All Rights Reserved. Copyright © 1979 by D. Reidel Publishing Company, Dordrecht, Holland.

(1,2,3). Alkylation of tris(cyclopentadienyl)uranium(IV)
chloride with lithium or Grignard reagents in diethyl ether or
tetrahydrofuran proceeds according to equation (1), to yield
very air sensitive, but thermally stable organometallics.

$$(\eta^5\text{-}C_5H_5)_3UCl \xrightarrow{\text{RMgX or RLi}} (\eta^5\text{-}C_5H_5)_3UR \qquad (1)$$

R = a wide variety of alkyl, aryl, and alkenyl groups

The colors of these monomeric compounds range from greenish for
the phenyl derivative through dark red for primary alkyls. The
proton NMR spectra of the $(C_5H_5)_3UR$ compounds exhibit large displacements of resonances from diamagnetic positions and narrow
linewidths. For a complete discussion of the NMR spectra of
U(IV) organometallic compounds, the reader is referred to
Chapters 10 and 11. Physical properties of the $(C_5H_5)_3UR$ derivatives are compiled in Table I. A representative molecular
structure is shown in Figure 1, while more extensive structural
remarks are presented in Chapter 8.

Other members of the $(\eta^5\text{-}C_5H_5)_3MR$ class of compounds are
the tris(cyclopentadienyl)Th(IV) derivatives (5). These compounds were prepared by alkylation of tris(cyclopentadienyl)-
thorium(IV) chloride with lithium or Grignard reagents in
toluene-diethyl ether mixtures according to equation (2); they
are diamagnetic, and highly air sensitive. The $(C_5H_5)_3ThR$ species possess even greater thermal stability than the $(C_5H_5)_3UR$
analogues (5). Physical properties of the tris(cyclopentadienyl)-

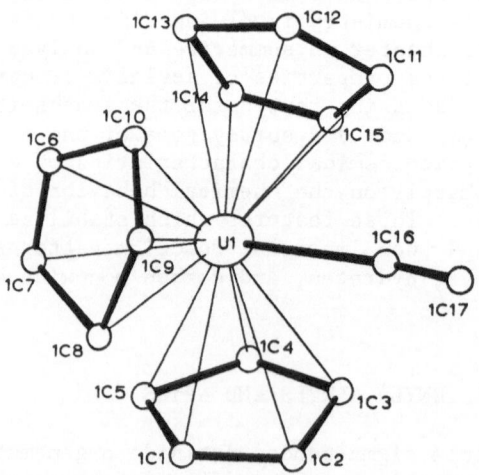

Fig. 1.   The molecular structure of $(\eta^5\text{-}C_5H_5)_3U(C\equiv CH)$ from
           reference 4.

Table I. Physical Properties of $U(C_5H_5)_3R$ Compounds

Compound	Color	Melting Point (°C)	μ_{eff} (B.M.)[a,b] 298°K	Proton NMR[e] R
$(C_5H_5)_3U(CH_3)$				
$(C_5H_5)_3U(i\text{-}C_3H_7)$			$(2934)^a$	19.3(6H,d,J=7Hz), 190.0(1H,sept,J=7Hz)^c
$(C_5H_5)_3U(n\text{-}C_4H_9)$	dark red	130(dec)	$3.36(2945)^a$	18.7(3H,t,J=7Hz), 27.6(2H,m,J=7Hz), 33.6(2H,m,J=7Hz), 200.0(2H,m,J=Hz)^c
$(C_5H_5)_3U(t\text{-}C_4H_9)$				23.2(9H,s)^c
$(C_5H_5)_3U(\text{neopentyl})$	dark red	148(dec)		22.1(9H,s), 192.0(2H,s)^c
$(C_5H_5)_3U(\text{ferrocenyl})$	brown			-13.46(2H,m),-30.59(2H,m)^f
$(C_5H_5)_3U(\text{allyl})$	dark brown	155(dec)		38.2(1H, quint, J=11Hz), 126.0(4H,br)^c
$(C_5H_5)_3U(2\text{-methyl allyl})$	reddish brown			
$(C_5H_5)_3U(\text{vinyl})$				-24.4(1H,dd,J=16.5Hz), 17.0(1H,dd,J=20.5Hz),^{c} 16.3(1H,dd,J=16.2Hz)
$(C_5H_5)_3U(C_6H_5)$	greenish	>165(dec)	$(2869)^a$	6.0(1H,t,J=7Hz), 15.0(2H,t,J=7Hz)^c 24.2(2H,d,J=7Hz)
$(C_5H_5)_3U(C_6F_5)$	dark brown	144(dec)		88.7(2F), 99.1(2F), 115.5(1F)^{c,d}
$(C_5H_5)_3U[p\text{-}C_6H_4U(C_5H_5)_3]$	red-orange			
$(C_5H_5)_3U(C_2H)$	yellow-green			14.74(1H,s)^f
$(C_5H_5)_3U(C_2C_6H_5)$	yellow-green	183-185(dec)	$2.88(3461)^b$	9.65(2H,t,J=8Hz), 14.7(2H,d,J=8Hz)^c
$(C_5H_5)_3U$ (p-methylbenzyl)	dark violet			214(s),29.41(d,J=8Hz),5.62(d,J=8Hz), 3.62(s)^c
$(C_5H_5)_3U(\text{benzyl})$	dark violet	200(dec)·		213(s), 29.37(d,J=8Hz), 4.65(t,J=8Hz)^c 10.12(t,J=8Hz)
$(C_5H_5)_3U(2\text{-}\underline{cis}\text{-2-butenyl})$				19.8(3H,s), 22.6(1H,quart,J=6Hz)^c 42.3(3H,d,J=6Hz)
$(C_5H_5)_3U(2\text{-}\underline{trans}\text{-2-butenyl})$			$(3032)^a$	-23.5(1H,quart,J=7Hz), 33.0(3H,d,J=7Hz)^c 33.6 (3H,s)
$(C_5H_4)_2Fe[U(C_5H_5)_3]_2$	green			

^{a}Value in parentheses is X_m x 10^6 (cgs) in benzene at 308°K.

^{b}298°K by Guoy method.

^{c}^{1}H NMR data at 25°C in ppm relative to internal benzene, (-) indicating shift to low field.

^{d}^{19}F data in ppm to high field of internal $C_6H_5CF_3$.

Table I. (cont.)

C_5H_5	Solvent	M.W.	Other data [g]	ref. [h]
10.0(15H,s)	C_6D_6		Ir(2,3),NMR in $C_6D_5CD_3$(3)	[2],3
10.9(15H,s)	C_6D_6		Ir	2
10.3(15H,s)	C_6D_6	512	Ir(2,3),NMR in $C_6D_5CD_3$(3).X-ray structure (73)	[2],3,73
11.4(15H,s)	C_6D_6		Ir	2
11.6(15H,s)	C_6D_6	510	Ir	2
-1.65(5H,s), -2.33(15H,s)	THF		Ir,M.S.	71
10.0(15H,s)	C_6D_6		Ir,low temp. NMR	2
			Ir, M.S., X-ray structure	72
9.3(15H,s)	C_6D_6		Ir	2
10.6(15H,s)	C_6D_6		Ir(2 1,3),NMR in THF (1), $C_6D_5CD_3$(3),M.S. (1)	[2],1,3
10.9(15H,s)	C_6D_6		Ir, M.S.	2
			Ir	71
-1.53(15H,s)	C_6D_6		Ir, X-ray structure (4)	[71],4
9.98(15H,s)	C_6D_6		Ir, NMR in THF, X-ray structure(74)	[1],74
9.95(s)	$C_6D_5CD_3$		Ir(3), NMR in THF(1), X-ray structure (73)	[3],1,73
9.90(s)	$C_6D_5CD_3$		Ir	3
10.6(15H,s)	C_6D_6		Ir	2
10.7(15H,s)	C_6D_6		Ir	2
			Ir	71

[e]Key: s=singlet; d=doublet; t=triplet; quart=quartet; quint=quintet; sept=septet; m=multiplet; br=broad.

[f]Pmr data in ppm based on TMS occurring at δ =0; a minus sign indicates an upfield shift.

[g]The value in parenthesis indicates the reference where the data can be found. M.S. = mass spectrum.

[h]When more than one reference is cited, data reported here was obtained from the reference in brackets.

thorium organometallics are compiled in Table II.

$$(\eta^5\text{-}C_5H_5)_3ThCl \xrightarrow{\text{RMgX or RLi}} (\eta^3\text{-}C_5H_5)_3ThR \qquad (2)$$

R = a wide variety of alkyl, aryl, and alkenyl groups

The related (indenyl)$_3$MR derivatives (where M = Th,U) have also been synthesized (6) by alkylation of the corresponding chlorides in tetrahydrofuran as shown in equation (3).

$$(\eta^5\text{-}C_9H_7)_3MCl \xrightarrow{\text{RLi}} (\eta^5\text{-}C_9H_7)_3MR \qquad (3)$$

R = CH$_3$, n-C$_4$H$_9$

The thermal stability of these latter compounds was found to be comparable to the corresponding cyclopentadienyl derivatives. Physical properties of the tris(indenyl) actinide organometallics are summarized in Table III. A ring-substituted cyclopentadienyl derivative, tris(methylcyclopentadienyl)Th(i-C$_3$H$_7$), has also been prepared (7). The only other reported σ-bonded actinide organometallic of this class is the tris(cyclopentadienyl)-neptunium(IV) n-butyl. This compound was prepared (8) by the procedure of reaction (4).

$$(\eta^5\text{-}C_5H_5)_3NpCl \xrightarrow[\text{diethyl ether}]{\text{n-C}_4\text{H}_9\text{Li}} (\eta^5\text{-}C_5H_5)_3Np(n\text{-}C_4H_9) \qquad (4)$$

Attempts to synthesize the phenyl derivative gave only a mixture of (C$_5$H$_5$)$_4$Np and (C$_5$H$_5$)$_3$Np (8).

III. BIS(CYCLOPENTADIENYL) ALKYLS

The central role which (C$_5$H$_5$)$_2$MX$_2$ compounds, A, play in the organometallic chemistry of early transition metals as precursors for a wide variety of sigma-bonded derivatives (9,10, 11,12) provided strong motivation for the synthesis of analogous actinide compounds. This new class of compounds would not only allow expansion of the organometallic chemistry of the actinides, but also permit development of meaningful chemical comparisons between organoactinide reaction patterns and those of analogous transition metal organometallics. Although (C$_5$H$_5$)$_2$UCl$_2$ (13) would seem to be the logical starting point for bis(cyclopentadienyl) uranium dialkyls, recent studies (14 and Chapter 7) have shown that this compound is actually a mixture of the ligand redistribution products (C$_5$H$_5$)$_3$UCl and (C$_5$H$_5$)UCl$_3$·2S, where S is tetrahydrofuran or $\frac{1}{2}$·1,2-dimethoxyethane (DME). Viable alternative approaches to stable bis(cyclopentadienyl) uranium precursors and their alkyls involve modification of the cyclopentadienyl ligands: linking the rings by a bridging group, peralkylating the rings, and fusing benzene

Table II.[5]

Physical Properties of $Th(C_5H_5)_3R$ compounds.

Compound	Color	C_5H_5	Proton NMR R [a,b]
$(C_5H_5)_3Th(neopentyl)$	white	6.09(15H,s)	1.41(9H,s), 1.28 (2H,s)
$(C_5H_5)_3Th(allyl)$	white	5.84(15H,s)	6.50(1H, quint, J=12Hz), 3.94 (4H, d, J=12Hz)
$(C_5H_5)_3Th(1-C_3H_7)$	white	5.95(15H,s)	1.89(6H, d, J=6.5Hz), 0.98(1H, m, J=6.5Hz)
$(C_5H_5)_3Th(2-trans-2-butenyl)$	white	6.16(15H,s)	6.95(1H, m, J=6Hz), 2.56(3H, m, 2.11(3H, d of m, J = 6Hz)
$(C_5H_5)_3Th(2-cis-2-butenyl)$	white	6.10(15H,s)	6.45(1H, m, J=6Hz), 2.18(3H, m), 2.05(3H, d of m, J=6Hz)
$(C_5H_5)_3Th(n-C_4H_9)$	white	6.00(15H,s)	1.39(9H,m)

a ^{1}H NMR data expressed in ppm relative to internal TMS.   b Key: s=singlet, d=doublet, m=multiplet, quint=quintet

Table III.[6]

Properties of $M(C_9H_7)_3R$, M = Th, U.

Compound	Color	Melting Point (°C)
$(C_9H_7)_3 Th(CH_3)$	yellow	213-215
$(C_9H_7)_3 U(CH_3)$	red-brown	211-213
$(C_9H_7)_3 Th(n-C_4H_9)$	yellow	125-126

functionalities to the rings (indenyl). These topics will be
treated sequentially in the discussion which follows. It
should also be noted that other routes to $(C_5H_5)_2UX_n$ $(n = 1,2)$
derivatives exist (15,16,17,18), but these have not been
applied to alkyl chemistry.

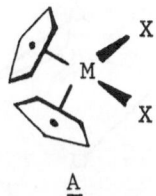

A

Ring-bridged bis(cyclopentadienyl) uranium halides can be
prepared by the procedure of equation (5) (19,20) and are isolated as dimeric tetrahydrofuran adducts.

$$2[A(C_5H_4)_2^{-2} \; 2Li^+] + 2UCl_4 \xrightarrow{\text{tetrahydrofuran}}$$

$$Li^+(THF)_2[A(\eta^5-C_5H_4)_2]_2U_2Cl_5^- + 3LiCl \qquad (5)$$

$$A = CH_2, \; (CH_3)_2Si, \; CH_2CH_2CH_2$$

A single crystal X-ray diffraction study of the A = CH_2 derivative revealed an unusual structure (shown in Figure 2) which
possessed a triple chlorine bridge and a "cryptated" lithium
ion. All evidence to date suggests that these dimers react as
$A(C_5H_4)_2UCl_2$ species (19,20). Monomeric nitrogenous base adducts can be obtained by cleaving the anionic dimers with 2,2'-
bipyridyl or 1,10-phenanthroline as shown in equation (6)
(19,20).

$$Li^+(THF)_2[A(\eta^5-C_5H_4)_2]_2U_2Cl_5^- \xrightarrow{N \frown N}$$

$$2A(\eta^5-C_5H_4)_2U\overset{N}{\underset{N}{\diagup}} \overset{}{\underset{Cl}{\diagdown}}_{Cl} + 2\,THF + LiCl \qquad (6)$$

The molecular structure of the A = CH_2 adduct with 2,2'-bipyri-
dyl is shown in Figure 3.

Alkylation of the ring-bridged bis(cyclopentadienyl) dimers
with stoichiometric quantities of lithium reagents at low
temperature in diethyl ether (equation (7)) gave, in the case of

$$Li^+(THF)_2[A(\eta^5-C_5H_4)_2]_2U_2Cl_5^- + 4RLi \xrightarrow{Et_2O}$$

$$2A(\eta^5-C_5H_4)_2UR_2 + 5LiCl \qquad (7)$$

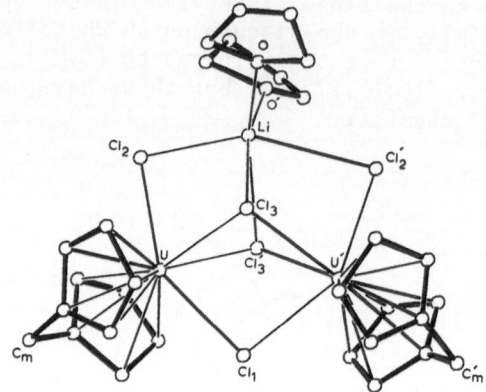

Fig. 2.   The molecular structure of $Li^+(THF)_2[CH_2(C_5H_4)_2]_2U_2Cl_5^-$ from reference 19.

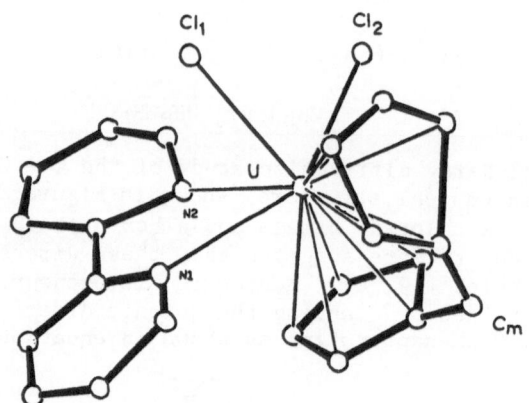

Fig. 3.   The molecular structure of $CH_2(C_5H_4)_2UCl_2(2,2'$-bipyridyl) from reference 20.

R = n-butyl, a dark red complex, which decomposed upon warming to room temperature to yield a <u>ca.</u> 60:40 mixture of butane and 1-butene. Preliminary studies of the R = CH_3, $CH_2Si(CH_3)_3$, and neopentyl derivatives indicate somewhat higher thermal stability (20,21). These results suggest that β-hydride elimination (22) is an important thermolysis pathway. In accord with these observations, alkylation products of the more coordinatively saturated 2,2'-bipyridyl adducts have even greater thermal stability (21). Further studies of these ring-bridged actinide alkyl systems are in progress.

A second synthetic approach to stable actinide bis(cyclo-pentadienyl) systems makes use of bulky substituents on the cyclopentadienyl ligand (for example, $C_5(CH_3)_5$ (23), $C_5(CH_3)_4$-(CH_2CH_3) (24)). In these cases, possibly for steric reasons, the corresponding trisubstituted compounds, $(\eta^5-C_5R_5)_3MCl$ (M = Th,U) are not formed. For R = CH_3, the corresponding bis(pentamethylcyclopentadienyl) actinide dihalides can be pre-pared by the reaction of the tetrahydrofuran adduct of penta-methylcyclopentadienyl Grignard reagent in toluene with the actinide tetrachloride according to equation (8) (23). For the red uranium derivative, cryoscopic molecular weight measure-

$$2[C_5(CH_3)_5]MgCl \cdot THF + MCl_4 \xrightarrow{\text{toluene}}_{100°}$$

$$[\eta^5-C_5(CH_3)_5]_2MCl_2 + 2MgCl_2 \cdot THF \tag{8}$$

$$M = Th, U$$

ments in benzene showed it to be monomeric. The colorless thorium derivative was insufficiently soluble for such deter-minations. The observed lack of association in solution and the relatively low coordination number (formally 8) stands in marked contrast to the associated $Li^+[A(\eta^5-C_5H_4)_2]_2U_2Cl_5^-$ species. Solution structure B is proposed for these compounds.

B

The analogous unsymmetrically substituted, bis(tetramethyl-ethylcyclopentadienyl)uranium (IV) dichloride was prepared by the reaction of the tin reagent $(C_5Me_4Et)(n-C_4H_9)_3Sn$, in toluene with uranium tetrachloride as shown in equation (9) (24).

$$2[C_5(CH_3)_4(CH_2CH_3)](n-C_4H_9)_3Sn + UCl_4 \xrightarrow{\text{toluene}}_{120°}$$

$$[\eta^5-C_5(CH_3)_4(CH_2CH_3)]_2UCl_2 + 2(n-C_4H_9)_3SnCl \tag{9}$$

Alkylation of the air sensitive bis(pentamethylcyclopenta-dienyl)thorium and uranium dichlorides with lithium reagents in diethyl ether proceeds according to equation (10) to yield the corresponding monomeric, air sensitive, but thermally stable dialkyls (23,25,26). Physical properties of alkyl-substituted bis(cyclopentadienyl)actinide compounds are set out in Table IV. Structure C is proposed for the bis(pentamethylcyclopentadienyl) actinide dialkyls. For R = CH_3, CH_2SiMe_3, and $CH_2C_6H_5$, both

Table IV. Physical Properties of $[\eta^5-C_5(CH_3)_5]_2MR_2$ and $[\eta^5-C_5(CH_3)_5]_2M(R)X$ Compounds (M = Th, U).

Compound	Color	Proton NMR[a,b]		Reference
		$C_5(CH_3)_5$	R	
$[C_5(CH_3)_5]_2U(CH_3)_2$	orange	5.03 (30 H, s)	-124 (6 H, s)	23
$[C_5(CH_3)_5]_2Th(CH_3)_2$	white	1.92 (30 H, s)	-0.19 (6 H, s)	23
$[C_5(CH_3)_5]_2U(CH_3)Cl$	red-orange	8.96 (30 H, s)	-154 (3 H, s)	23
$[C_5(CH_3)_5]_2Th(CH_3)Cl$	white	2.01 (30 H, s)	0.41 (3 H, s)	23
$[C_5(CH_3)_5]_2U[CH_2Si(CH_3)_3]_2$	orange	5.04 (30 H, s)	-6.14 (18 H, s, CH_3), -89 (4 H, s, CH_2)	25
$[C_5(CH_3)_5]_2Th[CH_2Si(CH_3)_3]_2$	white	1.98 (30 H, s)	0.31 (18 H, s, CH_3), -0.43 (4 H, s, CH_2)	25
$[C_5(CH_3)_5]_2U[CH_2Si(CH_3)_3]Cl$	red	9.48 (30 H, s)	-11.52 (9 H, s, CH_3), -124 (2 H, s, CH_2)	25
$[C_5(CH_3)_5]_2Th[CH_2Si(CH_3)_3]Cl$	white	2.04 (30 H, s)	0.45 (9 H, s, CH_3), 0.24 (2 H, s, CH_2)	25
$[C_5(CH_3)_5]_2Th[CH_2C(CH_3)_3]_2$	white	2.05 (30 H, s)	1.30 (18 H, s, CH_3), 0.15 (4 H, s, CH_2)	26
$[C_5(CH_3)_5]_2U(CH_2C_6H_5)_2$	black	7.62 (30 H, s)	{ 2.56 (4 H, t, meta H), -1.08 (2 H, t, para H), -16.2 (4 H, d, ortho H), -98.1 (4 H, s, CH_2) }	26
$[C_5(CH_3)_5]_2U(CH_2C_6H_5)Cl$	black	9.62 (30 H, s)	{ -2.74 (2 H, t, meta H), -5.15 (1 H, t, para H), -44.7 (2 H, d, ortho H), -110 (2 H, s, CH_2) }	26
$[C_5(CH_3)_5]_2Th(CH_2C_6H_5)_2$	white	1.87 (30 H, s)	1.40 (4 H, s, CH_2), 6.74-7.33 (10 H, m's, phenyl)	26
$[C_5(CH_3)_5]_2Th(CH_2C_6H_5)Cl$	white	1.92 (30 H, s)	1.86 (2 H, s, CH_2), 6.59-7.21 (5 H, m's, phenyl)	26
$[C_5(CH_3)_5]_2Th(C_6H_5)_2$	white	1.77 (30 H, s)	7.02-7.82 (10 H, m's, phenyl)	26
$[C_5(CH_3)_5]_2Th(C_6H_5)Cl$	white	1.90 (30 H, s)	7.05-7.74 (5 H, m's, phenyl)	26
$[C_5(CH_3)_5]_2U[C_4(C_6H_5)_4]$	brown	6.02 (30 H, s)	{ 5.72 (4 H, m), 4.52 (4 H, m), -0.27 (2 H, t), -1.25 (4 H, t), -33.0 (4 H, m) }	23
$[C_5(CH_3)_5]_2Th(CH_2CH_3)_2$	white	1.93 (30 H, s)	1.26 (6 H, t, CH_3), 0.21 (4 H, q, CH_2)	26

[a] ^{1}H NMR data expressed in ppm relative to TMS, a (-) sign indicating an upfield shift;
   spectra were recorded in C_6D_6 at 35°C.
[b] Key: s=singlet, d=doublet, t=triplet, q=quartet, m=multiplet.

$$[\eta^5\text{-}C_5(CH_3)_5]_2MCl_2 + 2RLi \xrightarrow{\text{diethyl ether}}$$

$$[\eta^5\text{-}C_5(CH_3)_5]_2MR_2 + 2LiCl \qquad\qquad (10)$$

$$M = Th, \; R = CH_3, \; CH_2SiMe_3, \; CH_2CMe_3, \; CH_2C_6H_5, \; C_6H_5$$
$$M = U, \; R = CH_3, \; CH_2SiMe_3, \; CH_2C_6H_5$$

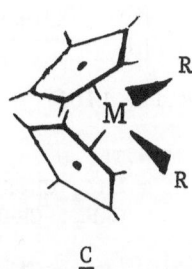

$$\underline{C}$$

the uranium and the thorium derivatives possess very high thermal stability. In toluene solution at 100°, the dimethyl derivative of thorium has a half-life of ∼ 1 week, while the corresponding half-life of the uranium dimethyl complex is ∼ 16 h. Curiously, the bis(neopentyl) derivatives show a lower thermal stability, and for M = U, it has not been possible to isolate the compound in an analytically pure state. The same situation holds for the bis(phenyl) uranium complex. A higher thermal stability for the thorium alkyls with respect to the corresponding uranium derivatives was also observed in the tris(cyclopentadienyl)actinide alkyl systems (5).

A uranium metallocycle has been obtained by the reaction of 1,4-dilithiotetraphenylbutadiene with $[C_5(CH_3)_5]_2UCl_2$ in diethyl ether as illustrated in equation (11)(23).

$$[\eta^5\text{-}C_5(CH_3)_5]_2UCl_2 \quad + \quad \begin{array}{c} C_6H_5 \\ Li \\ Li \end{array} \begin{array}{c} C_6H_5 \\ C_6H_5 \\ C_6H_5 \end{array} \longrightarrow$$

$$[\eta^5\text{-}C_5(CH_3)_5]_2U \begin{array}{c} C_6H_5 \\ C_6H_5 \\ C_6H_5 \\ C_6H_5 \end{array} + \quad 2LiCl \qquad (11)$$

For R = CH_3, CH_2SiMe_3, C_6H_5, and $CH_2C_6H_5$, the corresponding alkyl chlorides of thorium and uranium can be prepared by either mixing the dihalide with the corresponding dialkyl in toluene, as shown in equation (12) (23), or by alkylation of the dihalide with the stoichiometric amount of lithium reagent in diethyl ether (or THF) as shown in equation (13) (26). In equation (12)

$$[\eta^5\text{-}C_5(CH_3)_5]_2MCl_2 + [\eta^5\text{-}C_5(CH_3)_5]_2MR_2$$

$$\xrightarrow{\text{toluene}} 2[\eta^5\text{-}C_5(CH_3)_5]_2MRCl \qquad (12)$$

$$M = Th, U$$
$$R = CH_3, CH_2SiMe_3$$

the equilibrium lies 90-95% to the right. For the method used in equation (13) with R=CH$_3$, chloride-bromide exchange occurred

$$[\eta^5\text{-}C_5(CH_3)_5]_2MCl_2 + RLi \xrightarrow[\text{THF}]{\text{diethyl ether or}}$$

$$[\eta^5\text{-}C_5(CH_3)_5]_2MRCl + LiCl \qquad (13)$$

$$M = Th, U$$
$$R = CH_3, CH_2SiMe_3, CH_2CMe_3,$$
$$C_6H_5, CH_2C_6H_5$$

during the reaction because commercial methyl lithium contains lithium bromide. In contrast to the easy ligand redistribution which occurs in equation (12) for R = CH$_3$ and CH$_2$SiMe$_3$, no detectable reaction occurs for the neopentyl derivative of thorium at room temperature. The alkyl chloride can only be obtained, in low yield, after heating the reaction mixture to a temperature of $\sim 90°$, where competing thermal decomposition of the bis (neopentyl) compound also occurs. On the other hand, the procedure of equation (13) allows preparation of both the thorium and uranium neopentyl chloride derivatives. The thermal stability of the alkyl chlorides is even higher than that of the corresponding dialkyls.

Attempts to prepare the t-butyl chloride derivative of uranium by the method used in equation (13) gave a green dark solution instantaneously at -80°, from which the green tri-valent uranium complex, $\{[\eta^5\text{-}C_5(CH_3)_5]_2UCl\}_3$ (27), could be isolated in essentially quantitative yield. Analysis of the organic product(s) of this reaction has not yet been completed. The insoluble green monochloride compound, which is a trimer (27), dissolves easily in coordinating solvents such as THF and ether, with the formation of the corresponding adducts. Addi-tional routes to the synthesis of the monochloride are reduction of $[C_5(CH_3)_5]_2UCl_2$ with sodium amalgam in toluene, or diethyl ether (27) or hydrogenolysis of the alkyl chlorides (see Section VIC of this Chapter). Interestingly, the metallocycle shown in equation (11) can be obtained in 50% yield by a "disproportion-ation" reaction of $\{[C_5(CH_3)_5]_2UCl\}_3$ which reductively couples diphenylacetylene, as shown in equation (14) (27). The metallo-cycle can be obtained in even higher yield (70%) by reducing the bis(pentamethylcyclopentadienyl) dichloride with Na/Hg in the presence of diphenylacetylene.

$$\tfrac{2}{3}\left\{[\eta^5\text{-}C_5(CH_3)_5]_2UCl\right\}_3 + 2C_6H_5C_2 \longrightarrow$$

$$[\eta^5\text{-}C_5(CH_3)_5]_2U \underset{C_6H_5}{\overset{C_6H_5}{\bigvee}} \overset{C_6H_5}{\underset{C_6H_5}{}} + [\eta^5\text{-}C_5(CH_3)_5]_2UCl_2 \quad (14)$$

The diethyl derivative of bis(pentamethylcyclopentadienyl) thorium can be prepared in quantitative yield by the reaction of $\left\{[C_5(CH_3)_5]_2ThH_2\right\}_2$ and ethylene according (28) to equation (15) (bis(pentamethylcyclopentadienyl)actinide hydrides are discussed in Section V of this chapter). In an analogous manner, the

$$\left\{[\eta^5\text{-}C_5(CH_3)_5]_2ThH_2\right\}_2 + 4CH_2\text{=}CH_2 \xrightarrow{\text{toluene}}$$

$$2[\eta^5\text{-}C_5(CH_3)_5]_2Th(CH_2CH_3)_2 \qquad\qquad (15)$$

ethyl-chloride derivative can be made by the reaction of $\left\{[C_5(CH_3)_5]_2Th(H)Cl\right\}_2$ with ethylene as given in equation (16).

$$\left\{[\eta^5C_5(CH_3)_5]_2Th(H)Cl\right\}_2 + 2CH_2\text{=}CH_2 \xrightarrow{\text{toluene}}$$

$$2[\eta^5\text{-}C_5(CH_3)_5]_2Th(Cl)(CH_2CH_3) \qquad\qquad (16)$$

Alkyl derivatives of the bis(tetramethylethylcyclopentadienyl)uranium(IV) have not yet been reported, but similar trends can be expected.

A completely different approach to the synthesis of bis(cyclopentadienyl)actinide alkyls has also been achieved (29) by introducing pentahaptoindenyl ligands as shown in equation (17).

$$UCl_4 \xrightarrow[-70°]{3C_9H_7Na,\ THF} \xrightarrow[\text{2 days}]{RT} \xrightarrow[-70°]{RLi}$$

$$(\eta^5\text{-}C_9H_7)_2UR_2 \qquad\qquad (17)$$

$$R = CH_3,\ t\text{-}C_4H_9$$

Although a complete characterization of these dialkyls has not yet appeared, apparently reaction (17) capitalizes on a more favorable equilibrium for "$(C_9H_7)_2UCl_2$" with respect to disproportionation under the reaction conditions, than exists for the corresponding unsubstituted "$(C_5H_5)_2UCl_2$".

IV.   MONO(CYCLOPENTADIENYL) ACTINIDE ALKYLS

In contrast to the difficulties encountered in stabilizing
the bis(cyclopentadienyl) uranium dichlorides (see Section III)
the corresponding mono(cyclopentadienyl) uranium(IV) trichlorides
are well characterized compounds.   Thus, reactions such as
shown in equation (18) produce stable complexes, isolated as
adducts with Lewis bases (14,30).

$$UCl_4 + (C_5H_5)Tl \xrightarrow{S} (\eta^5\text{-}C_5H_5)UCl_3 \cdot 2S + TlCl \qquad (18)$$

$$S = THF, \tfrac{1}{2}DME$$

Recent work described in Chapter 7 has greatly elaborated the
variety of bases which can be utilized.   Judging from X-ray
diffraction results on $(CH_3C_5H_4)UCl_3 \cdot 2THF$ (14) and $(C_5H_5)UCl_3 \cdot$
$2OP(C_6H_5)_3$ (31), pseudooctahedral coordination geometries, $\underline{C}$,

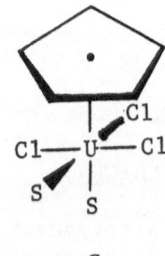

$$\underline{C}$$

will dominate mono(cyclopentadienyl) actinide structural chemistry.   In the case of mono(peralkylcyclopentadienyl) complexes,
the routes shown in equations (19) and (20) have been used to prepare ethyltetramethylcyclopentadienyl and pentamethylcyclopentadienyl derivatives (32,33).

$$UCl_4 + Sn[C_5(CH_3)_4(CH_2CH_3)](n\text{-}C_4H_9)_3 \xrightarrow{toluene}$$

$$[\eta^5\text{-}C_5(CH_3)_4(CH_2CH_3)]UCl_3 + (n\text{-}C_4H_9)_3SnCl \qquad (19)$$

$$MCl_4 + [C_5(CH_3)_5]MgCl \cdot THF \xrightarrow{THF}$$

$$[\eta^5\text{-}C_5(CH_3)_5]MCl_3 \cdot 2THF + MgCl_2 \qquad (20)$$

$$M = Th, U$$

The only alkyl complex known for the mono(cyclopentadienyl)
actinide series is a pentamethylcyclopentadienyl thorium trimethylsilylmethyl derivative (33).   It was prepared by reaction
of $[C_5(CH_3)_5]ThCl_3 \cdot 2THF$ in diethylether with one equivalent of
the corresponding lithium reagent, as shown in equation (21).

$$[\eta^5 \text{-} C_5(CH_3)_5]ThCl_3 \cdot 2THF + LiCH_2SiMe_3 \xrightarrow{\text{diethylether}}$$

$$[\eta^5 \text{-} C_5(CH_3)_5]ThCl_2CH_2SiMe_3 + LiCl + 2THF \qquad (21)$$

This compound, as yet of unknown molecularity, possesses a thermal stability comparable to some of the bis(pentamethylcyclopentadienyl)thorium alkyl derivatives. The reaction of $[C_5(CH_3)_5]ThCl_3 \cdot 2THF$ with three equivalents of methyllithium appears to give a reasonably stable product at room temperature, however, attempts to isolate it by removal of ether and extraction with pentane resulted in decomposition (33). This result suggests that the product will be stabilized by the coordination of Lewis bases.

V.   HOMOLEPTIC ALKYLS AND ARYLS

A.   Neutrally Charged Species

Early efforts to synthesize uranium tetraalkyls via approaches such as equation (22) were unsuccessful (34). The

$$UCl_4 + 4RLi \longrightarrow [UR_4] + 4LiCl \qquad (22)$$

R = various alkyl residues

products of these reactions, presumed to be uranium tetraalkyls, were unstable at room temperature and could not be isolated from the reaction mixtures for detailed characterization. It has, however, been possible to study the proton NMR spectra of these species at low temperature, and to analyze the thermal decomposition products (35). For R = CH_3, an isotropically shifted singlet at $\tau \approx 32.7$ was observed, but since coordination of other ligands such as halide or solvent could not be excluded, an explicit molecular structure was not proposed. The trimethylsilylmethyl derivative appears to be the most thermally stable member of the tetraalkyl uranium(IV) series (36). Thorium alkyls generated from $ThCl_4$ and lithium reagents by a procedure analogous to equation (22), are similarly unstable at room temperature (5,37). Tetrabenzyl derivatives of thorium (38) and uranium (39) have also been reported and are more thermally stable than the corresponding tetraalkyls. Thus, tetrabenzylthorium was synthesized by the route of equation (23) and is a pale-yellow, air-sensitive crystalline complex, stable at room temperature for brief periods of time.

$$ThCl_4 + 4C_6H_5CH_2Li \xrightarrow[-20°]{THF} Th(CH_2C_6H_5)_4 + 4LiCl \qquad (23)$$

The uranium analogue, prepared as shown in equation (24), is isolated as an $MgCl_2$ adduct.

$$UCl_4 \cdot 3THF + 2(C_6H_5CH_2)_2Mg \xrightarrow[-40^\circ]{THF}$$

$$U(CH_2C_6H_5)_4 \cdot MgCl_2 + MgCl_2 \qquad (24)$$

This species appears to be of even greater thermal stability than tetrabenzylthorium. The greater stability of the actinide benzyl derivatives vis-à-vis the alkyls may reflect additional coordinative saturation imparted by interaction of the metal ion with the π-system of the aromatic ring (D).

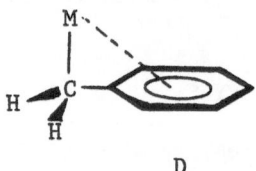

<u>D</u>

This effect has been noted in the structures of the zirconium and hafnium tetrabenzyls (40).

B.   Anionic Species

A number of anionic homoleptic uranium(IV) alkyls have been prepared (41) from the reaction of uranium tetrachloride with lithium reagents (equation (25)) in excess (to saturate the coordination sphere), yielding products formulated as hexa-alkyl dianions,

$$UCl_4 + \text{excess } RLi \xrightarrow{L} Li_2UR_6 \cdot L_n \qquad (25)$$

$R = CH_3, CH_2SiMe_3, C_6H_5, \underline{o}\text{-}C_6H_4CH_2NMe_2$
$L_n = 8(THF \text{ or } Et_2O) \text{ or } 7(tmed)$
tmed $= N,N,N',N'$-tetramethylethylenediamine

$[LiS_4]_2^{+2}[UR_6]^{-2}$, where S is a solvent molecule. The coordination geometry about uranium is believed to be octahedral as in <u>E</u>. The proposed hexaalkyl uranates decompose thermally near or

<u>E</u>

below room temperature. The methyl compounds are the least
stable, decomposing with gas evolution above -20°. The methyl-
trimethylsilyl, phenyl and dimethylaminomethylphenyl compounds
decompose rapidly in solution at room temperature, but could be
handled for very short periods of time at that temperature without substantial decomposition. The compounds could only be
characterized by infrared and proton NMR spectroscopy and by
measuring the Li: U: S: RH ratio after hydrolysis. The NMR spectra
were also used to calculate μ_{eff} values by the Evans method.
The same authors (41) also reported that the reaction of U(V)
pentaethoxide with lithium reagents, followed by dioxane precipitation, yields the thermally stable green octaalkyl uranium(V)
trianions (equation (26)). These products were characterized

$$U_2(OCH_2CH_3)_{10} + \text{excess RLi} \longrightarrow$$

$$Li_3UR_8 \cdot 3\text{dioxane} \qquad\qquad\qquad (26)$$

$$R = CH_3, \ CH_2SiMe_3, \ CH_2CMe_3$$

in a manner similar to the UR_8^{-2} species. The qualitative thermal stability of the pentavalent compounds decreases in the
order $CH_3 > CH_2SiMe_3 > CH_2CMe_3$, which follows increasing steric
crowding around the metal ion. Since in coordinately unsaturated
compounds, the approximate reverse trend is generally observed,
it was supposed that the uranium ion is coordinatively saturated
in these compounds. A molecular structure with eight alkyl
groups surrounding the uranium ion in a dodecahedral or bicapped
trigonal prismatic (or antiprismatic) geometry, with each lithium atom complexed by one dioxane and bridging a polyhedral face,
was considered to be consistent with the experimental data.
Physical properties of the anionic uranium (IV) hexaalkyls and
uranium (V) octaalkyls are presented in Table V.

VI. THE NATURE OF THE ACTINIDE-CARBON SIGMA BOND

A. Thermal Stability and Modes of Decomposition

Those electronic and molecular factors which govern whether
or not a particular organometallic compound will be isolable
and whether or not a certain reaction pathway is selected under
thermolysis conditions, are of prime importance in understanding
the fundamental properties of actinide-to-carbon sigma bonds.
For d-transition metal organometallics, it is becoming increasingly apparent that metal-carbon sigma bonds are not thermo-
dynamically unstable (22,42), but rather the isolability of compounds reflects the kinetic availability of reaction pathways
such as β-hydride elimination (equation (27)) (22,43). As in
the case of d-block complexes, the tendency of actinide alkyls

Table V. Physical Properties of Uranium(IV) and Uranium(V) Alkyl Anions.[41]

Compound	Color	Decomp. Pt. (°C)	μ_{eff} (B.M.)	Proton NMR[d,e,f] Data	Temperature (°C)
Li$_2$UMe$_6$·8 Et$_2$O	olive green	~-20		R: 10.51 (18 H, br, s, CH$_3$) L.B: 6.64 (32 H, q, OCH$_2$), 8.84 (48 H, t, CH$_3$)	-40
Li$_2$UMe$_6$·8THF	olive green	-15	2.9^b (-30°C)	R: 10.44 (18 H, br, s, CH$_3$) L.B: 6.32 (32 H, br, s, OCH$_2$), 8.22 (32 H, br, s, CH$_2$)	-40
Li$_2$UMe$_6$·7TMED	dark green	-5	2.9^b (-50°C)	R: 11.20 (18 H, br, s, CH$_3$) L.B: 7.71 (28 H, s, NCH$_2$), 7.91 (84 H, s, CH$_3$)	-40
Li$_2$U(CH$_2$SiMe$_3$)$_6$·8Et$_2$O	light green	30	2.7^b (0°C), 2.72^c(25°C)	R: 10.80 (54 H, SiC 3), 17.82 (12 H, Br, s, SiCH$_2$) L.B: 6.64 (32 H, q, OCH$_2$), 8.98 (48 H, t, CH$_3$)	0
Li$_2$U(CH$_2$SiMe$_3$)$_6$·8THF^a	light green	30	2.7^b (0°C), 2.72^c(25°C)	R: 11.22 (54 H, s, SiCH$_3$), 19.63 (12 H, br, s, SiCH$_2$) L.B: 6.51 (32 H, s, OCH$_2$), 8.25 (32 H, s, CH$_2$)	0
Li$_2$U(CH$_2$SiMe$_3$)$_6$·TMED^a	dark green	35	2.8^b (0 c), 2.77^c(25 c)	R: 10.50 (54 H, s, SiCH$_3$), 17.56 (12 H, br, s, SiCH$_2$) L.B: 7.53 (28 H, s, NCH$_2$), 7.75 (84 H, s, NCH$_3$)	0
Li$_2$UPh$_6$·8Et$_2$O	red	5	2.7^b (-10 c)	R: 2.90 (30 H, m, C$_6$H$_5$) L.B: 6.80 (32 H, q, OCH$_2$), 9.05 (48 H, t, CH$_3$)	-10
Li$_2$U(g-PhCH$_2$NMe$_2$)$_6$·8Et$_2$O	red	0	2.7^b (-10 c)	R: 3.05 (24 H, m, C$_6$H$_4$), 7.13 (12 H, s, NCH$_2$), 8.12 (36 H, s, NCH$_3$) L.B: 6.64 (32 H, q, OCH$_2$), 8.95 (48 H, t, CH$_3$)	-10
Li$_2$U(g-PhCH$_2$NMe$_2$)$_6$·TTMED	red	10	2.7^b (-10 c)	R: 2.98 (24 H, m, C$_6$H$_4$), 7.10 (12 H, s, NCH$_2$), 8.20 (36 H, s, NCH$_3$) L.B: 7.48 (28 H, s, NCH$_2$), 7.70 (84 H, s, NCH$_3$)	0
Li$_3$UMe$_6$·3dioxane	pale green	265-268		R: 8.70 (24 H, br, s, CH$_3$) L.B: 6.66 (24 H, s, OCH$_2$)	35
Li$_3$U(CH$_2$CMe$_3$)$_6$·3dioxane	olive green	120-122		R: 7.85 (16 H, br, s, CH$_2$), 9.32 (72 H, s, CCH$_3$) L.B: 6.68 (24 H, s, OCH$_2$)	35
Li$_3$U(CH$_2$SiMe$_3$)$_6$·3dioxane	olive green	150-154		R: 8.56 (16 H, br, s, CH$_2$Si), 10.2 (72 H, s, SiCH$_3$) L.B: 6.66 (24 H, s, OCH$_2$)	35

^{a} Infrared data available.

^{b} Magnetic moment determined by Evans NMR method.

^{c} Magnetic moment determined by Guoy method.

^{d} $_1$H NMR given as values in pyridine or ^{2}H$_5$-pyridine.

^{e} R=alkyl group, L.B. = Lewis Base.

^{f} Key: br = broad, s = singlet, t = triplet, q = quartet, m = multiplet

$$CH_2 = CHR$$

$$M-CH_2CH_2R \;\rightleftharpoons\; M-H \;\rightleftharpoons\; M-H + CH_2=CHR \qquad (27)$$

to undergo β-hydride elimination is strongly dependent on the degree to which the complex is coordinatively saturated. In all likelihood, coordinative congestion impedes other thermolysis processes as well. Qualitatively, the general order of thermal stability for the known sigma-bonded organoactinides can be summarized as shown below, where M is thorium or uranium in the

$$(C_5H_5)_3MR \sim (C_9H_7)_3MR > [C_5(CH_3)_5]_2MRCl > [C_5(CH_3)_5]_2MR_2 \geq$$

$$(C_9H_7)_2UR_2 \geq A(C_5H_4)_2UR_2 \geq [C_5(CH_3)_5]Th(R)Cl_2 >$$

$$[C_5(CH_3)_5]ThR_3 > Li_2UR_6 \cdot L_n > MR_4$$

+4 oxidation state.

Thus, the tris(cyclopentadienyl) alkyls and aryls with a formal coordination number of ten (assuming each cyclopentadienyl ligand formally occupies three coordination sites) are not only the most stable actinide-to-carbon σ-bonded organometallics, but also some of the most stable metal-to-carbon σ-bonded complexes. The stability of the n-butyl derivative of uranium, which has a half life in toluene solution at 97° of 1130 hr (2), is indeed remarkable. The next most stable alkyl complexes are the bis(cyclopentadienyl) derivatives, with a formal coordination number of 8. In contrast to the tris-(cyclopentadienyl) series, in which thermal stability is relatively insensitive to the presence or absence of β-hydrogen atoms in the alkyl group, different bis(cyclopentadienyl) actinide dialkyls exhibit a wide variation in thermal stability. Thus, for example, while the bis(pentamethylcyclopentadienyl) thorium(IV) dimethyl in toluene solution at 100° has a half-life of ~ 1 week, the diethyl derivative under the same conditions decomposes completely within a period of approximately 1 hr (44). Nevertheless, the bis(pentamethylcyclopentadienyl) actinide dialkyls possess a relatively high thermal stability, which is comparable to that found in the analogous bis(pentamethylcyclopentadienyl) derivatives of titanium and zirconium (45). It is interesting to note that a drop in stability is observed on going from the bis(pentamethylcyclopentadienyl) dialkyl derivatives of uranium to the ring-bridged cyclopentadienyl dialkyls. This fall in stability resembles that found in bis(pentamethylcyclopentadienyl) titanium(IV) dimethyl when compared to the corresponding ring-unsubstituted bis(cyclopentadienyl) (45). For both d- and f-block dialkyls, substitution of

an alkyl group by a halogen results in enhanced thermal stability
(6,46).

The organoactinide mono(cyclopentadienyl) alkyls are formally six-coordinate, and preliminary results suggest lower
thermal stability than found for bis(cyclopentadienyl) alkyls.
Thus, the product of the reaction of $[C_5(CH_3)_5]ThCl_3$ with methyllithium in ether decomposes when ether removal is attempted.
Likewise, the hexaalkyl uranium dianions, UR_6^{-2}, are six-
coordinate and unstable at room temperature. The formally four-
coordinate $[UR_4]$ species may be even less stable.

The mechanism of thermal decomposition has been studied in
greatest detail for $(C_5H_5)_3UR$ and $(C_5H_5)_3ThR$ compounds. On the
basis of product analysis, isotopic and stereochemical labelling,
and kinetic measurements it was concluded that thermolysis does
not take place _via_ β-hydride elimination, but rather the corresponding RH molecule is extruded and the source of the hydrogen
atom is the cyclopentadienyl ring (2,5). Hydrogen transfer was
shown to be intramolecular and stereospecific; there was no evidence for β-hydride elimination processes (equation (27))
occurring prior to RH elimination. In the case of thorium, the
organometallic thermolysis product was isolated in crystalline
form and found to have the structure shown in Figure 4. The

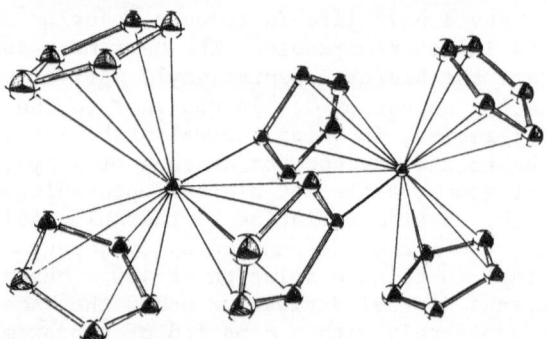

Fig. 4.   The molecular structure of $[(C_5H_5)_2Th(C_5H_4)]_2$ from
           reference 47.

hydrogen transfer mechanism can be discussed in terms of three
limiting transition states: homolytic bond scission (48) and
hydrogen abstraction within a highly constrained solvent cage
(<u>F</u>), a four-center concerted elimination (<u>G</u>) (49), and an
oxidative-addition reductive-elimination sequence (50) (<u>H</u>).
The unfavorability of β-hydride elimination in the $(C_5H_5)_3MR$
reasonably reflects the energetic inaccessibility of highly
crowded olefin-hydride intermediates or transition states such
as (<u>I</u>). In agreement with these ideas, the far less-saturated

thorium and uranium polyalkyls such as MR_4 (35), $A(C_5H_4)_2UR_2$
(19,20), and $[C_5(CH_3)_5]MR_2$ (23,44) give organic decomposition
products associated with β-hydride elimination in cases where
β-hydride atoms are present on R.   That MR_4 species without

β-hydride atoms are also unstable at room temperature suggests
that coordinative unsaturation also facilitates processes other
than β-hydride elimination.   After RH extrusion occurs on
$(C_5H_5)_3MR$ thermolysis, a carbene complex-cyclopentadienylide
species (J - L) is postulated to dimerize, forming (at least
in the case of thorium) the product shown in Figure 4.

Although $(C_5H_5)_3ThR$ and $(C_5H_5)_3UR$ complexes strongly
resist β-hydride elimination under thermolysis conditions,
photochemical excitation readily leads to β-hydride elimination

(51, 52).   On the basis of product analysis, deuterium labelling,
and frozen solution studies, the two-step mechanism illustrated
in equations (28) and (29) with the isopropyl derivatives has
been proposed.   Key processes are ligand-to-metal charge-transfer promotion of the complex to a less coordinatively saturated
mono- or trihapto excited state (M) (53), and reaction of the
metal hydride with the alkyl to produce alkane (98).   Photolysis
of the R = CH_3 derivatives proceeds in extremely low quantum
yield to produce primarily methane (and some ethylene) (52).
Photolysis of the

$$\text{(28)}$$

$$\text{(29)}$$

s-butyl compounds yields 1-butene as well as <u>cis</u>- and <u>trans</u>-2-butene (52). For the M = uranium compounds, some competition from the thermal decomposition (<u>i</u>. <u>e</u>., ring hydrogen atom abstraction) also typically occurs during photolysis.

B.   Reactions of Actinide-Carbon Sigma Bonds: Electrophiles

Alcoholysis of the tris(cyclopentadienyl) actinide alkyls cleaves both the monohapto and pentahapto ligands (equations (30) and (31) (2,5). In the case where M is thorium, the reaction is rapid and attack occurs first (exclusively) at the sigma

$$(\eta^5\text{-}C_5H_5)_3MR + R'OH \longrightarrow (\eta^5\text{-}C_5H_5)_3MOR' + RH \tag{30}$$

$$(\eta^5\text{-}C_5H_5)_3MR + R'OH \longrightarrow (\eta^5\text{-}C_5H_5)_2M(R)OR' + C_5H_6 \tag{31}$$

bond. Only then does cleavage of the cyclopentadienyl ring occur, probably as shown in equation (32). In the case of the uranium alkyls, the bonding is apparently less ionic (see

$$(\eta^5\text{-}C_5H_5)_3MOR' + R'OH \longrightarrow (\eta^5\text{-}C_5H_5)_2M(OR')_2 + C_5H_6 \tag{32}$$

Chapter 5 for similar observations in the $M(\eta^8\text{-}C_8H_8)_2$ series) and cleavage is far slower. Curiously, for uranium there is competitive cleavage of σ and π ligands (equations (30) and (31)).

C.   Reactions of Actinide-Carbon Sigma Bonds: Hydrogenolysis

As a consequence of the rather high coordinative saturation, the $(C_5H_5)_3ThR$ and $(C_5H_5)_3UR$ compounds are unreactive with respect to many interesting reagents. Thus, there is no reaction with hydrogen at room temperature in the course of many hours. At $90°$ in a sealed tube under 1 atm. of H_2, $(C_5H_5)_3Th(i\text{-propyl})$ is also unresponsive (54). In marked contrast to these observations, hydrogenolysis of bis(pentamethylcyclopentadienyl)actinide dialkyls occurs rapidly to yield the corresponding actinide dihydrides and the stoichiometric amount of alkane (equation (33)) (23). On the basis of molecular weight

$$2[\eta^5\text{-}C_5(CH_3)_5]_2MR_2 + 4H_2 \xrightarrow[25°]{\text{toluene}} \left\{[\eta^5\text{-}C_5(CH_3)_5]_2MH_2\right\}_2 + 4RH$$

$$M = Th, U \tag{33}$$

and spectral data, these first isolable organoactinide hydrides were proposed to have dimeric structures with both terminal and bridging metal-hydrogen bonds (23). This proposal has recently been confirmed by single crystal neutron diffraction studies (55), and the result is illustrated in Figure 5. The low frequencies of the M-H stretching modes in these compounds and the low

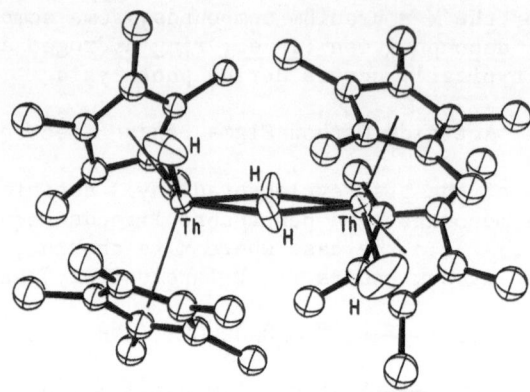

Fig. 5.   The molecular structure of $\left\{[C_5(CH_3)_5]_2ThH_2\right\}_2$ from
           reference 55.

field position of the hydride resonance in the $\{[C_5(CH_3)_5]_2-$
$ThH_2\}_2$ proton NMR are properties reminiscent of early transition
metal (d^0) hydrides (56).   The thorium hydride is far more
stable than the uranium hydride which although stable as a solid,
in solution readily (and reversibly) loses one-half equivalent
of H_2 per uranium.   The hydrogen elimination in the uranium complex is apparently due to the ready accessibility of the +3
oxidation state (57).

     Hydrogenolysis of the bis(pentamethylcyclopentadienyl)-
thorium alkyl chlorides produces the corresponding hydrochloride
according to equation (34).   The infrared spectrum of this complex suggests the presence of bridging hydrogen atoms and is
assigned a solid state structure similar to that found for the
dihydride but with terminal chlorine atoms instead of hydrogen
atoms (38).   In the case of the bis(pentamethylcyclopentadienyl)
uranium alkyl chlorides, the stability of the trivalent uranium

$$3[\eta^5\text{-}C_5(CH_3)_5]_2Th(R)Cl + 2H_2 \longrightarrow$$

$$\left\{\eta^5\text{-}C_5(CH_3)_5]_2Th(H)Cl\right\}_2 + RH \qquad (34)$$

oxidation state is evidenced again by the immediate reduction to
the uranium monochloride, as shown in equation (35) (27).

$$3[\eta^5\text{-}C_5(CH_3)_5]_2U(R)Cl \xrightarrow{H_2} \left\{[\eta^5\text{-}C_5(CH_3)_5]_2UCl\right\}_3 + 3RH \quad (35)$$

     The bis(pentamethylcyclopentadienyl) thorium and uranium
dihydrides catalyze the hydrogenation of olefins and the exchange of hydrogen atoms bound to arene and benzylic carbon
positions (58).   It is likely that both reaction patterns

involve both the creation and scission of actinide-to-carbon
sigma bonds.

D.   Reactions of Actinide-Carbon Sigma Bonds: Migratory
     Insertion of Carbon Monoxide.

     Migratory insertion of carbon monoxide (59) is a wide-
spread reaction pattern for transition metal alkyls and a fundamental component of catalytic processes such as hydroformyla-
tion (59,60).   The generally accepted mechanism for this reaction, which produces a monohaptoacyl, is shown in equation (36).

$$
\overset{\displaystyle CH_3}{\underset{\displaystyle M}{|}} + CO \rightleftharpoons \overset{\displaystyle CH_3}{\underset{\displaystyle M \leftarrow CO}{|}} \rightleftharpoons M - \overset{\displaystyle O}{\overset{\displaystyle \|}{C}}CH_3 \tag{36}
$$

Although the saturated $(C_5H_5)_3ThR$ and $(C_5H_5)_3UR$ complexes do not
appreciably react with carbon monoxide over a period of hours
(2,5) the bis(pentamethylcyclopentadienyl) thorium and uranium
dimethyls rapidly take up two equivalents of carbon monoxide at
low temperature to quantitatively yield the unusual dimeric
products illustrated in equation (37) (61).   Further metrical

$$\tag{37}$$

M= Th, U

95-100 %

information on the thorium insertion product is provided by
single crystal X-ray diffraction results, shown in Figure 6.
Thus, this insertion reaction has resulted in the coupling of
eight separate, carbon-containing fragments and the stereospecific formation of two carbon-carbon double bonds and four thorium-
oxygen bonds.

     The reaction pattern of equation (37) is remarkably simi-
lar to the behavior of several early transition metal systems.
Thus, bis(cyclopentadienyl) titanium, zirconium, and hafnium
alkyls reversibly add carbon monoxide (59,62) to form bihapto-
acyls such as N, which have been characterized crystallographi-
cally (59,62).   Even more relevant to the present results is the
observation (63) that bis(pentamethylcyclopentadienyl) zirconium
dimethyl reacts with two equivalents of carbon monoxide (the
first rapidly and reversibly, the second only very slowly and

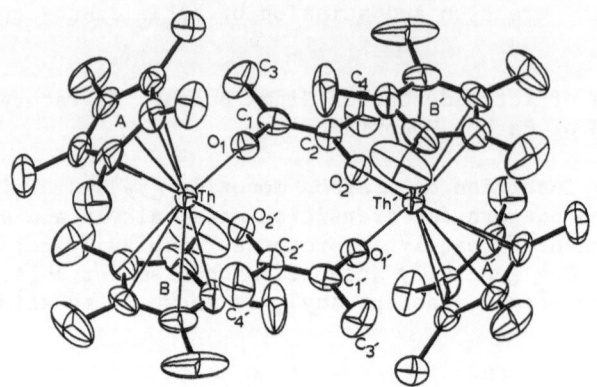

Fig. 6.   The molecular structure of $\{[C_5(CH_3)_5]_2Th\text{-}$
$[\mu\text{-}O_2C_2(CH_3)_2]\}_2$ from reference 61.

$\underline{N}$

irreversibly) to yield a monomeric complex of proposed structure $\underline{O}$.   It is known that a bihaptoacyl species is the product

$\underline{O}$

of the first carbon monoxide insertion (63).

Taken together, the above early transition metal and organoactinide observations provide important information on this
unique type of migratory insertion reaction, the mechanism of
which may be related to such important processes as Fischer-
Tropsch, methanol synthesis, ethylene glycol synthesis, methana-

tion, and isosynthesis chemistry (64). The greatly enhanced
reactivity of the organoactinides over the analogous transition
metal organometallics is especially noteworthy. This effect is
reasonably explained on the basis of the higher coordinative
unsaturation of the actinide ions (the same ligand array as zirconium but considerably larger ionic radii (65)) and the likelihood that the greater ionicity of the actinide-to-carbon sigma
bonds renders the methyl groups more nucleophilic in the alkyl
migration process (66). Whether bihaptoacyl species are actually
intermediates in the formation of the butenediolate products
(an assertion which seems reasonable but, especially in the case
of the more reactive actinide dimethyls, requires additional
support) and why the zirconium product is monomeric (0) while the
actinide analogues are dimeric (equation (37)) are important
questions. Further experiments with the organoactinides shed
some light on these issues.

The stabilization of the bihaptoacyl moiety reasonably
reflects the high affinity that early transition metal and actinide ions display for oxygen(67). With sufficient oxygen
affinity, resonance hybrid Q should become important, and

$$\underset{\underline{P}}{M-\overset{\overset{\displaystyle \cdot O}{\|}}{C}-R} \quad \longleftrightarrow \quad \underset{\underline{Q}}{M\leftarrow:\overset{\overset{\displaystyle O}{|}}{C}-R}$$

carbene-like chemistry (i.e., apparent intramolecular coupling)
could yield an olefinic species such as 0 (63,68). In an effort
to learn more about the reactivity of the actinide bihaptoacyl
species and to probe the factors which control whether monomeric
or dimeric products are formed in coupling, it was decided to
study the insertion process with bulkier alkyl residues.

The $[C_5(CH_3)_5]_2M(CH_2SiMe_3)_2$ systems, M = Th, U, react
rapidly with carbon monoxide and, interestingly, the system is
forced to form a monomeric coupling product analogous to 0
(equation (38)) (61). To introduce a further modification and in

(38)

M= Th, U   95-100 %

an attempt to isolate a bihaptoacyl, one equivalent of carbon
monoxide was reacted with the monoalkyl chlorides (62,26).   In
the case of trimethylsilylmethyl, the intermediate bihaptoacyl
can actually be detected spectroscopically, (^{1}H NMR), however,
it rearranges to a substituted enolate (equation (39)).   This

(Detected by ^{1}H NMR)

type of product is very strongly suggestive of carbene-like
migration chemistry (69).   In the case of the related thorium
neopentyl derivative, it is possible to isolate and completely
characterize the bihaptoacyl (equation (40)).   Upon heating in

solution, this compound slowly rearranges to a cis-t-butyl-
vinylate derivative.   It is expected in rearrangements involving
α-carbene centers that silicon will migrate more readily than
hydrogen (69).

Another variation on the above systems is the reaction of
carbon monoxide with the uranium metallocycle R (equation (41))

(26). In contrast to the reactions of the aforementioned actinide dialkyls, $\underline{R}$ reacts with only one equivalent of carbon

$$\underline{R} \xrightarrow[25^o]{CO} \underline{S} \qquad (41)$$

Ph = C_6H_5 95-100%

monoxide to yield a complex with proposed structure $\underline{S}$ (equation
(41)) (other possible resonance forms of $\underline{S}$ which involve delocalization of charge into the cyclopentadiene ligand ring are
not depicted). Support for this structural assignment is provided by the reaction of tetraphenylcyclopentadienone with
$\{[C_5(CH_3)_5]_2UCl\}_3$, which yields $\underline{S}$ in 50% yield (equation (42))
(26). A dimeric titanium complex related to $\underline{S}$ has been struc-

$$[\eta^5\text{-}C_5(CH_3)_5]_2UCl_2 + \underline{S} \qquad (42)$$

turally characterized (70). A zirconium metallocycle has been
proposed to react with carbon monoxide in an analogous manner
(63).

E.   Conclusions

At this stage it is becoming increasingly apparent that
actinide-to-carbon sigma bonds have a rich and varied chemistry.
It is remarkable that chemical reactivity can be manipulated to
such an extent by modification of the nature of the ligands in
the actinide coordination sphere, and that reaction patterns
with some combinations of ligands (and reactants) can be so
similar to those of early transition metals. The degree to
which this is so, and the degree to which such similarities as
well as differences can be exploited to learn and to develop
new chemistry, is a fascinating question for future research.

Acknowledgments

We are grateful to the National Science Foundation (grant CHE76-84494 A01) and NATO (grant 776) for generous support of this research. We thank Prof. V. W. Day and Dr. C. S. Day for recent stimulating structural collaboration. T. J. M. is a Camille and Henry Dreyfus Teacher-Scholar.

References

1. A. E. Gebala and M. Tsutsui, J. Am. Chem. Soc. , 95, 91 (1973).

2. T. J. Marks, A. M. Seyam, and J. R. Kolb, J. Am. Chem. Soc. , 95, 5529 (1973).

3. G. Brandi, M. Brunelli, G. Lugli, and A. Mazzei, Inorg. Chim. Acta, 7, 319 (1973).

4. J. L. Atwood, M. Tsutsui, N. Ely, and A. E. Gebala, J. Coord. Chem. , 5, 209 (1976).

5. T. J. Marks and W. A. Wachter, J. Am. Chem. Soc. , 98, 703 (1976).

6. J. Goffart, B. Gilbert, and G. Duyckaerts, Inorg. Nucl. Chem. Letters, 13, 186 (1977).

7. W. A. Wachter, Ph. D. thesis, Northwestern University, August 1976.

8. D. G. Karraker and J. A. Stone, Abstracts, 172nd National Meeting of the American Chemical Society, San Francisco, California, Sept. 1976, No. INOR 184, and Chapter 13 of this volume.

9. R. B. King, "Transition-Metal Organometallic Chemistry," Academic Press, N. Y. , 1969, Chapt. I and II.

10. P. C. Wailes, R. S. P. Coutts, and H. Weigold, "Organometallic Chemistry of Titanium, Zirconium, and Hafnium," Academic Press, New York, N. Y. 1974, Chapts. IVB, IVC, and VD.

11. R. R. Schrock and G. W. Parshall, Chem. Rev. , 76, 243 (1976).

12. J. Schwartz and J. A. Labinger, Angew. Chem. Int. Ed. , 15, 333 (1976).

13. P. Zanella, S. Faleschini, L. Doretti, and G. Faraglia, J. Organometal. Chem. , 26, 353 (1971).

14. R. D. Ernst, W. J. Kennelly, C. S. Day, V. W. Day, and T. J. Marks, _J. Am. Chem. Soc._, in press.

15. A. F. Reid and P. C. Wailes, _Inorg. Chem._, 5, 1213 (1966).

16. M. F. Brady and R. S. Marianelli, Abstracts, Eighth Regional Midwest A. C. S. Meeting, Columbia, Missouri, November, 1972, No. 235.

17. J. D. Jameson and J. Takats, _J. Organometal. Chem._, 78, C23 (1974).

18. P. Zanella, G. DePaoli, G. Bombieri, G. Zanotti, and R. Rossi, _J. Organometal. Chem._, 142, C21 (1977).

19. C. A. Secaur, V. W. Day, R. D. Ernst, W. J. Kennelly, and T. J. Marks, _J. Am. Chem. Soc._, 98, 3713 (1976).

20. C. S. Day, V. W. Day, R. D. Ernst, W. J. Kennelly, and T. J. Marks, manuscript in preparation.

21. R. D. Ernst and T. J. Marks, unpublished results.

22. a. P. J. Davidson, M. F. Lappert, and R. Pearce, _Chem. Rev._, 76, 219 (1976).
    b. M. C. Baird, _J. Organometal. Chem._, 64, 289 (1974).

23. J. M. Manriquez, P. J. Fagan, and T. J. Marks, _J. Am. Chem. Soc._, 100, 3939 (1978).

24. J. C. Green and O. Watts, _J. Organometal. Chem._, 153, C40 (1978).

25. J. M. Manriquez, P. J. Fagan, C. S. Day, V. W. Day, and T. J. Marks, _J. Am. Chem. Soc._, 100, 8112 (1978).

26. J. M. Manriquez, P. J. Fagan, and T. J. Marks, manuscript in preparation.

27. P. J. Fagan, J. M. Manriquez, T. J. Marks, C. S. Day, V. W. Day, and T. J. Marks, manuscript in preparation.

28. P. J. Fagan, J. M. Manriquez, and T. J. Marks, unpublished results.

29. A. M. Seyam and G. A. Eddein, _Inorg. Nucl. Chem. Lett._, 13, 115 (1977).

30. a. L. Doretti, P. Zanella, G. Faraglia, and S. Faleschini,
         J. Organometal. Chem., 43, 339 (1972).
     b. K.W. Bagnall and J. Edwards, J. Organometal. Chem.,
         80, C14 (1974).

31. G. Bombieri, G. de Paoli, and K.W. Bagnall, Inorg. Nucl.
     Chem. Lett., in press.

32. K.W. Bagnall in Chapter 7 of this volume.

33. J.M. Manriquez, P.J. Fagan, and T.J. Marks, unpublished
     results.

34. H. Gilman, R.G. Jones, E. Bindschadler, D. Blume,
     G. Karmas, G.A. Martin, Jr., J.F. Nobis, J.R. Thirtle,
     H.L. Yale, and F.A. Yoeman, J. Am. Chem. Soc., 78,
     2790 (1956).

35. T.J. Marks and A.M. Seyam, J. Organometal. Chem., 67,
     61 (1974).

36. R. Andersen, E. Carmona-Guzman, K. Mertis, E. Sirgudson,
     and G. Wilkinson, J. Organometal. Chem., 99, C19 (1975).

37. T.J. Marks and W.A. Wachter, unpublished results.

38. E. Köhler, W. Brüser, and K.-H. Thiele, J. Organometal.
     Chem., 76, 235 (1974).

39. K.-H. Thiele, R. Opitz, and E. Köhler, Z. Anorg. Allg.
     Chem., 435, 45 (1977).

40. G.R. Davies, J.A.J. Jarvis, and B.T. Kilbourn, J. Chem.
     Soc., Chem. Comm., 1511 (1971).

41. E.R. Sigurdson and G. Wilkinson, J. Chem. Soc., Dalton
     Trans., 812 (1977).

42. J.A. Connor, Topics Curr. Chem., 71, 71 (1977), and
     references therein.

43. a. G.M. Whitesides, J.F. Gaasch, and R.R. Stedronsky,
         J. Am. Chem. Soc., 94, 5258 (1972).
     b. H.C. Clark and C.S. Wong, J. Am. Chem. Soc., 96,
         7213 (1974).
     c. G. Wilkinson, Science, 185, 109 (1974).
     d. D.L. Reger and E.C. Culbertson, J. Am. Chem. Soc.,
         98, 2789 (1976).
     e. D.L. Reger and E.C. Culbertson, Inorg. Chem., 16,
         3104 (1977).

44. J.M. Manriquez, P.J. Fagan, and T.J. Marks, unpublished results.

45. J.E. Bercaw, R.H. Marvich, L.G. Bell, and H.H. Brintzinger, J. Am. Chem. Soc., 94, 1219 (1972).

46. M.R. Collier, M.F. Lappert, and R. Pearce, J. Chem. Soc., Dalton Trans., 445 (1973).

47. E.C. Baker, K.N. Raymond, T.J. Marks, and W.A. Wachter, J. Am. Chem. Soc., 96, 7586 (1974).

48. a. J. Evans, S.J. Okrasinski, A.J. Pribula, and J.R. Norton, J. Am. Chem. Soc., 99, 5835 (1977).

    b. G.M. Whitesides, E.J. Panek, and E.R. Stedronsky, J. Am. Chem. Soc., 94, 232 (1972).

49. a. R.R. Schrock and J.D. Fellmann, J. Am. Chem. Soc., 100, 3359 (1978).

    b. V. Malatesta, K.U. Ingold, and R.R. Schrock, J. Organometal. Chem., 152, C53 (1978).

50. a. C.S. Cundy, M.F. Lappert, and R. Pearce, J. Organometal. Chem., 59, 161 (1973).

    b. M.P. Brown, R.J. Puddephatt, and C.E.E. Upton, J. Organometal. Chem., 49, C61 (1973).

    c. A. Tamaki, S.A. Magennis, and J.K. Kochi, J. Am. Chem. Soc., 96, 6140 (1974).

    d. T.T. Tsou and J.K. Kochi, J. Am. Chem. Soc., 100, 1634 (1978).

    e. L. Abis, A. Sen, and J. Halpern, J. Am. Chem. Soc., 100, 2915 (1978).

51. D.G. Kalina, T.J. Marks, and W.A. Wachter, J. Am. Chem. Soc., 99, 3877 (1977).

52. D.G. Kalina and T.J. Marks, manuscript in preparation.

53. a. M. Ephritikhine and M.L.H. Green, J. Chem. Soc., Chem. Comm., 926 (1976).

    b. E. Vitz and C.H. Brubaker, Jr., J. Organometal. Chem., 104, C33 (1976).

    c. J.G.S. Lee and C.H. Brubaker, Jr., Inorg. Chim. Acta., 25, 181 (1977).

54. D.G. Kalina and T.J. Marks, unpublished observations.

55. R.W. Broach, A.J. Schultz, J.M. Williams, G.M. Brown,
     J.M. Manriquez, P.J. Fagan, and T.J. Marks, Science,
     in press.

56. a. J.C. Green and M.L.H. Green in Comprehensive
         Inorganic Chemistry, J.C. Bailar, Jr., H.J.
         Emeleus, R.S. Nyholm, and A.F. Trotman -
         Dickenson, eds., Pergamon Press, Oxford, 1973,
         Chapt. 48.

     b. G.L. Geoffroy and J.R. Lehman, Advan. Inorg. Chem.
         Radiochem., 20, 189 (1977).

57. L.J. Nugent in MTP International Review of Science,
     Inorganic Chemistry, Series Two, Vol. 7, K.W. Bagnall,
     ed., University Park Press, Baltimore, 1975, Chapt. 6.

58. J. Manriquez, N.L. Jones, P.J. Fagan, and T.J. Marks,
     Manuscript in preparation.

59. a. Wojcicki, Advan. Organometal. Chem., 11, 87 (1973).

     b. F. Calderazzo, Angew. Chem. Int. Ed., 16, 299 (1977).

     c. R. Eisenberg and D. Hendriksen, Advan. Catal.,
         in press.

60. a. G.W. Parshall, J. Mol. Catal., 4, 243 (1978).

     b. A. Stefani, G. Consiglio, C. Botteghi, and P. Pino,
         J. Am. Chem. Soc., 99, 1058 (1977) and references
         therein.

61. J.M. Manriquez, P.J. Fagan, T.J. Marks, C.S. Day, and
     V.W. Day, J. Am. Chem. Soc., 100, 7112 (1978).

62. a. G. Fachinetti, G. Fachi, C. Floriani, J. Chem. Soc.,
         Dalton Trans., 1946 (1977).

     b. G. Fachinetti, C. Floriani, H. Stoeckli-Evans,
         J. Chem. Soc., Dalton Trans., 2297 (1977).

63. a. J.M. Manriquez, D.R. McAlister, R.D. Sanner,
         J.E. Bercaw, J. Am. Chem. Soc., 100, 2716 (1978).

     b. J.M. Manriquez, D.R. McAlister, R.D. Sanner,
         J.E. Bercaw, J. Am. Chem. Soc., 98, 6733 (1976).

64. a.   H. Schulz, _Erdol_, _Kohle_, _Erdgas_, _Petrochem._, _30_,
          123 (1977).

     b.   M.A. Vannid, _Catal. Rev. Sci. Eng._, _14_, 153 (1976)
          and references therein.

     c.   G. Henrici-Olive, S. Olive, _Angew. Chem. Int. Ed._,
          _15_, 136 (1976).

     d.   G.C. Demitras and E.L. Muetterties, _J. Am. Chem.
          Soc._, _99_, 2796 (1977).

     e.   E.M. Cohn in "Catalysis," P.H. Emmet, ed. Reinhold,
          N.Y., Vol. 4, 1956.

65. a.   R.D. Shannon, _Acta. Cryst._, _A32_, 751 (1976).

     b.   Ionic radii (65a) for eight-coordination:
          Zr(IV), 0.84 Å; U(IV), 1.00 Å; Th(IV), 1.05 Å.

66. For a discussion of the potential importance of such
     factors in carbon monoxide insertion reactions, see
     reference 59c.

67. a.   D.L. Keppert, "The Early Transition Metals",
          Academic Press, N.Y., 1972, Chapt. 1.

     b.   R.G. Pearson, ed. "Hard and Soft Acids and Bases,"
          Dowden, Hutchinson, and Ross, Stroudsberg, PA, 1973.

     c.   A. Navrotsky in _MTP International Review of Science_,
          Inorganic Chemistry, _Series Two_, Vol. 5, D.W.A.
          Sharp, Ed., University Park Press, Baltimore,
          1975, Chapt. 2.

68. a.   W.J. Baron, M.R. DeCamp, M.E. Hendrick, M. Jones,
          Jr., R. Levin, and M.B. Sohn in "Carbenes," M.
          Jones, Jr., and R.A. Moss, eds., Wiley, New York,
          1973, Vol. 1, p. 128.

     b.   R.A. Moss in reference 68a, p. 280.

     c.   W. Kirmse, "Carbene Chemistry", Academic Press,
          New York, 1971, Chapt. 3, Section E.

     d.   R.W. Hoffmann, _Angew. Chem. Int. Ed._, _10_, 529
          (1971).

69. a.   C. Wentrup, _Top. Curr. Chem._, _62_, 173 (1976).

b.   Reference 68c, Chapter 12.

c.   J.H. Robson and H. Schechter, J. Am. Chem. Soc.,
      89, 7112 (1967).

70.   G. Fachinetti, C. Biran, C. Floriani, C. Floriani,
      A. Chiesi-Villa, and C. Guastini, J. Am. Chem. Soc.,
      100, 1921 (1978).

71.   M. Tsutsui, Neal Ely, and Allen Gebala, Inor. Chem., 14
      78 (1975).

72.   G.W. Halstead, E.C. Baker, and K.N. Raymond, J. Amer.
      Chem. Soc., 97, 3049 (1975).

73.   G. Perego, M. Cesari, F. Farina, and G. Lugli, Acts.
      Cryst., B32, 3034 (1976).

74.   J. L. Atwood, C.F. Haines, Jr., M. Tsutsui, and A.E.
      Gebala, J. Chem. Soc., Chem. Comm., 452 (1973).

PREPARATION AND CHEMISTRY OF URANOCENES

Andrew Streitwieser, Jr.

Department of Chemistry, University of California,
Berkeley, CA 94720 U.S.A.

INTRODUCTION

This summer of 1978 marks the tenth anniversary of the
discovery of uranocene, di-pi-cyclooctatetraene–uranium(IV), (1).
Before this event, organoactinide chemistry consisted of a mere
handful of papers (2-12). The ten years since that time have seen
a surge of varied organoactinide chemistry, based not only on
cyclooctatetraene (COT) but also on cyclopentadienyl and other
pi- as well as sigma-ligands. The subject has been reviewed
frequently in the last few years (13-27).

Uranocene was prepared as an expected f-orbital analog to
the d-transition metal metallocenes. In compounds such as
ferrocene it is now well established that a significant amount of
ring-metal covalency results from interaction of the highest
occupied e_{1g} MOs of two cyclopentadienyl (Cp) anions with
appropriately arranged vacant $d_{\pm 1}$ orbitals of Fe^{+2}, (Fig. 1).
A corresponding interaction with the highest occupied e_{2u} MOs
of two COT dianions requires $f_{\pm 2}$ atomic orbitals (Fig. 2) and
suggests the application to actinide elements. In one of his
first experiments to implement this suggestion, Dr. Ulrich
Müller-Westerhoff treated a THF solution of uranium tetrachloride
with cyclooctatetraene dianion and, in June, 1968, prepared
uranocene. The D_{8h} structure, I, was confirmed by X-ray
crystal analysis by Zalkin and Raymond (28, see also 29).

Nevertheless, the fact of the preparation based on the
concept of e_{2u} - f covalency does not by itself establish that
such interaction is actually important in ring-metal bonding in
uranocene. Extensive chemical and spectroscopic studies

149

T. J. Marks and R. D. Fischer (eds.), Organometallics of the f-Elements, 149–177.
All Rights Reserved. Copyright © 1979 by D. Reidel Publishing Company, Dordrecht, Holland.

 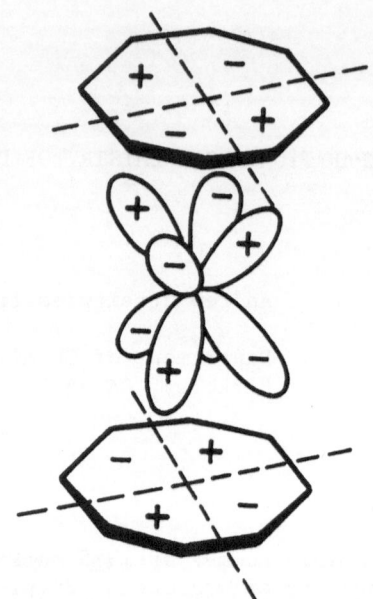

Figure 1: Ring-metal bonding in ferrocene; interaction between e_{1g} MOs of two cyclopentadienyls and $d_{\pm 1}$ AOs of iron.

Figure 2: Proposed ring-metal bonding in uranocene; interaction between e_{2u} MOs of two planar cyclooctatetraenes and $f_{\pm 2}$ AOs of uranium.

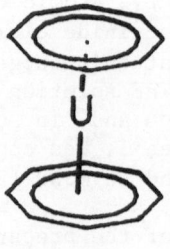

I

indicate that ring-metal bonding with uranocene compounds is significantly covalent and theoretical studies show that the f-orbital interaction postulated originally does contribute significantly to this covalency.

SUBSTITUTED URANOCENES

     To establish the chemistry and spectroscopic interpretations of the uranocene system, a number of substituted derivatives have been studied. Most common have been the 1,1'-disubstituted compounds, II, with the following substituents: methyl, ethyl, n-propyl, isopropyl, n-butyl, t-butyl, neopentyl, cyclopropyl, methoxy, ethoxy, benzyloxy, allyloxy, dimethylamino, trimethyl-ammonio, methoxycarbonyl, ethoxycarbonyl, benzyloxycarbonyl, vinyl, phenyl, p-methoxyphenyl, p-dimethylaminophenyl and others (30-36). The bis-disubstituted compounds include uranocenes prepared from 1,4-di-t-butylcyclooctatetraene, III, (37)

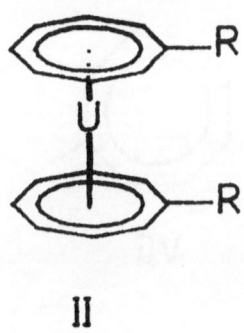

II

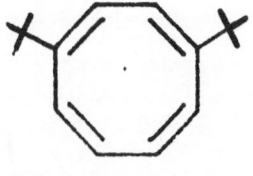

III

as well as the fused ring ligands, IV, V, VI and VII (38, 39).
Among the bis-tetrasubstituted uranocenes prepared are those
derived from 1,3,5,7-tetrasubstituted COTs, VIII, with methyl
and phenyl substitutents (40, 41).

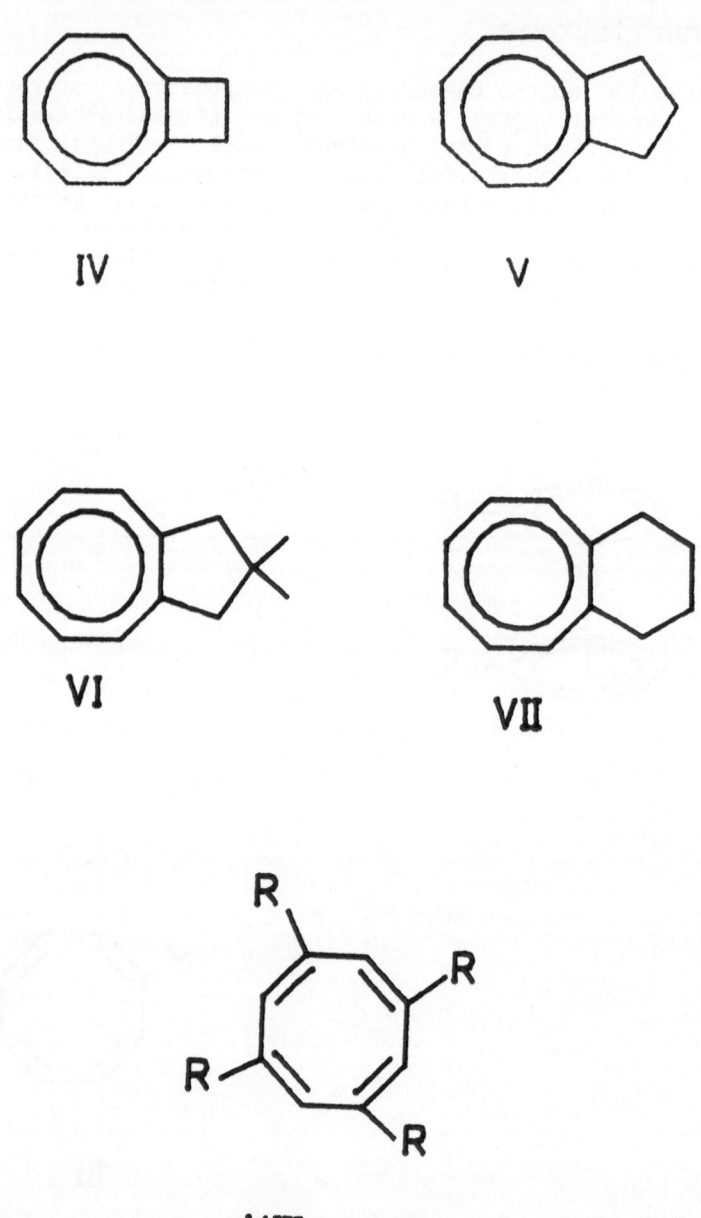

IV

V

VI

VII

VIII

Substituted uranocenes can be prepared by three types of methods:

1. Reaction of substituted $COT^=$ with UCl_4.
2. Reaction of the substituted COT with finely divided U°.
3. Functional group reactions of substituted uranocenes.

The first of these methods is also the most common. COTs react generally with potassium metal to give the corresponding dianion:

$$C_8H_8 \; + \; 2K \; = \; C_8H_8^= 2K^+$$

The potassium salts of COT itself (42) and of 1,3,5,7-tetra-methylcyclooctatetraene (43) have been shown by X-ray crystal analysis to involve a planar 8-membered ring having benzenoid C-C bond lengths of 1.40 A and solvated by potassium ions on opposite sides of the ring; i.e., the structure is that of a contact ion-triplet of a 10-pi-electron (4n +2) Huckel-aromatic ring. Details of the preparation of the parent uranocene are available (44, 45) and apply generally to many substituted COTs. Solutions of the dianions in THF react rapidly with a THF solution of UCl_4. Uranocene itself has low solubility and may be isolated by filtration and purified by Soxhlet extraction and by sublimation. Many substituted uranocenes have greater solubility and may be purified by recrystallization and by sublimation. Because most uranocenes are air-sensitive, all operations should be carried out with degassed solvents under inert atmosphere.

Some dianions have been prepared by deprotonation of a cyclooctatriene.

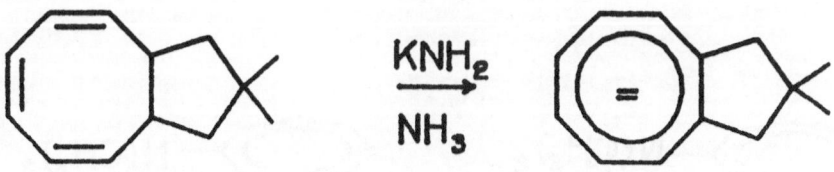

Some substituents are sensitive to elimination from the dianion; hence, the long reaction times involved in the normal preparation by reaction with potassium metal give rise to poor yields of the uranocene. For such cases an especially convenient approach is the treatment of the COT with sodium naphthalide in THF solution. The electron transfer to form the $COT^=$ is immediate and homogeneous and the reaction mixture can be treated rapidly with a solution of UCl_4 (33, 34).

$$2C_{10}H_8^- + C_8H_8 = 2C_{10}H_8 + C_8H_8^=$$

The naphthalene can be conveniently removed from the reaction product by sublimation.

Uranocene can also be prepared by reaction of COT with finely divided uranium metal (46). Cernia and Mazzei (22) reported briefly that this method can be applied to the colloidal suspension of uranium produced by allowing the low temperature reaction product of UCl_4 and butyllithium to warm to room temperature. We have confirmed this report and have shown that it can be applied to the preparation of some substituted uranocenes in good yield (37).

Subject to the sensitivity of the uranocene structure to strong acids and bases, some substituted uranocenes can be prepared by reactions of others. Some examples are:

SUBSTITUTED CYCLOOCTATETRAENES

The most common routes to substituted uranocenes start with
the corresponding substituted cyclooctatetraene. The preparation
and chemistry of cyclooctatetraenes have been reviewed (47, 48,
49). Many monosubstituted COTs are derived from bromocyclloocta-
tetraene which in turn is prepared by the careful bromination of
COT (50). Reaction of bromo COT with sodium or potassium
alkoxides in Me$_2$SO gives the alkoxy COT, undoubtedly by way of the
intermediate cyclooctatrienyne. The presence of dialkylamine in
the base reaction gives the dialkylamino COT. A convenient route
to alkylCOTs is via the reaction of BrCOT with the corresponding
dialkylcuprate (50, 51).

Some alkyl-substituted COTs can be obtained by reaction of
COT with the alkyllithium (52, 53). The reaction generally goes
in rather poor yield but was the basis of a convenient preparation
of 1,4-di-t-butylCOT (37). Paquette has summarized other routes
to some disubstituted COTs (48). Some 1,2-disubstituted COTs can
be prepared by co-cyclotetramerization in the presence of a nickel
catalyst of a disubstituted acetylene with excess acetylene.
As an example:

Others are available from the photolytic cycloaddition of
acetylene to benzene. An example is:

Annulated compounds are available by reaction of COT= with appropriate dihalides. The trienes can be deprotonated to the COT dianions. Examples are (38, 39, 55):

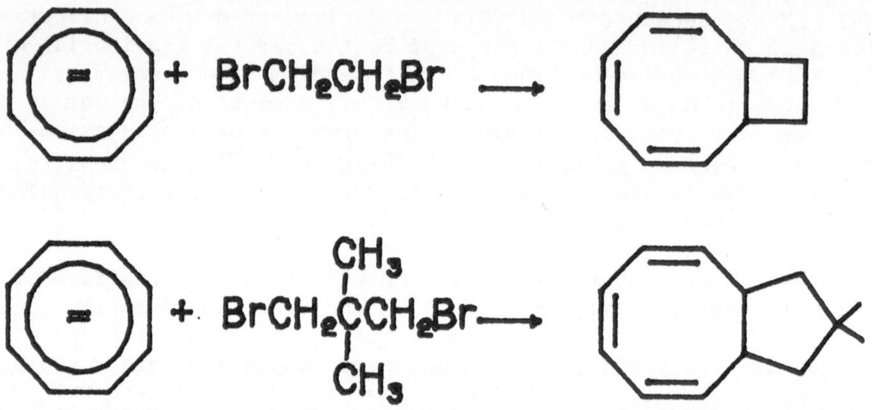

1,3,5,7-tetrasubstituted cyclooctatetraenes have special significance because of their symmetry. Their preparation involves special methods not subject to generalization. Examples are the tetramethyl, tetraphenyl and tetra-t-butyl compounds (56, 57, 37):

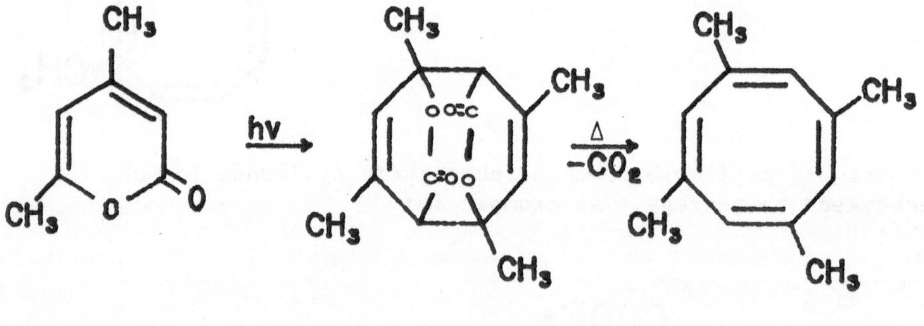

REACTIONS OF URANOCENES

Two of the important reactions of uranocenes are with oxygen and with water. Uranocene inflames on exposure to air but controlled air oxidation gives UO_2 and liberates the ligand quantitatively (44). Because of this air-sensitivity all operations with uranocenes require inert atmosphere conditions and carefully de-gassed solvents. The controlled oxidation provides a convenient structure proof. Of special interest is

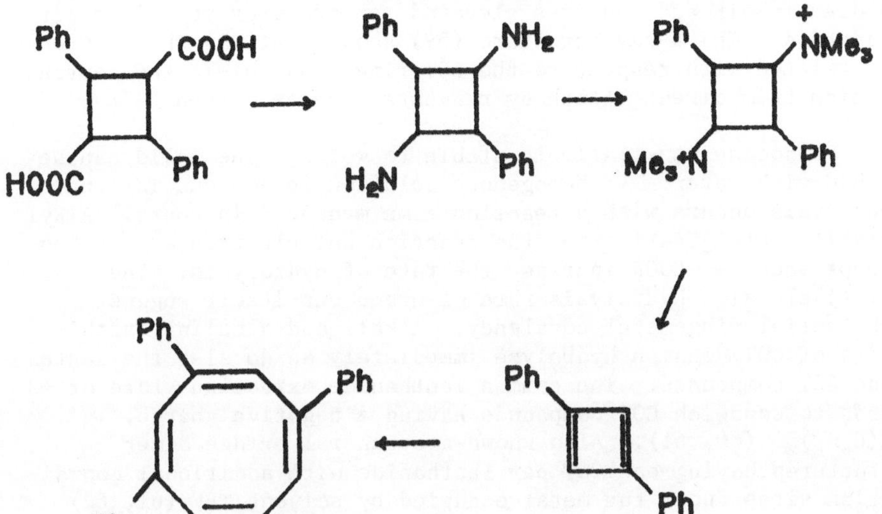
Ph
COOH
HOOC
Ph
Ph
NH_2
H_2N
Ph
Ph
$\overset{+}{N}Me_3$
$Me_3\overset{+}{N}$
Ph
Ph
Ph
Ph
Ph
Ph
Ph

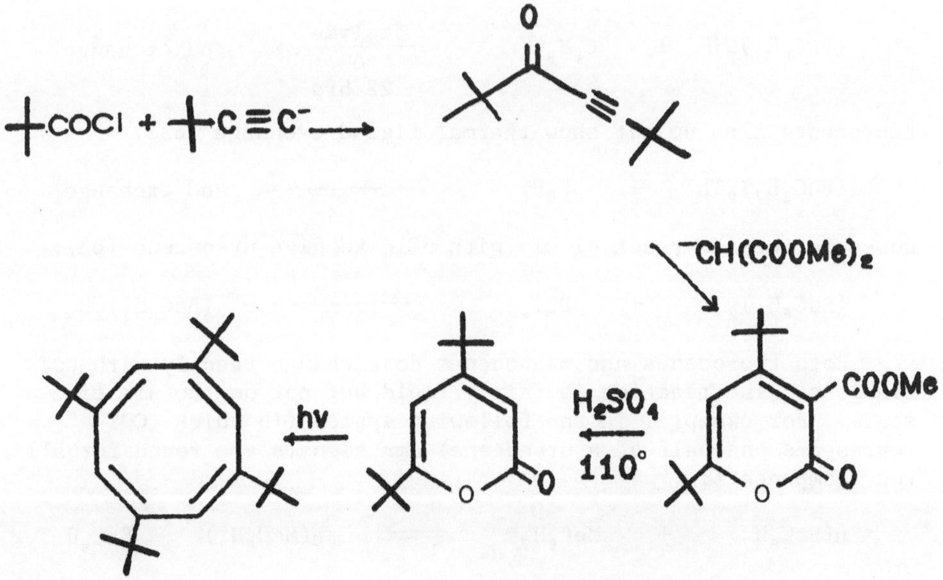
COCl
C≡C⁻
O
⁻CH(COOMe)₂
COOMe
O
O
H_2SO_4
110°
hν
O
O

the pronounced stability of bis-(1,3,5,7-tetraphenylcycloocta-tetraene)uranium towards oxygen (58). This compound can be handled in air and requires elevated temperatures to effect air oxidation. The X-ray structure (59) shows that the phenyl groups are twisted with respect to the COT rings and shield the central uranium from direct attack by reagents such as oxygen.

Uranocene is relatively stable to water. The solid can be washed with water. In homogenous solution in aqueous THF the hydrolysis occurs with a reaction time measured in hours. Alkyl substitutents greatly slow the reaction but electron-attracting groups such as -COOR increase the rate of hydrolysis. The relatively slow hydrolysis rate of uranocene itself suggest substantial ring-metal covalency. Alkali and alkaline earth salts of COT dianion hydrolyze immediately as do also the lantha-nide COT compounds. The common lanthanide oxidation state of +3 leads to sandwich COT compounds having a negative charge, $Ln(C_8H_8)_2^-$ (60, 61). Also known are central bridge dimer structures having one COT per lanthanide with additional coordi-nation sites about the metal occupied by solvent THF (61, 62). These lanthanide compounds of COT show essentially wholly ionic character in their ring-metal bonding. Not only do they hydrolyze readily but they react immediately with UCl_4 to give uranocene (61). The mono- and di-ring structures show apparently mobile ligand exchange.

By contrast, uranocenes do not show ligand exchange with COTs under thermal conditions (30, 63):

$$(BuC_8H_7)_2U \quad + \quad C_8H_8 \quad \xrightarrow[\substack{\Delta \\ 22\ hrs}]{diglyme} \quad no\ exchange$$

Thorocenes also do not show thermal ligand exchange (63):

$$(BuC_8H_7)_2Th \quad + \quad C_8H_8 \quad \xrightarrow{\hspace{3cm}} \quad no\ exchange$$

However, they do react slowly with UCl_4 to give uranocene (63):

$$(BuC_8H_7)_2Th \quad + \quad UCl_4 \quad \xrightarrow{\hspace{3cm}} \quad (BuC_8H_7)_2U$$

Both thorocenes and uranocenes do exchange readily with COT dianions. The reaction is fairly rapid but not on the nmr time scale. For example, in the following system (in which COT represents one-half of a uranocene) nmr signals are found for all four components:

$$u(EtC_8H_7) \quad + \quad MeC_8H_7^= \quad \rightleftharpoons \quad u(MeC_8H_7) \quad + \quad EtC_8H_7^=$$

A reasonable interpretation of these exchange reactions involves electron transfer to form the corresponding COT derivatives of the +3 actinide oxidation state:

$$(RC_8H_7)_2M \ + \ R'C_8H_7^= \ \rightleftharpoons \ (RC_8H_7)_2M^- \ + \ R'C_8H_7^-$$

$$\downarrow$$

ligand exchange

Karraker and Stone (64) have prepared the corresponding bis-(cyclooctatetraene)neptunium(III) and -plutonium(III) compounds as potassium salts; i.e., $KNp(C_8H_8)_2$ and $KPu(C_8H_8)_2$. These compounds are highly ionic and by analogy with the lanthanide compounds, may be expected to show facile ligand exchange. Although Karraker and Stone were not able to prepare a stable bis(cyclooctatetraene)uranium(III) derivative, its existance as a transient intermediate seems plausible by analogy.

Uranocenes are sensitive to strong acids. Reaction with acetic acid is no more rapid than with water but trifluoro-acetic acid, HCl, etc., give immediate decomposition. Lewis acids also give decomposition and no electrophilic aromatic-type substitution reactions in the COT ligand rings has been found. Aqueous potassium hydroxide added to a THF solution of uranocene gives immediate decomposition but uranocenes are stable to amines and even to cesium cyclohexylamide (44, 39). Uranocenes react slowly with butyllithium with liberation of the corresponding COT dianion but with no evidence of ring metallation. Alkoxy uranocenes and uranocyltrimethylammonium ions do undergo ring metallation reactions leading to functional group exchange (30, 33):

$$U(CH_3OC_8H_7)_2 \ + \ 2BuLi \ \longrightarrow \ (BuC_8H_7)_2U \ + \ 2CH_3OLi$$

$$U(C_8H_7N^+Me_3)_2 \ + \ 2PhLi \ \longrightarrow \ (PhC_8H_7)_2U \ + \ 2Me_3N \ + \ 2Li^+$$

The reaction is proposed to involve metallation next to the electronegative substitutent followed by loss of this substituent to give an intermediate uranium derivative of cyclooctatrienyne. The reactive triple bond reacts with the organolithium reagent to give a metallated uranocene which can next metallate another ring in a carbanion chain reaction (Chart 1).

The following experiment helps to confirm the postulated mechanism. Methyllithium does not react with 1,1'-dimethoxy-uranocene but it does react smoothly with the 1,1'-uranocyl-bis-trimethylammonium salt to give 1,1'dimethyluranocene. Methyl-lithium is known to be a generally weaker metallating agent than other alkyllithiums and is apparently too weak to metallate methoxyuranocene directly. The ammonium salt reaction, however, does show that methyllithium can react with any intermediates

Chart 1

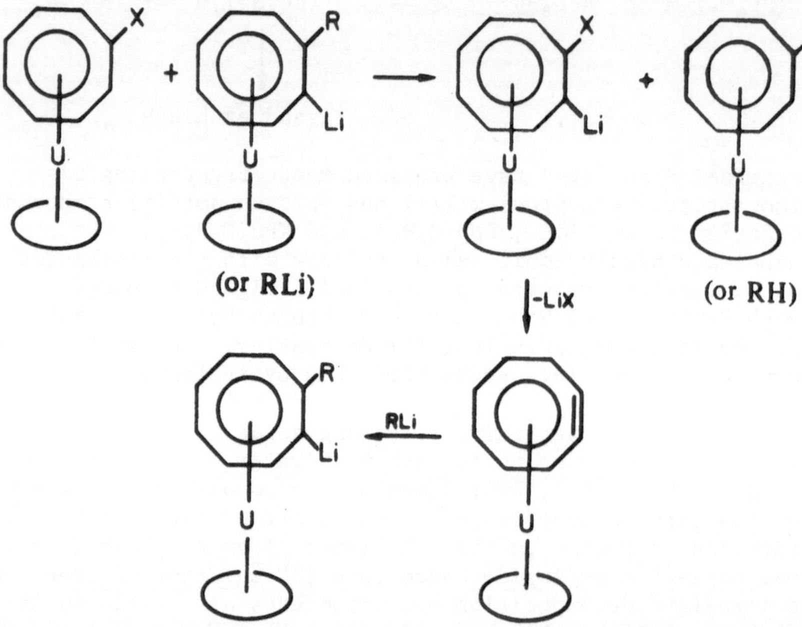

(or RLi)   -LiX   (or RH)

RLi

involved in the reaction sequence. When 1,1'-dimethoxyuranocene
was treated with a mixture of methyllithium and butyllithium a
mixture of uranocenes resulted which, on controlled air oxidation,
gave a mixture of methyl- and butylcyclooctatetraenes. This
result is consistent with the proposed mechanism in the following
way. Butyllithium can metallate the methoxyuranocene to give the
intermediate cyclooctatrienyne derivative. This intermediate can
now react with either methyllithium or butyllithium to give incorporation of both methyl and butyl groups.

Uranocenes are stable to normal catalytic hydrogenation
conditions; i.e., Pt/H. Thus, 1,1'-divinyluranocene can be
hydrogenated to 1,1'-diethyluranocene.

Electrochemical oxidation of uranocene gives evidence of a
monocation, di(cyclooctatetraene)uranium(V), which is short-lived
and reacts further to give an air-stable cluster (65). A number
of oxygen-containing compounds such as dimethyl sulfoxide, epoxides,
azoxybenzene, etc., do not react rapidly with uranocenes. Nitro
compounds of all types, however, do react with extreme rapidity
to liberate the COT ligand and precipitate UO_2. The nitro compounds are reduced generally to azo compounds and amines, often
in good yield (66). Electron transfer to give an intermediate

U(V) derivative associated with the nitro radical anion is not involved because thorocenes give the same reaction (63). Nitroso compounds are rapidly reduced by uranocenes and are allowable intermediates in the reaction sequence. Of special interest are the further observations that 1-nitro-2,4,6-tri-t-butylbenzene reacts measurably slowly with uranocenes and that the corresponding nitroso compound reacts faster (36, 37). These results strongly suggest that direct reaction occurs at uranium in a reaction of the type:

$$(C_8H_8)_2U \quad + \quad O_2NR \quad \longrightarrow \quad C_8H_8 \quad + \quad (C_8H_8)U{=}O \quad + \quad RNO$$

The intermediate $(C_8H_8)UO$ compound either reacts further or disproportionates.

The general nature of most reactions of uranocenes can be summarized as attack generally at the central metal. In terms of frontier orbital theory this generality can be rationalized by the theoretical consideration that both the HOMO and LUMO are uranium f-orbitals. The distance from the uranium to the plane of the COT ring is 1.92 A; i.e., the rings are 3.84 A apart. Reaction of another group at uranium must tilt or force the rings apart, thereby weakening the ring-metal bonding. In this connection it is significant to consider that the total coordination number afforded a U^{+4} ion by two COT dianions, 20, is less than the coordination number of 24 provided by the 4 Cp rings in UCp_4. Yet, with uranocene substituents such as dimethylaminomethyl or -ethyl there is no evidence for the type of internal coordination depicted in Fig. 3 (33). Uranocenes appear to be coordinatively saturated but for steric rather than electronic reasons.

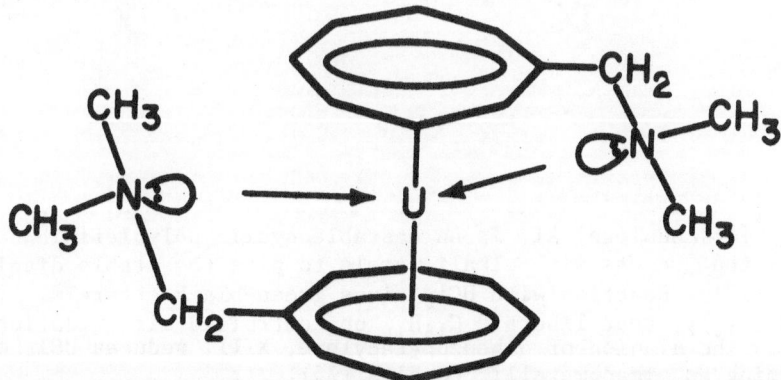

Fig. 3. Aminoalkyl and alkoxyalkyl uranocenes show no evidence of internal coordination of the type illustrated.

RELATED PI-SYSTEMS

The first organoactinide compounds known were of the cyclo-pentadienyl type, Cp_3UCl (2). The chemistry of such compounds is discussed in other papers in this collection. We mention here that these compounds react rapidly with COT dianions to give uranocenes (63).

$$Cp_3UCl \quad + \quad BuC_8H_7^= \quad \longrightarrow \quad (BuC_8H_7)_2U$$

$$Cp_3UBu \quad + \quad BuC_8H_7^= \quad \longrightarrow \quad (BuC_8H_7)_2U$$

The second reaction is slower than the first, consistent with the idea that the U-Bu bond is more covalent than U-Cl.

Other aromatic types of dianions are known and have been allowed to react with UCl . Pentalene forms a relatively stable dianion, IX (68). Reaction with UCl_4 gives a poorly characterized polymeric material stable only at low temperature (69). Similarly, fulvalene dianion, X, reacts with UCl_4 to give a complex product that appears to be more closely related to cyclopentadienyl-uranium compounds than to uranocenes (70).

IX   X

[16]Annulene, XI, is an unstable cyclic polyolefinic compound (71) that reacts with alkali metals to give the stable dianion XII (72). Reaction with UCl_4 gives green-black crystals, $U(C_{16}H_{16})_2$, that liberate $C_{16}H_{16}$ on controlled air oxidation (73, 74). The dianion of dibenzopyracylene, XIII, reduces UCl_4 without forming an organometallic complex (75).

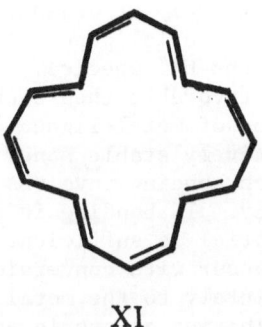

XI

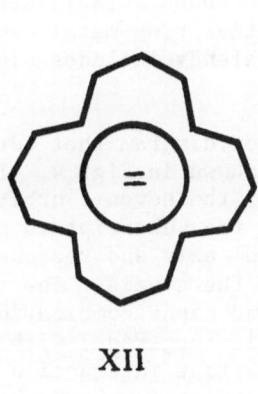

XII

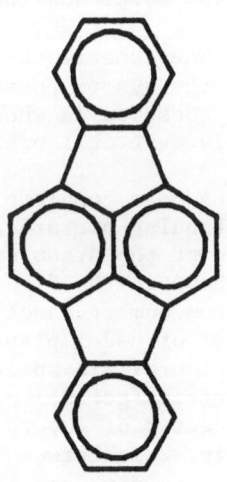

XIII

THEORY

The reactions with UCl_4 of organic pi-dianions having sufficient reduction potential generally involve some degree of electron transfer to the U^{+4} species. If steric and orbital overlap features are favorable then such electron transfer is partial with formation of metal-ligand covalent bonding and the production of a relatively stable sandwich-type compound. In different systems such bonding involves varying degrees of strength and covalency. If bonding is too weak and if the pi-anion reduction potential is sufficiently high, complete electron transfer to U^{+4} can occur with conversion to a lower valent U species or even completely to the metal, $U°$. COT apparently offers the best combination of steric and orbital overlap factors and, of the various pi-dianions examined, forms the most stable and covalent compound with uranium. Present evidence suggests that thorocene is more ionic and, indeed, uranium may form the most stable COT compounds among the actinides. With sufficient electron-donating groups, COT dianion becomes too strongly reducing and reduces the uranium without forming a uranocene. An example is 1,3,5,7-tetra-t-butylcyclooctatetraene (VIII, R=t-Bu). This compound forms a dianion which, on treatment with UCl_4 gives uranium metal and the neutral COT (37).

The general chemistry of uranocenes establishes convincingly that the system possesses substantial ring-metal covalency. The next question is whether such covalency includes significant involvement of f-orbitals.

A more complete MO interaction diagram that corresponds to the bonding picture of Fig. 2 is shown in Fig. 4. In the axial field of the uranocene D_{8h} system, the seven f-orbitals split as shown in the figure; that is, the subscript is an effective quantum number about the z (8-fold) axis and represents the number of nodal planes containing the z-axis. The two sets of COT pi-orbitals split into plus and minus combinations that are symmetric (gerade) or antisymmetric (ungerade) with respect to the center of inversion. The important interaction symbolized in Fig. 2 is between the $f_{\pm 2}$ atomic orbitals and the e_{2u} -combination to give the level scheme shown in Fig. 4. According to this scheme the first three ionization potentials of uranocene would correspond to the f-electrons, e_{2g} and e_{2u}, respectively. A semi-empirical LCAO MO treatment is in substantial agreement with this simple picture (76).

The photoelectron spectrum (PES) of uranocene shows the first three ionization potentials at 6.20, 6.90 and 7.85 eV (77, 78). The PES of thorocene is very similar except that the first peak, attributed to the 5f electrons of U^{+4}, is missing. These spectra were first interpreted in the manner discussed above,

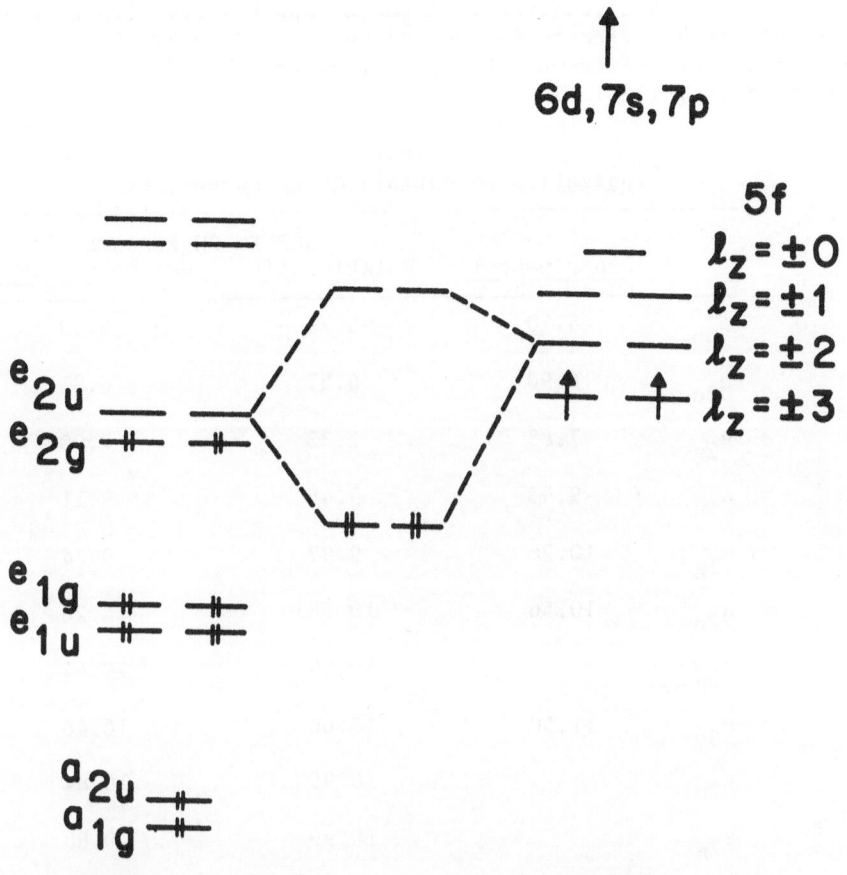

Fig. 4. Interaction diagram between π-MOs of two planar
         cyclooctatetraenes and uranium 5f orbitals in a D_{8h}
         configuration.

that is, I.P. (e_{2g}) > I.P. (e_{2u}). A subsequent reassignment
by Clark and Green (79), based on intensity changes resulting
from change in the energy of the ionizing light, placed I.P. (e_{2g})
< I.P. (e_{2u}) and suggests the significant involvement of metal
d-orbitals. The d-orbitals were specifically neglected in prior
semi-empirical MO treatments.

    A recent SCF-Xα scattered wave MO study has been reported for
uranocene and thorocene (80). The results give excellent agreement between calculated and experimental ionization potentials
(Table 1). The revised assignment of ionization potentials of
Clark and Green is confirmed. Moreover, a detailed study of the
two MO wave functions shows that f-orbital interaction as proposed

originally is indeed important but that d-orbital interaction is at least equally significant. A recent approximate incorporation of relativistic effects changes the calculated ionization potentials somewhat but leaves the overall agreement and conclusions unchanged (81) (Table 1).

Table 1
Ionization Potentials of Uranocene, eV

	Experimental	SCF Xα-SW Results Relativistic	Non-Relativistic
e_{3u}	6.20	5.73	6.30
e_{2u}	6.90	6.57	6.83
e_{2g}	7.85	7.33	7.00
e_{1u}	9.95	9.48	9.11
e_{1g}	10.28	9.97	9.74
a_{2u}	10.56	10.58	10.20
e_{3g}		10.58	10.44
e_{3u}	11.50	10.60	10.46
e_{2u}		10.99	10.84
e_{2g}		10.99	10.83
a_{1g}	12.37	11.69	10.99

PHYSICAL PROPERTIES

a. Visible Spectra

Uranocenes generally show a cascade of bands in the visible spectrum in the 600-700 nm region and with extinction coefficients of about 10^2-10^3. The most intense visible peak of uranocene itself is at 616 nm. Electron-donating groups shift this absorption to longer wavelengths whereas electron-attracting groups produce a hypsochromic shift (Table 2). This result suggests that the electronic transition involves charge transfer from ligand to metal. The corresponding transition in thorocene occurs at shorter wavelength but substituents have a closely comparable effect (63) (Table 2). It is interesting to note that the energy difference between the e_2 orbitals of uranocene and

Table 2
Absorption Spectra of
Bis-cyclooctatetraene actinide Compounds

| $(RC_8H_7)_2M$ | Central Metal M | | | |
R=	U	Th	Np^a	Pu^a
H	616^b	450	518	404
C_2H_5	619			408
$n-C_4H_9$	622	456^c	523	407
$i-C_3H_7$	622			
$CH=CH_2$	617			
C_6H_5	625			
OCH_3	634			
$OC(CH_3)_3$	632			
$N(CH_3)_2$	635			
$CH_2N(CH_3)_2$	619			
$COOC_2H_5$	599			
$1,3,5,7-(CH_3)_4{}^d$	650	480	546	429

(a) Ref. 32.

(b) This is the shortest wavelength and most intense band of a group of 4 bands. The four bands given in nm ($\varepsilon \times 10^{-3}$) in THF are: 616 (1.8), 643 (0.72), 661 (0.41), 671 (0.12).

(c) $\varepsilon = 64$.

(d) Bis-(1,3,5,7-tetramethylcyclooctatetraene)actinide-(IV).

the vacant metal f-levels as given by the SCF-Xα calculations is of the proper magnitude to correspond to such a charge-transfer transition; moreover, the corresponding energy difference in thorocene is sufficiently greater to account for the magnitude of the short wavelength shift of a comparable transition.

b. Magnetic Susceptibilities

Uranocenes are paramagnetic and have magnetic susceptibilities of the order of 2.2-2.7 BM at room temperature, consistent with 2 unpaired 5f electrons. The temperature dependency follows Curie-Weiss behavior generally to about 10K. Below this temperature a number of uranocenes have temperature independent magnetic susceptibilities although some are known that show Curie-Weiss behavior to the limit of the temperature measured, ~2K. This magnetic behavior is discussed in detail elsewhere in this collection.

c. Magnetic Resonance Spectra

As a normal closed-shell system, thorocenes have simple nmr spectra (Table 3). The results are generally consistent with lower effective charge in the ligand compared to the corresponding COT dianions. The ^{13}C magnetic resonance spectra (Table 4) are particularly indicative of substantial charge transfer from ligand to metal. Especially significant in this regard are the p-phenyl shifts in the phenyl-substituted compounds. The relative chemical shifts of the para-carbon and hydrogen have been shown to reflect closely the charge on a benzylic position (82).

Because of their paramagnetism, uranocenes have more complex nmr spectra. The proton magnetic resonances are somewhat broadened but still easily measurable. Their large chemical shifts can generally be interpreted in terms of contact and pseudocontact (dipolar) contributions. With the assumption that the magnetic moment lies entirely along the z- or D_{8h} axis of the uranocene structure (i.e., $\mu_x=\mu_y=0$), the dipolar contribution is completely specified by geometry,

$$\frac{\Delta H}{H_0} = \frac{K(3\cos^2 \theta-1)}{R^3}$$

where H is the shift at an applied magnetic field, H_0 , K is a temperature-dependent constant related to the magnetic susceptibility, θ is the angle between the principal magnetic axis and the vector from uranium to proton and R is the length of this vector.

Ring protons of uranocene itself resonate at about -35 ppm (upfield from TMS) at temperatures close to room temperature. In substituted uranocenes the different ring protons appear generally over the range -27 to-46 ppm (Table 5). The dipolar shift contribution can be estimated to be about -8 ppm and is approximately the same for all of the substituted uranoenes.

Table 3
Proton NMR Spectra of
Cyclooctatetraenes and Thorocenes

| COT Substituent | | δ, ppm from TMS | | |
		COT	Dianion	Thorocene
CH_3	ring:	5.6	5.6	6.5
	CH_3:	1.8	2.8	3.1
t-Bu	ring:	5.7	5.7	6.5
	CH_3:	1.1	1.5	1.7
1,3,5,7-$(CH_3)_4$	ring:	5.4	5.6	6.4
	CH_3:	1.7	2.8	3.0
n-Bu	ring:	5.7	5.7	6.5
	Bu:	2.1	2.9	3.2
		1.4	1.3	1.6
		1.0	0.9	1.0
Ph	ring:	6.0, 5.8	6.2, 5.8	6.8, 6.6
	Ph:	7.2	7.6	7.7
			7.1	7.5
			6.65(para)	7.3(para)

Table 4
^{13}C NMR Spectra of
Cyclooctatetraenes and Thorocenes

| COT Substitutent | | δ, ppm TMS | | |
		COT	Dianion	Thorocene
CH_3	C_1:	141	93	120
	ring:	136–128	92–89	109–106
	CH_3 :	24	34	33
n-Bu	C_1:	145	101	121
	ring:	134–126	92–89	109–106
	Subst:	38	50	49
		31	41	38
		22	24	23
		14	15	14
Ph	C_1:	142	103	121
	ring:	132	94–91	109–108
	p-Ph:	128	119	126

Table 5
Ring Proton NMR of
1,1'-Disubstituted Uranocenes

Substituent	δ, ppm TMS in THF at 39°	
	2H	4H
CH_3	34.0	32.1, 36.6, 41.0
C_2H_5	33.4	31.8, 35.0, 38.1
$CH(CH_3)_2$	35.6	32.2, 37.3, 39.6
$C(CH_3)_3$	39.1	32.6, 34.2, 36.0
OCH_3 [a]	30.2	27.5, 35.6, 43.7
$OC(CH_3)_3$ [a]	28.7	28.1, 36.2, 45.7
$COOC_2H_5$ [b]	36.0	30.4, 33.3, 42.8

(a) in C_6D_6

(b) in C_6D_6 at 27.5°

Hence, the contact shift for the ring protons in uranocene itself
is about -34 ppm relative to the position expected for a hypothetical non-paramagnetic analog (as suggested by thorocene). In
substituted uranocenes the corresponding contact shifts can vary
by ±10 ppm from the parent. The detailed interpretation of these
variations has not been completed but the gross direction of the
contact shift can be rationalized as shown in Fig. 5 (40). In
this figure the arrows indicate directions of magnetic moments.
For f-electrons the magnetic moment associated with the large
orbital angular momentum is alligned with the applied field in
the ground state and dominates the spin contribution which is
alligned against the field as shown in the figure. Electron
density transferred from a ligand pi-MO to any empty uranium
orbital will be spin-polarized so that the opposite moment is
induced in the pi-system. This pi-moment is relayed to ring
protons by a second spin-polarization in the C-H bond so that the
net moment near the proton is opposed to the applied field. A
higher applied field is necessary to achieve resonance, corresponding to an upfield shift.

Proton nmr spectra of substituent protons in some substituted uranocenes are summarized in Table 6. Protons in α-positions
as in methyl, ethyl, etc., uranocenes also experience contact
and pseudo contact contributions. Beyond the α-position proton

π -system

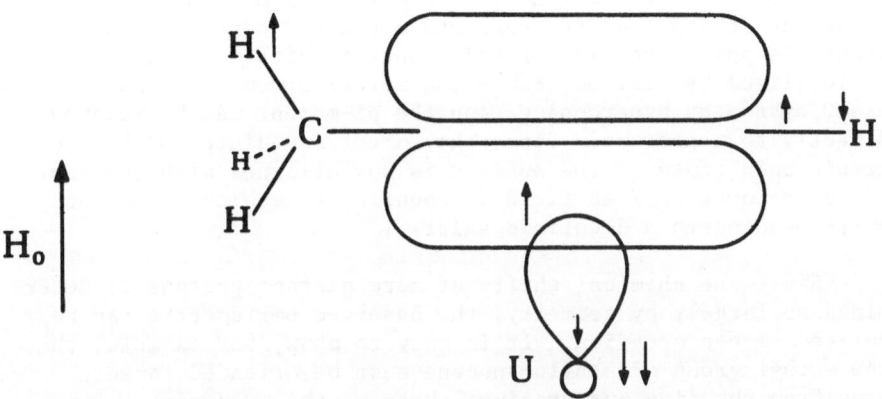

Fig. 5.   Illustrating the contact shift contributions to
ring and methyl proton nmr in uranocenes.

Table 6
Substituent Proton NMR of 1,1'-Disubstituted Uranocenes

Substituent	δ, ppm TMS in THF at 39°
CH_3	−6.52
CH_2CH_3	−15.88 (CH_2), −1.35 (CH_3)
$CH(CH_3)_2$	−13.68 (CH), −9.67 (CH_3, d, J=6Hz)
$C(CH_3)_3$	−10.97
$CH_2C(CH_3)_3$ [a]	−22.97 (CH_2), +3.43 (CH_3)
OCH_3 [a]	−3.73
$OC(CH_3)_3$ [a]	+2.08
$COOC_2H_5$ [b]	−4.39 (CH_2, q, J=7Hz), −6.19 (CH_3, t, J=7Hz)

[a] in C_6D_6.
[b] in C_6D_6 at 27.5°.

contact shifts generally have rapidly diminishing magnitude for saturated substitutents and the pseudocontact contributions dominate. For methyl groups the rotationally averaged pseudo-contact shift is estimated to be about -24 ppm. The observed resonance at about -7 ppm suggests a contact contribution of about +14 ppm. The sign of this contact shift can also be rationalized by Fig. 5. Since the methyl protons are part of the pi-MO system by hyperconjugation the pi-moment can be relayed directly to α-protons. Since the moment associated with electronic spin close to the nucleus is now alligned with the applied field, a lower applied field is required to achieve resonance corresponding to a downfield shift.

Since the chemical shifts of more distant protons is determined so largely by geometry, the observed nmr spectra can be related to conformation. It is easy to show, for example, that the methyl group of ethyluranocene must be oriented largely away from the ring and uranium; that is, the predominant conformation of RCH_2-groups is that shown in Fig. 6.

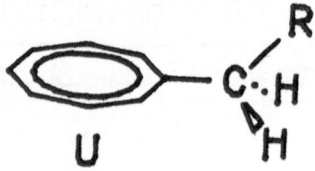

Fig. 6. Predominant conformation of RCH_2 groups in substituted uranocenes.

ACKNOWLEDGEMENTS

It has been my pleasure and good fortune to have been associated in this research with a number of excellent undergraduate, graduate and postdoctoral student researchers. They have contributed novel ideas, experimental skill and rewarding discussions. Most of this research has been supported by the National Science Foundation. Some of the work was carried out at the Lawrence Berkeley Laboratory and supported by the Basic Energy Sciences Division of the Department of Energy. The SCF Xα-SW calculations were supported in part by a NATO grant.

REFERENCES

1. A. Streitwieser, Jr. and U. Muller-Westerhoff, J. Am. Chem.
    Soc., 90, 7364 (1968).

2. L. T. Reynolds and G. Wilkinson, J. Inorg. Nucl. Chem., 2,
    246 (1956).

3. E. O. Fischer and Y. Hristidu, Z. Naturforsch, 17 b, 276 (1962).

4. E. O. Fischer and A. Treiber, Z. Naturforsch, 17b, 276 (1962).

5. R. D. Fischer, Theoret. Chim. Acta (Berl.), 1, 418 (1963).

6. G. L. Ter Haar and M. Dubeck, Inorg. Chem., 3, 1648 (1964).

7. C.-H. Wong, T.-M. Yen and T.-Y. Lee, Acta Cryst., 18, 340
    (1965).

8. F. Baumgartner, E. O. Fischer, B. Kanellakopulos and P.
    Laubereau, Angew. Chem., Internat. Edit., 4, 878 (1965).

9. F. Baumgartner, E. O. Fischer, B. Kanellakopulos and P.
    Laubereau, Angew. Chem., Internat. Edit., 5, 134 (1966).

10. D. G. Wilke, B. Bogdanovic, P. Hardt, P. Heimbach, W. Keim,
    M. Kroner, W. Oberkirch, K. Tanaka, E. Steinrucke, D. Walter,
    and H. Zimmerman, Angew. Chem., Internat. Edit., 5, 151 (1966).

11. A. F. Reid and P. C. Wailes, Inorg. Chem., 5, 1213 (1966).

12. F. Baumgartner, E. O. Fischer, B. Kanellakopulos and P.
    Laubereau, Angew Chem., Internat. Edit., 7, 634 (1968)

13. F. Calderazzo, Organometal. Chem. Rev. B., 6, 997 (1970).

14. R. G. Hayes and J. L. Thomas, Organometal. Chem. Rev. A,
    7, 1 (1971).

15. F. Calderazzo, Organometal. Chem. Rev. B, 9, 131 (1972).

16. G. T. Seaborg, Pure and Appl. Chem., 30, 539 (1972).

17. B. Kanellakopulos and K. W. Bagnall, "The Organometallic
    Chemistry of the Lanthanides and Actinides," MTP Interna-
    tional Review of Science, Inorg. Chem., Series One, 7,
    'Lanthanides and Actinides,' K. W. Bagnall (ed.),
    Butterworths, London, 1970.

18. F. Calderazzo, J. Organometal. Chem., 53, 173 (1973).

19. A. Streitwieser, Jr., "Uranocene - an f-Orbital Aromatic System?" Topics in Nonbenzenoid Aromatic Chemistry, 1, p. 221, T. Nozoe, R. Breslow, K. Hafner, S. Ito, I. Murata, eds., Hirokawa Publishing Co., Inc., Tokyo, 1971.

20. F. Calderazzo, J. Organometal. Chem., 79, 175 (1974).

21. T. J. Marks, J. Organometal. Chem., 79, 181 (1974).

22. E. Cernia and A. Mazzei, Inorgan. Chim. Acta, 10, 239 (1974).

23. T. J. Marks, J. Organometal. Chem., 95, 301 (1975).

24. T. J. Marks, J. Organometal. Chem., 119, 229 (1976).

25. Nachr. Chem., techn., 24, 313 (1976).

26. E. C. Baker, G. W. Halstead and K. N. Raymon, "The Structure and Bonding of 4f and 5f Series Organometallic Compounds," Structure and Bonding, J. D. Dunitz, P. Hemmerich, R. H. Holm, J. A. Ibers, C. K. Jorgensen, J. B. Neilands, D. Reinen and R. J. P. Williams, eds.; Springer-Verlag Berlin Heidelberg, New York, 1976.

27. T. J. Marks, Accts. Chem. Res., 9, 223 (1976).

28. A. Zalkin and K. N. Raymond, J. Am. Chem. Soc., 91, 5667 (1969).

29. A. Avdeef, K. N. Raymon, K. O. Hodgson and A. Zalkin, Inorg. Chem., 11, 1083 (1972).

30. C. A. Harmon and A. Streitwieser, Jr., J. Am. Chem. Soc., 94, 8926 (1972).

31. A. Streitwieser, Jr. and C. A. Harmon, Inorg. Chem., 12, 1102 (1973).

32. D. G. Karraker, Inorg. Chem., 12, 1105 (1973).

33. C. A. Harmon, D. P. Bauer, S. R. Berryhill, K. Hagiwara and A. Streitwieser, Jr., Inorg. Chem., 16, 2143 (1977).

34. D. G. Morrell, Dissertation, Univ. of Calif., 1973.

35. L. S. Hillard, unpublished results.

36. H. Burghard, unpublished results.

37. M. J. Miller, unpublished results.

38. S. R. Berryhill, A. Streitwieser, Jr., and W. D. Luke, Inorg. Chem., submitted.

39. W. D. Luke, unpublished results.

40. A. Streitwieser, Jr., D. Dempf, G. N. La Mar, D. G. Karraker and N. Edelstein, J. Am. Chem. Soc., 93, 7343 (1971).

41. A. Streitwieser, Jr. and R. Walker, J. Organometal. Chem., 97, C41 (1975).

42. J. H. Noordik, Th. E. M. van den Hark, J. J. Mooij and A. A. K. Klaasen, Acta Cryst., B30, 833 (1974).

43. S. Z. Goldberg, K. N. Raymond, C. A. Harmon and D. H. Templeton, J. Am. Chem. Soc., 96, 1348 (1974).

44. A. Streitwieser, Jr., U. Müller-Westerhoff, G. Sonnichsen, F. Mares, D. G. Morrell, K. O. Hodgson and C. A. Harmon, J. Am. Chem. Soc., 95, 8644 (1973).

45. A. Streitwieser, Jr., U. Müller-Westerhoff, F. Mares, C. B. Grant and D. G. Morrell, Inorg. Synth., in press.

46. D. F. Starks and A. Streitwieser, Jr., J. Am. Chem. Soc., 95, 3423 (1973).

47. G. Schröder, Cyclooctatetraene, Verlag Chemie, Weinheim (1965).

48. L. A. Paquette, Tetrahedron, 31, 2855 (1975).

49. G. I. Fray and R. G. Saxton, The Chemistry of Cyclooctatet-raene and its Derivatives, Cambridge Univ. Press, New York (1978).

50. J. Gasteiger, G. Gream, R. Huisgen, W. Konz and U. Schnegg, Chem. Ber., 104, 2412 (1971).

51. C. A. Harmon and A. Streitwieser, Jr., J. Org. Chem., 38, 549 (1973).

52. A. Cope and M. Kinter, J. Am. Chem. Soc., 73, 3424 (1951).

53. A. Cope and H. van Orden, J. Am. Chem. Soc., 74, 175 (1952).

54. A. C. Cope and H. C. Campbell, J. Am. Chem. Soc., 73, 3536 (1951).

55. S. W. Staley, G. M. Cramer and A. W. Orvedal, J. Am. Chem. Soc., 96, 7433 (1974).

56. P. de Mayo and R. W. Yip, Proc. Chem. Soc., 84 (1964).

57. E. H. White and H. C. Dunathon, J. Am. Chem. Soc., 86, 453 (1964).

58. A. Streitwieser, Jr. and R. Walker, J. Organometal. Chem., 97, C41 (1975).

59. L. K. Templeton, D. H. Templeton and R. Walker, Inorg. Chem., 15, 3000 (1976).

60. F. Mares, K. O. Hodgson and A. Streitwieser, Jr., J. Organometal. Chem., 24, C68 (1970).

61. K. O. Hodgson, F. Mares, D. F. Starks and A. Streitwieser, Jr., J. Am. Chem. Soc., 95, 8650 (1973).

62. F. Mares, K. O. Hodgson and A. Streitwieser, Jr., J. Organometal Chem., 28, C24 (1971).

63. C. LeVanda, unpublished results.

64. D. G. Karraker and J. A. Stone, J. Am. Chem. Soc., 96, 6885 (1974).

65. J. A. Butcher, Jr., J. Q. Chambers and R. M. Pagni, J. Am. Chem. Soc., 100, 1012 (1978).

66. C. B. Grant and A. Streitwieser, Jr., J. Am. Chem. Soc., 100, 2433 (1978).

67. S. Gabriel, unpublished results.

68. T. J. Katz and M. Rosenberger, J. Am. Chem. Soc., 84, 865 (1962); T. J. Katz, M. Rosenberger and R. K. O'Hara, ibid, 86, 249 (1964).

69. S. Mylonakis, unpublished results.

70. J. Cambray, Dissertation, Univ. of Calif., 1974.

71. F. Sondheimer and T. Gaoni, J. Am. Chem. Soc., 83, 4863 (1961).

72. J. F. Oth, G. Anthoine and J. M. Gilles, Tetrahedron Letters, 6265 (1968).

73. C. A. Harmon, Dissertation, Univ. of Calif., 1975.

74. S. R. Berryhill, Dissertation, Univ. of Calif., 1977.

75. C. M. Berke, unpublished results.

76. R. G. Hayes and N. Edelstein, J. Am. Chem. Soc., 94,
     8688 (1972).

77. J. P. Clark and J. C. Green, J. Organometal. Chem., 112,
     C14 (1976).

78. I. Gragala, G. Condorelli, P. Zanella and E. Tondello,
     VIIth Intern. Conf. Organometal. Chem., Sept. 1975, p. 3.

79. J. P. Clark and J. C. Green, J.C.S. Dalton Trans., 505 (1977).

80. N. Rösch and A. Streitwieser, Jr., J. Organometal. Chem.,
     145, 195 (1978).

81. N. Rösch and A. Streitwieser, Jr., unpublished results.

82. G. L. Nelson and E. A. Williams, Prog. Phys. Org. Chem.,
     12, 229 (1976).

KINETICALLY-STABLE LANTHANIDE METAL ALKYLS AND BRIDGING METHYLS

J. Holton and M. F. Lappert[*]

School of Molecular Sciences, University of Sussex,
Brighton BN1 9QJ, Great Britain

D. G. H. Ballard and R. Pearce

Imperial Chemical Industries Ltd., Corporate Laboratory,
P. O. Box 11, The Heath, Runcorn, Cheshire WA7 4QE

J. L. Atwood and W. E. Hunter

Department of Chemistry, The University of Alabama,
Alabama 45386, U. S. A.

SUMMARY

This paper presents results on three topics concerning synthetic,
structural, and chemical studies relating to three new classes of lanthanide
metal (M) alkyl: (i) methyl-bridged complexes involving the moiety
$M\langle^{Me}_{Me}\rangle AlMe_2$; (ii) methyl-bridged complexes involving the unit $M\langle^{Me}_{Me}\rangle M$;
and (iii) neutral, $[MR_3(thf)_2]$, and anionic, $[MR_4]^-$ or $[M(Cl)R'_3]^-$, silylmethyl
$[R = Me_3SiCH_2, R' = (Me_3Si)_2CH]$ complexes. Topic (iii) is not described
herein (but was included in the lecture) (the reader is referred to the
paper by J. L. Atwood, W. E. Hunter, R. D. Rogers, J. Holton, J. McMeeking,
R. Pearce, and M. F. Lappert, J. C. S. Chem. Comm., 1978, 140). (Also
covered in the lecture were results on alkyls of the early transition elements.)

Reaction of $[\{Mn(\eta-C_5H_5)_2Cl\}_2]$ with $Li[AlR_4]$ or in some cases
$Mg[AlR_4]_2$, affords the novel crystalline complexes $[M(\eta-C_5H_5)_2R_2AlR_2]$
(M = Sc, Y, Gd, Dy, Ho, Er, Tm, or Yb, with R = Me; or M = Sc, Y,
or Ho, with R = Et). The yttrium, unlike the scandium, tetra-alkylaluminates are fluxional at +40° C, but at -40° C bridging and terminal

---

[*]The lecture in Sogesta was by M. F. Lappert

T. J. Marks and R. D. Fischer (eds.), Organometallics of the f-Elements, 179–220.
All Rights Reserved. Copyright © 1979 by D. Reidel Publishing Company, Dordrecht, Holland.

alkyl groups give distinct n.m.r. signals; $\Delta \underline{G}^{\ddagger}_{392K}$ for site exchange in
$[Y(\eta-C_5H_5)_2Me_2AlMe_2]$ is 15.9 kcal mol^{-1}. A di-μ-alkyl-bridged structure
was confirmed by i.r. (bridging CH_3 band at 1250 and 1235 cm^{-1}), variable
temperature ^1H and ^{13}C n.m.r. (M = Sc or Y) and $\underline{X}$-ray (M = Yb) studies.
Additional data are given on the less stable $[Ti(\eta-C_5H_5)_2AlMe_4]$,
$[Ti(\eta-C_5H_5)_2AlMe_3Cl]$, and $[Ti(\eta-C_5H_5)_2H_2AlMe_2]$ (structure deduced in part
from e.s.r. spectra). A single crystal $\underline{X}$-ray analysis of $[Yb(\eta-C_5H_5)_2Me_2-$
$AlMe_2]$ has been carried out to $\underline{R}_1 = 0.036$ and $\underline{R}_2 = 0.042$; the complex has
an approximately tetrahedral Yb and Al environment (space group $\underline{Pn\,a\,2_1}$)
with the $YbMe_2Al$ unit strikingly similar to $AlMe_2Al$ in Al_2Me_6. Important
bond lengths (Å; t = terminal, b = bridge) and angles (^{0}) are: Yb-C
(cyclopentadienyl, average), 2.61(3); Yb-C_b, 2.59(3); Al-C_b, 2.14(5);
Al-C_t, 2.00(1), and Yb$\hat{C}$Al, 78.9(1.6); and C$\hat{A}$lC, 113.3(8).

The equimolar reaction of $[M(\eta-C_5H_5)_2Me_2AlMe_2]$ (M = Y, Dy, Ho, Er,
Tm or Yb) and pyridine gave a series of complexes $[\{M(\eta-C_5H_5)_2Me\}_2]$
(M = Y, Dy, Ho, Er, Tm, or Yb), assigned a dimeric structure on the
basis of ^1H and ^{13}C n.m.r. (M = Y) and i.r. data, and by crystal and
molecular structural determinations of $[\{M(\eta-C_5H_5)_2Me\}_2]$ (M = Y, Yb).
A similar reaction of $[Sc(\eta-C_5H_5)_2Me_2AlMe_2]$ with a Lewis base L gave
$[Sc(\eta-C_5H_5)_2(Me)L]$ (L = py or thf). Other reactions described are of
$[\{Y(\eta-C_5H_5)_2Me\}_2]$ with Lewis bases (an amine, phosphine, or phosphine
oxide) or with a Lewis acid ($Al_2Me_nCl_{6-n}$; n = 2, 4, or 6). A single
crystal $\underline{X}$-ray analysis of the isostructural $[\{M(\eta-C_5H_5)_2Me\}_2]$ has been
carried out to $\underline{R}_1 = 0.048$ (Y) or 0.066 (Yb) and $\underline{R}_2 = 0.055$ (Y) or
0.061 (Yb); the complexes have an approximately tetrahedral metal
environment (space group $\underline{P}2_1/\underline{n}$) with the YMe_2Y unit in the yttrium
compound remarkably similar to $AlMe_2Al$ in Al_2Me_6. Important
average bond lengths (Å) and angles for the yttrium compound (Yb in
parentheses) are: Y-C (cyclopentadienyl), 2.655(18) [2.613(13)];
Y-CH_3, 2.545(11) [2.511(35)]; Y$\hat{C}$Y, 87.7(3) [86.6(3)]; and $H_3C\hat{Y}CH_3$,
92.3(3) [93.4(4)].

The neutral $[M(CH_2SiMe_3)_3(thf)_2]$ (M = Tb, Er, or Yb; thf = tetrahydrofuran) and the ionic $[Li(L)_4][MR_4]$ [R = Me_3SiCH_2; M = Y, Er, or Yb, and $L_2 = Me_2NCH_2CH_2NMe_2$ (TMED); M = Y, L = thf], $[Li(thf)_4][M(Cl)R'_3]$ [M = Er or Yb, R' = $(Me_3Si)_2CH$], and $[LiL_2]$-$[M(\eta-C_5H_4SiMe_3)_2Cl_2]$ (L = $\frac{1}{2}$TMED or thf) are reported; their structures are deduced from i. r., ^{13}C, and ^1H n. m. r. (M = Y) spectra, conductivity, and by a single crystal $\underline{X}$-ray analysis, [l(Yb-C) (av.) = 2.38 $\overset{\circ}{A}$] of $[Li(thf)_4][Yb(Cl)R'_3]$.

DI-η-CYCLOPENTADIENYLMETAL(III) TETRA-ALKYLALUMINATES $[M(\eta-C_5H_5)_2R_2AlR_2]$ (M = Ti, Sc, Y, Gd, Dy, Ho, Er, Tm, or Yb, WITH R = Me; or M = Sc, Y, or Ho, WITH R = Et), AND THE CRYSTAL AND MOLECULAR STRUCTURE OF $[Yb(\eta-C_5H_5)_2Me_2AlMe_2]$

Structural inorganic chemistry at the molecular level is conventionally described in terms of the central metal ion(s) and the associated ligands. The latter are generally held to occupy either terminal or, in di- or poly-nuclear complexes, bridging positions. Hydride or hydrocarbyl (R$^-$) ligands, being free from electrons in non-bonding orbitals, may only enter into bridging situations by means of electron-deficient bonding, $\underline{i.\,e.}$, where the number of constituent atomic orbitals exceeds the number of available electrons. For this reason, electron-deficient molecules are often en-countered among hydrides or alkyls of the $\underline{s}$-block and main group 3 elements, which have available energetically low-lying vacant $\underline{p}$-orbitals.

The formation of electron-deficient alkyl bridges in di- or poly-nuclear metal complexes is well-established for the light $\underline{s}$-block (Li, Be, Mg), aluminium, and (in the solid state) the heavier group 3b elements.[1] That this is not an exclusive property of these metals has been made clear in some recent reports of transition-metal complexes which contain alkyl bridges between adjacent metal atoms. Indeed, such a chemistry may become general within the transition-metal series. The relevant bonding problems, in $\underline{e.\,g.}$, $[M_2L_6]$ bridged-metal complexes, have been discussed.[2]

The present collaboration derives from the complementary interests of our three groups in polymerisation catalysts (Runcorn) and the chemistry (Sussex) and structural characterisation (Alabama) of transition-metal alkyls. The initial impetus was provided by the discovery of the novel methyl-bridged titanium- and yttrium-aluminium compounds $[M(\eta-C_5H_5)_2MeAlMe_2]$.[3a] This new class of electron-deficient, early transition-metal alkyl has been much extended to include scandium and the later lanthanide metals and an X-ray analysis of two compounds (M = Y or Yb) has established the alkyl-bridged structure.[3b]

Stimulated by this discovery and of the first structurally-authenticated electron-deficient transition-metal alkyl, $[\{Cu(CH_2SiMe_3)\}_4]$,[4] we considered that such compounds may well feature for a wider range of d- and f- block metal complexes, because these metals also have energetically-accessible vacant atomic orbitals. A principal objective of this series is to examine this hypothesis.

We recognise electron-deficient alkyl bridges to be of two types: either single- (1) or double- (2), where, respectively, one or two alkyl groups join a pair of adjacent metal atoms. Such bonding may be homo- (M = M') or hetero- (M ≠ M'). Bridges of higher order, as in $[(LiMe)_4]$, where a methyl group spans three metal atoms,[5] have not been reported in the transition-metal series. Examples of

M—R—M'   M⟨R⟩M'

(1)   (2)

singly-bridged alkyls are authenticated (X-ray) for $[\{Cu(CH_2SiMe_3)\}_4]$,[4] but may be present in $[Re(CH_2SiMe_3)_4]$ if this [6a] proves to have a cluster structure. Double bridges occur more extensively. Examples of homometallic complexes are to be found in the well-characterised $[\{M(\eta-C_5H_5)_2Me\}_2]$ (M = Y, Dy, Ho, Er, or Yb),[7] $[(MnR_2)_x]$ (R = Me_3CCH_2, x ≈ 4; R = $PhMe_2CCH_2$, x ≈ 2; or R = Me_3SiCH_2, x ≈ n),[6b] $[Cr_2(CH_2SiMe_3)_4(PMe_3)_2]$,[6c] and $[\{Ni(\eta-allyl)Me\}_2]$ (allyl = C_3H_5, C_4H_7, or C_5H_9);[8] and may occur in the less well-characterised $[\{Zr(\eta-C_5H_5)_2Me\}_2]$.[9] Examples of heterometallic bridges are in $[M(\eta-C_5H_5)_2-\mu-R_2AlR_2]$ (M= Ti, Sc, Y, Gd, Dy, Er, Ho, Tm, or Yb, and R ≈ Me; M ≈ Sc or Y, and R = Et),[3] and $[Ni(\eta-allyl)Me_2AlMe_2]$ (allyl = C_3H_5, C_4H_7, or C_5H_9).[8]

Doubly-bridged species containing only one alkyl group (e.g.,
μ-halogeno-μ-alkyl) are exceedingly rare, but may occur in [Ti(η-C$_5$H$_5$)$_2$AlMe$_3$Cl]
(this work and ref. 10) and [Ni(η-allyl)MgMe$_2$Cl(OEt$_2$)],[8b] and are commonly
proposed as intermediates in halide/alkyl exchange reactions.[11]   Other bridges,
e.g., di-μ-halide or -alkoxy, are, of course, well-known,[12] but may be
described as electron-precise, in contrast to the electron-deficient (cf., also
H$^-$ as a bridging ligand) types under discussion.

Bridges involving hydrocarbyl groups other than alkyl are known
also: these have included the singly-bridging (i) aryl ligand, e.g.,
the homometallic [{Cu(C$_6$H$_4$CH$_2$NMe$_2$-o)}$_4$][13] and the heterometallic
[M$_2$Li(C$_6$H$_4$CH$_2$NMe$_2$-o)$_4$] (M = Cu,[13b] Ag,[13c] or Au[13d]) or [Au$_2$Zn$_2$Ph$_6$];[13e]
(ii) another arene-derived ligand, e.g., the bridging benzyne in
[Os$_3$(CO)$_7$(PPh$_2$)C$_6$H$_4$)],[14] (iii) a cyclopentadienyl-derived ligand, e.g.,[15]
in [Fe(η-C$_5$H$_5$)(η-C$_5$H$_4$)Au$_2$(PPh$_3$)$_2$]$^+$; and doubly-bridged complexes,
e.g., μ-(aryl)$_2$ as in [{Ti(η-C$_5$H$_5$)$_2$Ph}$_2$],[16] or μ-(alkynyl)$_2$ as in
[Sc(η-C$_5$H$_5$)$_2$C$\equiv$CPh$_2$].[17]

There has been much speculative discussion about bridges of these
types in compounds considered to be intermediates in Ziegler-Natta catalysis.
They may serve a number of possible roles: alkylating, stabilising, or
solvating the metal centre.[18]

In this paper we give full details (see ref. 3 for preliminary
communications) of the preparation and properties of [M(η-C$_5$H$_5$)$_2$(μ-R)$_2$AlR$_2$]
(M = Ti, Sc, Y, Gd, Dy, Ho, Er, Tm, or Yb) and some related Ti^{III}
complexes and X-ray structural data for the complexes where M = Yb
(the Y structure is discussed elsewhere[19]). These transition-metal-
aluminium species, especially where M = Ti, serve as useful models
for hypothetical doubly-alkyl-bridged intermediates in Ziegler-Natta
catalyst systems.

Results and discussion

Preparation and properties. The reaction of lithium tetra-alkylaluminates with the appropriate chlorodi-η-cyclopentadienyl-lanthanide, in toluene, gave high yields of the group 3 metal tetra-alkylaluminate, equation (1).

$$[\{M(\eta-C_5H_5)_2Cl\}_2] + 2Li[AlR_4] \longrightarrow 2[M(\eta-C_5H_5)_2R_2AlR_2] + 2LiCl \qquad (1)$$

(M = Sc, Y, Gd, Dy, Ho, Er, Tm, or Yb, with R = Me; or
M = Sc, Y, or Ho, with R = Et)

The air-sensitive complexes were, in many cases, highly coloured (see Table 1) and were obtained as needles from toluene-hexane mixtures at -30^0 C.   They were soluble in toluene, benzene, or methylene chloride, but insoluble in saturated hydrocarbon solvents (the ethyl complexes were slightly soluble).   The early lanthanide complexes (Sm and Gd) were found to be insoluble in toluene and, for this reason, are thought to be more ionic.   An increase in ionicity from right to left across the lanthanide series may be related to the effects of the lanthanide contraction.   With a decrease in size of the lanthanide(III) ion from left to right there is an increase in the polarising power (polarising power α $\frac{\text{charge}}{\text{radius}}$; charge = 3+) and hence an increase in metal-ligand covalent character.

[Gd(η-C$_5$H$_5$)$_2$Me$_2$AlMe$_2$] was recrystallised from methylene chloride, although only as a microcrystalline material.   The yellow samarium compound was insoluble in methylene chloride and was not isolated free of LiCl; hence it has not been fully characterised.

The tetramethylaluminates were thermally very stable, melting, in most cases, without decomposition (Table 1).   The exceptions are [Tm(η-C$_5$H$_5$)$_2$Me$_2$AlMe$_2$], which decomposed above 130^0 C, and [Gd(η-C$_5$H$_5$)$_2$Me$_2$AlMe$_2$], which did not melt but slowly decomposed above 170^0 C.   The high value for the latter may again be a consequence of an increase in ionicity.   The tetra-ethylaluminates were much less

Table 1

Analytical data, yields, colour, and melting points for [M(η-C₅H₅)₂AlR₄] and [Ti(η-C₅H₅)₂AlMe₃Cl]

Compound	Yield (%)	Colour	M.p. (°C)	Found (required) (%)			
				C	H	M	Al
[Sc(η-C₅H₅)₂AlMe₄]	67	Pale-yellow	108-110 [a]	63.8 (64.1)	8.4 (8.5)	17.3 (17.4)	10.1 (10.3)
[Y(η-C₅H₅)₂AlMe₄]	78	Colourless	143 [b]			28.6 [c] (29.0)	9.0 (8.8)
[Gd(η-C₅H₅)₂AlMe₄]	45	White	> 170 dec	44.4 (44.9)	6.2 (5.9)	42.9 (42.8)	6.9 (7.1)
[Dy(η-C₅H₅)₂AlMe₄]	65	Pale-yellow	145-146	43.6 (44.3)	5.8 (5.8)	43.7 (43.2)	6.8 (7.1)
[Ho(η-C₅H₅)₂AlMe₄]	78	Straw	142-143	44.0 (44.0)	5.9 (5.8)	43.6 (43.5)	7.0 (7.0)
[Er(η-C₅H₅)₂AlMe₄]	76	Pink	133-135	43.2 (43.7)	5.8 (5.8)	43.6 (43.5)	7.0 (7.0)
[Tm(η-C₅H₅)₂AlMe₄]	60	Pale-green	> 130 dec	43.6 (43.5)	5.9 (5.7)		
[Yb(η-C₅H₅)₂AlMe₄]	72	Orange-red	133-135	42.8 (43.1)	5.6 (5.7)	45.1 (44.3)	6.8 (6.9)
[Sc(η-C₅H₅)₂AlEt₄]	62	Pale-yellow		69.8 (67.9)	9.5 (9.5)		
[Y(η-C₅H₅)₂AlEt₄]	70	Colourless		59.9 (59.7)	8.4 (8.35)	24.2 (24.5)	6.8 (7.45)
[Ho(η-C₅H₅)₂AlEt₄]	73	Straw		49.2 (49.3)	7.0 (6.9)		
[Ti(η-C₅H₅)₂AlMe₄]	58	Deep-green				17.8 [c] (18.1)	9.9 (10.15)
[Ti(η-C₅H₅)₂AlMe₃Cl]	46	Deep-green				16.5 [d] (16.8)	9.6 (9.4)

[a] Sublimation 100°C/0.1 mmHg.   [b] Sublimation 120°C/0.05 mmHg.   [c] Li and Cl were absent

[d] Additionally, Found: Cl, 12.4%.   Required: Cl 12.4%.

stable, slowly decomposing at room temperature, in solution or in
the solid state.   The scandium or yttrium compounds $[M(\eta-C_5H_5)_2Me_2AlMe_2]$
can be purified by sublimation at 100^0 C/0.1 mmHg and 120^0 C/0.05 mmHg,
respectively.   By contrast, attempts to purify $[Gd(\eta-C_5H_5)_2Me_2AlMe_2]$ in
a similar manner resulted in isolation of $[Gd(\eta-C_5H_5)_3]$.

Attempts to prepare $[Y(\eta-C_5H_5)_2Me_2InMe_2]$ by a similar procedure to
that shown in equation (1) were unsuccessful, $[\{Y(\eta-C_5H_5)_2Me\}_2]$ being isolated.
This is presumably formed by loss of $InMe_3$ from the tetramethylindate.
Unlike aluminium, the other group 3 elements are not prone to forming
strong electron-deficient bridges in their alkyls, e.g., BMe_3 and $GaMe_3$ are
both monomeric; and $InMe_3$, although a weakly-bound tetramer in the
solid state, is monomeric in solution.[1]   Although boron alkyls do not
contain electron-deficient bridges, they differ in this respect from many
boron hydrides; it is interesting, therefore, that recently several
lanthanide tetrahydroborates $[M(\eta-C_5H_5)_2BH_4(thf)]$ (M = Sm, Er, or Yb)
and $[M(\eta-C_5H_5)_2BH_4]$ (M = Er or Yb) have been prepared.[20]

The titanium(III) complex, $[Ti(\eta-C_5H_5)_2Me_2AlMe_2]$, was prepared
in a similar manner using either $[Li(AlMe_4)]$ or $[Mg(AlMe_4)_2]$ as alkylating
agent.   It was markedly less stable than the lanthanide complexes,
decomposing slowly in the solid state to an unidentified purple oil, or
rapidly in toluene to a purple solution at ambient temperature.   Somewhat
surprisingly, stability in solution was enhanced by addition of Al_2Me_6;
at ca. 0^0 C there was little apparent decomposition over 1 h in the presence
of 0.5 to 5 mol of Al_2Me_6 per mol of the titanium compound.   This con-
trasts with other observations on transition-metal alkyls where thermal
stability decreased in the presence of organoaluminium compounds.[21]
The origin of this stabilisation may lie in affecting the equilibrium between
the tetramethylaluminate and its presumed dissociation products
$[\{Ti(\eta-C_5H_5)_2Me\}_2]$ and Al_2Me_6.   The homonuclear complex $[\{Ti(\eta-C_5H_5)_2Me\}_2]$
is probably much less stable than $[Ti(\eta-C_5H_5)_2Me_2AlMe_2]$ and decomposes
irreversibly at the temperatures studied.   There is evidence that this

methyl-titanium(III) compound exists at low temperature: thus a soluble green species is formed at -70^0 C in the reaction of $[\{Ti(\eta-C_5H_5)_2Cl\}_2]$ with methyl-lithium in diethyl ether, but attempts at its isolation have been singularly unsuccessful.[22]

Interaction of $[\{Ti(\eta-C_5H_5)_2Cl\}_2]$ and Al_2Me_6 (slight excess) in toluene gave the adduct $[Ti(\eta-C_5H_5)_2AlMe_3Cl]$ which may be identical to a complex of the same formula reported earlier by Natta et al.[10] Its stability was similar to that of the Ti^{III} tetramethylaluminate.

Attempts to prepare the related $[Y(\eta-C_5H_5)_2AlMe_3Cl]$, for which ^{1}H and [13]C n. m. r. data would be obtainable and from which an assignment of the μ-chloro-μ-methyl or di-μ-methyl structure would be possible, were unsuccessful; $[Y\{(\eta-C_5H_5)_2Cl\}_2]$ was recovered unchanged after heating to ca. 100^0 C with Al_2Me_6 (excess) in toluene; and interaction of $[\{Y(\eta-C_5H_5)_2Me\}_2]$ and $Al_2Me_4Cl_2$ in benzene produced the insoluble $[\{Y(\eta-C_5H_5)_2Cl\}_2]$.[23] Lack' of success in the preparation of the yttrium complex may merely reflect the relative insolubility of the yttrium(III) chloro- complex. Experiments are in hand to examine the preparation of species containing mixed bridges, e. g., μ-halide-μ-alkyl, using cyclopentadienyl complexes with solubilising substituents on the cyclopentadienyl ring.

Reaction of $[\{Ti(\eta-C_5H_5)_2Cl\}_2]$ with $Na[AlH_2Me_2]$ in toluene proceeded less readily than with $[Li(AlMe_4)]$. After stirring overnight at ambient temperature, a purple solution was obtained, which contained a mixture of products, the major component of which was identified by e. s. r. (vide infra) as $[Ti(\eta-C_5H_5)_2H_2AlMe_2]$. Pure material was not obtained.

A comparative reaction between the titanium(IV) complex $[Ti(\eta-C_5H_5)_2Cl_2]$ and $Li[AlMe_4]$ (excess) was examined. Free Al_2Me_6 and the known[24] $[Ti(\eta-C_5H_5)_2Me_2]$ were obtained. Thus $[Li(AlMe_4)]$ acts as a simple alkylating agent and shows no tendency to act as a reducing agent to give a Ti^{III} product. Transient methyl-bridged titanium(IV)-aluminium

complexes may be present in solution, but it is to be expected that the coordinatively saturated $[Ti(\eta-C_5H_5)_2Me_2]$ will show a lesser tendency to
form methyl bridges to aluminium than the unsaturated titanium(III)
complex, $[\{Ti(\eta-C_5H_5)_2Me\}_n]$.

A number of reactions of $[Li(AlMe_4)]$ or $[Mg(AlMe_4)_2]$ with the cyclo-
pentadienyl-free $[Ti(CH_2Ph)_3Cl]$ and $[Hf(\eta-C_3H_5)_3Br]$ were studied. These
reactions appeared to give a mixture of products; however, from
$[Hf(\eta-C_3H_5)_2Br]$ and $[Mg(AlMe_4)_2]$ the major product was tentatively
identified as $[Hf(\eta-C_3H_5)_3Me]$.

<u>Spectroscopic characterisation.</u> Trivalent lanthanide complexes,
except those of lanthanum ($\underline{f}^0$) and lutetium ($\underline{f}^{14}$), are paramagnetic, often
giving large contact shifts in the n.m.r. spectra.[25] ^1H and ^{13}C n.m.r.
investigations were therefore concentrated on the diamagnetic scandium and
yttrium complexes.

N.m.r. data for solutions of the scandium or yttrium compounds are
consistent with the presence of a μ-dialkyl bridge between aluminium and
the group 3a metal (Table 2). The complex $[Y(\eta-C_5H_5)_2Me_2AlMe_2]$ and the
ethyl congener are fluxional at 40^0 C, the ^1H n.m.r. spectra showing for the
methyl complex a broad singlet and for the ethyl analogue two broad peaks
in the high field region. On cooling to -40^0 C, a well-resolved spectrum
is obtained, showing two different alkyl environments, bridging and terminal.
The bridging-alkyl groups are observed coupled to yttrium (^{89}Y, monoisotopic,
spin = $\frac{1}{2}$), characterised as a doublet for the methyl complex $[^2\underline{J}(^{89}Y-C^1H_3)$,
5.0 Hz] (Figure 1) and an eight line multiplet for the ethyl analogue due to
coupling to the methyl protons as well as yttrium $[^2\underline{J}(^{89}Y-C^1H_2)$ 4.0 Hz]
(Figure 2). The scandium complexes $[Sc(\eta-C_5H_5)_2R_2AlR_2]$ (R = Me or Et)
are non-fluxional at room temperature and two distinct alkyl environments
are observed. The bridge methyl and methylene protons were, in each
case, broad due to unresolved coupling to the quadrupolar scandium ($I = \frac{7}{2}$).

From line-shape ^1H n.m.r. studies at different temperatures for
$[Y(\eta-C_5H_5)_2Me_2AlMe_2]$ a value for the free energy of activation for the

Table 2

^{1}H n.m.r. data[a] on $[M(\eta-C_5H_5)_2R_2AlR_2]$

Compound	η-C₅H₅ (τ)	Assignment MR₂Al (τ)	AlR₂ (τ)
[Sc(η-C₅H₅)₂Me₂AlMe₂]	3.88 (s, 10H)	10.29 (broad s, 6H)	10.84 (s, 6H)
[Y(η-C₅H₅)₂Me₂AlMe₂][b]	3.82 (4.05)[e] (s, 10H)	10.32 (10.20)[d] (d, 6H)[c]	10.98 (10.09)[e] (s, 6H)
[Sc(η-C₅H₅)₂Et₂AlEt₂]	3.81 (s, 10H)	10.58 (q, 4H) 8.52 (t, 6H)	10.01 (q, 4H) 8.85 (t, 6H)
[Y(η-C₅H₅)₂Et₂AlEt₂][b]	3.80 (s, 10H)	10.36 (8 line multiplet, 4H) 8.70 (t, 6H)	10.19 (q, 4H) 8.99 (t, 6H)

a In CD_2Cl_2 at room temperature, unless otherwise stated.

b At −45°C.

c $^2\underline{J}(^{89}Y-C^1H_3)$, 5.0 Hz.

d In toluene-$\underline{d}_8$.

e $^2\underline{J}(^{89}Y-C^1H_2)$, 4.0 Hz.

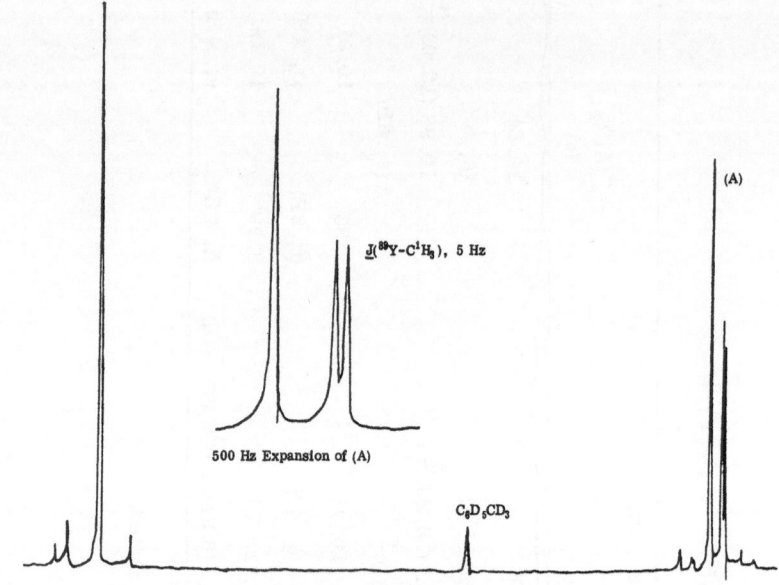

Figure 1   The ^{1}H n. m. r. spectrum of $[Y(\eta-C_5H_5)_2Me_2AlMe_2]$ in $C_6D_5CD_3$ at -40^0 C

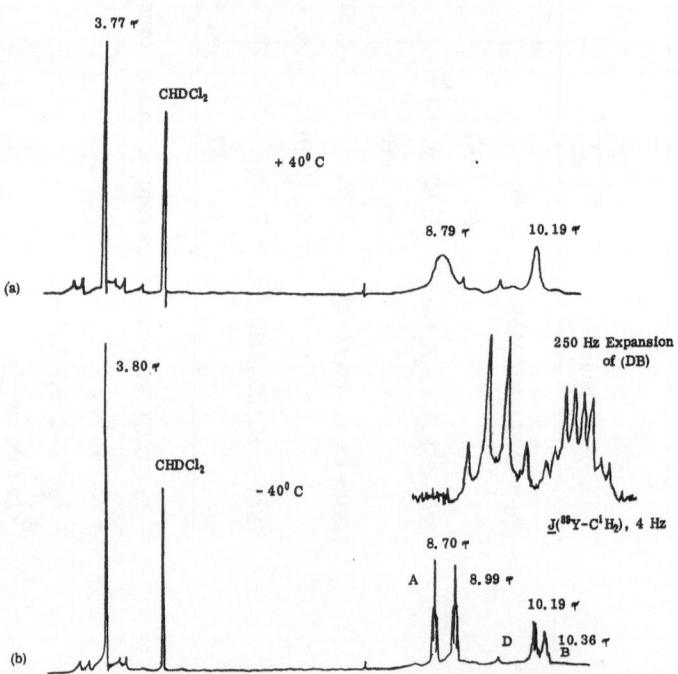

Figure 2   The ^{1}H n. m. r. spectra of $[Y(\eta-C_5H_5)_2Et_2AlEt_2]$ in CD_2Cl_2:
(a) at $+ 40^0$ C, (b) at $- 40^0$ C

bridge-terminal site-exchange process was obtained, $\Delta G \ddagger_{392 K}$ = 15.9 kcal mol^{-1}.
This may be compared with $\Delta G \ddagger$ = 11.0 kcal mol^{-1} for Al$_2$Me$_6$ in cyclopentane;[26]
thus the bridge in the yttrium complex is considerably less labile than that in
trimethylaluminium.   Further data are required before any interpretation
as to the mechanism of site-exchange processes in the yttrium-aluminium
complexes can be made.   There has been much discussion in the literature
as to the mechanism of bridge-terminal site-exchange processes in aluminium
alkyls, but to date distinctions between alternative processes have not been
fully-resolved.[27] Above 100^{0}C the two methyl resonances in [Sc(η-C$_5$H$_5$)$_2$Me$_2$AlMe$_2$]
collapse to a singlet, again demonstrating rapid site-exchange.   It is
interesting that the activation energy for site-exchange should be appreciably higher for scandium than yttrium.

Variable temperature ^{13}C n.m.r. spectra give similar structural
results (Table 3).   The limiting ^{13}C n.m.r. spectra of the yttrium complexes
allowed the first measurement of the yttrium-carbon coupling constant
[$^1\underline{J}(^{89}$Y-^{13}C), 12.2 Hz].

The i.r. spectra of the various methyl-bridged complexes, as
paraffin or hexachlorobutadiene mulls, were essentially identical, as
were those of the ethyl complexes.   This provided a very useful diagnostic
technique in the identification of new alkyl-bridged lanthanide complexes.
The cyclopentadienyl frequencies were observed (at 3100m, 1446m, 1020s,
and 800bs cm^{-1}) in all the complexes and are consistent with other cyclopentadienyl complexes.[28]   In [M(η-C$_5$H$_5$)$_2$Me$_2$AlMe$_2$] bands at 1440, 1360
(asym CH$_3$ bend), 1250, 1235 (sym CH$_3$-bridge bend) and 1190 cm^{-1}
(sym CH$_3$-terminal bend) were assigned by comparison with Al$_2$Me$_6$.[29] In a
similar manner for [M(η-C$_5$H$_5$)$_2$Et$_2$AlEt$_2$], bands at 2790s, 2730s (C-H
stretch), 1470s (asym, CH$_3$ bend), 1412s, 1395m, 1380m (CH$_2$-Al bend),
1365 (sym CH$_3$ bend), 1235 (CH$_2$ bend), 1210m, 1190m (CH$_2$-Al bend),
990s, 960s, 930m, and 905 m (C-C stretch) cm^{-1} were assigned by comparison with Al$_2$Et$_6$.[30]   The skeletal modes and rocking frequencies in
the 770-300 cm^{-1} region were not assigned.

Table 3

^{13}C n. m. r. data[a] on $[M(\eta-C_5H_5)_2R_2AlR_2]$

Compound	$\eta-C_5H_5$ (p.p.m.)	Assignment[b]		AlR_2 (p.p.m.)
		MR_2Al (p.p.m.)		
$[Sc(\eta-C_5H_5)_2Me_2AlMe_2]$	113.2s	20.7s		−6.3s
$[Y(\eta-C_5H_5)_2Me_2AlMe_2]$[c]	112.2s	7.86d[d]		−7.9s
$[Sc(\eta-C_5H_5)_2Et_2AlEt_2]$	112.1s	34.9 (br, CH$_2$)	15.5 (s, CH$_3$)	1.5 (br, CH$_2$) / 10.4 (s, CH$_3$)
$[Y(\eta-C_5H_5)_2Et_2AlEt_2]$[c]	111.7s	20.75 (d, CH$_2$)[d]	13.0 (s, CH$_3$)	−0.34 (s, CH$_2$) / 10.4 (s, CH$_3$)

[a] In CD_2Cl_2 at 35° C, unless otherwise stated.

[b] Chemical shifts are quoted as downfield from $SiMe_4$.

[c] At −45° C.

[d] $^1\underline{J}(^{89}Y-^{13}C)$, 12.2 Hz.

With the exception of the scandium complex, only a weak parent ion was observed in the mass spectrum of $[M(\eta-C_5H_5)_2Me_2AlMe_2]$, the ion at highest m/e being $[M(\eta-C_5H_5)_2AlMe_3]^+$. Generally the fragmentation followed a similar course for all the methyl complexes: loss of one methyl group and then rapid loss of $AlMe_3$ leading to the dominant metal-containing peak $[M(\eta-C_5H_5)_2]^+$. In most cases, no ions of the type $[M(\eta-C_5H_5)_2AlMe_2]^+$ or $[M(\eta-C_5H_5)_2AlMe]^+$ were noted. An ion due to $[M(\eta-C_5H_5)_3]^+$ was also observed in some cases, probably from rearrangement. Similar rearrangement processes have been observed in a large number of cyclopentadienyl compounds.[31]

The e.s.r. spectra of $[Ti(\eta-C_5H_5)_2Me_2AlMe_2]$ and $[Ti(\eta-C_5H_5)_2AlMe_3Cl]$ in toluene at ca. -40^0 C showed a strong broadened signal at g = 1.977, without resolved fine structure. These spectra were similar to those of the complexes $[Ti(\eta-C_5H_5)_2Cl_2AlR_2]$ notably in peak shape; for the latter these were interpreted in terms of broadening via coupling to ^{27}Al $(I = \frac{5}{2})$ giving a 1 : 1 : 1 : 1 : 1 : 1-sextet.[32] A di-μ-methyl structure for $[Ti(\eta-C_5H_5)_2AlMe_4]$ appears reasonable, but is not firmly established. For $[Ti(\eta-C_5H_5)_2AlMe_3Cl]$, either a di-$\mu$-methyl or a μ-chloro-μ-methyl structure is possible; the data do not allow a more precise assignment.

The e.s.r. spectrum of the solution obtained by reaction of $[\{Ti(\eta-C_5H_5)_2Cl\}_2]$ with $Na[AlH_2Me_2]$ showed the presence of two dominant species, the lesser of which was destroyed by heating to ca. 60^0 C for a few minutes. The remaining signal (Figure 3) is assigned to the complex $[Ti(\eta-C_5H_5)_2(\mu-H)_2AlMe_2]$ by analogy with data on other di-μ-hydrido-titanium(III)-aluminium complexes.[33]

The Molecular Structure of $[Yb(\eta-C_5H_5)_2Me_2AlMe_2]$. The molecular structure and atom numbering scheme are shown in Figure 4, while Figure 5 presents a stereoscopic view of the unit cell packing. Structural parameters are found in Table 4. The range of compounds which exhibit three-centre two-electron bridge bonding of type (2) is not extensive. Structural data, compiled in Table 5, show that this type of bonding produces two geometrical manifestations: (i) the metal-carbon bridge bond

Figure 3 The e. s. r. spectrum of [Ti(η−C₅H₅)₂H₂AlMe₂] in toluene at ca. 20⁰ C: g, 1. 992; a(H), 4. 65 G; a(Al), 6. 9 G.

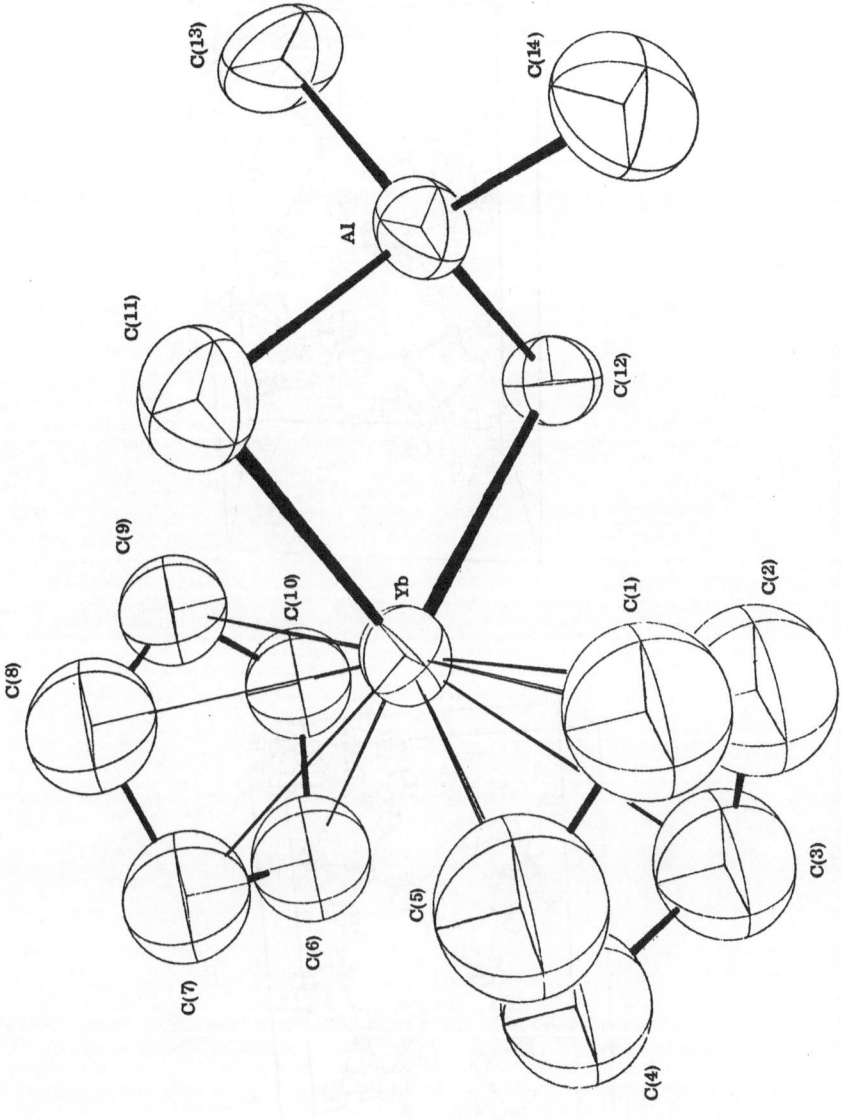

Figure 4 Diagram of [Yb{η-C$_5$H$_5$}$_2$Me$_2$AlMe$_2$] showing the numbering system used, and anisotropic thermal motion (ellipsoids are sealed to enclose 50% probability) for Yb, Al, and methyl C atoms

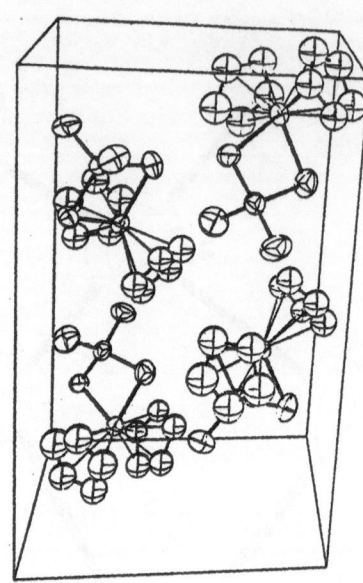

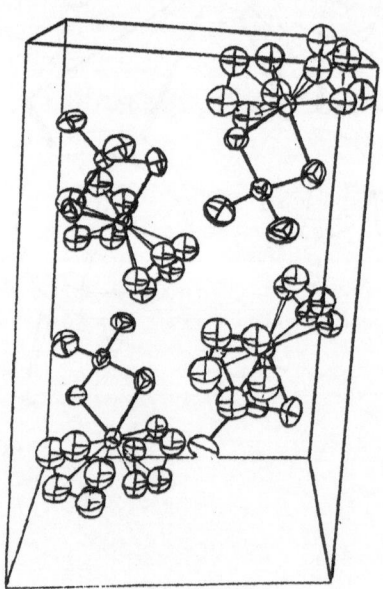

Figure 5 Stereoscopic view of the unit cell packing in [Yb(η-C_5H_5)$_2$Me$_2$AlMe$_2$]

Table 4

Bond lengths (Å) and angles (^{0}) with standard deviations
in parentheses.

Yb–C(1)	2.603(24)	Yb–C(6)	2.570(29)
Yb–C(2)	2.626(24)	Yb–C(7)	2.577(26)
Yb–C(3)	2.623(28)	Yb–C(8)	2.596(26)
Yb–C(4)	2.620(32)	Yb–C(9)	2.593(24)
Yb–C(5)	2.673(28)	Yb–C(10)	2.591(24)
Yb–C(11)	2.609(23)	Yb–C(12)	2.562(18)
Al–C(11)	2.165(22)	Al–C(12)	2.096(18)
Al–C(13)	2.014(25)	Al–C(14)	1.991(24)
C(1)–C(2)	1.41(3)	C(6)–C(7)	1.44(3)
C(2)–C(3)	1.47(3)	C(7)–C(8)	1.31(3)
C(3)–C(4)	1.44(4)	C(8)–C(9)	1.43(3)
C(4)–C(5)	1.33(4)	C(9)–C(10)	1.33(3)
C(5)–C(1)	1.36(4)	C(10)–C(6)	1.32(3)
Yb–Al	3.014(6)		
Yb–C(11)–Al	77.7(7)	Yb–C(12)–Al	80.0(6)
C(11)–Yb–C(12)	87.1(6)	C(11)–Al–C(12)	113.3(8)
Cent 1–Yb–Cent 2	133.1	C(13)–Al–C(14)	118(1)
Cent 1–Yb–C(11)	107.8	C(13)–Al–C(11)	101(2)
Cent 1–Yb–C(12)	107.1	C(13)–Al–C(12)	107(1)
Cent 2–Yb–C(11)	104.5	C(14)–Al–C(11)	109(1)
Cent 2–Yb–C(12)	107.7	C(14)–Al–C(12)	107(1)
C(1)–C(2)–C(3)	104(3)	C(6)–C(7)–C(8)	106(2)
C(2)–C(3)–C(4)	105(2)	C(7)–C(8)–C(9)	107(2)
C(3)–C(4)–C(5)	110(3)	C(8)–C(9)–C(10)	108(2)
C(4)–C(5)–C(1)	109(3)	C(9)–C(10)–C(6)	109(2)
C(5)–C(1)–C(2)	111(3)	C(10)–C(6)–C(7)	108(2)

Table 5

Comparison of structural parameters for main group organometallic compounds with electron-deficient bridging groups

Compound	Bridging group	M–C (bridge) (Å)	M–C (σ) (Å)	M–$\hat{\text{C}}$–M (°)	Ref.
[{Be(Me₂)}₂]	Me	1.93		66.0	a
[{Be(Me)(C: CMe)NMe₃}₂]	–C≡CMe	1.87	1.75	77.2	b
[{Mg(Me₂)}ₙ]	Me	2.24		75.0	c
[{Mg(Et₂)}ₙ]	Et	2.26		72.0	d
[Mg(AlMe₄)₂]	Me	2.21(Mg), 2.10(Al)	1.96	77.7	e
[Li(AlEt₄)]	Et	2.30(Li), 2.02(Al)		77.2	f
[(AlMe₃)₂]	Me	2.123	1.952	75.7	g
[(AlPh₃)₂]	Ph	2.18	1.96	76.5	h
[Al₂Me₅(NPh₂)]	Me	2.142	1.945	78.9	i
[Yb(η-C₅H₅)₂Me₂AlMe₂]	Me	2.18 (Al), 2.59 (Yb)	2.00 (Al)	78.8	This study
[Y(η-C₅H₅)₂Me₂AlMe₂]	Me	2.10 (Al), 2.58 (Y)	1.98 (Al)	80.8	j

[a] A. I. Snow and R. E. Rundle, Acta Cryst., 1951, 4, 348. [b] B. Morosin and J. Howatson, J. Organometallic Chem., 1971, 29, 7. [c] E. Weiss, J. Organometallic Chem., 1964, 2, 314. [d] E. Weiss, J. Organometallic Chem., 1965, 4, 101. [e] J. L. Atwood and G. D. Stucky, J. Amer. Chem. Soc., 1969, 91, 2538. [f] R. L. Gerteis, R. E. Dickerson, and T. L. Brown, Inorg. Chem., 1964, 3, 872. [g] J. C. Huffman and W. E. Streib, J. C. S. Chem. Comm., 1971, 911. [h] J. F. Malone and W. S. McDonald, J. C. S. Chem. Comm., 1967, 444. [i] V. R. Magnuson and G. D. Stucky, J. Amer. Chem. Soc., 1969, 91, 254. [j] G. R. Scollary, Austral. J. Chem., in the press.

length is significantly longer than a normal metal-carbon sigma bond distance, and (ii) the metal-carbon-metal bond angle is small. For example, for trimethylaluminium,[34] the bridge bond length is 2.124(2) Å compared to the aluminium-carbon terminal distance of 1.952(4) Å, while the Al-$\hat{C}$-Al bond angle is 75.7(1)^0. For the complex [Yb$(\eta$-C$_5$H$_5)_2$Me$_2$AlMe$_2$] quite similar values are found: 2.14(2) Å for the Al-C$_b$ bond length, and 78.8^0 for the Yb-$\hat{C}$-Al bond angle. The Yb-C$_b$ length of 2.58(3) Å is much longer than the only known Yb-C σ-bond length, 2.38(1) Å in [Li(thf)$_4$]-[Yb$\{$CH(SiMe$_3)_2\}_3$Cl].[35]

Even though the uncertainty in the metal-carbon bond length is large because of the high thermal motion of the carbon atoms, the structural parameters of the units (3) and (4) are strikingly similar.

(3)   (4)

We infer, therefore, that the bonding in the two systems is of a like nature. This point is discussed in some detail towards the end of this paper.

The Yb-C(cyclopentadienyl) average distance, 2.61(3) Å may be compared with 2.585(8) η in [$\{$Yb$(\eta$-C$_5$H$_5)_2$Cl$\}_2$],[35a] and 2.611(13) Å in [$\{$Yb$(\eta$-C$_5$H$_5)_2$Me$\}_2$].[7] The Al-C$_t$ average of 2.00(1)Å is normal, as is C$_t$-$\overset{\wedge}{Al}$-C$_t$ [118(1)^0].

DI-η-CYCLOPENTADIENYLMETAL(III) METHYLS $[\{M(\eta-C_5H_5)_2Me\}_2]$
(M = Y, Dy, Ho, Er, Tm, or Yb) AND THE CRYSTAL AND
MOLECULAR STRUCTURE OF $[\{M(\eta-C_5H_5)_2Me\}_2]$ (M = Y or Yb)

We now turn from the heterometallic complexes, exemplified above
by the bis(η-cyclopentadienyl) metal(III) tetra-alkylaluminates, to the
homometallic compounds illustrated in this paper by some of the corresponding methyls $[\{M(\eta-C_5H_5)_2Me\}_2]$. A preliminary communication
has appeared.[7]

In considering strategy towards the planned synthesis of these novel
complexes we noted several pertinent facts. Firstly, using aluminium
compounds as model, it seemed striking that Al forms not only many μ-Me$_2$-
bridged but also numerous μ-Cl$_2$-Al$_2$ analogues. Secondly, Me$_2$-bridged
heterometallic complexes appeared to be more obviously realiseable than
corresponding homometallic transition-metal complexes; we noted that
the known Mg[AlMe$_4$]$_2$ has μ-Me$_2$-heterometallic bridges. Thirdly, by
choosing as reagents a Cl$_2$-bridged-binuclear transition-metal complex
and Li[AlMe$_4$], a clear synthetic route was at hand; this concept was
exemplified above.: $[\{M(\eta-C_5H_5)_2Cl\}_2] \longrightarrow [M(\eta-C_5H_5)_2(\mu-Me_2)AlMe_2]$.
Fourthly, just as Cl$_2$-bridged bimetallic complexes are cleaved by bases,
so there was the expectation (with precedent in organoaluminium chemistry)
that MMe$_2$Al complexes would behave similarly; this is an important theme
of the present paper: $[M(\eta-C_5H_5)_2(\mu-Me_2)AlMe_2] \longrightarrow [\{M(\eta-C_5H_5)_2Me\}_2]$.
The results are also of significance as a contribution to the organometallic
chemistry of the group 3a elements, which is as yet sparsely documented.

The organometallic chemistry of the group 3a elements and the
lanthanides has, until recently, been a relatively neglected area.[36] This
is particularly true for metal-carbon σ-bonded complexes. Stable or well-
characterised materials are only obtained by a careful choice of ligand.[37]

The simple methyls, e.g., [YMe$_3$], appear to be polymeric materials,
soluble only in donor solvents and have not been isolated in a pure state.
With the more bulky neopentyl or trimethylsilyl ligands discrete stable
complexes have, however, been obtained. These include the homoleptic

$[Y\{CH(SiMe_3)_2\}_3]$ [38] and $[Sc(CH_2SiMe_2C_6H_4-\underline{o})_3]$, [39] and the heteroleptic $[MR_3(thf)_2]$ [thf = tetrahydrofuran; M = Sc or Y; R = Me_3SiCH_2, [39] $(Me_3Si)_2CH$, [38] or Me_3CCH_2; [34] and M = Yb, R = Me_3SiCH_2 [23]]. Additionally, the hexa co-ordinate metal complexes $[MR_3]$ [with the chelating ligand $Me_2P(CH_2)_2$ = R, and M = La, Pr, Nd, Sm, Gd, Ho, Er, or Lu] have been described. [40] Lanthanide Grignard analogues were believed to be formed by reacting the metal (Eu, Yb, or less convincingly, La, Ce, or Sm) with an alkyl or aryl iodide. [41] The simple metal(III) phenyls $[MPh_3]$ (M = Sc or Y), [4] like the methyls, are somewhat ill-defined. However, the corresponding '-ate' complexes $[MAr_4]^-$ (M = La or Pr, Ar = Ph; [4] or M = Yb or Lu, Ar = $2,6-Me_2C_6H_3$) have been prepared and are more fully-characterised, including a single crystal $\underline{X}$-ray diffraction analysis of $[Li(thf)_4][Lu(C_6H_3Me_2-2,6)_4]$. [42] A metal(II) complex $[Yb(C_6F_5)_2(thf)_4]$, has been prepared $\underline{via}$ the mild arylmercury(II) reagent. [43] Two alkynyl complexes, $[Sc(C{\equiv}CPh)_3]$ [37] and $[Eu(C{\equiv}CMe)_2]$, [44] are known.

A more extensive range of alkyl or aryl complexes exists with the 'stabilising'cyclopentadienyl ligand. [These include the metal(III) complexes described in our preliminary communication] [7] [$\{M(\eta-C_5H_5)_2R\}_n$] [M = Y, Gd, Py, Ho, Er, or Yb, with R = Me, and n = 2; [7] M = Gd, Er, or Yb, with R = Ph; M = Sc, Gd, Ho, Er, or Yb, with R = PhC≡C; [17, 45] M = Sc, with R = $(CH_2)_2PPh_2$ [46]], $[Ho(\eta-C_5H_5)(C{\equiv}CPh)_2]$, [45] $[M(\eta-C_5H_5)_2(\mu-R)_2AlR_2]$ (M = Sc, Y, Gd, Dy, Ho, Er, Tm, or Yb, with R = Me; or M = Sc, Y, or Ho, with R = Et) (see above), and the cerium(IV) complexes $[Ce(\eta-C_5H_5)_3R]$ and $[Ce(indenyl)_2R_2]$ (R = Me, Et, Ph, or $PhCH_2$). [47]

Above, results were described relating to the preparation of the double methyl-bridged-metal(III) tetra-alkylaluminates $[M(\eta-C_5H_5)_2(\mu-R)_2AlR_2]$. We now give details ($\underline{cf.}$ ref. 7) of the preparation and characterisation of the corresponding homometallic group 3a metal complexes [$\{M(\eta-C_5H_5)_2Me\}_2$] (M = Y, Dy, Ho, Er, Tm, or Yb); the crystal and molecular structures of the Y and Yb complexes definitively establish the double methyl-bridge.

During the course of this work three of the compounds $[\{M(\eta-C_5H_5)_2Me\}_n]$ (M = Gd, Er, or Yb) were reported independently;[45] they were obtained by an alternative procedure [from the metal(III) chloride and LiMe], and their structures were discussed in terms of a monomeric pseudo-7-co-ordinate metal environment.[48] Using the same procedure we have now prepared the ytterbium complex and find it to be essentially identical to that obtained from the tetramethylaluminate (see below).

It is interesting that while $[Ti(\eta-C_5H_5)_2Me_2AlMe_2]$ is significantly less stable than the group 3A metal analogues, attempts to prepare $[\{Ti(\eta-C_5H_5)_2R\}_n]$ (R = Me or Et) have failed;[49] by contrast, $[Ti(\eta-C_5H_5)_2-(CH_2CMe_3)]$ is stable and monomeric, perhaps because of the presence of the bulkier alkyl ligand.

Results and discussion

Preparation. The complexes $[\{M(\eta-C_5H_5)_2Me\}_2]$ (M = Y, Dy, Ho, Er, Tm, or Yb) were prepared according to equation (1), using an equi-molar quantity of the base pyridine (py). In principle, the substrate $[M(\eta-C_5H_5)_2(\mu-Me)_2AlMe_2]$ has two potential acceptor sites, M or Al, with equation (1) showing the latter as the more electrophilic.

$$[(\eta-C_5H_5)_2M \underset{Me}{\overset{Me}{\diagdown}} Al \underset{Me}{\overset{Me}{\diagdown}}] \xrightarrow{py} [(\eta-C_5H_5)_2M \underset{Me}{\overset{Me}{\diagdown}} M(C_5H_5-\eta)_2]$$

$$+ AlMe_3(py) \quad (1)$$

The di-η-cyclopentadienyl(methyl) metal(III) complexes were isolated in high yield (ca. 80%) as air-sensitive crystalline solids (Table 6). They are soluble in dichloromethane, hot toluene, or benzene, and partially soluble in cold toluene or benzene. They are insoluble in saturated hydrocarbon solvents. The compounds are thermally-stable up to ca. 150^0 C, whereafter they slowly decompose without melting. The mode of decomposition is under investigation.

Table 6

Data on the lanthanide methyl complexes

Compound	Colour	Yield (%)	M.p. (°C)	Magnetic moment B.M. (calc)	Analysis: Found (required) (%)			
					C	H	M	Al
[{Y(η-C$_5$H$_5$)$_2$Me}$_2$]	Colourless	87	>158 (dec)	–	56.4 (56.4)	5.7 (5.7)	37.7 (38.0)	0.0
[{Dy(η-C$_5$H$_5$)$_2$Me}$_2$]	Pale-yellow	78	>165 (dec)	9.9 (10.6)	43.1 (43.0)	4.25 (4.3)	54.5 (52.8)	–
[{Ho(η-C$_5$H$_5$)$_2$Me}$_2$]	Straw	84	>160 (dec)	10.0 (10.6)	42.8 (42.6)	4.2 (4.2)	–	–
[{Er(η-C$_5$H$_5$)$_2$Me}$_2$]	Pink	82	>159 (dec)	9.5 (9.6)	42.2 (42.3)	3.9 (4.2)	51.9 (53.5)	0.0
[{Tm(η-C$_5$H$_5$)$_2$Me}$_2$]	Pale-green	82	>160 (dec)	7.5 (7.6)	42.4 (42.1)	4.2 (4.2)	–	–
[{Yb(η-C$_5$H$_5$)$_2$Me}$_2$]	Orange-red	81	>165 (dec)	4.0 (4.5)	41.8 (41.5)	4.4 (4.1)	53.7 (54.4)	0.0
[Sc(η-C$_5$H$_5$)$_2$(Me)py]$^-$ a	Cream	71	–	–	69.8 (71.4)	6.5 (6.7)	–	
[Sc(η-C$_5$H$_5$)$_2$(Me)(thf)]	Cream	85	–	–	68.9 (68.7)	8.1 (8.1)	–	

a Found: N, 4.5. Calc.: N, 5.2%

Reaction of $[Sc(\eta-C_5H_5)_2(\mu-Me)_2AlMe_2]$ with the Lewis base (L)
py or tetrahydrofuran (thf) took a different course [equation (2)],
whence Sc rather than Al is the more electrophilic centre.

$$[(\eta-C_5H_5)_2Sc\underset{Me}{\overset{Me}{\diagdown}}Al\underset{Me}{\overset{Me}{\diagup}}] \xrightarrow{\ L\ } [(\eta-C_5H_5)_2Sc\underset{L}{\overset{Me}{\diagup}}] + \tfrac{1}{2}Al_2Me_6 \quad (2)$$

The pale-yellow adducts $[Sc(\eta-C_5H_5)_2(Me)(L)]$ were obtained in
high yield (ca. 70%) and were purified by crystallisation from toluene.
For the pyridine product (L = py), the base L was readily removed
in vacuo. After prolonged evacuation, a material was obtained that
did not correspond to the expected $[\{Sc(\eta-C_5H_5)_2Me\}_2]$ and has not yet
been identified. The thf adduct was more stable in vacuo, but attempted
sublimation ($160^0 C/10^{-4}$ mmHg) resulted in decomposition.

From equations (1) and (2), it appears that in a related series
of complexes, Lewis acidity increases in the order: lanthanide $\approx$
yttrium < aluminium < scandium.

Spectroscopic characterisation. The ^{1}H n. m. r. spectrum of the
diamagnetic yttrium complex $[\{Y(\eta-C_5H_5)_2Me\}_2]$ was characteristic of
a doubly-bridged methyl structure (Table 7). The spectrum showed a
triplet (10. 81 τ) for the bridge-methyl protons and was invariant
between -40^0 C and 40^0 C. The methyl protons are found as a triplet
due to coupling to two equivalent yttrium atoms, $^1J(^{89}Y-C^1H_3) = 3.6$ Hz,
(^{89}Y, spin = $\tfrac{1}{2}$, 100%). It is interesting to compare the ^{1}H n. m. r.
spectrum of $[\{Y(\eta-C_5H_5)_2Me\}_2]$ with those of Al_2Me_6 [50] and $[\{Ni(\eta-allyl)Me\}_2]$,[51]
where the bridge-methyl protons are found at 10. 67 τ and 12. 1 τ,
respectively. The unusually-strong shielding in the nickel case was
ascribed to interaction with the non-bonding d-electrons. This now
seems likely, because in $[\{Y(\eta-C_5H_5)_2Me\}_2]$ and Al_2Me_6, where no
d-electrons are available for this type of interaction, the bridging-
alkyl protons occur at similar lower field values.

The scandium complexes $[Sc(\eta-C_5H_5)_2(Me)(L)]$ (L = py or thf) showed broad singlets at 10.6 τ and 9.97 τ, respectively, attributed to the methyl protons, and peaks corresponding to the ligands in the expected lower field region (Table 7).

The $^{13}C-\{^1H\}$ n.m.r. spectrum of $[\{Y(\eta-C_5H_5)_2Me\}_2]$ is consistent with a dimeric structure. The bridging methyl carbon atoms occur at 23.0 p.p.m. as a triplet due to coupling to two equivalent yttrium atoms with $^1\underline{J}(^{89}Y-^{13}C) = 25.0$ Hz (Table 7).

The i.r. spectra of the methyl-bridged complexes were essentially identical. The cyclopentadienyl frequencies were assigned as for (see above) $[M(\eta-C_5H_5)_2R_2AlR_2]$ at 3100m, 1447m, 1020s, and 790s cm^{-1}. The methyl deformations were found at 1368m and 1195s cm^{-1}, assigned to the asymmetric CH_3 bend and the symmetric CH_3 bend, respectively. The strong band at ~ 1195 cm^{-1} is characteristic of these compounds and of a methyl attached to a metal.[52] For instance, in the polymeric complexes $[(MgMe_2)_n]$ or $[(BeMe_2)_n]$ similar symmetric deformations occur at 1193 and 1209 cm^{-1}, or 1250 and 1260 cm^{-1}, respectively;[53] this may be compared with mononuclear complexes; e.g., $SnMe_4$ has absorptions at 1198 and 1205 cm^{-1},[54] and trans-$[Pt(Me)(I)(PR_3)_2]$ has the band at 1217 cm^{-1}.[55]

The i.r. spectra of $[Sc(\eta-C_5H_5)_2(Me)(L)]$ (L = py or thf) showed significant differences from the methyl-bridged species. The characteristic strong band at ca. 1195 cm^{-1} was missing, the methyl deformations occurring as weak bands at 1365, 1350 (asymmetric CH_3 bend), 1190, and 1182 (symmetric CH_3 bend) cm^{-1}. Bands were also observed due to the ligand: for L = py at 1609, 1590, 1492, 1449, 1224, 1160, 1076, 1048, 767, 710, and 640 cm^{-1}; and for L = thf at 925m, 869s, and 850s cm^{-1}; these compare reasonably with other py-[56] or thf-[57] metal-complexes.

The mass spectra of the ytterbium and erbium complexes did not show a parent molecular ion, the ion at highest m/e being (parent-Me)$^+$. Peaks were observed due to cyclopentadienyl rearrangement, e.g.,

Table 7

N. m. r. data[a] on $[Y(\eta-C_5H_5)_2Me]_2$ and $[Sc(\eta-C_5H_5)_2(Me)L]$

Compound	Technique	$\eta-C_5H_5$	Me	L
$[Y(\eta-C_5H_5)_2Me]_2$	^1H (τ)	3.79s	10.81t[b]	–
	^{13}C (p.p.m.)	111.3	23.0[c]	–
$[Sc(\eta-C_5H_5)_2(Me)py]$	^1H (τ)	4.15s	10.6s	1.8, 2.65, 2.75, 2.95
$[Sc(\eta-C_5H_5)_2(Me)(thf)]$	^1H (τ)	3.86s	9.97s	6.75m, 8.8m

[a] At room temperature in dichloromethane

[b] $^2\underline{J}(^{89}Y-C^1H_3) = 3.6$ Hz

[c] $^1\underline{J}(^{89}Y-^{13}C) = 25.0$ Hz

$[M(\eta-C_5H_5)_3]^+$, with the most intense metal-containing peak being $[M(\eta-C_5H_5)_2]^+$.

The electronic spectra in the range 350-900 nm and the magnetic moments of the highly coloured complexes $[\{M(\eta-C_5H_5)_2Me\}_2]$ (M = Dy, Ho, Er, or Tm) were recorded as dichloromethane solutions and were diagnostic of the respective lanthanide(III) ion. [58, 59]

Reactions of $[\{Y(\eta-C_5H_5)_2Me\}_2]$. -(a) With Lewis Bases.

Although the scandium adducts with thf or py, $[Sc(\eta-C_5H_5)_2(Me)(L)]$, are stable and isolatable, the corresponding yttrium complexes appear to be less robust and have not been fully characterised. By observing the bridge protons in the ^{1}H n.m.r. spectrum, the reaction with base (and, indeed, other reactions) can be monitored; the triplet (10.81 τ) collapsed to a doublet (usually broad and unresolved) at lower field upon bridge-breaking. The complex $[\{Y(\eta-C_5H_5)_2Me\}_2]$ appears to react readily with hard donors such as amines, thf, or phosphine oxides e.g., PPh$_3$(:O), but less readily with soft donors such as phosphines, e.g., PPh$_3$ or PEt$_3$. Neither the pyridine nor triphenylphosphine oxide adducts were isolated. ^{1}H n.m.r. experiments showed loss of the triplet due to the bridging methyl protons and the pyridine complex showed a broad lower field resonance (9.67 τ) attributed to the non-bridging methyl group. The low thermal stability of the yttrium-pyridine adduct, compared with the scandium analogue, again points to Y^{III} being a weaker Lewis acid than Sc^{III} [cf., equations (1) and (2)].

(b) With aluminium alkyls or chloro-alkyls. The complex $[\{Y(\eta-C_5H_5)_2Me\}_2]$ reacted readily with an aluminium alkyl or chloro-alkyl to give di-μ-alkyl- or di-μ-chloro- bridged species [equations (3) and (4)]; the mixed bridge compound was considerably less stable.

$$[\{Y(\eta-C_5H_5)_2Me\}_2] + Al_2Me_2Cl_4 \longrightarrow 2[(\eta-C_5H_5)_2Y\langle^{Cl}_{Cl}\rangle AlMe_2] \qquad (3)$$

$$[\{Y(\eta\text{-}C_5H_5)_2Me\}_2] \; + \; Al_2Me_6 \; \longrightarrow \; 2\,[(\eta\text{-}C_5H_5)_2Y \overset{\displaystyle Me}{\underset{\displaystyle Me}{\diagdown\!\!\diagup}} AlMe_2] \qquad (4)$$

There are numerous well-established examples of compounds with di-μ-chloro- bridges and di-μ-alkyl analogues, while less common, are well documented (see above and ref. 1). Evidence for the existence of μ-alkyl-μ-chloro- bridges is sparse, but they are often postulated as a structural feature of reactive intermediates, e.g., in exchange reactions.[1]   Only [Ti(η-C$_5$H$_5$)$_2$AlMe$_3$Cl] has been isolated and characterised by analysis and e. s. r., although the nature of the bridge was not determined.[46]   Tetranuclear [Ru$_2$M$_2$] complexes (M = Zn or Mg) were believed to contain   (μ-Cl, Me) bridges.[60]

In the light of the prior art, the instability of [Y(η-C$_5$H$_5$)$_2$AlMe$_3$Cl], equation (5), is not unexpected.

$$[Y(\eta\text{-}C_5H_5)_2Me]_2 \; + \; Al_2Me_4Cl_2 \; \longrightarrow \; 2\,[(\eta\text{-}C_5H_5)_2Y \overset{\displaystyle Cl}{\underset{\displaystyle Me}{\diagdown\!\!\diagup}} AlMe_2]$$

$$\text{not isolated}$$

$$\Big\downarrow \qquad\qquad (5)$$

$$[\{Y(\eta\text{-}C_5H_5)_2Cl\}_2] \; + \; Al_2Me_6$$

Failure to isolate [Y(η-C$_5$H$_5$)$_2$AlMe$_3$Cl] may be a consequence of the insolubility of [$\{$Y(η-C$_5$H$_5$)$_2$Cl$\}_2$] in solvent toluene.   This would favour its disproportionation to the symmetrical dimers.   Separate experiments demonstrated that no complex could be isolated from the reverse reaction, between [$\{$Y(η-C$_5$H$_5$)$_2$Cl$\}_2$] and excess of Al$_2$Me$_6$.   Experiments are in hand to increase the solubility of a related yttrium(III) chloride dimer, by using various substituted cyclopentadienyl ligands; this may lead to more stable mixed-bridge systems.

The crystal structure of [$\{$Yb(η-C$_5$H$_5$)$_2$Me$\}_2$] unambiguously demonstrates that the complex is dimeric in the solid state, rather than seven-co-ordinate and hence monomeric, as suggested by Tsutsui and Ely.[48]

As the two samples were prepared by different methods, equation (1) or (6),[48] it was important to show that the samples were identical.

$$[\{Yb(\eta\text{-}C_5H_5)_2Cl\}_2] + 2\,LiMe \longrightarrow [\{Yb(\eta\text{-}C_5H_5)_2Me\}_n] + 2\,LiCl \quad (6)$$

Accordingly, we repeated the earlier synthesis, based on equation (6), and obtained orange-red crystals from benzene and showed that they were identical (i. r. and analysis) with those obtained from reaction according to equation (1).

The Molecular Structure of $[\{Y(\eta\text{-}C_5H_5)_2Me\}_2]$ and $[\{Yb(\eta\text{-}C_5H_5)_2Me\}_2]$. The molecular structure and atom numbering scheme for the yttrium complex is shown in Figure 6, and a stereoscopic view of the unit cell packing of the ytterbium analogue is presented in Figure 7. From the crystal data it is evident that the two compounds are isostructural. In the preliminary communication, only the structure of the ytterbium analogue was reported.[7] Although it was inferred that the two ytterbium atoms were linked together via an electron-deficient methyl-bridge (5), situation (6), could not be definitely ruled out. Because of high thermal motion, the presence of the seventy electron ytterbium atom, or an absorption problem, it was not possible to locate the hydrogen atoms of the methyl group. However, this ambiguity has now been resolved with the structure of the yttrium complex. (The metal has only thirty-nine

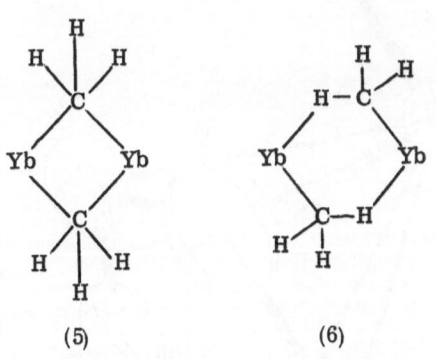

(5)   (6)

electrons, and the linear absorption coefficient is only 60% of that of $[\{Yb(\eta\text{-}C_5H_5)_2Me\}_2]$.) Structure (5) is the correct choice.

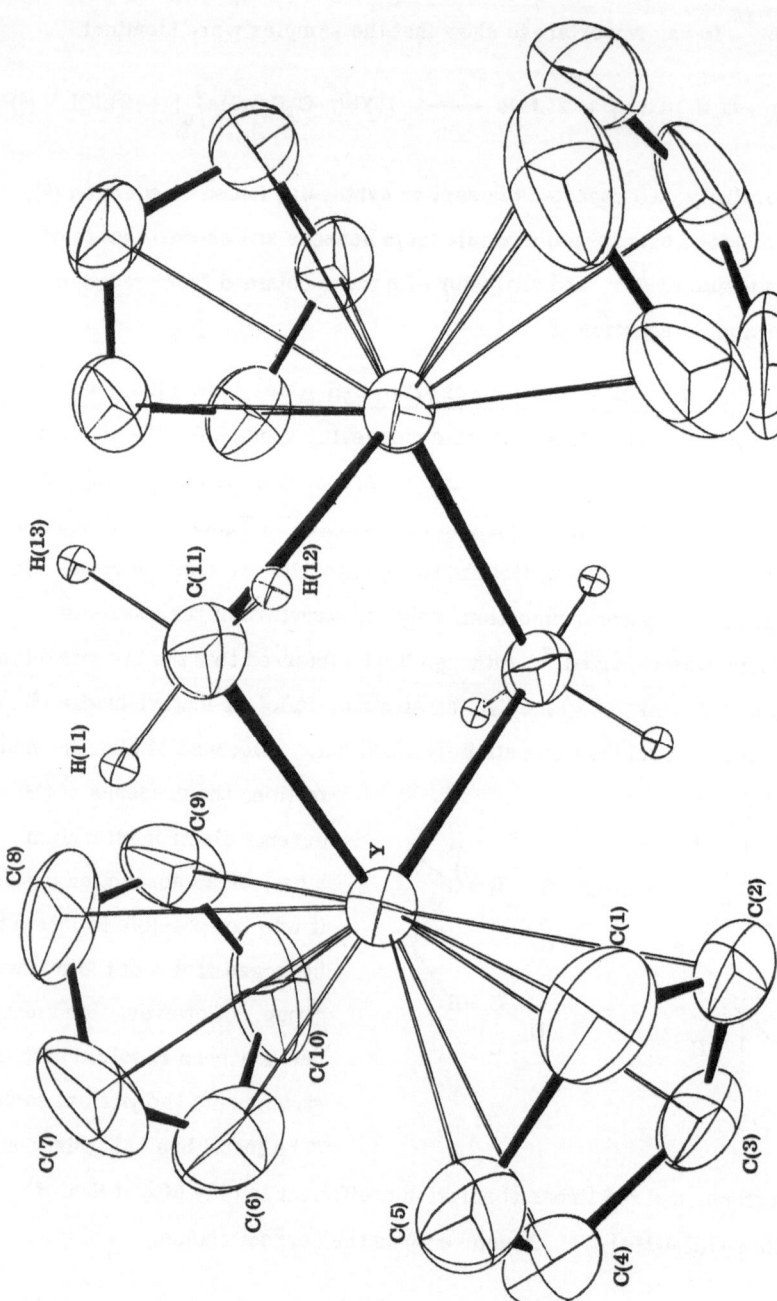

Figure 6 Diagram of [Ȳ(η-C₅H₅)₂Me]₂] showing the numbering system used, with the non-hydrogen atoms represented by their 50% probability ellipsoids for thermal motion

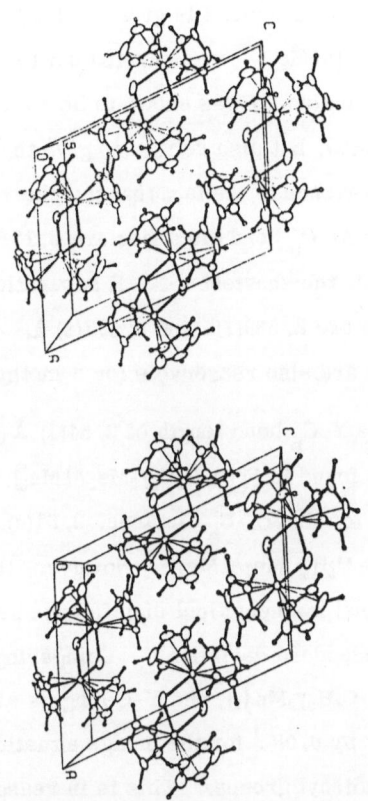

Figure 7 Stereoscopic view of the unit cell packing in $[\{Yb(\eta-C_5H_5)_2Me\}_2]$

The methyl hydrogen atoms were clearly visible on the difference Fourier map, and although their positional parameters were not refined, substantiation for the placement is given in Table 8. The low temperature X-ray structure of $[AlMe_3]_2$[34] was carried out to resolve just such a bonding dilemma, and there also (1) was shown to be correct. The hydrogen atoms were not only located, but also refined to give the values listed in Table 8. Of particular importance is the fact that the shortest Al$\cdots$H approach is 2.17(3) Å, and the Al-C_b bond distances are 2.125(2) and 2.123(2) Å. For $[\{Y(\eta-C_5H_5)_2Me\}_2]$, the shortest Y$\cdots$H separation is 2.54 Å, while the Y-C_b bond lengths are 2.553(10) and 2.537(9) Å. The C-H bond lengths and H-$\hat{C}$-H angles are also reasonable for a methyl group.

The average Y-C_b bond length of 2.54(1) Å (Table 9) is shorter than the 2.58(3) Å found in $[Y(\eta-C_5H_5)_2Me_2AlMe_2]$,[61] and the same holds (see above) for the Yb-C_b distance, 2.51(4) Å, compared to 2.58(1) Å in $[Yb(\eta-C_5H_5)_2Me_2AlMe_2]$. However, the differences are of only marginal mathematical significance because of the large associated standard deviations. Comparing the two compounds $[\{M(\eta-C_5H_5)_2Me\}_2]$, the Y-C lengths are longer than the Yb-C distance by 0.03 Å for the bridge situation, and 0.04 Å for the cyclopentadienyl groups. This is in reasonable agreement with the commonly reported difference in the M^{3+} radii, 0.02[12] to 0.06 Å.[62]

Table 8

Comparison of the bonding parameters of the bridging methyl groups in $[\{Y(\eta-C_5H_5)_2Me\}_2]$ and $[AlMe_3]_2$[a]

Bond (lengths in Å)	M = Al	M = Y
C(11)-H(11)	0.88(3)	0.93
C(11)-H(12)	0.94(4)	0.86
C(11)-H(13)	0.96(4)	1.15
M-C(11)	2.125(2)	2.553(10)
M-C(11)	2.123(2)	2.537(9)
M...H(11)	2.17(3)	2.63
M...H(12)	-	3.04
M...H(13)	-	3.29
M...H(11)'	-	3.45
M...H(12)'	-	2.54
M...H(13)'	-	2.63
C(11)...C(11)'	3.355(4)	3.67(2)
H(11)-Ĉ(11)-H(12)	102(3)^0	111^0
H(11)-Ĉ(11)-H(13)	100(3)	95
H(12)-Ĉ(11)-H(13)	100(3)	118

[a] Ref. 34

Table 9

Comparison of structural parameters for the yttrium and ytterbium compounds $[\{M(\eta-C_5H_5)_2CH_3\}_2]$[a]

Bond (lengths in Å)	M = Yb	M = Y
M–C(11)	2.536(17)	2.553(10)
M–C(11)'	2.486(17)	2.537(9)
M–C (electron-deficient) av.[a]	2.511(35)	2.545(11)
M–C(1)	2.628(16)	2.661(8)
M–C(2)	2.622(16)	2.683(9)
M–C(3)	2.622(18)	2.677(10)
M–C(4)	2.594(17)	2.640(10)
M–C(5)	2.604(18)	2.642(9)
M–C(6)	2.621(19)	2.660(12)
M–C(7)	2.622(18)	2.669(10)
M–C(8)	2.621(19)	2.653(9)
M–C(9)	2.622(20)	2.621(10)
M–C(10)	2.621(22)	2.641(11)
M–C (cyclopentadienyl), av.[b]	2.613(13)	2.655(18)
C(1)–C(2)	1.34(3)	1.40(1)
C(2)–C(3)	1.38(3)	1.40(1)
C(3)–C(4)	1.36(3)	1.43(1)
C(4)–C(5)	1.39(3)	1.38(1)
C(5)–C(1)	1.40(3)	1.40(1)
C(6)–C(7)	1.36(3)	1.37(2)
C(7)–C(8)	1.36(3)	1.36(2)
C(8)–C(9)	1.40(3)	1.36(2)
C(9)–C(10)	1.35(3)	1.38(2)
C(10)–C(6)	1.39(3)	1.40(2)

Table 9 (Cont.)

Angle (°)		
M–C(11)–M'	86.6	87.7(3)
C(11)–M–C(11)'	93.4	92.3(3)
Cent1–M–Cent2	128.2	128.9
Cent1–M–C(11)	105.9	106.1
Cent1–M–C(11)'	110.0	110.2
Cent2–M–C(11)	06.?	105.3
Cent2–M–C(11)'	107.4	107.8
C(5)–C(1)–C(2)	108(1)	108(1)
C(1)–C(2)–C(3)	110(1)	109(1)
C(2)–C(3)–C(4)	106(1)	107(1)
C(3)–C(4)–C(5)	110(1)	107(1)
C(4)–C(5)–C(1)	106(1)	109(1)
C(10)–C(6)–C(7)	109(1)	106(1)
C(6)–C(7)–C(8)	108(1)	109(1)
C(7)–C(8)–C(9)	108(1)	109(1)
C(8)–C(9)–C(10)	109(1)	108(1)
C(9)–C(10)–C(6)	106(1)	108(1)

References

[1]Cf. , (a) G. E. Coates and K. Wade in 'Organometallic Compounds', Vol. 1, 3rd. Ed. , G. E. Coates, M. L. H. Green, and K. Wade (Eds), Methuen, 1967; (b) T. Mole and E. A. Jeffery, 'Organoaluminium Compounds', Elsevier, 1972; (c) P. J. Davidson, M. F. Lappert, and R. Pearce, Chem. Rev. , 1976, 76, 219.

[2]For a discussion of the bonding in complexes of this type, see R. Mason and D. M. P. Mingos, J. Organometallic Chem. , 1973, 50, 53; R. Mason, Pure Appl. Chem. , 1973, 33, 513; R. H. Summerville and R. Hoffmann, J. Amer. Chem. Soc. , 1976, 98, 7240.

[3](a) D. G. H. Ballard and R. Pearce, J. C. S. Chem. Comm. , 1975, 621; (b) J. Holton, M. F. Lappert, G. R. Scollary, D. G. H. Ballard, R. Pearce, J. L. Atwood, and W. E. Hunter, ibid. , 1976, 425.

[4]J. A. J. Jarvis, R. Pearce, and M. F. Lappert, J. C. S. Dalton, 1977, J. A. J. Jarvis, B. T. Kilbourn, R. Pearce, and M. F. Lappert, J. C. S. Chem. Comm. , 1973, 475.

[5]E. Weiss and E. A. C. Lucken, J. Organometallic Chem. , 1964, 2, 197.

[6](a) K. Mertis, A. F. Masters, and G. Wilkinson, J. C. S. Chem. Comm. , 1976, 858; (b) R. A. Andersen, E. Carmona-Guzman, J. F. Gibson, and G. Wilkinson, J. C. S. Dalton, 1976, 2204; (c) R. A. Andersen, R. A. Jones, G. Wilkinson, M. B. Hursthouse, and K. M. A. Malik, J. C. S. Chem. Comm. , 1977, 283.

[7]J. Holton, M. F. Lappert, D. G. H. Ballard, R. Pearce, J. L. Atwood, and W. E. Hunter, J. C. S. Chem. Comm. , 1976, 480.

[8](a) K. Fischer, K. Jonas, P. Misbach, R. Stabba, and G. Wilke, Angew. Chem. Internat. Ed. Engl. , 1973, 12, 943; (b) H. Schenkluhn, Thesis, University of Bochum, 1971.

[9]P. C. Wailes, H. Weigold, and A. P. Bell, J. Organometallic Chem. , 1972, 34, 156.

[10]G. Natta, G. Mazzanti, U. Giannini, and S. Cesca, Angew. Chem. , 1960, 72, 39.

[11] Cf., R.J. Puddephatt and P.J. Thompson, J.C.S. Dalton, 1975, 1810; J.C.S. Chem. Comm., 1975, 841; J. Organometallic Chem., 1976, 120, C51; G.W. Rice and R.S. Tobias, ibid., 1975, 86, C37. M.H. Chisholm and M.W. Extine, J. Amer. Chem. Soc., 1976, 98, 6393; D.B. Carr and J. Schwartz, ibid., 1977, 99, 638; M. Yoshifuji, M.J. Loots, and J. Schwartz, Tetrahedron Letters, 1977, 1303.

[12] Cf., F.A. Cotton and G. Wilkinson, 'Advanced Inorganic Chemistry', Interscience, New York, 3rd ed., 1972.

[13] (a) J.M. Guss, R. Mason, I. Sotofte, G. van Koten, and J.G. Noltes, J.C.S. Chem. Comm., 1972, 446; (b) G. van Koten and J.G. Noltes, J.C.S. Chem. Comm., 1972, 940; (c) A.J. Leusink, G. van Koten, J.W. Marsman, and J.G. Noltes, J. Organometallic Chem., 1973, 55, 419; (d) G. van Koten and J.G. Noltes, ibid., 1974, 82, C53; P.W.J. de Graaf, J. Boersma, and G.J.M. van der Kerk, ibid., 1977, 127, 391.

[14] C.W. Bradford, R.S. Nyholm, G.J. Gainsford, J.M. Guss, P.R. Ireland, and R. Mason, J.C.S. Chem. Comm., 1972, 87.

[15] V.G. Andrianov, Yu.T. Struchkov, and E.R. Rossinskaya, J.C.S. Chem. Comm., 1973, 338.

[16] Cf., R.S.P. Coutts and P.C. Wailes, Adv. Organometallic Chem., 1970, 9, 135.

[17] R.S.P. Coutts and P.C. Wailes, J. Organometallic Chem., 1970, 25, 117.

[18] Cf., J. Boor Jr., Macromol. Rev., 1967, 3, 115; J.R. Jones, J. Chem. Soc. (C), 1971, 1171; V.A. Korner, V.A. Vasiliev, N.A. Kal'nicheva, and O.I. Belgorodskaya, J. Polymer Sci., Polymer Chem. Ed., 1973, 11, 2557; P. Pino, G. Consiglio, and H.J. Ringger, Annalen, 1975, 509; W.P. Long and L.S. Breslow, ibid., p.463; M. Zenbayashi, K. Tamao, and M. Kumada, Tetrahedron Letters, 1975, 1719; D.R. Armstrong, P.G. Perkins, and J.J.P. Stewart, J.C.S. Dalton, 1972, 1972.

[19] G.R. Scollary, Austral. J. Chem., in the press.

[20] T.J. Marks and G.W. Grynkewich, Inorg. Chem., 1976, 15, 1302.

[21] T. Yamamoto and A. Yamamoto, J. Organometallic Chem., 1973, 57, 127.

[22](a) T. Chivers and E. D. Ibrahim, J. Organometallic Chem., 1974, 77, 241; (b) cf., P. C. Wailes, R. S. P. Coutts, and H. Weigold, 'Organometallic Chemistry of Titanium, Zirconium, and Hafnium', Academic Press, 1974; F. W. van der Weij. H. Scholtens, and J. H. Teuben, J. Organometallic Chem., 1977, 127, 299.

[23]J. Holton, D. Phil. Thesis, University of Sussex, 1976.

[24]T. S. Piper and G. Wilkinson, J. Inorg. Nuclear Chem., 1956, 3, 104; H. C. Beachell and S. A. Butler, Inorg. Chem., 1965, 4, 1133.

[25]Cf., D. R. Eaton and W. D. Phillips, Adv. Mag. Res., (Ed.) J. S. Waugh, 1965, 1, 103.

[26]K. C. Ramey, J. F. O'Brien, I. Hasegawa, and A. E. Bochert, J. Phys. Chem., 1965, 64, 3418.

[27]T. B. Stanford and K. L. Henold, Inorg. Chem., 1975, 14, 2426, and references therein.

[28]Cf., H. P. Fritz, Adv. Organometallic Chem., 1962, 1, 240.

[29]T. Ogawa, Spectrochim. Acta, 1968, 24A, 15.

[30]E. G. Hoffmann, Z. Elektrochem., 1960, 64, 616.

[31]Cf., M. R. Litzow and T. R. Spalding, 'Mass Spectrometry of Inorganic and Organometallic Compounds', Elsevier, Amsterdam, 1973, p. 519.

[32]H. J. M. Bartelink, H. Bos. J. Smidt, C. H. Vrinssen, and E. A. Adema, Rec. Trav. chim., 1962, 81, 225, and references therein.

[33]J. G. Kenworthy, J. Myatt, and M. C. R. Symons, J. Chem. Soc. (A), 1971, 1020.

[34]J. C. Huffman and W. E. Streib, J. C. S. Chem. Comm., 1971, 911.

[35]J. L. Atwood, W. E. Hunter, R. D. Rogers, J. Holton, J. McMeeking, R. Pearce, and M. F. Lappert, J. C. S. Chem. Comm., 1978,

[35a]E. C. Baker, L. D. Brown, and K. N. Raymond, Inorg. Chem., 1975, 14, 2680.

[36] Cf. , H. Gysling and M. Tsutsui, Adv. Organometallic Chem. , 1970, 9, 361; R.G. Hayes and J. L Thomas, Organometallic Rev. , 1971, 7A, 1; M. Tsutsui, N. Ely, and A. E. Gebala, Ann. New York Acad. Sci. , 1974, 239, 160; M. Tsutsui, N. Ely, and R. Dubois, Accounts Chem. Res. , 1976, 9, 217.

[37] F.A. Hart, A.G. Massey, and M.S. Saran, J. Organometallic Chem. , 1970, 21, 147.

[38] G.K. Barker and M.F. Lappert, J. Organometallic Chem. , 1974, 76, C45.

[39] M.F. Lappert and R. Pearce, J.C.S. Chem. Comm., 1973, 126.

[40] H. Schumann and S. Hohmann, Chem. Zeitung, 1976, 100, 336.

[41] D.F. Evans, G.V. Fazakerly, and R.F. Phillips, J. Chem. Soc. (A), 1971, 1931.

[42] S.A. Cotton, F.A. Hart, M.B. Hursthouse, and A.J. Welch, J.C.S. Chem. Comm., 1972, 1225.

[43] G.B. Deacon and D.G. Vince, J. Organometallic Chem. , 1976, 112, C1.

[44] E. Murphy and G.E. Toogood, J. Inorg. Nuclear Chem. Letters, 1971, 7, 755.

[45] M. Tsutsui and N.M. Ely, Inorg. Chem. , 1975, 14, 2680.

[46] L.E. Manzer, Inorg. Chem. , 1976, 15, 2567.

[47] B.L. Kalsotra, R.K. Multani, and B.D. Jain, J. Inorg. Nuclear Chem. , 1973, 35, 311.

[48] M Tsutsui and N. Ely, J. Amer. Chem. Soc. , 1975, 97, 3551.

[49] F.W. van der Weij, H. Scholtens, and J.H. Teuben, J. Organometallic Chem. , 1977, 127, 299.

[50] E.G. Hoffman, Trans. Faraday Soc. , 1962, 58, 642.

[51] K. Fischer, K. Jonas, P. Misbach, R. Stabba, and G. Wilke, Angew. Chem. Intern. Ed. Engl. , 1973, 12, 943.

[52] K. Nakamoto, 'Characterisation of Organometallic Compounds', Interscience, 1969, Ch. 3.

[53] P. Krohmer and J. Goubeau, Z. anorg. Chem., 1969, 369, 238.

[54] N. Sheppard, Trans. Faraday Soc., 1955, 51, 1465.

[55] D. M. Adams, J. Chatt, and B. L. Shaw, J. Chem. Soc., 1960, 2047.

[56] N. S. Gill, R. H. Nuttal, D. E. Scaife, and D. W. A. Sharp,
J. Inorg. Nuclear Chem., 1961, 18, 79.

[57] J. Lewis, J. R. Miller, R. L. Richards, and A. Thompson,
J. Chem. Soc. (A), 1965, 5850.

[58] L. Holleck and L. Hartinger, Angew. Chem., 1955, 67, 648.

[59] J. H. Van Vleck and A. Frank, Phys. Rev., 1929, 34, 1494, 1625.

[60] D. J. Cole-Hamilton and G. Wilkinson, J.C.S. Dalton, 1977, 797.

[61] G. R. Scollary, Austral. J. Chem., 1978, 31, 44.

[62] M. C. Day and J. Selbin, 'Theoretical Inorganic Chemistry', Reinhold,
1969, p. 118.

LABILITY AND STABILITY IN F-TRANSITION METAL ORGANOCOMPOUNDS: SOME APPLICATIONS OF UNUSUAL LIGANDS AND TECHNIQUES TO THEIR SYNTHESIS

Kenneth W. Bagnall

Department of Chemistry, University of Manchester, Manchester M13 9PL, England

I. INTRODUCTION

The lability of lanthanide and actinide complexes is a factor which is sometimes overlooked by preparative chemists who are not familiar with the field. When this occurs, the results can be a publication which is, at best, misleading. The aim of this chapter is to draw attention to factors of this kind by discussing briefly the underlying reasons for labile behaviour, and then critically to consider the lability of lanthanide and actinide systems with π-bonded ligands, together with some preparative implications of behaviour of this kind. Finally, some aspects of the recent work on σ-bonded lanthanide and actinide species is discussed.

II. LIGAND FIELD SPLITTING IN LANTHANIDE AND ACTINIDE SYSTEMS

The magnitudes of the ligand field splittings in lanthanide ion systems are only about 1% of those observed for d-transition metal ions, and are of the order of $100cm^{-1}$ ($\sim kT$).[1] It is difficult to assess the values of these splittings for f^1, f^2 and f^3 actinide ions in the earlier part of the actinide series, but it seems probable that spin-orbit coupling in these ions is of the same order as the ligand field splittings, and the latter are certainly larger for the actinide ions than for the corresponding lanthanide ions. For example[2], for the U^{4+} ion (f^2) in an octahedral field, the splitting caused by spin-orbit interaction is more than twice that for the isoelectronic lanthanide ion, Pr^{3+}. Thus in the early part of the actinide series there are noticeable, but not very large, ligand field effects in the electronic spectra of Pa(IV), U(IV) and U(V)

T. J. Marks and R. D. Fischer (eds.), Organometallics of the f-Elements, 221–248.
All Rights Reserved. Copyright © 1979 by D. Reidel Publishing Company, Dordrecht, Holland.

systems, whereas in the lanthanide systems such effects are
much less easily discerned. Further along the actinide series
these effects become appreciably less marked and the situation
then becomes similar to that which is observed for the lanthanide
ions.

The change in behaviour across the actinide series can be
explained as being the result of the more diffuse nature of the
5f orbitals in the early part of the series than is the case
with the lanthanides. This spatial extension diminishes very
markedly with increasing atomic number as a result of the rapid
increase in effective nuclear charge across the series. The
rate of this latter increase is greater than that found for the
lanthanides, as can be seen from a consideration of the ratios
of the radii of isoelectronic ions in the same oxidation state,
and with the same C.N., for $r_{Ac}3+/r_{La}3+$ is 1.07 and this ratio
has fallen to 1.04 for $r_{Cf}3+/r_{Dy}3+$.

Because of the absence of extensive interactions between
4f or 5f orbitals and the ligand orbitals, ligand field
stabilisation effects are at best minimal and contribute very
little to the kinetic stability of the complexes formed by these
elements, and such complexes should be regarded as kinetically
labile. Thus there is a considerable degree of flexibility in
the coordination geometries adopted by the complexes and in
ligand exchange reactions because there is little or no gain or
loss of ligand field stabilisation energy involved in an intra-
or intermolecular rearrangement.

III. π-BONDED LIGANDS

A.   Cyclopentadienyl Complexes

The labile behaviour of $\eta^5-C_5H_5$ complexes of the lanthanides
and actinides is shown in the formation of[3a] $[(\eta^5C_5H_5)_2LnCl]_2$
from $(\eta^5C_5H_5)_3Ln$ and $LnCl_3$ (Ln = lanthanide element) in THF
(equation 1):

$$4(\eta^5C_5H_5)_3Ln + 2LnCl_3 \rightarrow 3[(\eta^5C_5H_5)_2LnCl]_2 \quad -------- \quad (1)$$

and in the analogous formation[3b] of $(\eta^8C_8H_8)LnCl$ from
$[(\eta^8C_8H_8)_2Ln]^-$ and $LnCl_3$. In these cases it would be interesting
to react a given tris cyclopentadienyl or bis cyclooctatetraenyl
lanthanide with the trichloride of a different lanthanide element,
a reaction which should produce three possible products in
solution; for example, the two different $[(\eta^5C_5H_5)_2LnCl]_2$ and
$[(\eta^5C_5H_5)Ln(\mu Cl_2)Ln'(\eta^5C_5H_5)_2]$. The equilibrium constants for
the exchange might well be estimated by spectroscopic means.

Similarly, in the actinide series $(\eta^5C_5H_5)_3ThCl$ has been obtained by subliming $(\eta^5C_5H_5)_4Th$ through a layer of thorium tetrachloride[4] (equation 2);

$$3(\eta^5C_5H_5)_4Th + ThCl_4 \rightarrow 4(\eta^5C_5H_5)_3ThCl \qquad \text{-------- (2)}$$

It is therefore obvious that great care must be exercised in characterising the products of any reaction which involves the possibility of ligand exchange of this type; conventional analyses are inadequate for this purpose, and [1]H-n.m.r. spectra may also mislead.

An example which illustrates this problem is the reported[5] formation of a compound of composition "$(\eta^5C_5H_5)_2UCl_2$", prepared by treating uranium tetrachloride with the stoicheiometric quantity of thallium(I) cyclopentadienide in dimethoxyethane (DME), followed by evaporation of the filtrate, so that the analysis of the product merely represents the quantities of reagents used. In 1974 Kanellakopulos and co-workers[6] showed that the chemical behaviour of this product was consistent with the formulation $[(\eta^5C_5H_5)_3UCl]_2[UCl_4(DME)_2]$ or $[(\eta^5C_5H_5)_3U]_2^+[UCl_6]^{2-}$, and also reported evidence for the complex $[(\eta^5C_5H_5)_3U]^+[UCl_5]^-$. A product of composition $(\eta^5C_5H_5)UCl_3 \cdot DME$ had also been reported[7], but without full characterisation. The complex $(\eta^5C_5H_5)_3UCl$ is sterically rather crowded as compared with a monomeric compound of composition $(\eta^5C_5H_5)_2UCl_2$, so that the latter would be expected to form complexes with reasonably small oxygen or nitrogen donor ligands, so changing the symmetry of the environment of the uranium atom from that in $(\eta^5C_5H_5)_3UCl$. Thus the u.v./visible spectra should be easily differentiated, and the same should apply to complexes of the type $(\eta^5C_5H_5)UCl_3L_2$, where L is a monodentate ligand. U.v./visible spectroscopy should then make it possible to identify the genuine compounds of these compositions and to follow ligand re-arrangement reactions leading to mixtures of disproportionation products.

In a study of this kind[8a] in the author's laboratory, the u.v./visible spectra of complexes of the type $[(\eta^5C_5H_5)UCl_3L_2]$ (L = $CH_3CON(CH_3)_2$, DMA; $(CH_3)_3CCON(CH_3)_2$, DMPVA; $(C_6H_5)_3PO$, TPPO and $\frac{1}{2}[(C_6H_5)_2P(O)(CH_2)_2P(O)(C_6H_5)_2]$, DPPOE) in the 1000-1300 nm region (Table 1) were found to be essentially identical, with the most intense band (solid reflectance) at 1116nm (L_2 = DPPOE). Overall, the spectrum is somewhat similar to that[9] of $[UCl_4(CH_3CONH(iC_3H_7))_4]$, although shifted a little as might be expected for the change in ligand. It should be noted that the solution spectrum of $(\eta^5C_5H_5)_3UCl$ (Table 1 and Figure 1) exhibits six maxima in the 1000-1300nm region, the strongest of which are at wavelengths at which the complexes $[(\eta^5C_5H_5)UCl_3L_2]$ (Figure 1) do not absorb, so that disproportionation of any species of composition $(\eta^5C_5H_5)_2UCl_2L_x$ can be recognised easily. A study

Table 1.   The u.v./visible spectra* (1000–1500nm) of
           $[(\eta^5C_5H_5)UCl_3(DPPOE)]_2$ (1), $[(\eta^5C_5H_5)_2UBr_2(DPPOE)]_2$ (2)
           and $[(\eta^5C_5H_5)_3UCl]$ (3) [8a]

	THF solution	Solid reflectance
1	1116, 1184	1116
2	1114, 1145**	1107, 1134, 1290, 1470
3	1004, 1025, 1045,	1036,
	1127, 1205, 1290	1139, 1199, 1295

*   Most intense bands underlined
** CH_2Cl_2 solution; solubility in THF is very low

of the $(\eta^5C_5H_5)ThCl_3$ complex systems with oxygen and other donor
ligands has also been carried out[8b], but here the f^0 system does
not give rise to u.v./visible spectra and one is restricted to
[1]H- and ^{13}C-n.m.r. spectroscopy for identification purposes.
The results are less clear cut than in the uranium(IV) system;
similar studies of the neptunium(IV) and plutonium(IV) systems
are now beginning in the author's laboratory.

The reaction of complexes of the type $[(\eta^5C_5H_5)UCl_3L_2]$,
where L is a monodentate ligand, with the stoicheiometric
quantity of thallium(I) cyclopentadienide invariably yielded
products of the analytical stoicheiometry $(\eta^5C_5H_5)_2UCl_2L_x$ on
evaporation of the filtrate from the reaction, but in all cases
the u.v./visible spectra indicated that disproportionation had
occurred, for features assignable to $(\eta^5C_5H_5)_3UCl$ were predominant in them.

This disproportionation presumably involves an equilibrium
of the type shown in equation 3:

$$2[(\eta^5C_5H_5)_2UCl_2L_2] \rightleftharpoons [(\eta^5C_5H_5)_3UCl] + [(\eta^5C_5H_5)UCl_3L_2] + 2L$$
$$----(3)$$

The same conclusion has been reached in an extensive study of
"$[(\eta^5C_5H_5)_2UCl_2]$" by infrared, u.v. visible and [1]H-n.m.r.
spectroscopy, as well as by X-ray powder diffraction[10b].

Some of the complexes of the monocyclopentadienyl trichloride
have a relatively low solubility in solvents such as THF,
examples being those with bidentate nitrogen donor ligands, such
as 2,2'-bipyridyl and 1,10-phenanthroline[10b] and the triphenylphosphine oxide complex, $[(\eta^5C_5H_5)UCl_3(TPPO)_2]$[8a]. It has now
been shown [8a] that reaction of $[(\eta^5C_5H_5)_3UCl]$ with $[UCl_4(TPPO)_2]$
and the free ligand in THF yields crystals of this monocyclo-

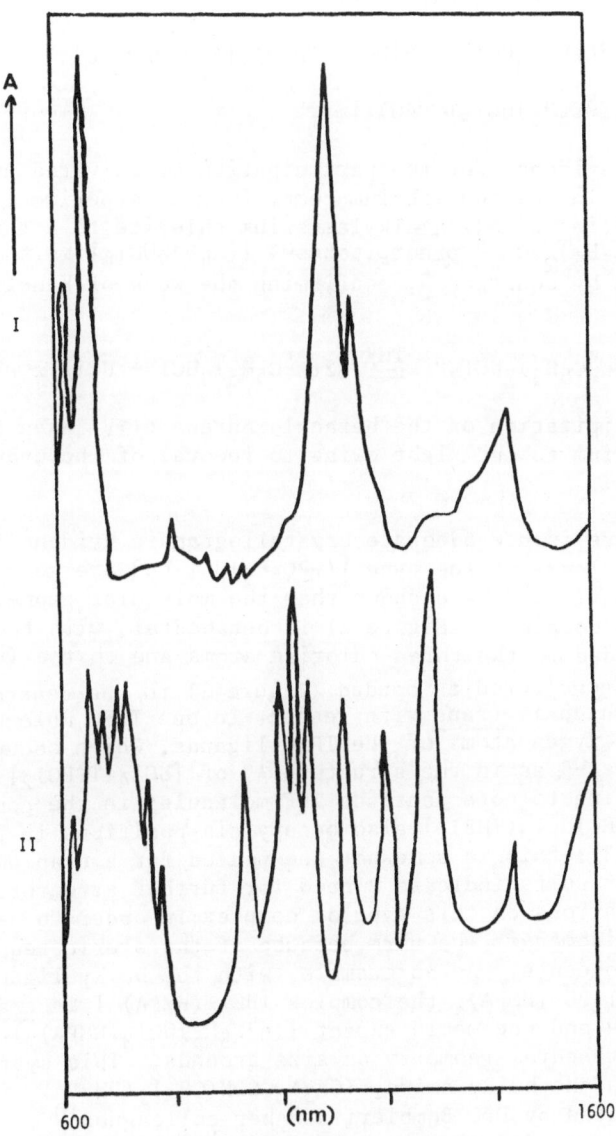

Figure 1.   The u.v./visible spectra (600-1600nm) of the
            complexes $(\eta^5C_5H_5)UCl_3(THF)_2$ (I) and $(\eta^5C_5H_5)_3UCl$
            (II) in THF solution.

pentadienyl complex, while the u.v./visible spectrum of the
supernatant shows that small amounts of the other two uranium(IV)
complexes are present, confirming that the process is also an
equilibrium one (equation 4):

$$[(\eta^5C_5H_5)_3UCl] + 2[UCl_4(TPPO)_2] + 2TPPO \rightleftharpoons$$

$$3[(\eta^5C_5H_5)UCl_3(TPPO)_2] \qquad\qquad \text{-------- (4)}$$

Further evidence for the participation of solvated uranium tetrachloride in the equilibrium comes from an experiment in which
the addition of a tetraalkylammonium chloride to a thf solution
of "$(\eta^5C_5H_5)_2UCl_2$" precipitates[8c] $[(R_4N)_2UCl_6]$ in the proportion
required by equation 5, confirming the work of Kanellakopulos
et al[6]:

$$3"(\eta^5C_5H_5)_2UCl_2" \underset{}{\overset{THF}{\rightleftharpoons}} 2(\eta^5C_5H_5)_3UCl + UCl_4 \cdot xTHF \quad \text{---(5)}$$

the precipitation of the hexachlorouranate(IV) then shifts the
equilibrium to the right owing to removal of the uranium tetrachloride.

There is now adequate crystallographic evidence[10] to show
that complexes of the type $[(\eta^5C_5H_5)UCl_3L_2]$ are genuine; structure
determinations[10] have shown that the molecular geometry of the
bis TPPO complex : (Figure 2) is octahedral, with the uranium
atom bonded to the three chlorine atoms and to the $(\eta^5C_5H_5)$
ligand (considered as bonded (Figure 2) to the centre of the
ring), which is trans with respect to one TPPO molecule, and to
the two oxygen atoms of the TPPO ligands, which occupy cis
positions[10a] as in the structure[11a] of $[UCl_4(TPPO)_2]$. It is
interesting to note that the THF molecules in the complex
$[(\eta^5CH_3C_5H_4)UCl_3(THF)_2]$ also occupy cis positions in the molecule[10b]
(Figure 3); this is somewhat unexpected for a uranium(IV) complex
and the results indicate a need for further structure
determinations on this type of complex in order to see whether
the cis isomer is the norm because of the steric requirements
of the molecule. As an example, with the bulky ligand
$[(CH_3)_2N]_3PO$ (HMPA), the complex $[UCl_4(HMPA)_2]$ is trans octahedral[11b] and one would expect $[(\eta^5C_5H_5)UCl_3(HMPA)_2]$ to retain
trans octahedral geometry on size grounds. This system, and
that with the bulky amide, $(CH_3)_3CCON(CH_3)_2$ (DMPVA), is being
investigated by Dr. Bombieri and her colleagues[10c]. Clearly
there is a good deal of scope for preparative studies of this
kind with other π-bonding ligands, using the known oxygen or
nitrogen donor complexes of the actinide(IV) halides as starting
materials instead of the more usual tetrachloride, and for
extensions of such work to the analogous neptunium(IV) and
plutonium(IV) systems, in which the smaller radii of the
actinide(IV) centres may lead to changes in the stoicheiometry

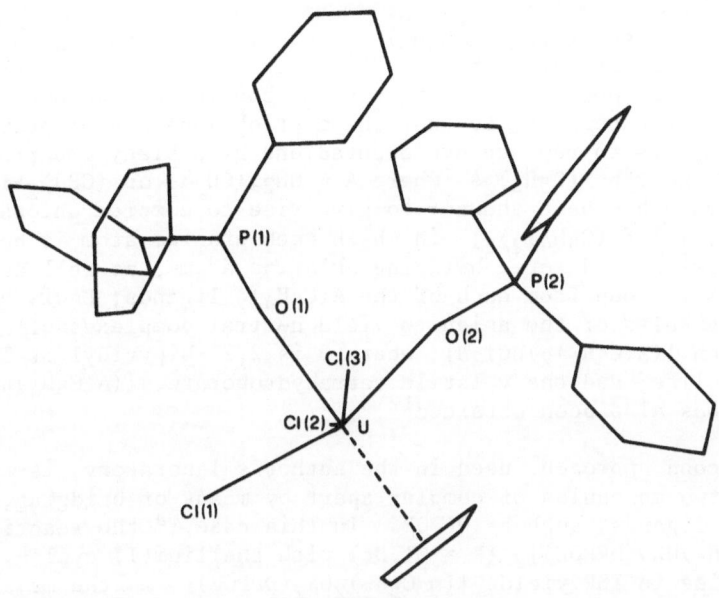

Fig. 2.   The structure of $[(\eta^5C_5H_5)UCl_3(PPh_3O)_2]$.[10a]

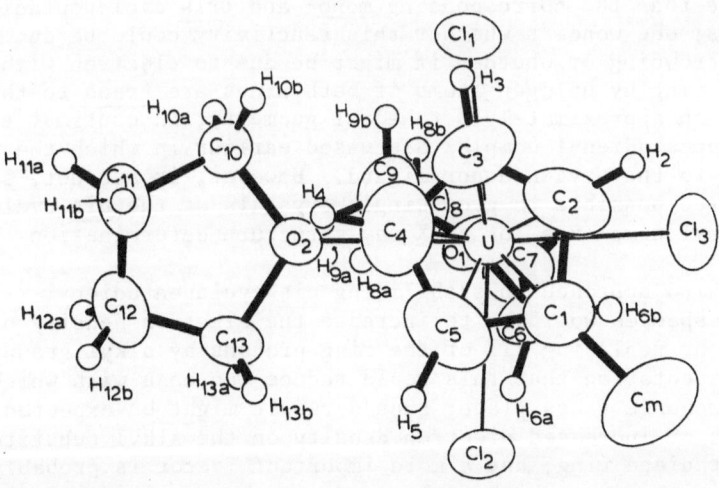

Fig. 3.   The structure of $[(\eta^5CH_3C_5H_4)UCl_3(THF)_2]$[10b]

and coordination geometry adopted by the complexes formed.

The exchange of the $(\eta^5 C_5 H_5)$ ligand almost certainly takes place via an intermediate bimolecular complex in which the η^5 bonded rings of one molecule are also η^1 bonded to a second molecule of the complex, so that the most obvious way of preventing the exchange is to replace cyclopentadiene by a biscyclopentadienyl-alkane of the type $(C_5 H_4)_2 A$, where $A = CH_2, (CH_2)_3$ or $(CH_3)_2 Si$. This approach has been shown[12] to give rise to complex anions of the type $[U_2 Cl_5 (A(C_5 H_4)_2)_2]^-$ in which each uranium atom is bonded to one terminal and three bridging chlorine atoms, as well as to one $(\eta^5 C_5 H_4)$ group from each of the $A(C_5 H_4)_2$ ligands; Lewis bases react with salts of the anion to yield neutral complexes of composition $[(A(C_5 H_4)_2)UCl_2 B]$, where B is 2,2'-bipyridyl or 1,10-phenanthroline, and the volatile tetrahydroborate, $[(A(C_5 H_4)_2)U(BH_4)_2]$ has also been obtained[12].

A second approach, used in the author's laboratory, is to hold the two molecules of complex apart by means of bridging, bidentate ligands, such as DPPOE. In this case,[8a] the reaction of $[(\eta^5 C_5 H_5)UX_3 (DPPOE)]_2$ (X = Cl,Br) with thallium(I) cyclo-pentadienide in THF yields $[(\eta^5 C_5 H_5)_2 UX_2 (DPPOE)]_2$ as the major product, contaminated with only small amounts of $[(\eta^5 C_5 H_5)_3 UX]$, detected by examination of the u.v./visible spectra of the products (Figure 4). The spectrum of this type of complex differs markedly from those of $[(\eta^5 C_5 H_5)UX_3 L_2]$ and $[(\eta^5 C_5 H_5)_3 UX]$ (see Table 1 and Figures 1 and 4), so that identification is relatively easy. It is interesting to note that these bis cyclopentadienyl complexes are much more air and moisture-sensitive than the corresponding mono- and tris cyclopentadienyl complexes; one wonders whether this reactivity could be due to lack of crowding or whether it might be due to electron withdrawal from the ring by halogen atoms if both rings are trans to these atoms in an approximately octahedral geometry, in contrast to the monocyclopentadienyl complex discussed earlier in which the ring is trans to the oxygen donor ligand. However, it has not, so far, proved possible to grow single crystals of the bis cyclo-pentadienyl complexes for an X-ray structure determination.

A third approach to stabilising bis cyclopentadienyl actinide species would be to increase the electron density on the ring by replacing all of the ring protons by alkyl groups, in the expectation that this would reduce the ease with which ligand exchange takes place; such a result might be expected on the basis of increased electron density on the alkyl substituted cyclopentadiene ring, but a more important factor is probably the steric crowding which would ensue in the resulting tris cyclopentadienyl complex. If this is too great for the tris complex to be formed, then the bis complex cannot disproportionate.

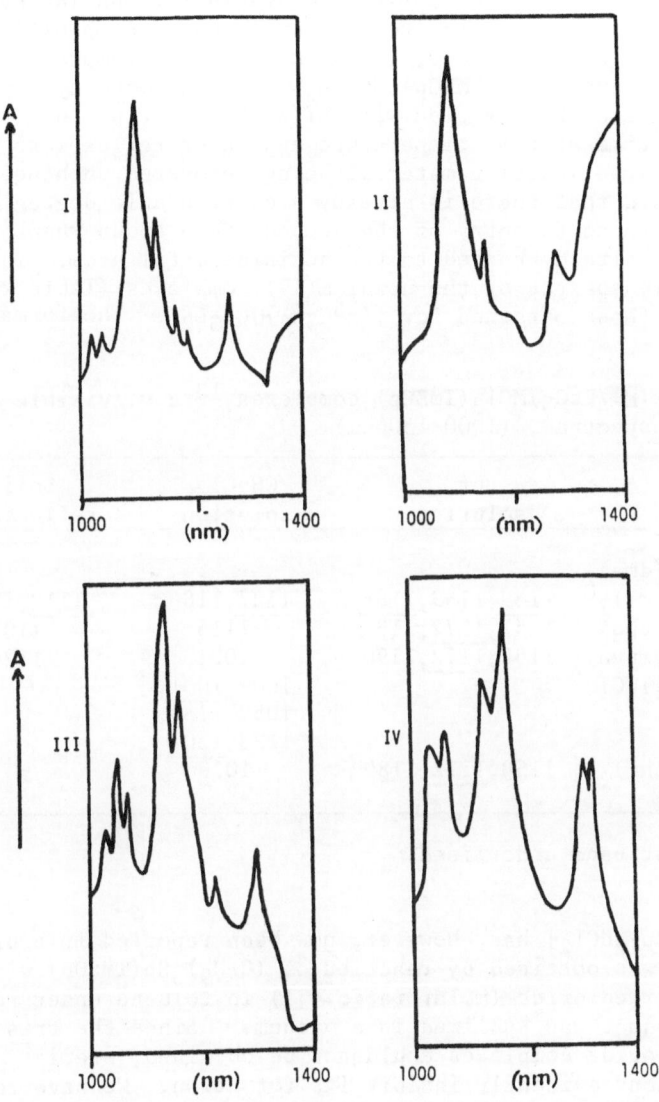

Figure 4. The u.v./visible spectra (1000–1400nm) of the complexes $[(\eta^5C_5H_5)UCl_3(DPPOE)]_2$ (I), $[(\eta^5C_5H_5)UBr_3(DPPOE)]_2$ (II), $[(\eta^5C_5H_5)_2UCl_2(DPPOE)]_2$ (III) and $[(\eta^5C_5H_5)_2UBr_2(DPPOE)]_2$ (IV).

An approach of this type has been attempted using $[(CH_3)_4(C_2H_5)C_5]^-$ (TMECp) in place of (C_5H_5) in the author's laboratory[13] for the thorium(IV) and uranium(IV) halide systems, but although it is easy to prepare complexes of composition $[(\eta^5TMECp)MCl_3L_2]$ (M = Th,U; L = RCONR$_2'$), all attempts to prepare the bis complexes, $[(\eta^5TMECp)_2MCl_2L_x]$ from the mono cyclopentadienyl complexes by reaction with Li(TMECp) over prolonged periods of time at room temperature and under reflux were unsuccessful, the starting material being recovered unchanged. This suggests that there is already a considerable degree of steric hindrance to entry of the second TMECp group when other ligands are already bonded to the actinide metal atom.   The u.v. visible spectra of the uranium(IV) complexes (Table 2) are similar to those obtained for $[(\eta^5C_5H_5)UCl_3L_2]$.   The formation

Table 2. $(Me_4EtC_5)MCl_3$(TMECp) complexes,[13] u.v./visible spectra*, (1000-1500nm)

	thf solution	CH_2Cl_2 solution	Solid reflectance
TMECpThCl$_3$(dma)$_2$	–	–	–
TMECpUCl$_3$(dma)$_2$	1139,1169,<u>1188</u>	1137,<u>1186</u>	<u>1112</u>,1167
TMECpUCl$_3$(tppo)$_2$	<u>1157</u>,1172,<u>1192</u>	<u>1115</u>	<u>1107</u>
TMECpUCl$_3$(dmpva)$_2$	1155,<u>1172</u>,1190	1081	1120
(TMECp)(Cp$_2$)UCl	–	1009,<u>1031</u>, 1058,<u>1133</u>, 1198,<u>1302</u>	–
TMECpUCl$_3$(thf)$_x$	1158,<u>1172</u>,1189	1079	–

*   Strongest band underlined.

of $[(\eta^5TMECp)_2UCl_2]$ has, however, now been reported in a brief note[14]; it was obtained by reaction of $(C_4H_9)_3Sn$(TMECp) with uranium tetrachloride (molar ratio 4:1) in toluene under reflux and the product was sublimed in a vacuum.   Since the tris TMECp uranium chloride complexes could not be obtained, steric considerations evidently inhibit its formation. We have found that the reaction of the mono TMECp complexes, $(\eta^5TMECp)UCl_3L_2$, with Tl(C$_5$H$_5$) results in disproportionation, the products including $[(\eta^5TMECp)(\eta^5C_5H_5)_2UCl]$ and the starting material[13]. The formation of the bis pentamethylcyclopentadienyl (PMCp) complexes, $(PMCp)_2MCl_2$ (M = Th,U) has now been reported by Marks and co-workers[15], and they have found that these give rise to bis alkyl species which are remarkably stable to heat[15].   Some aspects of their chemical behaviour will be mentioned later in this chapter and some structural studies are described elsewhere in this volume.

Another approach to stabilising $[(\eta^5 C_5 H_5)_2 UX_2]$ entity would be by replacement of one of the halogen atoms by a multidentate anionic ligand in the expectation that the need to break more than one bond between the uranium atom and a ligand of this type would provide a kinetic barrier to the disproportionation. In one series of such experiments, β-diketones were used as possible stabilising ligands[16]; however, $[(\eta^5 C_5 H_5)UCl_2(acac)]$ (Hacac = pentane-2,4-dione) reacts with thallium(I) cyclopentadienide in THF to yield a mixture of disproportionation products which includes $(\eta^5 C_5 H_5)_3 UCl$, U(acac)$_4$ and, possibly, UCl$_2$(acac)$_2$. This is not very surprising, for β-diketonates are known to be quite labile, as shown by experiments on the exchange of the bound ligand in mixtures of uranium(IV) β-diketonates, such as U(acac)$_4$ and U(dpm)$_4$ (Hdpm = $\underline{t}C_4 H_9 COCH_2 CO\underline{t}C_4 H_9$, 2,2,6,6-tetramethylheptane-3,5-dione); the ^{1}H-n.m.r. spectra of these mixtures in CDCl$_3$ shows all possible combinations of mixed chelate[18].

However, $[(\eta^5 C_5 H_5)UCl_2(acac)]$ is evidently coordinatively unsaturated, for with TPPO the complex $[(\eta^5 C_5 H_5)UCl_2(acac)(TPPO)_2]$ is formed and its u.v. visible spectrum is almost identical to that of $[(\eta^5 C_5 H_5)UCl(acac)_2]$ (Table 3 and Figure 5), another complex which is stable with respect to disproportionation.

Table 3. Cyclopentadienyl-acetylacetonato and (pyrazol-1-yl) borato complexes, u.v./visible spectra* (1000-1500nm)[16] (Cp = $\eta^5 C_5 H_5$)

CpUCl$_2$(acac)	1117, <u>1157</u>, 1190
CpUCl$_2$(acac)(TPPO)$_2$	1110, <u>1183</u>
CpUCl(acac)$_2$	1108, <u>1166</u>, 1188
Cp$_2$U(acac)$_2$	<u>1133</u>, 1193
Cp$_2$UCl(acac)(TPPO)$_2$	1141, <u>1184</u>
CpUCl$_2$(HBPz$_3$)	1157, <u>1175</u>
CpUCl(HBPz$_3$)$_2$	1077, <u>1123</u>, 1172
Cp$_2$UCl(HBPz$_3$)	<u>1171</u>
Cp$_2$U(HBPz$_3$)$_2$	<u>1175</u>
CpUCl$_2$(H$_2$BPz$_2$)	<u>1159</u>, 1183
CpUCl(H$_2$BPz$_2$)$_2$(tppo)$_2$	<u>1128</u>, 1159
CpUCl$_2$(H$_2$BPz$_2$)(tppo)$_2$	<u>1112</u>, 1149, 1183
CpThCl$_2$(HBPz$_3$)L$_x$**	–
CpThCl$_2$(H$_2$BPz$_2$)Ly**	–

*
**   Solvent; thf or CH$_2$Cl$_2$. Strongest band underlined
     x = 1, L = dmpva (Me$_3$CCONMe$_2$); x = 1.5, L = dma (MeCONMe$_2$);
     y = 1.5, L = dma.

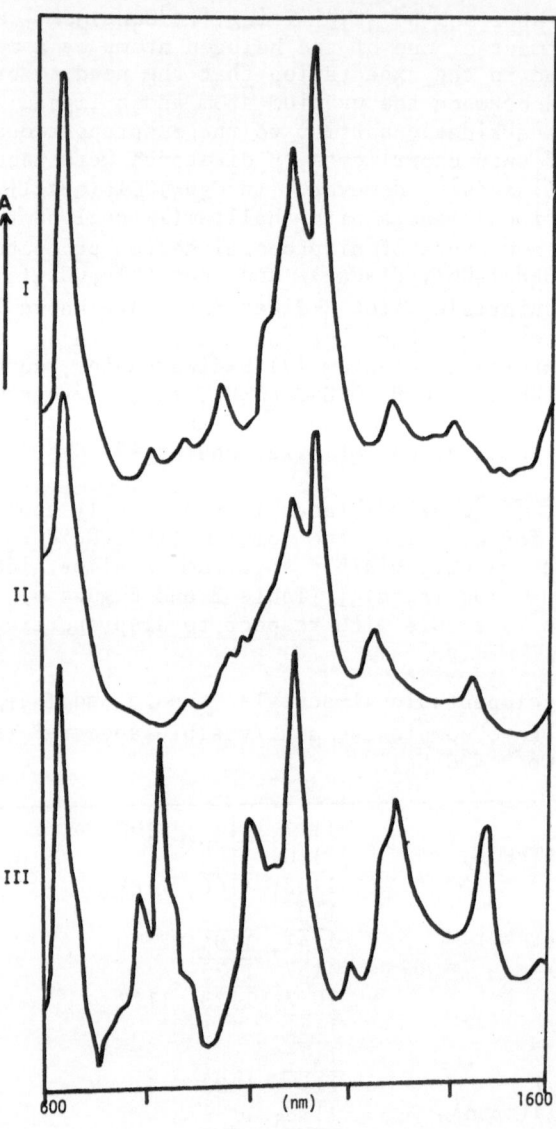

Figure 5.   The u.v./visible spectra of $[(\eta^5C_5H_5)UCl(acac)_2(TPPO)_2]$
(I) $[(\eta^5C_5H_5)_2UCl(acac)(TPPO)_2]$ (II) and
$[(\eta^5C_5H_5)_2U(acac)_2]$ (III) in THF solution.

Reaction of this bis TPPO complex with Tl(C_5H_5) in THF then yields
[($\eta^5C_5H_5$)$_2$UCl(acac)(TPPO)$_2$], the u.v./visible spectrum of which
is almost identical to that of [($\eta^5C_5H_5$)$_2$(acac)$_2$] (Table 3)
prepared from [($\eta^5C_5H_5$)UCl(acac)$_2$] and Tl(C_5H_5).[16]   The
preparation of [($\eta^5C_5H_5$)$_2$U(acac)$_2$] has previously been reported
at an A.C.S. meeting.[17c]   Both of these bis cyclopentadienyl-
acac complexes are stable with respect to disproportionation[16],
a result which is probably due to steric crowding. It should be
noted that [($\eta^5C_5H_5$)$_3$U(acac)] has not yet been obtained; when
[($\eta^5C_5H_5$)$_3$UCl] is treated with the stoicheiometric quantity of
Na(acac), the bis cyclopentadienyl complex, [($\eta^5C_5H_5$)$_2$U(acac)$_2$]
results. This last is one of the few products obtained in this
work which is soluble in hydrocarbons such as n-pentane[16]. It is
possible that the reaction of [($\eta^5C_5H_5$)$_3$UCl] with Ag(acac) in
toluene may lead to [($\eta^5C_5H_5$)$_3$U(acac)], but the results of this
experiment are not yet available[16].

Another example of a stable bis cyclopentadienyl uranium(IV)
complex is the borohydride, [($\eta^5C_5H_5$)$_2$U(BH$_4$)$_2$] prepared by
reaction of UCl$_2$(BH$_4$)$_2$ with Tl(C_5H_5) or of "($\eta^5C_5H_5$)$_2$UCl$_2$" with
sodium borohydride in dimethoxyethane; the product was isolated
by vacuum sublimation at 80°C/10^{-2} Torr and a preliminary X-ray
investigation suggests an approximately tetrahedral arrangement
of the ligands about the uranium atom.[17a] The potentially
tridentate ligand, [HBPz$_3$]$^-$ (Pz = the pyrazole group, $C_3H_3N_2$),
might provide more of a kinetic barrier to disproportionation
than a β-diketone,and it is known that the complex [($\eta^5C_5H_5$)UCl$_2$
(HBPz$_3$))], in which the HBPz$_3$ ligand is tridentate, is thermally
stable and can be sublimed in a vacuum.[17b] This complex reacts[16]
very slowly with Tl(C_5H_5) in THF to yield [($\eta^5C_5H_5$)$_2$UCl(HBPz$_3$)],
and it is also possible to obtain [($\eta^5C_5H_5$)$_2$U(HBPz$_3$)$_2$] in a
similar manner; both of these bis cyclopentadienyl complexes are
stable with respect to disproportionation[16] and some details of
their spectra are given in Table 3.

These results on the cyclopentadienyl uranium(IV) systems
inevitably raise some doubts as to the validity of the reports
of a number of other cyclopentadienyl and π-complex species which
have appeared in recent years. Thus the preparation[19] of
"($\eta^5C_5H_5$)$_2$U(C_6H_5)$_2$" from "($\eta^5C_5H_5$)$_2$UCl$_2$" and C_6H_5Li in THF at
-10°C must be regarded as very doubtful in view of the above
information on the nature of the starting material used for the
preparation and it is probable that the product is a mixture of
($\eta^5C_5H_5$)$_3$U(C_6H_5) and, perhaps, solvated Li$_2$U(C_6H_5)$_6$. The nature
of ($\eta^5C_5H_5$)$_2$U(NR$_2$)$_2$, prepared by Takats and co-workers from
U(NR$_2$)$_4$ and the stoicheiometric quantity of cyclopentadiene[20a],
is much less uncertain, for this compound should be genuine on
three counts. Firstly, U(NEt$_2$)$_4$ is dimeric, the structure[21]
consisting of two trigonal bipyramidal units sharing an edge, so
that the two uranium atoms are held apart; it would be interesting

to see whether $(\eta^5 C_5H_5)_2 U(NEt_2)_2$ is similar in structure.
Secondly, there may be some degree of $p\pi$ - $f\pi$ interaction in the
U-N bonds, leading to a stronger terminal U-N bond, which might
increase the activation energy required for the ligand exchange
reaction; however, the non-bridging U-N bond lengths are[21] 2.22(2)
Å. Thirdly, the steric crowding about the uranium atom must be
considerable and it is very doubtful whether there is space for
an additional σ-bonded C_5H_5 ring to form the disproportionation
reaction intermediate. This last consideration is probably
paramount.

There is also chemical evidence for the genuineness of the
bis dialkylamido complex; Takats et al[20b] have shown that the
compound undergoes insertion reactions with CS_2, COS and CO_2 to
yield the corresponding bis carbamates. The dithio- and oxo-
thiocarbamates are monomers in benzene whereas the dioxocarbamate
is polymeric, and variable temperature ^{1}H-n.m.r. studies indicate
dynamic solution behaviour. However, no u.v./visible spectra have
been reported. In other work[20c] Jamerson and Takats have shown
that the formation of $[(\eta^5 C_5H_5)_3 U(NEt_2)]$ from reaction of $U(NR_2)_4$
with an excess of cyclopentadiene is extremely slow, an indication
of the steric crowding, and that formation of $(\eta^5 C_5H_5)_4 U$ from
the diethylamide does not occur even at 80°C, although it does
occur with the corresponding dimethylamide, again a reflection of
the steric crowding in the ethyl analogue.

In connection with this type of alkylamide complex it would
also be interesting to see whether the trimeric complex[22]
$[U_3 (MeNCH_2CH_2NMe)_6]$ retains the U-U-U backbone on reaction with
C_5H_6 and whether the distorted tetrahedral monomer[23], $U\{N(C_6H_5)_2\}_4$,
reacts with cyclopentadiene to yield a stable bis cyclopentadienyl
complex; the U-N bond lengths in $U\{N(C_6H_5)_2\}_4$ range from 2.21(2)
to 2.35(2)Å (average 2.27Å)[23]; in this case either the steric
crowding about the uranium atom or the U-N bond breaking step,
mentioned above, may provide a barrier to the disproportionation
of $(\eta^5 C_5H_5)_2 U\{N(C_6H_5)_2\}_2$. These systems are being investigated
by Dr. G. Bombieri[10c].

Apart from the reaction with cyclopentadiene, dialkylamides
do not appear to have attracted very much attention for the
synthesis of organo-lanthanide or actinide complexes. The proton
affinity of the $-NR_2$ group is extremely high and reactions with
other organic species of lower proton affinity are obviously
possible. A number of products obtained by reaction of $(\eta^5 C_5H_5)_2$
$U(NR_2)_2$ with thiols and dithiols have been reported by Takats
et al[20a,c] who have found that by the reaction with thiophenol
yields $(\eta^5 C_5H_5)_3 U(SC_6H_5)$ as a result of disproportionation, whereas
with the bulky t-butylthiol much less disproportionation was
observed, presumably because of steric considerations. It is of
interest that $[(\eta^5 C_5H_5)_2 U(S\underline{t}C_4H_9)_2]$ is soluble in n-pentane. In

the corresponding reaction with o-mercaptophenol
(H_2omp), the mass spectrum of the product showed the presence of
the ion $[(\eta^5C_5H_5)_2U(omp)]_2^+$, suggesting a dimeric formulation for
the parent bis cyclopentadienyl compound, and the same result was
observed with the catechol reaction product, whereas with toluene-
3,4-dithiol (H_2tdt) the mass spectrum of the reaction product
showed the presence [20c] of the monomeric ions $[(\eta^5C_5H_5)_2U(tdt)]^+$
and $[(\eta^5C_5H_5)U(tdt)]^+$

In the lanthanides, the preparation of $[(\eta^5C_5H_5)_3CeCl]$, and
a number of derivatives obtained by replacement of the halogen
atom, have been described[71]; the preparative route was the reaction
of $(C_5H_5NH)_2CeCl_6$ with NaC_5H_5 in THF, a method applied also for
the preparation[28] of $[(\eta^5C_5H_5)_4Ce]$; the Ce(IV) centre is slightly
smaller than U(IV) in comparable compounds, so that the Ce(IV)
centre will be even more crowded than in the latter case, but the
main problem is the oxidising nature of Ce(IV) in the hexachloro-
cerate(IV) which, in the author's experience, is so strong an
oxidant that a Ce(III) product would be more likely. For this
reason these reactions should be regarded with reserve and re-
investigated. The preparation[72], of $[(\eta^5C_5H_5)_3Ce(OiC_3H_7)]$ from
the less oxidising $[Ce(OiC_3H_7)_4]$ and $Mg(C_5H_5)_2$ illustrates the
utility of alkoxides as a starting material for syntheses of
complexes of this type, and the procedure should be applicable to
plutonium(IV) systems.

B. Indenyl Complexes

The situation in the case of the bis indenyl uranium(IV)
complexes also requires clarification; the complexes
$[(\eta^5C_9H_7)_2UR_2]$ (R = CH_3, tC_4H_9) have been reported[26] to be formed
by reaction of the dichloride (prepared in situ from $Na(C_9H_7)$ and
UCl_4 in THF) with RLi at -70°C, but they have not been fully
characterised; other work[27a] on the uranium(IV)/indenyl system
has indicated that $[(\eta^5C_9H_7)_3UCl]$ can be prepared and its
structure[27a] is known, so that there is a possible disproportionation route, and there is experimental evidence[27b] for exchange
of the indenyl ligand between this compound and uranium tetrachloride. However, $[(\eta^5C_9H_7)UCl_3]$ does not appear to exist[27c],
and this may inhibit disproportionation of the bis indenyl
complex. Further work on these systems is being undertaken[27b].
The reported[28] preparation of $(\eta^5C_9H_7)_4Ce$ from $(C_5H_5NH)_2CeCl_6$
is very doubtful.

C. Allyl Complexes

The reported bis allyl complex, $(\eta^3C_3H_5)_2UCl_2$, prepared[24]
by reaction of $[(\eta^3C_3H_5)_4U]$ with hydrogen chloride in ether at

-40°C, is almost certainly a mixture of disproportionation products, although the bis allyl bis alkoxides [$(\eta^3C_3H_5)_2U(OR)_2$], prepared by reaction of the tetraallyl with the alcohol in hexane (R = C_2H_5 or iC_3H_7) or diethyl ether (R = tC_4H_9), are genuine and a crystal structure of the isopropoxide has shown the compound is a dimer with the two uranium(IV) centres bridged by two alkoxy groups[25]. Two of these alkoxides (R = C_2H_5,iC_3H_7) decompose slowly at room temperature and are stored at -20°C, but the t butoxide (which is pale green, whereas the others are brown) does not decompose at room temperature under nitrogen[25]. It would be interesting to have u.v./visible spectra for these compounds and there is certainly a good deal of scope for further research on the allyl systems.

Although the two known tetraallyl actinide complexes, [$(\eta^3C_3H_5)_4M$] (M = Th[44], U[45]), are thermally very unstable, cyclopentadienyl lanthanide[46] and actinide[47] allyl complexes are very much more stable in this respect. In the uranium(IV) complex [$(\eta^5C_5H_5)_3U(C_4H_7)$]($C_4H_7$ = 2-methylallyl) the allyl group is σ-bonded with a U-C bond length of 2.48(3)Å, appreciably larger than that in the corresponding phenylacetylide[48], [$(\eta^5C_5H_5)_3U$ (C≡CC$_6$H$_5$)],which is 2.33(2)Å.

D.   Other Systems

The synthetic utility of alkoxides as starting materials is shown by the formation of [$(\eta^8(C_8H_8)_3Ce$] and [$(\eta^8C_8H_8)_3Ce_2$] by reduction of Ce(OiC_3H_7)$_4$ in the presence of cyclooctatetraene[72]. However the reported preparation of the tetrafluorency complex, $(\eta^5C_{13}H_9)_4Ce$, from $(C_5H_5NH)_2CeCl_6$ and Na($C_{13}H_9$) in THF[29] must be very doubtful indeed both on steric grounds and on oxidative power of the [$CeCl_6$]$^{2-}$ anion, and the bis cycloheptatrienyl cerium(IV) dichloride claimed by the same authors[30] would, if genuine, involve a considerable degree of ionic bonding, yet it is reported[30] to be unaffected by water or dilute acid. The preparation of lanthanide and actinide fluorenyl complexes is certainly worth investigation, for on steric grounds one would expect the latter to form [$(\eta^5C_{13}H_9)_2MX_2$] (X = halogen) which should be stable with respect to disproportionation.

E.   Metal Atomisation Techniques

Any survey of π-bonded lanthanide and actinide systems would be incomplete without a mention of metal atomization techniques as a synthetic method. Such techniques have been developed for the synthesis of low or zero oxidation state transition metal complexes, such as the bis arene niobium compounds[31], Nb(arene)$_2$ (arene = benzene, toluene or mesitylene) by co-condensation of

niobium vapour with the aromatic ·compounds at 77K. The technique
does not yield the bis arene uranium compounds[32], but it has been
successfully applied to the synthesis of $Nd_2(\eta^8 C_8 H_8)_3$ from the
metal and cyclooctatetraene[33] and in an investigation of the
reaction of the lanthanide metals with 1,3-butadiene and 2,3-
dimethyl butadiene[34] it has been shown that with the former the
main product is $Ln(C_4 H_6)_3$ (Ln = Sm,Nd,Er) while with the latter the
main product is $Ln(C_6 H_8)_2$ (Ln = La,Er). These compounds are very
unstable with respect to oxidation or hydrolysis, the latter process
yielding 2-butene as the major product, together with 1-butene,
octadiene, etc. The bonding in these compounds is not ionic
(i.e. not ¯C-C=C-C¯) since polymerisation does not occur in the
presence of an excess of the diene, and is probably not of the
form

since butadiene is not obtained on hydrolysis; the alternative

would require a rather high formal ‚charge on the lanthanide
metal[34].

    In another extremely interesting contribution from the same
laboratory, a simple reductive method is reported for the
conversion of cis,1,5-cyclooctadiene to the cyclooctatetraenyl
dianion; in this, anhydrous $PrCl_3$ is reduced by potassium in THF
under reflux, and the filtered product is reacted with the diene
at room temperature[75]. The function· of the lanthanide element
in this useful synthesis is uncertain, but reductive techniques
of this kind are certainly worth further investigation.

IV.   LANTHANIDE AND ACTINIDE-CARBON SIGMA BONDED COMPLEXES

A.   Carbonyls and Cyanides

    During and since the Manhattan project (1940-5) period,
several unsuccessful attempts were made to prepare uranium
carbonyls, using the standard technique of reductive carbonyl-
ation. Later work was based on the premise that filled f_{xz}^2 or
f_{yz}^2 orbitals, or the $fy(3x^2 - y^2)$ and $fx(x^2 - 3y^2)$ orbitals, would
be of the correct symmetry for back donation into a π* orbital
on the CO carbon atom, but· no products were obtained. However,

a matrix isolation technique, in which uranium metal was condensed
in a dilute CO-Ar matrix at 4K, has provided infrared evidence
for the existence of a uranium carbonyl species, for at 4K a CO
stretching mode has been observed at 1961cm^{-1} (cf $Cr(CO)_6$, 1990cm^{-1};
$Mo(CO)_6$, 1992cm^{-1}), but this carbonyl species was stable only
below 30K.[35]   Using a similar technique, evidence for the
existence of lanthanide (Pr,Nd,Eu,Gd,Ho) carbonyls in argon
matrices at 8-12K has also been observed, but these are also
unstable above these low temperatures[36]. Evidently the overlap
between the 4f and 5f orbitals and the carbon 2p* orbitals is
rather poor. The formation of lanthanide complexes with d
transition metal carbonyls, in which the carbon atom of the CO
molecule is bonded to the d transition metal, and the oxygen atom
to the f-transition metal, underlines the difference between f
and d transition metals; in one example of this kind, species of
the type $(\eta^5C_5H_5)_2Ln(CO)_3M(\eta^5C_5H_5)$ have been obtained (Ln = Dy,
Ho,Er,Yb; M = Mo,W).[37]

Somewhat surprisingly, there appears to have been no attempt
to prepare PF_3 or substituted PF_3 complexes of the lanthanides or
actinides, for here one would expect a better chance of overlap
between filled 4f or 5f orbitals and the vacant phosphorus 3d
orbitals. A possible preparative route would be reaction of
matrix-isolated lanthanide or actinide carbonyls with PF_3 at low
temperature. It might also be of interest to attempt the
preparation of CS complexes of these elements, although there is
no very obvious preparative route available.

In view of the difficulty experienced in obtaining f
transition metal carbonyls, it is not surprising to find that the
few known cyano-complexes are exceedingly susceptible to hydrolysis.
There are reports of the formation of solvated lanthanide cyanides,
$M(CN)_3(THF)_2$ (M = Pr-Lu,Y), precipitated when solutions of the
tribromides in THF are treated with lithium cyanide[38], but no
infrared spectroscopic or other data have been reported for these
products. The dicyanides, $M(CN)_2$ (M = Eu,[39,40] Yb[40]) have been
prepared by reaction of the metal with ammonium cyanide in
anhydrous liquid ammonia, a method which yields a mixture of the
di- and tri-cyanide; the latter can be leached out with absolute
ethanol. In an extension of this method, a number of lanthanide
tricyanides (Ce,Pr,Sm,Ho) have been prepared by an electrolytic
method in which a concentrated solution of ammonium cyanide in
liquid ammonia is the electrolyte and the electrodes are a platinum
foil cathode and a platinum rod/lanthanide metal anode.[40] These
tricyanides are reported to be thermally stable to 600°C; the
cyanide stretching modes, ν_{CN}, are in the range 2095-2180cm^{-1} and
the CN groups are not ionic. An acid-base system such as the
NH_4CN/NH_3 one used here should be applicable to the preparation of
a variety of lanthanide or actinide cyanides from the metal amides,
$M(NH_2)_x$, or imides, $M(NH)_y$ (equation 6):

$$M(NH_2)_x + xNH_4CN \rightarrow M(CN)_x + 2xNH_3 \qquad \text{--------(6)}$$

but this procedure does not appear to have been used.

Reports of the actinide cyanides are few; the complex $[Cl_3U(CN)(NH_3)_4]$ is precipitated when the tetrachloride is treated with sodium cyanide in liquid ammonia[41] and reaction of $(\eta^5C_5H_5)_nU$ (n = 3 or 4) with hydrogen cyanide in an inert solvent yields[43] $(\eta^5C_5H_5)_{n-1}U(CN)$; in both cases the infrared spectrum indicates that the CN group is not ionic. The corresponding cerium(IV) compounds, $(\eta^5C_5H_5)_3Ce(CN)$ and $(\eta^5C_9H_7)_2Ce(CN)_2$ are reported[42] to be formed from the corresponding chlorides by heating them with potassium cyanide in THF under reflux at 70-75ºC/3-4 hours; the CN group in these products is not ionic (ν_{CN} at 2115 and 2155cm^{-1} respectively) but the compounds need verification.

B.   Alkyls and Aryls, $(\eta^5C_5H_5)_nMR$

All of the known σ-bonded alkyl or aryl lanthanide compounds of the types[49] $(\eta^5C_5H_5)_2LnR$ (R = CH_3, $C\equiv CC_6H_5$, C_6H_5, etc) and $(\eta^5C_5H_5)LnR_2$ (R = $C\equiv C.C_6H_5$) are remarkably stable to heat, with decomposition temperatures in excess of 100ºC. The bonding in the π or polyhapto organolanthanides appears to be essentially ionic[50], which would be in accord with the view that the 4f orbitals do not extend sufficiently far from the nucleus to contribute to covalent bonding to any great extent, and in these compounds the electron density must be spread over several carbon atoms. In the σ- bonded species, however, the monohaptocarbanions, such as CH_3^- or $(C_6H_5C\equiv C)^{-1}$ will have the electron density concentrated on a single carbon atom and there will be a greater possibility of a localised interaction of some kind between the metal and the carbon atoms; for example the f_{x^3} orbital can, for certain molecular symmetries, be of the same symmetry as the carbon p_z orbital, and there is obviously the possibility of orbital overlap.

Similar considerations apply to the actinide complexes of the type $(\eta^5C_5H_5)_3UR$[51,52], and for the corresponding thorium complexes[53]; for these elements the 5f orbitals are more extended spatially than the 4f orbitals, so that even better overlap of the type envisaged for the lanthanide species would be expected. When decomposition of these compounds takes place, the alkane is eliminated by way of proton abstraction from a C_5H_5 ring, giving rise to complexes such as[54] $[(\eta^5C_5H_5)_2(\eta^5,\eta^1C_5H_4)Th]_2$. The failure of the actinide compounds to undergo decomposition by β-elimination is probably due to coordination saturation or steric crowding[51a], but it is interesting that the hydride $(\eta^5C_5H_5)_3UH$, was not formed[51a]. Such a species would be highly reactive. However, Kanellakopulos and co-workers have reported evidence for the

formation of $(\eta^5C_5H_5)_3UH$ and the corresponding deuteride by
reaction[51b] of $[(\eta^5C_5H_5)_3UCl]$ with CaH_2 or CaD_2 and Marks et al[15]
have shown that the bis $((CH_3)_5C_5)$ (PMCp) thorium and uranium
dimethyls react rapidly in toluene with hydrogen at room temperature
to form the doubly bridged dihydrides (equation 7):

$$2[(PMCp)_2M(CH_3)_2] + 4H_2 \rightarrow [(PMCp)_2MH_2]_2 + 4CH_4 \qquad -------- (7)$$

The mechanism of this reaction should certainly prove to be
interesting, for although one can envisage oxidative addition of
hydrogen as a first step for the uranium compound, this cannot
apply to the thorium one. The thorium dihydride is stable at 80°C
in toluene solution, but the uranium compound loses hydrogen
reversibly at room temperature, a binuclear reductive elimination[15]

The photolysis of $(\eta^5RC_5H_4)_3Th(iC_3H_7)$ (R = H,CH$_3$) yields a
mixture of propane (53%) and propene (47%), which must occur via
β- elimination, in contrast to thermolysis in toluene at 170°C
which gives propane and the dimeric bridged complex mentioned
above[54]. The thorium product remaining from the photolysis
reaction is dark green and is of composition $(RC_5H_4)_3Th$[55a],
whereas the deep violet $(\eta^5C_5H_5)_3Th$ is obtained by sodium
reduction of $(\eta^5C_5H_5)_3ThCl$ in THF/naphthalene at 40°C/4 days;[56]
the analogous reactions of $[(\eta^5C_9H_7)_3ThR]$[55b] and $[(\eta^5C_9H_7)_3ThCl]$[27b] yield dark red and yellow products respectively.
It would be interesting to have further structural information on
these products.

Lanthanide-alkyl aluminium complexes of composition
$(\eta^5C_5H_5)_2LnR_2AlR_2$ (R = CH$_3$, Ln = Y[57], Sc,Gd,Dy,Ho,Er,Tm,Yb[58];
R = C$_2$H$_5$, Ln = Sc,Y[58]) are obtained by reaction of $[(\eta^5C_5H_5)_2LnCl]_2$ with $LiAlR_4$ in toluene and these are interesting because
the structures[58] involve double alkyl bridges; the yttrium
compound (R = C$_2$H$_5$) exhibits fluxional behaviour at room
temperature[58]. Reaction of these compounds (R = CH$_3$, Ln = Y,Dy,
Ho,Er,Yb) with the equimolar quantity of pyridine in toluene
yields $[(\eta^5C_5H_5)_2Ln(CH_3)]_2$ in which a symmetrical double methyl
bridge is present[59], established from ^{1}H-n.m.r. spectroscopy
(Ln = Y) and a single crystal X-ray structure determination
(Ln = Yb). The bridges involve electron-deficient 3-centre
bonds, which is rare for transition metal complexes[59]. It would
be interesting to have information on the analogous reactions
with tricyclohexylaluminium, in which the cyclohexyl ring could
bridge in other ways.

Although one might expect lability of the M-C σ-bonded
groups in the lanthanide and actinide complexes in this group,
there do not appear to have been any studies involving exchange
of the alkyl groups R with, say, LiR'. In most cases there could

well be steric problems which would inhibit the formation of the
exchange intermediate, but the systems are worth examination.
Marks et al[15] have, however, recorded one instance of alkyl
exchange with halogen to form the compounds $[(PMCp)_2M(Cl)(CH_3)]$
(M = Th,U) (equation 8):

$$[(PMCp)_2MCl_2] + [(PMCp)_2M(CH_3)_2] \rightarrow 2[(PMCp)_2M(Cl)(CH_3)] \quad ---(8)$$

In this reaction ^{1}H-n.m.r. spectroscopy indicates that the reaction
equilibrium lies 90-95% to the right of equation 8; the uranium
compound is monomeric in benzene whereas the thorium one is
dimeric, probably by way of chlorine bridging, a reflection of
the larger thorium(IV) radius.[15]   Similarly, ^{1}H-n.m.r.
spectroscopic studies of $(\eta^5C_5H_5)_3UR$ (R = C_2H_5,nC_4H_9,C_6H_5,C_3H_5
(allyl) or OC_2H_5) and AlR'_3 (R' = CH_3,C_2H_5,iC_4H_9) systems have
provided evidence for alkyl exchange, probably by way of a double
alkyl bridged intermediate, whereas there is no exchange in the
system[51a] $(\eta^5C_5H_5)_3UR + (\eta^5C_5H_5)_3UR'$ even at 413K.

C.   Simple and Anionic Alkyls or Aryls, MR_n and Li_xMR_{x+y}

     The only simple alkyl or aryl species to possess reasonable
stability at room temperature are the Grignard type lanthanide
compounds[60], RLnI (Ln = Sm,Eu,Yb), the compounds LnR_3 where R
is bulky[61] and the actinide(IV) tetrabenzyls, $Th(CH_2C_6H_5)_4$ [62]
and $U(CH_2C_6H_5)_4MgCl_2$.[63]   The lanthanide compounds RLnI are
obtained by reaction of the metal with alkyl or aryl iodides in
THF, samarium metal reacting less readily than the alkaline earth-
like europium and ytterbium metals. There is also some evidence
for the formation of an ethyl cerium iodide from a similar
reaction.[60]   The scandium and yttrium silylmethyl and neopentyl
compounds, $Ln[CH_2M'(CH_3)_3]$.2THF (M' = C or Si), are obtained by
the conventional reaction of $LnCl_3$ with RLi in a mixture of hexane,
ether and THF at [61] 0°C and their stability is probably a
reflection of the steric crowding about the Sc and Y atoms.
Thorium tetrabenzyl, which decomposes above 85°C, has been
prepared from the tetrachloride and $(C_6H_5CH_2)Li$ in THF below -20°C.
The same preparative method at room temperature gave a product of
approximate composition $(C_6H_5CH_2)_3Th(THF)$. This yields toluene
and THF only on hydrolysis and the absence of hydrogen in the
hydrolysis products indicates that this is a thorium(IV) compound
which is probably polymeric, presumably involving bridging benzyl
groups (for example, σ-bonded to two thorium atoms via the ring
and the methylene carbon atom)[62] Although the reaction of $ThCl_4$
.3THF with a benzyl Grignard or with dibenzyl magnesium did not
yield identifiable products, the reaction of UCl_4.3THF with the
latter in THF at -40°C yielded a reddish-brown product of
composition $(C_6H_5CH_2)_4U.MgCl_2$ (possibly $Mg[(C_6H_5CH_2)_4UCl_2]$) which
decomposed above 130°C; $(C_6H_5CH_2)_4U$ itself was not obtained[63], but

one would expect the above uranium(IV) compound to yield complexes on treatment with Lewis bases such as pyridine.

Trifluoromethyls, $U(CF_3)_4$ and $U(CF_3)_6$ have been reported, and it is claimed that they are formed by reaction of the metal halides with CF_3 radicals (e.g. from C_2F_6) in a glow discharge[64], but few details are available to substantiate this claim.

In contrast to the complexes $(\eta^5C_5H_5)_3AnR$, the actinide tetraalkyls decompose with the formation of the alkene as the major product when the group R includes a β- hydrogen atom, leaving the actinide (Th,U) in the metallic form. Since the $(\eta^5C_5H_5)_3AnR$ compounds are always much more stable to heat than the compounds AnR_4, it seems reasonable to conclude that the latter are coordinatively unsaturated as compared to the former, in which coordination saturation or steric crowding hinders the possible thermolysis pathway of β-hydride elimination.[66]

Somewhat surprisingly, R_2Zn compounds have not been used for preparative work in the actinide series, where reactions of the halides (e.g. UCl_5,R_4NUCl_6) might be expected to yield products analogous to those found in the $NbCl_5/(CH_3)_2Zn$ system,[67] $(CH_3)NbCl_4$ and $(CH_3)_2NbCl_3$.

The stabilisation of simple lanthanide alkyls by using bulky ligands, as mentioned above, to produce sterically hindered molecules is even more effective for the production of quite stable anionic species. Anionic phenyls, $Li[Ln(C_6H_5)_4]$, have been known for some years,[68] and the structure of a complex anion derived from m-xylene, in the compound $[Li(THF)_4]$ $[(2,6(CH_3)_2C_6H_3)_4 Lu]$, is approximately tetrahedral, and this is certainly sterically crowded.[69] In the case of the analogous compound with an even bulkier alkyl group, $[Li(THF)_4][Yb(CH(Si(CH_3)_3)_2)_3Cl]$, the crowding is too severe for a fourth alkyl group and the remaining space in the tetrahedron is occupied by a chloride ion.[70]

The synthetic utility of the actinide alkoxides has been demonstrated in the preparation of a series of uranium(V) alkyl complexes, $Li_3UR_8(dioxan)_3$ ($R = CH_3,CH_2C(CH_3)_3,CH_2Si(CH_3)_3$) from the alkoxide, $U_2(OC_2H_5)_{10}$ and an excess of RLi. These compounds are reported to be remarkably stable to heat, the methyl derivative decomposing at 265-268°C, and the neopentyl and trimethylsilyl compounds at 120-122°C and 150-154°C respectively. This reflects the increasing steric crowding about the uranium atom and the implication is that the uranium atom is coordinatively saturated.[74] The uranium(IV) alkyl species, $Li_2UR_6(solvent)_x$ (solvent = Et_2O or THF, x = 8; TMED, x = 7) are much less stable to heat and the reverse trend in thermal stability is observed, with the bulkier ligand ($R = CH_2SiMe_3$) stable to 30-35°C, and the methyl derivative to -5°C (TMED) to -20°C (Et_2O). Here the uranium(IV) centre is clearly coordinatively unsaturated, but all attempts to prepare compounds of the type Li_4UR_8 were unsuccessful.[74] Somewhat surprisingly, the uranium(IV) compounds with the potentially bidentate ligand derived from benzyldimethylamine[73,74] were also thermally unstable. This could be due to excessive distortion in the five membered chelate ring which might be alleviated in the case of the potentially 6-membered ring of 2-methyl benzyl-dimethylamine, which lithiates at the 2-methyl group. It is interesting to note that the corresponding reaction with uranium hexaisopropoxide only yielded adducts with lithium, magnesium or aluminium alkyls in which the Li,Mg or Al atoms are bonded to the oxygen atoms of the alkoxide groups.[74]

D.   Conclusions

The main point which comes over quite strongly in all of the alkyl systems is simply that steric crowding appears to be essential for thermal stability in these systems. It will be more difficult to study alkyl exchange reactions in these systems than is the case with the corresponding exchange reactions of π- bonded ligands, for the intermediates will necessarily be derived from coordinatively unsaturated species, which means low temperature studies. Nevertheless such exchange reactions should be investigated, and it would also be of interest to have information on the reactivity of the metal-carbon bond in these systems. Similarly, in the systems with π bonded ligands, exchange leading to disproportionation can also be inhibited by steric crowding or by holding the metal centres apart, so preventing access to the intermediate necessary for disproportionation. Finally, in systems which can give rise to u.v./visible spectra, spectroscopy is a valuable diagnostic technique which is all too often overlooked.

References

1.   B.N. Figgis, "Introduction to Ligand Fields", Interscience, New York, 1966, p.8.

2.   R.A. Satten, D.J. Young and D.M. Gruen, J. Chem. Phys., 33, 1140 (1960).

3.(a)R.E. Maginn, S. Manastyrskyj and M. Dubeck, J. Amer. Chem. Soc., 85, 672 (1963).   (b) F. Mares, K. Hodgson and A. Streitwieser, J. Organometallic Chem., 28, C24 (1971).

4.   P. Laubereau (personal communication to Prof. B. Kanellakopulos (1968)).

5.   P. Zanella, S. Faleschini, L. Doretti and G. Faraglia, J.
        Organometallic Chem., 26, 353 (1971).

6.   B. Kanellakopulos, C. Aderhold and E. Dornberger, J.
        Organometallic Chem., 66, 447 (1974).

7.   L. Doretti, P. Zanella, G. Faraglia, and S. Faleschini, J.
        Organometallic Chem., 43, 339 (1972).

8.(a)   K.W. Bagnall, J. Edwards and A.C. Tempest, J. Chem. Soc.,
        Dalton, 295 (1978). (b) K.W. Bagnall, A. Beheshti and F.
        Heatley, J. Less-Common Metals (in press). (c) K.W. Bagnall
        and J. Edwards (unpublished work).

9.   K.W. Bagnall, J.G.H. du Preez, J. Bajorek, L. Bonner, H.
        Cooper and G. Segal, J. Chem. Soc., 2682 (1973).

10.(a)G. Bombieri, G. de Paoli, and K.W. Bagnall (submitted to
        Inorg. Nuclear Chem. Letters). (b) R.D. Ernst, W.J.
        Kennelly, C.S. Day, V.W. Day and T.J. Marks (to be
        published). (c) G. Bombieri (personal communication).

11.(a)G. Bombieri, D. Brown and R. Graziani, J. Chem. Soc.,
        Dalton, 1873 (1975).

   (b)J.F. de Wet and S.F. Darlow, Inorg. Nuclear Chem. Letters,
        7, 1041 (1971).

12.   C.A. Secaur, V.W. Day, R.D. Ernst, W.J. Kennelly and
        T.J. Marks, J. Amer. Chem. Soc., 98, 3713 (1976).

13.   K.W. Bagnall, A. Beheshti and A.C. Tempest (to be published).

14.   J.C. Green and O. Watts, J. Organometallic Chem., 153,
        C40 (1978).

15.   J.M. Manriquez, P.J. Fagan and T.J. Marks, J. Amer. Chem. Soc.,
        100, 3939 (1978).

 6.   K.W. Bagnall and A.C. Tempest (to be published).

17.(a)P. Zanella, G. de Paoli, G. Bombieri, G. Zanotti and R.
        Rossi, J. Organometallic Chem., 142, C21 (1977). (b) K.W.
        Bagnall and J. Edwards, J. Organometallic Chem., 80, C14
        (1974). (c) M.F. Brady and R.S. Marianelli, Abstracts,
        8th. Midwest Regional A.C.S. meeting, Columbia, Missouri,
        Nov. 1972, No. 235.

18.   T. Siddall, Tert., and W.E. Stewart, J.C.S., Chem. Commun.,
        922 (1969).

19.   G. Lugli and G. Brandi, German patent 2,140,698 (1972)
        (according to Chem. Abs., 77, 19808 (1972).

20.(a) J.D. Jamerson and J. Takats, J. Organometallic Chem.,
        78, C23 (1974). (b) A.L. Arduini, J.D. Jamerson and
        J. Takats (to be published). (c) J.D. Jamerson and J.
        Takats (personal communication).

21.   J.G. Reynolds, A. Zalkin, D.H. Templeton, N.M. Edelstein
        and L.K. Templeton, Inorg. Chem., 15, 2498 (1976).

22.   J.G. Reynolds, A. Zalkin, D.H. Templeton and N.M.
        Edelstein, Inorg. Chem., 16, 599 (1977).

23.   J.G. Reynolds, A. Zalkin, D.H. Templeton and N.M.
        Edelstein, Inorg. Chem., 16, 1090 (1977).

24.   S. Poggio, G. Lugli and A. Mazzei, German patent
        2,257,787 (1973).

25.   M. Brunelli, G. Perego, G. Lugli and A. Mazzei (to be
        published).

26.   A.M. Seyam and G.A. Eddein, Inorg. Nuclear Chem.
        Letters, 13, 115 (1977).

27.(a) J.H. Burns and P.G. Laubereau, Inorg. Chem., 10, 2789
        (1971). (b) J. Goffart (personal communication). (c)
        A.M. Seyam (personal communication).

28.   B.L. Kalsotra, S.P. Anand, R.K. Multani, J. Organometallic
        Chem., 28, 87 (1971).

29.   B.L. Kalsotra, R.K. Multani and B.D. Jain, J. Inorg.
        Nuclear Chem., 34, 2679 (1972).

30.   B.L. Kalsotra, R.K. Multani and B.D. Jain, J. Organo-
        metallic Chem., 31, 67 (1971).

31.   F.G.N. Cloke, M.L.H. Green and D.H. Price, J.C.S., Chem.
        Commun., 431 (1978).

32.   M.L.H. Green (personal communication).

33.   S.R. Ely, T.E. Hopkins and C.W. De Kock, J. Amer. Chem.
        Soc., 98, 1624 (1976).

34.   W.J. Evans, S.C. Engerer and A.C. Neville, J. Amer. Chem.
        Soc., 100, 331 (1978).

35.   J.L. Slater, R.K. Sheline, K.C. Lin and W. Weltner, Jr.,
        J. Chem. Phys., 55, 5129 (1971).

36.   J.L. Slater, T.C. DeVore and V. Calder, Inorg. Chem.,
        13, 1808 (1974).

37.   A.E. Crease and P. Legzdins, J. Chem. Soc., Dalton, 1501
        (1973).

38.   K. Rossmanith, Monatsh, 96, 1407 (1966); 97, 1698 (1967).

39.   I. Colquhoun, N.N. Greenwood, I.J. McColm and G.E.
        Turner, J. Chem. Soc., Dalton, 1337 (1972).

40.   I.J. McColm and S. Thompson, J. Inorg. Nuclear Chem.,
        34, 3081 (1972).

41.   K.W. Bagnall and J.L. Baptista, J. Inorg. Nuclear Chem.,
        32, 2283 (1970).

42.   B.L. Kalsotra, R.K. Multani and B.D. Jain, J. Inorg.
        Nuclear Chem., 34, 2265 (1972).

43.   B. Kanellakopulos, E. Dornberger and H. Billich, J.
        Organometallic Chem., 76, C42 (1974).

44.   G. Wilke, B. Bogdanovic, P. Hardt, P. Heimbach, W. Keim,
        M. Körner, W. Oberkirch, K. Tanaka, E. Steinrücke,
        D. Walter and H. Zimmerman, Angew. Chem., 78, 157 (1966).

45.   G. Lugli, W. Marconi, A. Mazzei, N. Paladino and U.
        Pedretti, Inorg. Chim. Acta, 3, 253 (1969).

46.   M. Tsutsui and N. Ely, J. Amer. Chem. Soc., 97, 3551
        (1975).

47.   G.W. Halstead, E.C. Baker and K.N. Raymond, J. Amer.
        Chem. Soc., 97, 3049 (1975).

48.   J.L. Atwood, C.F. Hains, Jr., M. Tsutsui and A.E. Gebala,
        J.C.S., Chem. Commun., 452 (1973).

49.   N.M. Ely and M. Tsutsui, Inorg. Chem., 14, 2680 (1975).

50.   e.g. L.J. Nugent, P.G. Laubereau, G.K. Werner and K.L.
        Vander Sluis, J. Organometallic Chem., 27, 365 (1971).

51.(a) T.J. Marks, A.M. Seyam and J.R. Kolb, J. Amer.Chem.Soc.,
        95, 5529 (1973). (b) B. Kanellakopulos (personal
        communication).

52.   A.E. Gebala and M. Tsutsui, J. Amer. Chem. Soc., 95, 91
        (1973).

53.   T.J. Marks and W.A. Wachter, J. Amer. Chem. Soc., 98,
        703 (1976).

54.   E.C. Baker, K.N. Raymond, T.J. Marks and W.A. Wachter,
        J. Amer. Chem. Soc., 96, 7586(1974).

55.(a) D.G. Kalina, T.J. Marks and W.A. Wachter, J. Amer. Chem.
        Soc., 99, 3877 (1977). (b) T.J. Marks (personal
        communication).

56.   B. Kanellakopulos, E. Dornberger and F. Baumgärtner, Inorg.
        Nuclear Chem. Letters, 10, 155 (1974).

57.   D.G.H. Ballard and R. Pearce, J.C.S., Chem. Commun., 621
        (1975).

58.   J. Holton, M.F. Lappert, G.R. Scollary, D.G.H. Ballard,
        R. Pearce, J.L. Atwood and W.E. Hunter, J.C.S., Chem.
        Commun., 425 (1976).

59.   J. Holton, M.F. Lappert, D.G.H. Ballard, R. Pearce, J.L.
        Atwood and W.E. Hunter, J.C.S., Chem. Commun., 480 (1976).

60.   D.F. Evans, G.V. Fazakerley and R.F. Phillips, J. Chem.
        Soc., A, 1931 (1971).

61.   M.F. Lappert and R. Pearce, J.C.S., Chem. Commun., 126
        (1973).

62.   E. Kohler, W. Bruser and K.-H. Thiele, J. Organometallic
        Chem., 76 235 (1974).

63.   K.-H. Thiele, R. Opitz and E. Kohler, Z. anorg. allgem.
        Chem., 435, 45 (1977).

64.   R.J. Lagow, L.L. Gerchman and R.A. Jacob, U.S. patents
        3,954,585 and 3,992,424 (1976), according to Chem. Abs.,
        85, 160324 (1976) and 86, 72887 (1976).

65.   T.J. Marks and A.M. Seyam, J. Organometallic Chem., 67,
        61 (1974).

66.   T.J. Marks, Adv. Chem. Series (Ed. R.B. King), 150,
        232 (1976).

67.   G.W.A. Fowles, D.A. Rice and J.D. Wilkins, J. Chem. Soc.,
        Dalton, 2313 (1972).

68.   F.A. Hart, A.G. Massey and M.S. Saran, J. Organometallic
        Chem., 21, 147 (1970).

69.   S.A. Cotton, F.A. Hart, M.B. Hursthouse and A.J. Welch,
        J.C.S., Chem. Commun., 1225 (1972).

70.   J.L. Atwood, W.E. Hunter, R.D. Rogers, J. Holton, J.
        McMeeking, R. Pearce and M.F. Lappert, J.C.S., Chem. Commun.,
        140 (1978).

71.   B.L. Kalsotra, R.K. Multani and B.D. Jain, Israel J. Chem.,
        9, 569 (1971).

72.   A. Greco, S. Cesca and G. Bertolini, J. Organometallic
        Chem., 113, 321 (1976).

73.   R. Andersen, E. Carmona-Guzman, K. Mertis, E. Sigurdson
        and G. Wilkinson, J. Organometallic Chem., 99, C19
        (1975).

74.   E. Sigurdson and G. Wilkinson, J. Chem. Soc., Dalton,
        812 (1977).

75.   W.J. Evans, A.L. Wayda, Chia-Wun Chang and W.M. Cwirla,
        J. Amer. Chem. Soc., 100, 333 (1978).

76.   V.K. Vasil'ev, V.N. Sokolov and G.F. Kondratenkov,
        J. Organometallic Chem., 142, C7 (1977).

THE STRUCTURE AND BONDING OF $4f$ AND $5f$ SERIES ORGANOMETALLIC
COMPOUNDS

Kenneth N. Raymond

Department of Chemistry, University of California,
Berkeley, California 94720

I.   INTRODUCTION

This paper will focus on the molecular structures of organoactinides and –lanthanides (as determined primarily by x-ray diffraction). To the extent that these structures or comparison of
structural parameters tell us something about the bonding, the
details of these structure results will be discussed in this light.
Fortunately the science of x-ray diffraction has advanced to the
stage where it can be used as a general structural probe in the
investigation of <u>series</u> of compounds in addition to the characterization of individual examples. In the area of organoactinides
and –lanthanides this has been a particularly important tool for
the characterization of the basic chemistry of these species. Unlike the classic d transition metal organometallic compounds there
appeared no early, unifying law such as the 18 electron rule.
However it became clear that the chemistry of these compounds was
more extensive and complex than would be expected of simple ionic
salts. In several cases the structures found have been completely
unanticipated — no stronger statement can be made about the role
of crystallography in the understanding that we have today of this
area of chemistry.

In part this paper will be a review of the structure and
bonding of the organoactinides and –lanthanides. However the em-
phasis will be on the correlation of structural results and their
interpretation rather than a comprehensive inclusion of all available results. As a basis for comparison, non-f-series organometallic compounds will often be included in the discussion.

Although the simple ionic model cannot be expected strictly

T. J. Marks and R. D. Fischer (eds.), Organometallics of the f-Elements, 249–280.

to apply to many (and perhaps all) of the compounds described,
nevertheless it is a very useful format in which to present and
compare structures. In coordination chemistry, this model is often
more effective than it has any right to be, given the assumptions
made by the model. A brief review of ionic radii and their use
will therefore be given.

In simple ionic salts where the cation and anion charges are
±1, it is found that the distances, r_0, between ions in different
salts are constant factors for each ion — as seen below.

Interatomic Distances (r_0) in Halides (Å)

	Cl^-	Br^-	I^-
Na^+	2.57	2.98	3.23
K^+	2.81	3.29	3.53
Rb^+	3.14	3.43	3.66

These differences can be summarized as:

$K^+ \approx Na^+ + .32$ Å $Br^- \approx Cl^- + .16$ Å
$Rb^+ \approx K^+ + .14$ $I^- \approx Br^- + .25$ Å

Pauling[1] made the original assumption that in addition to writing:

$$r_0 = r_+ + r_- \tag{1}$$

the ratio of radii (based essentially on a hydrogen atom radial
wave function) would be determined by the "effective nuclear
charge," Z^* on each ion:

$$r_+/r_- = Z_-^*/Z_+^* \tag{2}$$

This gives rise to the so-called "univalent" ionic radii. When
ions of higher charge (e.g. Mg^{2+} or O^{2-}) are involved, some correction has to be made for the fact that the extra lattice energy
from the attraction of the more highly charged ions will compress
the structure to a lower value of r_0 than would be the case for
the structure in which the charges are all ±1. The resultant
"crystal radii" for a crystal in which the cation has a charge
+i and the anion a charge -j is given by

$$r_{ij} = r_{11}\left(\frac{1}{ij}\right)^{1/n-1} \tag{3}$$

The integer n is the Born exponent[1] [5, 7, 9, 10 and 12 for ions
with He, Ne, Ar(Cu^+), Kr(Ag^+), or Xe(Au^+) electronic configurations, respectively]. Thus for MgO

$$\left(r_{\text{crystal}}/r_{\text{univalent}}\right) = \left(\tfrac{1}{4}\right)^{1/6} = 0.794$$

In comparing ionic radii and structural parameters for different compounds, not only may different charges be involved but also different coordination numbers. For a given metal ion and ligand anion, the metal-ligand bond length will increase as the coordination number increases since the ligand-ligand repulsion forces increase rapidly as more ligands are placed around the metal. Again, a simple Born repulsion model predicts for two coordination numbers I and II

$$r_{II}/r_{I} = \left(\frac{CN_{II}}{CN_{I}}\right)^{1/n-1} \tag{4}$$

For 4-coordinate and 6-coordinate Zn^{2+} the ratio

$$r_6/r_4 = \left(\tfrac{6}{4}\right)^{1/8} = 1.052$$

The most complete and useful set of ionic radii today are those of Shannon and Prewitt.[2] These authors make the assumptions: (1) Additivity of both cation and anion radii to reproduce interatomic distances is valid if one considers coordination number, electronic spin, and other factors. (2) With these limitations, radii are independent of structure type. (3) Both cation and anion radii vary with coordination number. With these assumptions Shannon and Prewitt have produced a self-consistent set of ionic radii from 900 different structures. Their values differ significantly from those originally presented by Pauling but are certainly closer to "true" ionic radii.

The steric constraints imposed by organometallic ligands such as Cp^- and COT^{2-} are as important as formal coordination number in determining coordination geometries. Any covalent bonding effects are an additional perturbation. By coordination number will be meant the number of electron pairs involved in ligand-to-metal coordination.

The effects of covalent bonding on structural parameters can be seen in the following table. Note in particular that although the ionic radius for Mg^{2+} is 0.11 Å greater than low-spin Fe^{2+}, the Mg-C bond lengths are .26 Å greater than in ferrocene. The additional decrease of 0.15 Å in the Fe-C bonds can be attributed to the increased bond order of ferrocene, which is confirmed by many other physical properties, including direct thermochemical measurement.

Table 1a.   Gas Phase Electron Diffraction Structural Data for 3d
            Series Metallocenes

	R(M-C) Å	I.R.[a] (Å)	R(M-C)—(I.R.)	Ref.
Cp_2V	2.280(5)	.79	1.49	3
Cp_2Cr	2.169(4)	.73	1.44	3
Cp_2Mn	2.383(3)	.83	1.55	4
$(MeCp)_2Mn$ (L.S.)	2.144(12)	.67	1.47	5
$(MeCp)_2Mn$ (H.S.)	2.433(8)	.83	1.60	5
Cp_2Fe	2.064(3)	.61	1.45	6
Cp_2Co	2.119(3)	.65	1.47	7, 8
Cp_2Ni	2.196(4)	.69	1.51	9
Cp_2Mg	2.339(4)	.72	1.62	15

[a]From ref. 2.

Table 1b.   Single Crystal X-ray Diffraction Data for 3d Series
            Metallocenes

	R(M-C) Å	I.R.[a] (Å)	R(M-C)—(I.R.)	Ref.
Cp_2V	2.24	.79	1.45	10
Cp_2Cr	2.14	.73	1.41	10
Cp_2Mn	2.41[b]	.83	1.58	11
Cp_2Fe	2.045(4)	.61	1.44	12
Cp_2Co	2.096(8)	.65	1.45	13
Cp_2Ni	2.15	.69	1.46	10
Cp_2Mg	2.304(8)	.72	1.58	11
$[(MeCp)_2Fe]I_3^-$	2.05(2)	.55	1.50	14

[a]From ref. 2.

[b]Polymeric.

II. CYCLOPENTADIENYL COMPLEXES

A. Lanthanide Complexes

1. __Complexes of the type Ln(Cp)₃.__ The first well-characterized organometallic compounds of the lanthanides or actinides
were the tricyclopentadienides of various lanthanide ions synthesized by Birmingham and Wilkinson in 1954.[16] Anhydrous metal
chlorides were reacted with sodium cyclopentadienide in tetrahydrofuran as follows:

$$MCl_3 + 3\ NaC_5H_5 \xrightarrow{THF} M(C_5H_5)_3 + 3\ NaCl$$

[M = Sc, Y, La, Ce, Pr, Nd, Sm, Gd, Dy, Er, Yb]

The product was shown to be the THF adduct, but a later modification of the procedure by E. O. Fischer and H. Fischer, using benzene or diethyl ether in place of THF and the potassium salt in place of the sodium salt of cyclopentadiene, produced solvent-free complexes of Tb, Ho, Tm and Lu. The remaining tricyclopentadienides have since been made. All of the complexes are very air- and moisture-sensitive, are stable to heat, and sublime, except for the europium derivative, at elevated temperatures. Although the preparation of the lanthanide tricyclopentadienides was followed by considerable synthetic activity, for several years nothing was known regarding their structures. The chemical and physical properties of the compounds suggested ionic behavior: addition of water or acid gives cyclopentadiene, addition of ferrous chloride gives ferrocene, and the magnetic susceptibilities are all close to the free ion values. Thus the structure originally suggested was that the $C_5H_5^-$ ions would lie at the vertices of an equilateral triangle and early interpretations of visible spectra was based on D_{3h} molecular symmetry. However it was found that $Lu(C_5H_5)_3$ had a dipole moment of 0.85 ± .09 D, ruling out a symmetrical planar structure. The first x-ray structure analysis was of $Sm(C_5H_5)_3$, which showed a disordered structure with two types of molecules and at least three types of metal-carbon bonds in a complex polyhedral arrangement.[17] However, the analysis apparently was based on an incorrect unit cell, for subsequently the structure of the scandium complex, $Sc(C_5H_5)_3$, was determined[18] and the unit cell found to be the same as one-half that of the samarium compound.

The molecular structure of $Sc(C_5H_5)_3$ (Fig. 1) exhibits both η^5- and bridging η^1-cyclopentadienyl rings. Each scandium is

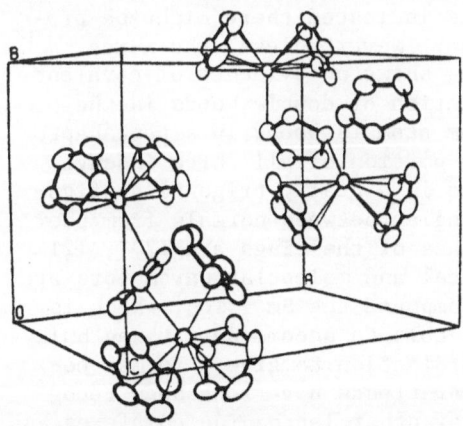

Fig. 1. The crystal and molecular structure of $Sc(C_5H_5)_3$, from Ref. (18).

pentahapto bound to two terminal C_5H_5 groups and monohapto bound
to two bridging groups, leading to a polymeric chain arrangement.
A summary of the crystal and molecular parameters appears in
Table 2. There is no structural evidence for any electric distortion of the bridging ring, but it is appreciably closer to one
scandium than the other. The bridging and terminal rings have
identical geometries. The interaction of the bridging ring with
the scandium is through only one carbon, since no other carbon
atom in the ring is within 3 Å of the metal. Although it was suggested that this preferred orientation of the metal relative to
the ring is an indication of some covalent bond character, it
would appear more likely that three η^5-cyclopentadienyl rings
(formally nine electron-pair bonds) are too much for the coordination sphere of the small Sc^{3+} ion, whereas two η^5 rings (six
bonds) are not enough. Thus bridging rings of lower hapticity are
formed to saturate the metal coordination, with the exact geometry
being dictated by steric requirements, largely ion packing in
character.

This point is borne out by the structure of tris indenyl
samarium.[19] An earlier report of the nmr spectrum was interpreted
as evidence of covalent bonding in the tetrahydrofuran adduct of
samarium triindenide.[20] Indenyl anion, is isoelectronic with

$C_5H_5^-$ if the five-membered indenyl ring is considered as an isolated unit. However, unlike the completely delocalized $C_5H_5^-$
system there is a localization of charge at the C_1 position in the
indenyl anion, and the nmr results indicated there might be preferred bonding through that carbon. Nevertheless, the x-ray
structure analysis of $Sm(indenyl)_3$ shows no evidence of covalent
bonding since there is no localization of double bonds in the
five-membered rings. The samarium atom is bound in a pentahapto
fashion to the five-membered ring portion of all three indenyl
groups, and the rings are oriented in a nearly trigonal configuration about the samarium. The angles between normals from the
samarium to the least squares planes of the rings are 120°, 121°
and 116°, respectively. The crystal and molecular parameters are
summarized in Table 2. In this compound the Sm^{3+} ion, which is
considerably larger than Sc^{3+}, is able to accommodate three bulky
ligands in a fully pentahapto coordination to give a formal coordination number of nine. Similar trends have long been recognized in the coordination number of other lanthanide complexes as
a function of ionic radius.[21]

Table 2. Summary of Molecular Parameters for Lanthanide Cyclopentadienide Complexes

	LnCp₃ type			LnCp₃X		LnCp₂X dimers		
	$Sc(C_5H_5)_3$	$Sm(indenyl)_3$	$Nd[C_5H_4(CH_3)]_3$	$Pr(C_5H_5)_3(CNC_6H_{11})$	$[YbCp_3]_2pyz$	$[Sc(C_5H_5)_2Cl]_2$	$[Yb(C_5H_4(CH_3))_2Cl]_2$	$[YbCp_2(CH_3)]_2$
Reference	18	19	22	23	26	25	24	28
Ionic radius of metal ion (Å)[a]	0.87	1.13*	1.19*	1.21*	1.08	0.87	0.98	0.98
Approximate coordination number in crystal	8	9	10	10	10	8	8	8
Average η^5 M–C bond length, Å	2.49(2)	2.75(5)	2.79(5)	~2.77(2)	2.68(1)	2.46(2)	2.585(8)	2.613(13)
M–σ bond lengths, Å[b]	2.52, 2.63	–	2.98(1)	2.68	2.61(1)	2.57	2.63	2.51(3)
Mean C–C bond, Å	1.40(3)	1.42(2)	1.40(3)	1.35(3)	1.387(5)	1.35(4)	1.41(1)	1.63
Average η^5 M–C bond length minus ionic radius	1.62	1.62	1.60	1.60	1.60	1.59	1.60	1.63

[a]The ionic radii [from Ref. (2)] have been corrected for coordination number by interpolation when marked *.

[b]This bond length is for a metal-carbon bond in the LnCp₃ compounds and a metal-adduct atom bond in the other complexes.

One other structure of a lanthanide tricyclopentadienide provides a further demonstration of the correlation between ionic
size and coordination. Neodymium tris(methylcyclopentadienide)
crystallizes as a tetramer (Fig. 2).[22] The Nd^{3+} ion (which is
slightly larger than Sm^{3+}) is pentahapto bound to three cyclopentadienyl rings and monohapto bound to a fourth ring. This
fourth ring is in turn η^5-bonded to another Nd^{3+} ion, until the
tetramer is generated. The distances between tetramers are those
expected for van der Waals contact. The crystal and molecular
parameters are compared with the other tris cyclopentadienyl complexes in Table 2.

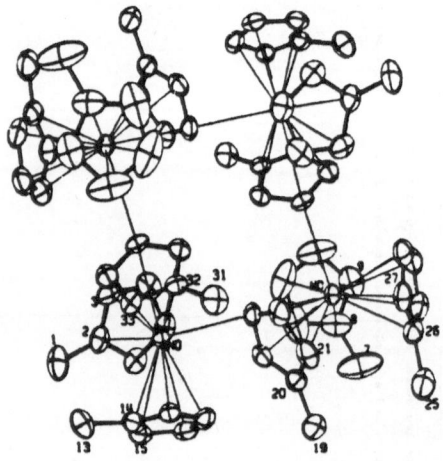

Fig. 2.   The crystal and molecular structure of tris(methylcyclo-
         pentadienyl)neodymium(III), from Ref. (22).

Viewed as a group, all of the lanthanide cyclopentadienyl
complexes show the behavior expected if the metal-ring interaction
were purely electrostatic in nature. Their chemical behavior
reflects a lability of the $C_5H_5^-$ ligand. Their physical properties, including magnetic susceptibility, nmr spectra and visible
spectra, show no evidence of significant covalent interaction
and their crystal structures are consistent with purely ionic
interactions in which the geometry observed is that which maxi-
mizes the coordinate interactions consistent with both the metal
ionic radius and the size and shape of the anion. The total
formal coordination with the anions depends on the size of the
metal ion (see Table 2) and the metal-carbon distances for these
complexes correlate very well with ionic radii.

    2.   LnCp₃ donor complexes. Lanthanide tricyclopentadienides
act as strong Lewis acids. Although the strongest coordination

is usually from oxygen and nitrogen, some complexes with more
polarizable bases are also known. Adducts are formed with triphenylphosphine, tetrahydrofuran, ammonia or cyclohexylisonitrile
and have been known for some time.[16]

It has been suggested that the source of the strong Lewis
acidity of the lanthanide tricyclopentadienides could result from
covalency involving the empty $5d_{z^2}$ orbital.[27] However, the accu-
mulated evidence in both the solid state and in solution points
instead to ionic bonding. For the lanthanide tricyclopentadienides, completion of the coordination sphere usually requires
polymerization or adduct formation with another Lewis base. Thus
the complex coordination of the LnCp$_3$ compounds in the crystalline
state is exhibited when the compound is made solvent-free, either
by using a non-coordinating solvent or through vacuum sublimation;
the polymeric structures are broken up and the η^1-bond displaced
upon addition of a Lewis base.

The structure of the cyclohexylisonitrile adduct of praseodymium tricyclopentadienide[23] has three pentahapto cyclopentadienyl rings forming the base of a trigonal pyramid and the isonitrile carbon at the apex. The isonitrile carbon is 2.68 Å from
the metal and the cyclopentadienyl carbons range in distance from
2.75 to 2.84 Å. The isonitrile C≡N stretch in Pr(C$_5$H$_5$)$_3$(CNC$_6$H$_{11}$)
is displaced upward 70 cm^{-1} upon coordination and this has been
interpreted as suggesting that the isonitrile acts as a σ donor
and not as a π acceptor. In similar complexes of actinide +3 ions
the shift of the C≡N stretch is less, which suggests, in this
model, back donation and a more covalent complex. However, in the
absence of force constants from a normal coordinate analysis,
little confidence can be placed in conclusions based on frequency
shifts of a few tens of wavenumbers.

The strong Lewis acidity of Ln(C$_5$H$_5$)$_3$ complexes, the recent
use of pyrazine as an effective electron transfer agent in transition metal chemistry, and our desire for an organolanthanide complex with a continuous π-bridging ligand system to complement
previous studies suggested to us the possibility of a pyrazine-
bridged dimer. Such a complex would place the metal atoms far
enough apart to eliminate through-space magnetic interactions so
that any electron exchange would have to take place through the
ligand π system. Furthermore, the anticipated approximate C$_{3v}$
site symmetry at the metal center would allow structural comparison with the (C$_5$H$_5$)$_3$U-X complexes of uranium(IV). The reaction
of Yb(C$_5$H$_5$)$_3$ with pyrazine under inert atmosphere conditions in
benzene gives the dimeric product desired.[26] Slow sublimation
under vacuum begins (remarkably!) at 75°C and gives green-brown
crystals. The molecular unit is a dimer located about a crystallographic inversion center (Fig. 3). Two ytterbium atoms, each
with three η^5-cyclopentadienide rings, are nearly linearly bridged

by a pyrazine ring coordinated through its nitrogens. The
ytterbium-nitrogen distance is 2.61 Å and the average ytterbium-
carbon distance is 2.68(1) Å. The coordination about the
ytterbium is nearly C_{3v} in symmetry. The magnetic susceptibility
has been measured in the range 3-100°K and shows linear Curie-
Weiss behavior with C = 1.51(4), θ = 1.3(6)° and μ_{eff} = 3.48 BM.
There is no evidence of any magnetic interaction between metal
centers or reduction of magnetic moment due to f-orbital covalency.
This lack of interaction and the consistency of the MCp₃ bonding
parameters in both lanthanide and actinide compounds (vide infra)
make an ionic formulation of the bonding most appropriate.

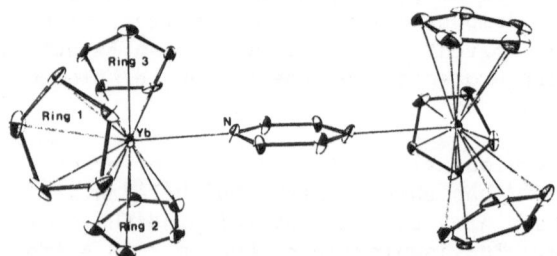

Fig. 3.   A perspective view of the complex $(C_5H_5)_3Yb(NC_4H_4N)$-
           $Yb(C_5H_5)_3$. The thermal ellipsoids are drawn at the 50%
           probability level, from Ref. (26).

3.   <u>Complexes of the type Ln(Cp)₂X</u>. By controlling the
stoichiometry of the reaction between lanthanide trichlorides and
sodium cyclopentadienide it is possible to replace the chloride
ions stepwise. Equilibria are rapidly established, so the addition of $Ln(C_5H_5)_3$ to one or two equivalents of $LnCl_3$ will produce
$Ln(C_5H_5)_2Cl$ and $Ln(C_5H_5)Cl_2$, respectively.

The compounds $Ln(C_5H_5)_2Cl$ also have been made only with the
lanthanides above samarium. A structure analysis of the ytterbium
member of this series has been completed.[24] The crystal and
molecular parameters of this and related complexes are compared
in Table 2. The crystal structure consists of discrete dimers
with C_i site symmetry (Fig. 4). The two ytterbium atoms, each
with two pentahapto-bound methylcyclopentadiene rings, are nearly
symmetrically bridged by the two chlorine atoms. The bridging
unit is required by symmetry to be planar and is nearly square,
with a Yb-Cl-Yb angle of 97.95(5)° and Yb-Cl bond lengths of
2.628(2) and 2.647(2) Å. The coordination of each ytterbium atom
is distorted tetrahedral, with the centroids of the cyclopentadienide rings and the chlorines forming the apices of the tetrahedron atom is 126.7°. These parameters have been compared[16] with
those for Cp_2ZrF_2, Cp_2ZrI_2, Cp_2TiCl_2 and $Cp_2Hf(CH_3)_2$ which are
127.8, 126.3, 121.5 and 132°, respectively.

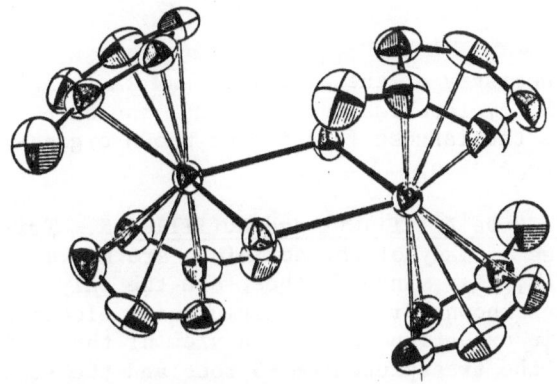

Fig. 4.   The molecular structure of [Yb(C$_5$H$_4$(CH$_3$))$_2$Cl]$_2$, from Ref. (24).

Two other dimeric complexes which are related in structure to [Yb(C$_5$H$_4$(CH$_3$))$_2$Cl]$_2$ are [Sc(C$_5$H$_5$)$_2$Cl]$_2$, [YbCp$_2$(CH$_3$)]$_2$ and [Ce(C$_8$H$_8$)Cl·2 THF]$_2$ (vide infra). The methyl bridged dimer[28] has essentially the same geometry as the chloro bridged compounds. The Yb-CH$_3$ bond lengths are 2.49 and 2.54 Å.   The non-bonded Yb-Yb and methyl-methyl distances are 3.45 and 3.66 Å, respectively.   (The latter is greater than the non-bonded contact distance, while the Yb-Yb distance is substantially shorter than that previously observed.)   The CH$_3$-Yb-CH$_3$ angle is 93.4°.

Replacing the chloride by other bridging ligands also gives dimeric complexes in non-coordinating solvents.   Crystals of Yb(C$_5$H$_4$(CH$_3$))$_2$(ϕCO$_2$) grown from benzene show a dimeric structure with a metal-metal distance of nearly 5 Å.[16]   This is much longer than the metal-metal distance in carboxylate-bridged d-transition metal dimers such as copper acetate, where there is considerable covalent bonding to give the structure:

$$
\begin{array}{c}
\diagup \mathrm{C} \diagdown \\
\mathrm{O} \qquad\ \ \mathrm{O} \\
|\ \qquad\quad | \\
\mathrm{M} \qquad\ \ \mathrm{M}
\end{array}
$$

The metal-metal distance is then nearly the same as the O-O distance.   For the lanthanide dimer the bridging unit exists as the structure

$$
\begin{array}{c}
\diagup \mathrm{C} \diagdown \\
\mathrm{O} \qquad \mathrm{O} \\
\diagup \qquad\quad \diagdown \\
\mathrm{Ln} \qquad \mathrm{Ln}
\end{array}
$$

which is consistent with predominantly ionic bonding.

B.   Actinide Complexes

The organometallic chemistry of the actinides and lanthanides began with complexes of the cyclopentadienide ligand, and such compounds still constitute the largest fraction of known organo-actinide and -lanthanide complexes.

1.   Complexes of the type Ac(III)Cp₃ and Ac(III)Cp₂X.   Tri-cyclopentadienide complexes of many of the actinides are known (Ac = Th, U, Pu, Am, Cm, Bk, Cf).   Indeed, these are the only cyclopentadienide complexes known for the transplutonium elements, where +3 is the most stable oxidation state.   In view of the simi-lar size and chemistry of the transplutonium +3 ions and the +3 lanthanides, it is not surprising that powder pattern data indi-cate that the Pu, Am, Cm, Bk and Cf complexes are isostructural with the lanthanide complexes $Pr(C_5H_5)_3$, $Sm(C_5H_5)_3$ and $Gd(C_5H_5)_3$.[29] A by-product of the preparation of $Bk(C_5H_5)_3$ has also been iso-lated and shown by mass spectral analysis to be the dimer [Bk-$(C_5H_5)_2Cl]_2$.[30] This complex is isostructural with the correspon-ding lanthanide dimer, $[Sm(C_5H_5)_2Cl]_2$, and undoubtedly has the same chloride-bridged structure as the ytterbium dimer, $[Yb(C_5H_4-(CH_3))_2Cl]_2$.

The tris cyclopentadienide complex of uranium is known, and like the lanthanide(III) tris cyclopentadienyl complexes, $U(C_5H_5)_3$ is a strong Lewis acid.

Two preparations of thorium(III) tricyclopentadienide have been reported.   The preparation of Kanellakopulos et al.[31] is violet with $\mu = 0.4$ μ_B.   A recent preparation by Marks et al.[32] is dark green with $\mu = 2.1$ μ_B.   Unfortunately nothing is known as yet about the structures of these very interesting materials.

2.   Complexes of the type AcCp₃X.   The first structure anal-ysis of an organoactinide or -lanthanide was carried out for $U(C_5H_5)_3Cl$.   The molecular parameters are listed in Table 3.   Un-fortunately, the positions of the carbon atoms were poorly deter-mined and provide little information regarding the hapticity of the rings.   Subsequently, the trisindenyl uranium and thorium chlorides were synthesized.   Since the structure determination of $U(C_5H_5)_3Cl$ was not accurate enough to establish the hapticity of the cyclopentadienyl rings, the indenyl structure was studied to resolve this question.   The bonding parameters of $U(C_9H_7)_3Cl$ are summarized in Table 3.   The coordination polyhedron is a distorted tetrahedron in which the centers of the five-membered rings occupy three apices and the chloride ion occupies the fourth.   The C-C bond lengths for the five- and six-membered ring of the indenide anion average 1.43 and 1.41 Å, respectively.   The individual ura-nium to carbon distances of bridging carbon and non-bridging carbon atoms as well as the averages, 2.86 Å (bridging) and 2.72 Å (non-

Table 3. Summary of Molecular Parameters for Actinide (IV) π-Cyclopentadienide Complexes and Bond Lengths Predicted by Ionic Radii

	$U(C_5H_5)_3Cl$	$U(C_5H_5)_3F$	$U(C_5H_4CH_2\phi)_3Cl$	$U(C_9H_7)_3Cl$	$U(C_5H_5)_3C_2H$	$U(C_5H_5)_3(C\equiv C\phi)$
Reference	42	43	33	44	35	34
M-X(σ) bond, Å	2.559(16)	2.11(1)	2.627(2)	2.593(3)	2.36(2)	2.33(2)
Calculated U-X bond	2.61	2.15	2.61	2.61	2.39	2.39
Average M-C distance, Å (π-cyclopentadienyl)	2.74[a]	2.74	2.733(1)	2.78[b]	2.73(5)	2.68
Calculated M-Cp bond	2.70	2.70	2.70	2.70	2.70	2.70
Formal coordination number	10	10	10	10	10	10

	$U(C_5H_5)_3(p\text{-xylyl})$	$U(C_5H_5)_3(n\text{-butyl})$	$U(C_5H_5)_3(CH_2-C(CH_3)=CH_2)$	$U(C_5H_5)_4$	$[Th(C_5H_5)_2(C_5H_4)]_2$
Reference	36	36	39	45	41
M-X(σ) bond, Å	2.54(2)	2.43(3)	2.48(3)	-	2.55
Calculated U-X bond	2.53	2.53	2.53	-	2.52
Average M-C distance, Å (π-cyclopentadienyl)	2.71(1)	2.73(1)	2.74(1)	2.81(2)	2.83
Calculated M-Cp bond	2.70	2.70	2.70	2.76	2.75
Formal coordination number	10	10	10	12	10

[a] The positions of the carbon atoms were not refined in this structure.

[b] U-C average distance is for the five membered ring of the indenide anion.

bridging), differ significantly. Possibilities for bonding in this complex therefore could include a 1,2,3-trihapto mode and a 1,2,3,8,9-pentahapto mode for the indenyl group. The former corresponds to bonding by the isolated allyl anion (carbons 1, 2, 3) and an isolated aromatic ring (carbons 4 through 9). Pentahapto bonding seems more likely with the observed distortion induced by the short C···Cl contacts of the three six-membered rings.

The question of the mode of bonding of the cyclopentadienyl rings in complexes of the type $U(C_5H_5)_3X$ was settled by the structural determination of tris(benzylcyclopentadienide)uranium(IV) chloride.[35] The structure has essentially the same coordination geometry as the LnCp$_3$X adducts, such as [Cp$_3$Yb]$_2$ pyrazine (Fig. 3). Weighted least-squares planes through each of the three cyclopentadienide rings show all atoms are within 0.02 Å of the plane. The U-C distances range from 2.68 through 2.81 Å and average 2.733(1) Å. These results are consistent only with three π-bonded pentahapto cyclopentadienyl rings attached to the uranium atom. The substituents of the five-membered rings are disposed so as to point toward the chlorine atom [as in $U(C_9H_7)_3Cl$].

The physical properties of the uranium triscyclopentadienyl chloride, bromide and iodide are all very similar. The mass spectra of the F, Cl, Br and I complexes show they are all monomers in the gas phase. The crystal data and molecular parameters for $U(C_5H_5)_3F$ are summarized in Table 3. The unit cell contains $U(C_5H_5)_3F$ monomers having C_{3v} crystallographic site symmetry. The coordination angles are virtually identical with those reported for $U(C_5H_5)_3Cl$ and $U(C_5H_4CH_2\phi)_3Cl$. The U-F bond distance of 2.11 Å can readily be accounted for by subtracting the difference in the ionic radii between fluorine and chlorine (0.48 Å) from the known U-Cl distances for $U(C_5H_5)_3Cl$ [2.56(2) Å] or $U(C_9H_7)_3Cl$ [2.593(3) Å]. Although stable homoalkyls of the actinides have only recently been made and, except for very sterically hindered ligand groups are of limited thermal stability, the actinide complexes AcCp$_3$R are relatively stable for Th and U.

Infrared and nuclear magnetic resonance data for the $U(C_5H_5)_3R$ complexes are consistent with a metal-carbon σ bond, and this fact has been confirmed by the structural analysis of compounds for which R = C_2Ph,[34] C_2H,[35] p-xylyl[36] and n-butyl.[36] The molecular geometry about the uranium is the same as that described earlier for the other Ac(Cp)$_3$X structures. The molecular parameters of the tris(cyclopentadienyl)uranium alkyls and aryls are compared in Table 3. The U-C σ bond lengths are considerably shorter than the U-C π cyclopentadienide distances. The uranium-carbon-carbon bond angle for the phenylacetylide groups is 175(2)°, as expected for sp-hybridized carbon. The $U(C_5H_5)_3X$ structure appears to be relatively insensitive to the ligand X. As expected, the U-C σ bond length for an sp^3-hybridized carbon is greater

than that for the analogous sp bond length. Apparently Tsutsui
et al. are not aware that the bond radius for sp hybridized carbon
is smaller than that for sp^3 by about 0.14 Å.[37] In their structure reports[34] and later[35] they ascribe this shorter (than alkyl)
distance to "enhanced covalency in the σ bond." In fact, the
U–C distances of the acetylides are entirely consistent with the
other values in Table 3. However, there appears to be a discrepancy between the U–C σ bond lengths found in $U(C_5H_5)_3$(p-xylyl) and
$U(C_5H_5)_3$(n-butyl). This discrepancy probably can be ascribed to
experimental errors due to x-ray absorption.

The allyl group is able to form both σ- and π-bonded complexes with the actinides. The limiting modes of bonding in
metal allyl complexes and the ratio of PMR intensities from magnetically equivalent protons are illustred in Fig. 5.

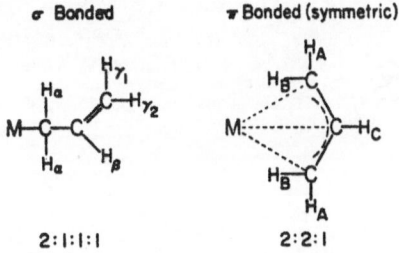

σ Bonded   π Bonded (symmetric)

2:1:1:1   2:2:1

Fig. 5. Bonding in metal allyl complexes.

Tetraallyluranium compounds have been prepared by the reaction of the Grignard reagents with uranium tetrachloride in
diethyl ether. The PMR spectra exhibited by tetraallyluranium
is typical of the symmetrical π-allyl structure in Fig. 5. The
spectrum of tetra(2-methylallyl)uranium(IV) is also consistent
with a symmetrical π-bonded structure in which the ligands are
presumably tetrahedrally arrayed about the uranium atom. Thorium
also forms a tetraallyl complex, and the low temperature PMR spectrum implies a π-allylic structure. The π-allylic form is therefore the most stable structure for the actinide homoallyls.

The temperature dependence of the PMR spectrum of $U(C_5H_5)_3$-
(allyl) has shown the compound to be fluxional at room temperature
with a σ⇌π⇌σ interconversion.[38] At lower temperatures, the
A_2BCD pattern characteristic of a monohapto allyl linkage is
frozen out. Using the rate constant for exchange found from the
temperature-dependent nmr spectrum, it is estimated that the monohapto (σ) species is 8-9 Kcal/mol in energy lower than the trihapto (π) allyl.

In the solid state only the σ-bonded allyl species is observed. The molecular structure of tris(η^5-cyclopentadienyl) (2-methylallyl)uranium(IV)[39] (Fig. 6) clearly shows the σ bonded nature of the 2-methylallyl group, which has a U–C σ-bond length of 2.48(3) Å. Coordination about the uranium is the same as the previously described U(C$_5$H$_5$)$_3$R and U(C$_5$H$_5$)$_3$X structures (Table 3). The localized double bond associated with the allylic σ-bonding mode is not observed in either terminal C–C bond length due to a crystallographic disorder.

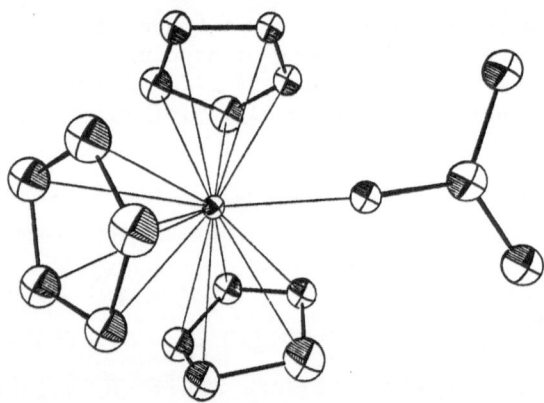

Fig. 6.   The structure of tris(cyclopentadienyl) (2-methylallyl)-
          uranium(IV), from Ref. (39).

The monohapto coordination of the allyl group in U(C$_5$H$_5$)$_3$-(C$_3$H$_5$) and U(C$_5$H$_5$)$_3$(C$_4$H$_7$) might be considered surprising in comparison with the structure of U(C$_5$H$_5$)$_4$. In U(C$_5$H$_5$)$_4$, four η^5 (π) cyclopentadienide rings are tetrahedrally coordinated to the uranium atom with an average U–C bond length of 2.81(2) Å. Knowing that the π-bonded allyls are lower in energy in U(allyl)$_4$ and U(2-methylallyl)$_4$ and that all four rings in U(C$_5$H$_5$)$_4$ are π-bonded, one might expect that the trihapto (π) form in U(C$_5$H$_5$)$_3$(allyl) would be lower in energy than the monohapto (σ) form.

The average U–C bond length in U(C$_5$H$_5$)$_4$ is significantly greater (2.81 vs 2.73 Å) than that in the U(C$_5$H$_5$)$_3$R and U(C$_5$H$_4$-CH$_2$C$_6$H$_5$)$_3$Cl structures, and this increase must be due to steric hindrance caused by the addition of a fourth π-bonded cyclopentadienide ring to the coordination sphere. Steric hindrance is also probably responsible for the observed lability of the fourth C$_5$H$_5$ ligand in UCp$_4$. The steric repulsion is not large enough, however, to counteract the energy gained in bonding the additional η^5 cyclopentadienide ring to the uranium atom. The opposing factors, steric repulsion and increase in coordinate

bonds, which determine the structure of the complex, are depicted
for both U(C₅H₅)₄ and U(C₅H₅)₃ (2-methylallyl) in Fig. 7. The
difference in energy between the σ- and π-limiting structures is
viewed as due largely to steric rearrangement energy. The difference in bonding between the σ and π allyl is formally the energy
associated with *one* coordinate bond (η^1 to η^3) while that for a σ
and π cyclopentadienide ring is *two* coordinate bonds (η^1 to η^5).
The steric requirements for the π-bonded structures are similar,
since the 2-methylallyl group and the cyclopentadienide anion
occupy approximately the same area on the surface of the coordination polyhedron. Thus the steric rearrangement energy required
for the σ to π conversion must be nearly the same. Apparently
this energy is greater than that released by one coordinate bond
but less than two. Thus the ground state for the allyl complex
is σ and for cyclopentadienyl it is π.

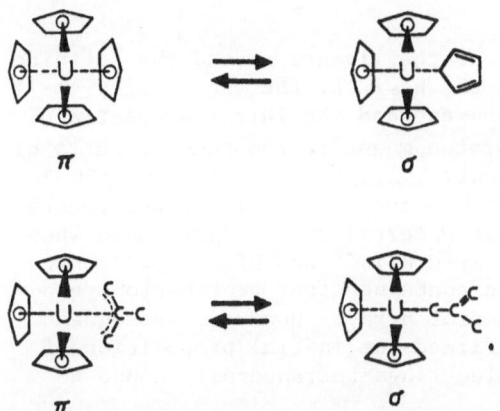

Fig. 7.   Interconversion of the σ- and π-bonded modes of allyl
          and cyclopentadienyl in the U(Cp)₃R structure.

The corresponding lanthanide allyl, Sm(C₅H₅)₂(allyl), has
recently been reported and preliminary indications, based upon
the absence of infrared absorptions in 1610-1640 cm⁻¹ region, are
consistent with a π-bonded structure.[40] Since the π-bonded structure would be formally eight coordinate and the σ-bonded structure
only six coordinate, this would be the predicted ground state.

The product [(C₅H₅)₂ThC₅H₄]₂, which results from the pyrolysis
of Th(C₅H₅)₃(n-butyl),[41] is a dimeric complex μ-di(η^5:η^1-cyclo-
pentadienyl)-dithorium(IV) in which each thorium is π (η^5) bonded
to two cyclopentadienyl rings. The third ring is also π (η^5)
bonded but has undergone a metallation reaction such that a ring
proton has been replaced by a σ (η^1) bond to the second thorium

atom. The coordination about the thorium is approximately the
same as in other actinide $Ac(C_5H_5)_3X$ structures. The centers of
the rings and the σ-bonded carbon occupy the apices of an approximate tetrahedron with the inter-ring angles ranging from 115 to
122°. The terminal C_5H_5 rings are equidistant from the metal,
with an average Th–C distance of 2.83 Å while the Th–C σ bond
length is significantly shorter, 2.55 Å. The bridging C_5H_4 ring
is not as well defined and shows greater fluctuation in the Th–C
distances.

3. Complexes of the type $Ac(IV)Cp_4$. Since the reduction
potential of the +4 state of the actinides rises rapidly beyond
uranium and since cylcopentadienide anion is a good reducing
agent, it is reasonable that only Th, Pa, U and Np form stable
tetravalent cyclopentadienides. While Pa, U, Np, Pu and Am have
stable oxidation states greater than 4, only complexes involving
oxygen or halogen donors are known for these higher oxidation
states.

A tetrahedral arrangement of the ligands around the actinide
atom was initially proposed on the basis of the vanishingly small
dipole moment of $U(C_5H_5)_4$ in benzene and the infrared spectra of
$Th(C_5H_5)_4$ and $U(C_5H_5)_4$. The proton magnetic resonance spectra of
$U(C_5H_5)_4$ and $Th(C_5H_5)_4$ showed only a single resonance (at +20.36
ppm and 1.10 ppm upfield from C_6H_6, respectively) for the twenty
protons in each complex.[46] Some uncertainty was introduced when
the structures of $Ti(C_5H_5)_4$,[47] $Zr(C_5H_5)_4$[48] and $Hf(C_5H_5)_4$[49] were
determined and showed that none contained four pentahapto cyclopentadienide rings in a tetrahedral array. However, subsequent
x-ray diffraction analysis confirmed the initial proposition of
four pentahapto cyclopentadienide rings tetrahedrally bound to
the uranium ion.[45]

The individual molecules of $U(C_5H_5)_4$ have point symmetry S_4
(Fig. 8). Disorder is present in the structure and is associated
with the crystallographic site symmetry. The refinement was based
upon a model in which each molecular site contained either the
molecule pictured in Fig. 8 or its enantiomorph, producing an
average structure containing about equal quantities of each
molecule with a site symmetry of D_{2d}. Coordination about the
uranium atom has the cyclopentadienide rings placed at the apices
of a tetrahedron. The cyclopentadienide ring is planar with an
average C–C distance of 1.39(1) Å and a range of 1.40(3)–1.37(2)
Å. The average U–C bond distance is 2.81(2) Å, with values ranging from 2.78 to 2.83 Å. Comparison with other bond lengths in
Table 3 reveals that this distance is significantly longer than
those found in the ten-coordinate $U(C_5H_5)_3X$ complexes. An examination of the carbon-carbon nonbonded contact distances in conjunction with the U–C distances in $U(C_5H_5)_3X$, $U(C_5H_5)_3R$ and
$U(C_5H_5)_4$ structures reveals considerable crowding in $U(C_5H_5)_4$.

The $C \cdots C$ distances between the three rings in $U(C_5H_4CH_2C_6H_5)_3Cl$ and $U(C_5H_5)_3(C_4H_7)$ range from 3.04 to 3.17 Å. The $C \cdots C$ distances between the various rings in $U(C_5H_5)_4$ are all 2.94(3) Å. This crowding results in the average U–C bond length in $U(C_5H_5)_4$ of 2.81(2) Å compared to 2.74 Å for the various $U(C_5H_5)_3X$ and $U(C_5H_5)_3R$ molecules.

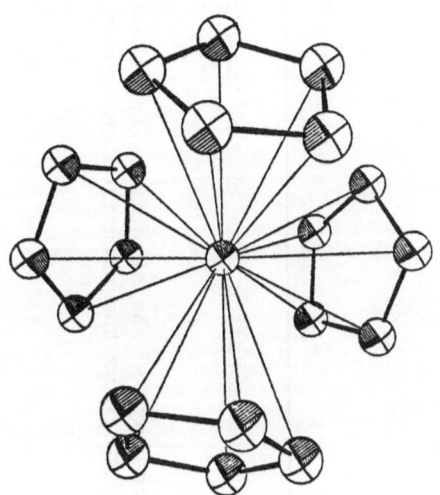

Fig. 8.   Molecular structure of $U(C_5H_5)_4$ viewed down the S_4 molecular axis, from Ref. (45).

Comparison of the structure of $U(C_5H_5)_4$ with those of the homologous Ti, Zr and Hf complexes is made in Table 4.   Unfortunately, the accuracy of the Zr and Hf structures is not sufficient to prove the exact nature of bonding.   It is evident from the table, however, that the total hapticity increases smoothly as the ionic radius of the metal ion increases.   The two limiting structures, $Ti(C_5H_5)_4$ and $U(C_5H_5)_4$, represent a transition from a structure in which two rings are pentahapto–bound and two are monohapto to a tetrahedral array of four pentahapto cyclopentadienide rings.

C.   Bonding in the Lanthanide and Actinide Cyclopentadienyl π Complexes

As noted earlier, the chemical properties indicate ionic bonding.   In general the physical evidence also suggests a high degree of ionic character[16] and the mass of the crystallographic evidence supports an ionic formulation of the bonding.   The pentahapto coordination of the $C_5H_5^-$ rings is consistent with either covalent or ionic bonding.   However, the observed mixtures

Table 4. Comparison of $M(C_5H_5)_4$ Complexes.

Metal	Ref.	Number of η^1-C_5H_5 Rings	Number of η^5-C_5H_5 Rings	Average π-C_5H_5 M-C Distance (Å)	Average σ M-C Distance	Metal Ionic Radius[a]
Ti	47	2	2	2.38	2.33	0.605
Hf	48	2[b]	2	2.50	2.34	0.71
Zr	49	1[b]	3	2.64	2.47	0.72
U	45	0	4	2.81	-	0.97

[a] The ionic radii [from Ref. (2)] are, for purposes of comparison, all chosen for a coordination number of six.

[b] These assignments have been disputed. The low accuracy of the structures and the small changes in structure required to change the ligand hapticity leave the situation ambiguous.

of η^5-coordination with bonding of lower hapticity so as to maximize the number of contacts consistent with the size of the metal
ion is clearly associated with ionic behavior. Furthermore, in
none of the structures is the aromatic geometry of the $C_5H_5^-$ ring
affected, as might be expected if any charge localization were
occurring. There is a clear correspondence of coordination
number and metal-carbon bond lengths with ionic radius.

The calculated bond lengths in Table 3 use ionic radii of
1.06 and 1.45 Å for F^- and Cl^-. These are average values derived
from those of Shannon and Prewitt[2] when corrected to a coordinate
number for the anion of 1, as described in the Introduction. The
"ionic value" for Cp anion, 1.61 Å, is based on the average values
here and in Table 2. The radius for sp^3 carbon, 1.44 Å, is based
on the Mg-C distances observed in several Grignard structures and
the sp carbon radius, 1.30 Å, is based on the .14 Å difference in
radii of sp and sp^3 carbon.[37] As can be seen in Table 3, the
ionic radii reproduce closely the observed distances. This con-
trasts sharply with the parameters in Table 1, where strong covalent effects are evident in the parameters.

III. CYCLOOCTATETRAENE COMPOUNDS

A.   Actinide Complexes

Interest in the organometallic chemistry of the lanthanides
and actinides was renewed with the synthesis of uranocene, $U(C_8H_8)_2$
in 1968 by Streitwieser and Mueller-Westerhoff. This was the first
example of an f-transition metal containing the cyclooctatetraene
dianion, and it was the suggestion of increased f-orbital participation in the bonding, resulting from the unique orbital symmetry
properties of the cyclooctatetraene dianion, that was the impetus
for the first studies.[16] In particular it was proposed that the
e_2 orbital of the cyclooctatetraene ring could overlap with the
$f_{\pm 2}$ orbitals of the metal just as the e_1 orbitals of cyclopentadiene can overlap with $d_{\pm 1}$ orbitals of transition metals to give
sandwich compounds. Hence the trivial name "uranocene" was proposed. The subsequent crystal structure determination confirmed a
π-sandwich structure with D_{8h} molecular symmetry.[16] Following the
characterization of $U(COT)_2$, $Th(COT)_2$, $Pu(COT)_2$, $Np(COT)$, and
$Pa(COT)_2$ have been prepared and their x-ray powder patterns have
shown them all to be isostructural with $U(COT)_2$.[16]

The molecular structure of $U(C_8H_8)_2$ and $Th(C_8H_8)_2$ consists of
a central metal atom symmetrically π-bonded to two $[C_8H_8]^{2-}$ rings
which are related by the inversion center at the heavy atom (Fig.
9). In uranocene the uranium-carbon bonds are equal within experimental error and average 2.647(4) Å. The thorium-carbon bonds are
slightly longer and average 2.701(4) Å. The mean C-M-C angle for

adjacent carbons in the dianion ring is 30.5(3)° in uranocene and 29.7(2)° in thoracene. These angles and distances along with the planarity of the dianion ring establish almost exact D_{8h} molecular symmetry.

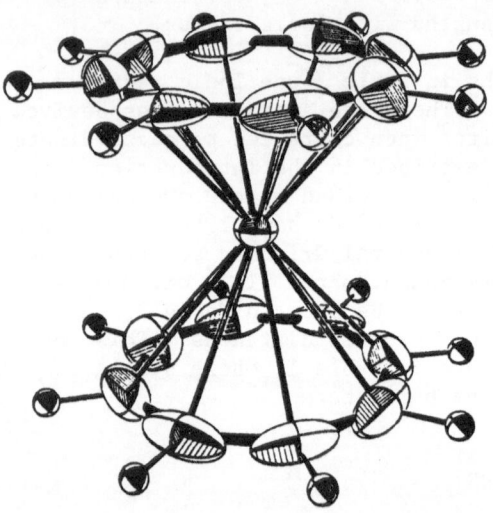

Fig. 9. Molecular structure of U(C₈H₈)₂ and Th(C₈H₈)₂, from Ref. (63).

It is interesting to compare these actinide (IV) cycloocta-tetraene complexes with similar compounds of the group IVB transition elements Ti, Zr and Hf. Bis(cyclooctatetraene) complexes of all three are known although structural data are only available for the first two. All would appear to involve both planar and nonplanar COT rings and to exhibit a "slipped" sandwich structure rather than the true sandwich structure of uranocene.

The compound Ti(C₈H₈)₂ can be prepared from TiCl₄ and Na₂C₈H₈ and in the presence of (C₂H₅)₂AlH it can be converted to Ti₂(C₈-H₈)₃.[50] The structures of both compounds are similar,[51,52] involving one symmetrical η⁸-coordinated COT ring and one non-planar COT ring of lower hapticity per titanium. The Ti₂(COT)₃ complex uses the non-planar ring to bridge the two titanium atoms. The zirconium complex[53-55] is similar except that a coordinated THF molecule is also present.

This failure to adopt a uranocene-type structure can be explained in two ways. One way is to note that the actinide ions are substantially larger than the group IVB ions and thus require more ligands to saturate the coordination sphere. Uranium complexes are often 9- or 10-coordinate. The COT dianion contains

10 π electrons, or is formally a 5-coordinate donor, making $U(C_8-H_8)_2$ a 10-coordinate complex. The smaller transition elements cannot accommodate so many coordination sites and so the second COT ring slips to one side, making the complex only 7 or 8 co-ordinate. The other explanation observes that the two COT di-anions contribute 20 π electrons to a M^{+4} center, thus violating the effective atomic number rule. Lanthanide and actinide com-plexes seldom follow this rule, but *d*-transition complexes often do and in particular Group IVB organometallic complexes almost invariably have 16 to 18 valence electrons.[56] Thus the second COT ring slips to one side to reduce the number of valence electrons. It is interesting that the mixed compound $Ti(C_8H_8)(C_5H_5)$ has 17 valence electrons, is formally eight coordinate and does exist in a true π-sandwich configuration.[57]

The eclipsed conformation of the rings in uranocene and tho-racene is not required in order to maximize ligand and *f*-orbital overlap and does not appreciably change ligand-ligand repulsive interactions. The structure of $U[C_8H_4(CH_3)_4]_2$ contains two types of molecules, type A with the methyl groups nearly eclipsed and type B with them nearly staggered. While neither molecule is con-strained to have any symmetry higher than 1, the angles and dis-tances within and between the dianion rings establish that the molecular coordination of the uranium atoms in both molecules A and B is virtually identical with that in the parent compound, with the exception of the rotomeric conformation of the ligands. An interesting feature of the $U[C_8H_4(CH_3)_4]_2$ structure is that all of the methyl groups in both types of molecules are bent in toward the uranium from the plane of the dianion ring. The angle of this bend averages 4.1°, corresponding to a distance of 0.10 Å from the appropriate least-squares plane. Van der Waals' attraction cannot account for this convex bending, as molecule A already has intra-molecular methyl-methyl contacts 0.3 Å shorter than the sum of the van der Waals' radii of two methyl groups, nor are there inter-molecular forces which would force all the methyl groups inward. The explanation apparently lies in the electronic structure, although the mechanism remains unknown. An unusual alternation of bond angles in this and the phenyl substituted analogue $U[C_8H_4-(Ph)_4]_2$ has been noted[58] but the origin of this, too is unknown.

In addition to these actinide(IV) compounds, the increasing stability of the +3 oxidation state for the trans-uranium elements has led to the preparation of compounds of formula $K[M(C_8H_8)_2]$ where M = Np or Pu. In their chemical behavior these compounds are similar to the corresponding lanthanide complexes (<u>vide</u> <u>infra</u>) and their x-ray powder patterns suggest they have the same structure.

B.   Lanthanide Complexes

Several lanthanide cyclooctatetraene complexes have been synthesized. The divalent metal complexes $Eu(C_8H_8)$ and $Yb(C_8H_8)$[59] were prepared by direct reaction of the metal with cyclooctatetraene in liquid ammonia and no structural data exist for them but they probably involve some kind of bridging interaction.

The trivalent complexes $[K(diglyme)] [Ln(C_8H_8)_2]$ were prepared from the anhydrous metal trichlorides and COT^{2-}. The cerium complex has been the subject of a structural investigation.[60] The structure consists of a contact ion pair formed by the $[Ce(C_8H_8)_2]^-$ anion and diglyme-coordinated potassium cation, as shown in Fig. 10. The anion consists of a central cerium atom, lying on a crystallographic mirror plane, which is symmetrically π-bonded to the two COT rings, the average Ce–C bond length being 2.742(8) Å. The molecular geometry is very close to D_{8d}, the rotomeric configuration which corresponds to an eclipsed geometry as opposed to the D_{8h} symmetry in uranocene. Otherwise the general features of the bonding are substantially the same as $U(C_8H_8)_2$ and $Th(C_8H_8)_2$ and are compared with them in Table 5.

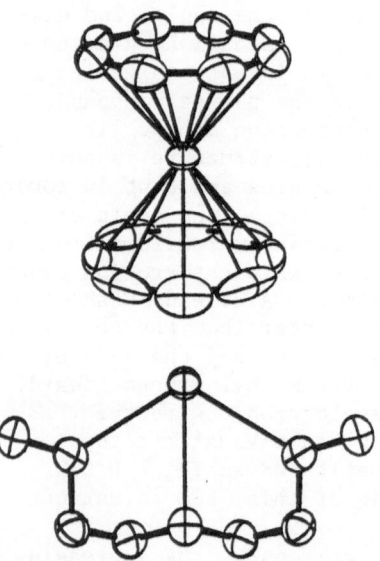

Fig. 10.   The molecular structure of $[K(diglyme)] [Ce(COT)_2]$, from Ref. (60).

A dimeric compound of formula $[Ce(C_8H_8)Cl \cdot 2\ THF]_2$ also has been prepared and structurally characterized. The molecular structure consists of two cerium atoms which are asymmetrically bridged by the two chloride ions with Ce–Cl distances of 2.855(2)

Table 5. Comparison of M-C Bond Lengths in Several Cyclooctatetraene Complexes

Compound	Ref.	Average M-C Bond Length (Å)	Formal Metal Ion Coordination Number [a]	Metal Ionic Radius (Å) [b]	Effective COT Ionic Radius (Å)
$U(C_8H_8)_2$	63	2.647(4)	10	1.08*	1.57
$U(C_8H_4(CH_3)_4)_2$	64	2.658(4)	10	1.08*	1.58
$Th(C_8H_8)_2$	63	2.701(4)	10	1.12*	1.58
$[K(diglyme)][Ce(C_8H_8)_2]$	60	2.742(8)	10	1.17*	1.57
$[Ce(C_8H_8)Cl \cdot 2THF]_2$	65	2.710(2)	9	1.15	1.56
$[Nd(COT)(THF)_2][Nd(COT)_2]$ [c]	61	2.68(1) / 2.79(1) / 2.68(1)	10 / 10 / 9	1.14* / 1.14* / 1.12	1.54 / 1.65[d] / 1.56
$Zr(C_8H_8)_2 \cdot THF$	54	2.461(7)	9	0.90*	1.56
$Ti(C_8H_8)(C_5H_5)$	57	2.323(4)	8	0.77*	1.55
$[K(diglyme)]_2[C_8H_4(CH_3)_4]_4$	66	3.003(8)	7	1.46	1.54
$K_2(COT)(diglyme)$ [c]	67	2.98(2) / 3.05(2)	6 / 7	1.38 / 1.46	1.60 / 1.59
$Rb_2COT(diglyme)$ [c]	68	3.10(1) / 3.15(1)	6 / 7	1.52 / 1.56	1.58 / 1.59

[a] Defined as the number of coordinating electron pairs.

[b] For the metal ion with this oxidation state and coordination number. When values for the particular coordination number have been obtained by interpolation from data of other coordination numbers they are marked *.

[c] There are two types of coordination sites in this compound.

[d] This average Nd-COT interaction is for a bridging ring.

and 2.935(2) Å. The Ce-Ce distance is 4.642(3) Å. The COT ring
is symmetrically π-bonded to the cerium (Fig. 11). The coordination about the cerium is approximately cubic. The COT dianion occupies one face of the cube and the opposite face has the two
chloride ions and two THF oxygen atoms on adjacent corners. (It
is interesting that the slipped sandwich complex $Zr(C_8H_8)_2 \cdot THF$
mentioned previously, which also has a single planar η^8 COT ring,
has a similar coordination.)[16] Related complexes of neodymium
have been prepared by Karraker containing bromide and iodide in
place of chloride. While their chemical properties are similar to
the dimeric chloride compound their powder patterns suggest they
may have different structures. Since they also have increasing
amounts of solvent, the bromide containing three THF molecules and
the iodide four, these may be complexes in which the halide bridge
is broken by addition of another solvent molecule to give a monomer such as $[Ln(COT)X \cdot 3THF]$.

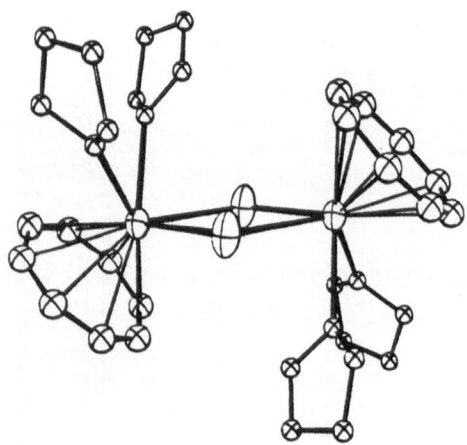

Fig. 11. The molecular structure of the dimeric complex,
          $[Ce(COT)Cl \cdot 2THF]_2$, from Ref. (65).

The mixed complex $[Nd(COT)(THF)_2][Nd(COT)_2]$[61] has a structure
(Fig. 12) which combines the features of both $Ce(COT)_2^-$ (Fig. 10)
and the dimer $[Ce(COT)Cl \cdot 2THF]_2$. This material is almost certainly the solvate of the same series of compounds $Ln_2(COT)_3$ reported
by Cesca et al.[62]

Mixed sandwich compounds of the type $M(C_8H_8)(C_5H_5)$ [M = Nd,
Sm, Ho, Er] are now known. They were prepared by adding sodium
cyclopentadienide to $[Ln(C_8H_8)Cl \cdot 2THF]_2$ or the addition of K_2COT
to $Ln(C_5H_5)Cl_2$. There are no structural data for these materials

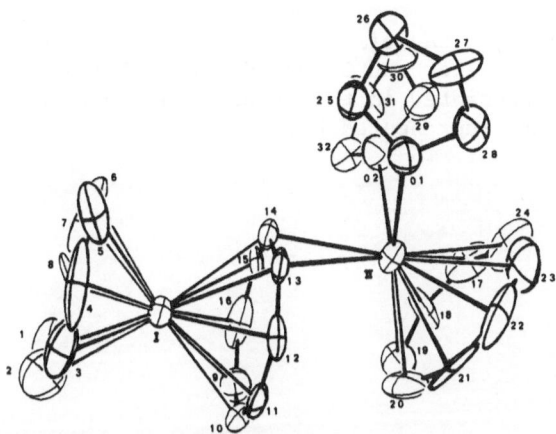

Fig. 12. The molecular structure of [Nd(COT)(THF)$_2$][Nd(COT)$_2$], from Ref. (61).

but they presumably resemble the corresponding titanium compound Ti(C$_8$H$_8$)(C$_5$H$_5$), whose structure is known. For this complex the carbons of the COT^{2-} ring lie some 0.03 Å closer to the metal than the carbons of the C$_5$H$_5^-$ ring, as expected from the difference in charge. This is also found for the corresponding metal-carbon bond lengths of the lanthanides.

A comparison of the metal-carbon bond lengths, ionic radii and formal coordination numbers of these compounds is summarized in Table 5. The formalism used in establishing coordination number assumes that a η^8-cyclooctatetraene ligand is a 5 electron-pair donor. The ionic radii have been adjusted for both the charge of the central metal and coordination number as described earlier. It will be seen that the difference in metal-carbon bond lengths corresponds very closely to the difference in ionic radii if co- ordination numbers are taken into account. The effective ionic radius for the COT dianion averages 1.57 Å. For the lanthanides and actinides, as well as other relatively ionic COT complexes, metal-carbon bond lengths can be predicted within .01 or .02 Å by using this effective ionic radius. This is even closer agreement than was seen in the cyclopentadiene compounds. Thus the struc- tural features of these compounds are consistent with ionic bonding. Other physical properties of uranocene and related compounds seem to imply a significant amount of covalent character in the bonding. Unless this is a dominant feature of the bonding it cannot be expected to appear as a structural effect.

Acknowledgment

I am pleased to acknowledge the efforts of: Charles Eigenbrot and John Robbins and Drs. Ted Baker and Gordon Halstead for help in assembling the material presented in this paper; and many other past co-workers whose work is represented here.

References

1. L. Pauling, "The Nature of the Chemical Bond," 3rd ed., Cornell University Press, Ithaca, N.Y., 1960, pp. 537-540.

2. R. D. Shannon, Acta Cryst., A 32, 751 (1976).

3. E. Gard, A. Haaland, D. P. Novak, and R. Seip, J. Organometal. Chem., 88, 181-189 (1975).

4. A. Almenningen, A. Haaland, and T. Motzfeldt, in "Selected Topics in Structure Chemistry," Universitets-forlaget, Oslo, 1967, p. 105.

5. A. Almenningen, A. Haaland, and Svein Samdal, J. Organometal. Chem., 149, 219-229 (1978).

6. A. Haaland and J. E. Nilsson, Acta Chem. Scand, 22, 2653 (1968).

7. A. Almenningen, E. Gard, A. Haaland, and J. Bunvoll, J. Organometal. Chem., 107, 273-279 (1976).

8. A. K. Hedberg, L. Hedberg, and K. Hedberg, J. Chem. Phys., 63, 1262-1266 (1975).

9. L. Hedberg and K. Hedberg, J. Chem. Phys., 53, 1228-1234 (1970).

10. P. J. Wheatley, "Perspectives in Structural Chemistry," New York, 1967.

11. W. Bünder and E. Weiss, J. Organometal. Chem., 92, 1-6 (1975).

12. J. Dunitz, L. E. Orgel, and A. Rich, Acta Cryst., 9, 373 (1956).

13. W. Bünder and E. Weiss, J. Organometal. Chem., 92, 65-68 (1975).

14. J. W. Bats, J. J. deBoer, and D. D. Bright, Inorg. Chim. Acta, 5, 605 (1971).

15. A. Haaland, J. Lusztyk, D. P. Novak, J. Brunvoll, and K. B. Sarovieysky, J. Chem. Soc., Chem. Commun., 1974, 54.

16. For a review of the structural studies cited to 1975 and pertinent references see, E. C. Baker, G. W. Halstead, and K. N. Raymond, Struct. Bonding (Berlin), 25, 23 (1976).

17. C. H. Wong, T. Lee, and T. Lee, Acta Cryst. Sect. B, 25, 2580 (1969)

18. J. L. Atwood and K. D. Smith, J. Am. Chem. Soc., 95, 1488 (1973).

19. Atwood, J. L., J. H. Burns, and P. G. Laubereau, J. Am. Chem. Soc., 95, 1830 (1973).

20. M. Tsutsui and A. J. Gysling, J. Am. Chem. Soc., 91, 3175 (1969).

21. T. Moeller, "Comprehensive Inorganic Chemistry," Vol. 4, p. 1, Oxford, 1973.

22. J. H. Burns, W. H. Baldwin, and F. H. Fink, Inorg. Chem., 13, 1916 (1974).

23. J. H. Burns and W. H. Baldwin, J. Organomet. Chem., 120, 361 (1976).

24. E. C. Baker, L. D. Brown, and K. N. Raymond, Inorg. Chem., 14, 1376 (1975).

25. J. L. Atwood and K. D. Smith, J. Chem. Soc., Dalton Trans., 1973, 2487.

26. E. C. Baker and K. N. Raymond, Inorg. Chem., 16, 2710 (1977).

27. R. G. Hayes and J. L. Thomas, Organometal. Chem. Rev. A, 7, 1 (1971).

28. J. Holton, M. F. Lappert, D. G. H. Ballard, and R. Pearce, J.C.S. Chem. Comm., 1976, 480.

29. P. G. Laubereau and J. H. Burns, Inorg. Chem., 9, 1091 (1970); Inorg. Nucl. Chem. Letters, 6, 59 (1970).

30. P. G. Laubereau, et al., Inorg. Nucl. Chem. Letters, 6, 611 (1970).

31. B. Kanellakopulos, E. Dornberger, and F. Baumgärtner, Inorg. Nucl. Chem. Letters, 10, 155 (1974).

32. D. G. Kalina, T. J. Marks, and W. A. Wachter, J. Am. Chem. Soc., 99, 3877 (1977).

33. J. Leong, K. O. Hodgson, K. N. Raymond, Inorg. Chem., 12, 1329 (1973).

34. J. L. Atwood, C. F. Hains, M. Tsutsui, and A. E. Gebala, J.
     Chem. Soc. Chem. Commun., 1973, 452; J. L. Atwood, M. Tsutsui,
     N. Ely, and A. E. Gebala, J. Coord. Chem., 5, 209 (1976).

35. M. Tsutsui, N. Ely, and R. Dubois, Accts. Chem. Res., 9, 217
     (1976).

36. G. Parego, M. Cesari, F. Forina, and G. Jugli, Gazz. Chim.
     Ital., 105, 643 (1975); Acta Cryst., B32, 3034 (1976).

37. "Tables of Interatomic Distances," Special Publication No. 11,
     The Chemical Society, London, 1958.

38. T. J. Marks, A. Seyam, and J. R. Kolb, J. Am. Chem. Soc., 95,
     5539 (1973).

39. G. W. Halstead, E. C. Baker, and K. N. Raymond, J. Am. Chem.
     Soc., 97, 3049 (1975).

40. M. Tsutsui and N. Ely, J. Am. Chem. Soc., 97, 3551 (1975).

41. E. C. Baker, K. N. Raymond, T. J. Marks, and W. A. Wachter,
     J. Am. Chem. Soc., 96, 7586 (1974).

42. C. Wong, T. Yen, and T. Lee, Acta Cryst., 18, 340 (1965).

43. R. R. Ryan, R. A. Penneman, and B. Kanellakopulos, J. Am. Chem.
     Soc., 97, 4258 (1975).

44. J. H. Burns and P. G. Laubereau, Inorg. Chem., 10, 2789 (1971).

45. J. H. Burns, Organometal. Chem., 69, 225 (1974).

46. R. von Ammon, B. Kanellakopulos, and R. D. Fischer, Chem. Phys.
     Letters, 2, 513 (1968).

47. J. L. Calderon, F. A. Cotton, B. G. DeBoer, and J. Takats, J.
     Am. Chem. Soc., 95, 3592 (1971).

48. V. I. Kulishov, N. G. Bokii, and Yu. T. Struchkov, J. Struct.
     Chem. (USSR), 11, 646 (1970).

49. V. I. Kulishov, N. G. Bokii, and Yu. T. Struchkov, Zh. Strukt.
     Khim., 13, 1110 (1972).

50. H. Breil and G. Wilke, Angew. Chem. Intern. Ed. Engl., 5, 898
     (1966).

51. H. Dietrich and M. Soltwisch, Angew. Chem. Intern. Ed. Engl., 8,
     765 (1969).

52. H. Dierks and H. Dietrich, Acta Cryst., B, 24, 58 (1968).

53. H. J. Kablitz, R. Kallweit, and G. Wilke, J. Organometal.
     Chem., 44, C 49 (1972).

54. D. J. Brauer and C. Krüger, J. Organometal. Chem., 42, 129
     (1972).

55. H. J. Kablitz and G. Wilke, J. Organometal. Chem., 51, 241
     (1973).

56. C. A. Tolman, Chem. Soc. Rev., 3, 337 (1972).

57. P. A. Kroon and R. B. Helmhodt, J. Organometal. Chem., 25,
     451 (1970).

58. L. K. Templeton, D. H. Templeton, and R. Walker, Inorg.
     Chem., 15, 3000 (1976).

59. R. G. Hayes and J. L. Thomas, J. Am. Chem. Soc., 91, 6876
     (1969).

60. K. O. Hodgson and K. N. Raymond, Inorg. Chem., 11, 3030 (1972).

61. S. R. Ely, T. E. Hopkins, and C. W. DeKock, J. Am. Chem. Soc.,
     98, 1624 (1976); C. W. DeKock, S. R. Ely, T. E. Hopkins, and
     M. A. Brault, Inorg. Chem., 17, 625 (1978).

62. A. Greco, S. Cesia, and G. Bertolini, J. Organometal. Chem.,
     113, 321 (1976).

63. A. Avdeef, K. N. Raymond, K. O. Hodgson, and A. Zalkin, Inorg.
     Chem., 11, 1083 (1972)

64. K. O. Hodgson and K. N. Raymond, Inorg. Chem., 12, 458 (1973).

65. K. O. Hodgson and K. N. Raymond, Inorg. Chem., 11, 171 (1972).

66. B. G. Goldberg, K. N. Raymond, C. A. Harmon, and D. H.
     Templeton, J. Am. Chem. Soc., 96, 1348 (1974).

OPTICAL SPECTROSCOPY OF f-ELEMENT COMPOUNDS

W. T. Carnall

Chemistry Division, Argonne National Laboratory,
Argonne, Illinois 60439, U.S.A.

I. INTRODUCTION

    The energies and intensities of transitions observed in the
optical spectra of lanthanide (Ln) and actinide (An) compounds
can typically be measured with a high degree of accuracy. The
observed transitions can then be directly represented as upper
state energy levels where the structure is induced by the environment. We discuss here the systematic theoretical interpretation
of these transitions both in terms of energy level structure and
transition probability. Particularly for the trivalent lanthanides and actinides, the detail to which the interpretation can
be carried is unique in the periodic table. Although we will emphasize the electronic structure of organometallic lanthanides
and actinides in the present discussion, it will be clear that
this type of ligand does not present any unique interpretive
problems. The basic framework of the interpretation is not dependent upon the specific ionic environment. On the other hand,
organometallic compounds represent a particularly interesting
group in which to study excited state relaxation.

    As lanthanide and actinide metal atoms are progressively ion-
ized, the observed energy level structure undergoes profound
changes. We are able to make accurate general statements about
this structure because systematic trends have been deduced from
an existing large but still incomplete reservoir of data from
atomic spectroscopy. Our interest will normally center on the
one or two lowest-energy electronic configurations at any given
state of ionization. The energy region for optical spectroscopy
will be taken to extend to ~ 50000 cm^{-1} or ~ 6 eV, the beginning of
the vacuum ultraviolet range. In the lanthanide series, the

T. J. Marks and R. D. Fischer (eds.), Organometallics of the f-Elements, 281–307.

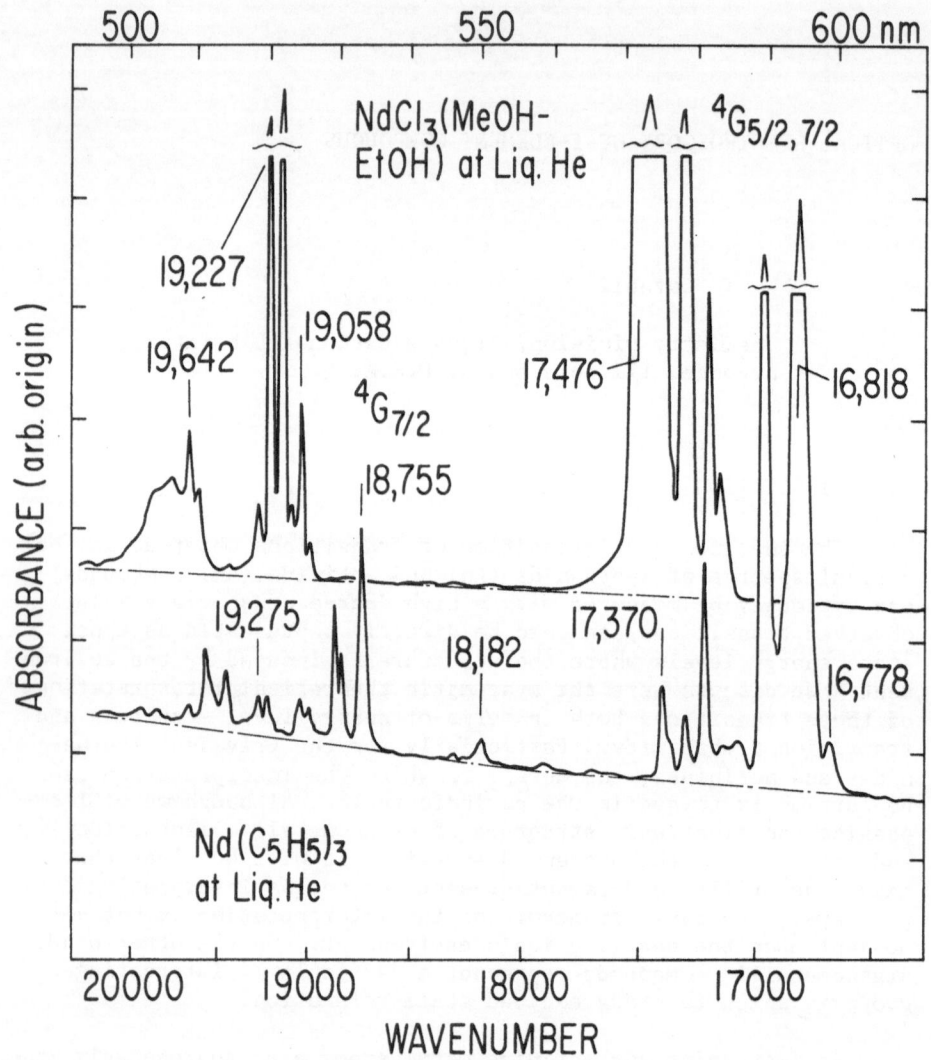

Fig. 1. Absorption Spectra of NdCl$_3$ (MeOH – EtOH) and NdCp$_3$ at 4°K

electronic structure consists of the Xe-core with outer 5d and/or 6s electrons. The $4f^N$-shell is located interior to and shielded by the filled $5s^2 5p^6$ shells. A similar situation occurs in the actinide series which is built on the Rn-core. The characteristic lack of sensitivity of the f^N-series to the environment compared to, for example, members of the d^N-transition series, is ascribed to the inner (shielded) character of the f-electron shell.

Normally the singly ionized states of the f-elements are of lesser interest to chemists because of the difficulty of stabilizing them in compounds. However, all the valence states from +2 through +7 have been characterized in one or another actinide or lanthanide compound. In several instances unusual oxidation states have been stabilized in organometallic compounds. Although our principal purpose is to examine the nature of the electronic structure revealed in optical spectra, and to show how this is related to transition probability theory, insights into bonding character will also be mentioned.

II. NATURE OF THE TRANSITIONS OBSERVED IN ACTINIDE AND LANTHANIDE ABSORPTION SPECTRA

In order to interpret the various features observed in the spectra of f-element organometallics, we need to be aware of the types of transitions that can occur. Consider first the basic electronic structure associated with atoms as they are progressively ionized. Two types of electronic transitions may be distinguished: those transitions that occur between different configurations of electrons and those that occur between states within a particular configuration.

Taking Eu^{2+} as an example, the lowest energy configuration is $4f^7$, with the next higher energy configurations being $4f^6 5d$ and $4f^6 6s$, respectively, Brewer (1971). Thus with the ground state of $4f^7 (^8S_{7/2})$ as the reference energy, promotion of an f-electron to a d shell ($4f^7 \rightarrow 4f^6 5d$) requires ~ 4.3 eV (34500 cm^{-1}), and the intensity of the absorption band observed as a result of this process would be representative of a normal "allowed" electric dipole transition. Electronic transitions can in principle also occur at <34500 cm^{-1} in Eu^{2+}, but the corresponding absorption bands, characterized as due to parity forbidden transitions between states within the $4f^7$-configuration ($4f^7 \rightarrow 4f^7$), would be $\sim 10^6$-fold weaker than the f$\rightarrow$d bands. Typical sharply defined absorption bands ($f^3 \rightarrow f^3$) in neodymium tricyclopentadienide, (NdCp$_3$), observed at $\sim 4°K$ are shown in Fig. 1, Pappalardo (1965a), where they are compared to the same group in NdCl$_3$ (dissolved in CH$_3$OH-C$_2$H$_5$OH and cooled to form a glass).

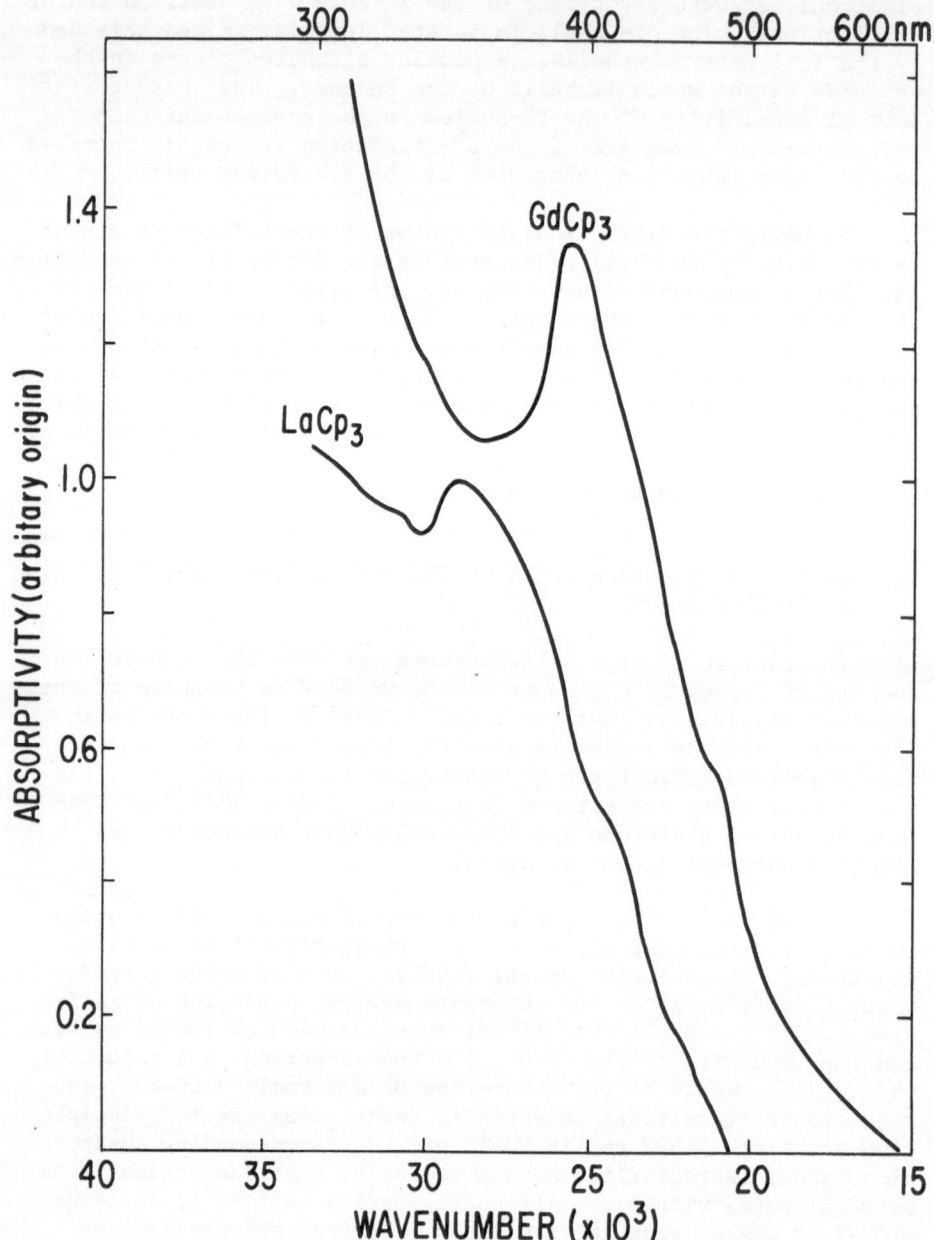

Fig. 2. Diffuse Transmission Spectra of LaCp₃ and GdCp₃ at ~78°K

Examination of the structure in the region of an f→f transition frequently reveals the presence of weak sharp bands that cannot be correlated with the f→f electronic transitions. These extra lines are usually interpreted as vibronic (lattice) states coupled to the actual f→f states, and have been observed, for example, strongly coupled to the $^3F_3$ and $^3F_2$ states of TmCp$_3$, Pappalardo (1969). In some cases detailed analysis in terms of characteristic vibrations have been made, Cohen and Moos (1967).

In addition to the foregoing, both charge transfer transitions and spectra solely characteristic of the ligands (π→π*) are observed in the systems of interest here. The process of charge (electron) transfer, usually from filled molecular orbitals principally concentrated on the ligands, to the partly filled f-shell gives rise to relatively intense broad bands. Energy correlations with the redox potential of the trivalent f-elements have been made, Jørgensen (1969), Nugent (1975). Since Eu^{3+} is the most oxidizing of the trivalent lanthanides, the electron transfer bands would be expected to occur at the lowest wave number. Pappalardo and Jørgensen (1967) pointed out the surprising result that intense bands characterized as resulting from charge transfer transitions were identified in YbCp$_3$ at a slightly lower energy than in EuCp$_3$. The broad adsorption bands in ThCp$_3$ at ~18000 and 25000 cm^{-1}, Kanellakopulos, et al. (1974), are somewhat reminiscent of the spectrum of YbCp$_3$ in the same energy range.

Since attention will focus on the LnCp$_3$ compounds, the optical properties associated with the Cp-ligand are important to recognize. In both LaCp$_3$ and GdCp$_3$ the transmission spectrum of a thin film, Fig. 2, reveals strong ultraviolet absorption associated with the Cp ligand, Pappalardo and Losi (1965). Excitation at 27300 cm^{-1} yielded strong fluorescence near 20000 cm^{-1} which could be interpreted as a transition from essentially triplet states of the C_5H_5 to a singlet ground state. Strong absorption in the ultraviolet range is an expected characteristic of organolanthanide and actinide compounds and will, in general, severely limit the range in which f-electron transitions can be observed. In $(\eta^5-C_9H_7)UI$ the cut-off is already observed near 14000 cm^{-1}, Goffart and Duyckaerts (1978), and the same problem occurs in Ln^{3+} complexes with the cyclooctatetraene dianion, Hodgson et al. (1973).

III. INTERPRETATION OF THE ENERGY LEVEL STRUCTURE OBSERVED IN THE OPTICAL SPECTRA OF THE TRIVALENT LANTHANIDES AND ACTINIDES

The purpose of this section is to briefly summarize the extensive theoretical model presently being used to interpret f-electron spectra and to show some examples of the correlation between theory and experiment. The examples will be taken from trivalent systems where the most extensive experimental work is available. There is an intrinsic interest in developing insights into the electronic structure of actinides and lanthanides in different environments; however, for present purposes our interest centers on obtaining the best available description of the electronic states (eigenvectors) in a particular environment for use in intensity calculations to be discussed later. It will be shown that there is considerable similarity between the well characterized energy level schemes for f-elements in certain crystal hosts and what can be established of the level schemes in particular organometallic compounds. The similarity in part justifies the use of the eigenvectors derived for better characterized systems in connection with calculations involving the organometallic compounds. Of particular relevance to this discussion is the availability of published spectra recorded at low temperature and moderately high resolution for practically all of the $LnCp_3$ and the corresponding $AnCp_3$ from Th^{3+} through Cf^{3+}.

III.1. Energy level analysis

Extensive analysis of trivalent lanthanide and actinide spectra has been possible because the crystal or ligand-field splitting of states is small compared to the principal electrostatic and magnetic interactions which determine the structure of the f^N-configuration. As a consequence many electronic states are relatively isolated in energy and selection rules governing transitions can be related in detail to the experimental data. The similarities in the energies of states in $LnCp_3$ and LnX_3, where X is a halide, has led to the characterization of the cyclopentadienides as highly ionic in their bonding, Nugent et al. (1971), Hayes and Thomas (1971). Interpretation of cyclopentadienide spectra is however, limited by the strong ultraviolet absorption already discussed, Fig. 2, which restricts the energy range over which f→f transitions can be observed.

The model interactions that make up the total Hamiltonian or representation of the energy of the system are treated in detail in the literature, Judd (1963), Wybourne (1965), Dieke (1968), and Carnall et al. (1977). As shown in the lecture on electronic structure, the basic single configuration model for interactions within the f^N-configuration can be written:

$$E = E_F + E_{CF} \tag{3.1}$$

$$\text{where } E_F = \sum_{k=0}^{6} F^K(nf,nf) f_K + \zeta_f A_{so} \quad (k \text{ even}) \tag{3.2}$$

Thus the energy of any state is the sum of atomic (free-ion) interactions (E_F) and those characteristic of the crystal (ligand)
field (E_{CF}). $F^K(nf,nf)$ and ζ_f represent the radial parts of the
electrostatic and spin-orbit interactions between the f-electrons,
respectively; f_K and A_{so} are the corresponding angular interactions whose matrix elements can be computed.

In proceeding with the energy level analysis of an ion such
as Nd^{3+}:LaF$_3$, the problem is to evaluate radial integrals such as
those in (3.2). In practice this is done by a data fitting process in which the integrals are treated as parameters; however, the
values which these parameters may assume and still retain physical
significance is limited. Extensive Hartree-Fock (HF) calculations
of the F^K and ζ, (3.2), have been carried out, but the resulting
HF-integrals are larger than those determined by a fit to experimental data. This discrepancy simply indicates that the ab initio
results do not include the effects of "configuration interaction";
they assume as does (3.2) that the ground configuration is purely
f^N in character. In practice, the f^N-electrons can be thought of
as spending some time in higher-lying configurations where they
move in larger orbits and thus interact less than in the f^N-
configuration. The fitted parameters reflect these effects of
other configurations. On the other hand, the ratios of F^4/F^2 and
F^6/F^2 found in the ab initio calculations, while typically somewhat
smaller than those deduced from experiment, serve as a useful
first approximation.

It has been known for many years that analysis of experimental data for the trivalent f^N-spectra based solely on (3.2), results in large deviations between observed and computed energy
levels, Rajnak and Wybourne (1963). In response to this deficien-
cy, additional effective operators have been added to the Hamiltonian in order to take explicit account of the effects of configuration mixing. To the extent that such a procedure is successful, it should result in a lesser distortion of the purely f^N-
character of the F^K and ζ obtained from fitting experimental data.

The effects of the environment on the electronic orbitals of
Ln^{3+} or An^{3+} are significant, but as noted earlier, $E_F \gg E_{CF}$. Consequently, efforts have concentrated on an expansion of the atomic
Hamiltonian to permit the accurate description of observed group-
ings of crystal-field levels. The crystal-field interaction itself, (3.1), appears at present to be adequately described in
terms of the single particle model

$$E_{CF} = \sum_{k,q,i} B_q^k (C_q^k)_i \qquad (3.3)$$

discussed in the companion lecture.

Rajnak and Wybourne (1963) showed that both two and three-body terms must be considered in correcting E_F for the effects of configuration mixing. The interacting configurations will be those of the same parity as f^N. Opposite parity configurations interact with f^N via the crystal-field. Although the configurations of interest can be classified as weakly coupled to f^N by the Coulomb field, their inclusion in the model was essential to recent developments in energy level analysis.

In second-order perturbation theory, there are ℓ two-body scalar terms required to describe the two-electron excitations in the ℓ^N-configuration, i.e., for f^N there are three terms. The interactions in question for f^3 would include, for example, $4f^3$ with $4f5d^2$, $4f5f^2$, or $4f6p^2$. In the development by Rajnak and Wybourne (1963) the additional terms are of the form

$$\alpha\ L(L+1) + \beta\ G(G_2) + \gamma\ G(R_7) \tag{3.4}$$

where α, β, and γ are parameters. The eigenvalues of $G(G_2)$ and $G(R_7)$ have been tabulated, Wybourne (1965).

The overt effects of (3.4) on the terms of a configuration are not simple to describe, but clearly the $L(L+1)$ coefficient of α provides for a shift in energy of the higher with respect to the lower orbital angular momentum terms. There is some similarity in the effects of α and β on a configuration since both are functions of L. They should be varied together where this can be justified by the data base, or one should be assigned an approximate fixed value while the other is varied. One effect of γ is to provide for a shift in terms of differing multiplicity. However, $G(R_7)$ has very large matrix elements for those few states of different seniority number than the N-value of the f^N-configuration being studied. Thus γ can move a single state several hundred cm^{-1} with respect to the rest of the configuration. The importance of this effect has been well documented.

Perturbations of a pure ℓ^N-configuration with N>2 are also described by three-body operators, Rajnak (1965), Judd (1966a). Typical interactions of f^3 with the class of same parity configurations differing from it in the quantum numbers of one electron and described by three-body operators would be $4f^2nf$ (i.e., $4f^25f$), $4f^2np$, or $n'p^64f^3 \leftrightarrow n'p^54f^4$. We have used Judd's formulation which results in six new terms added to the Hamiltonian, $T^i t_i$ (i = 2, 3, 4, 6, 7, 8) where T^i are the integrals treated as parameters. As a very rough estimate for $4f^N$, the two-body operators represent an approximate 4% contribution to the Hamiltonian, while the three-body terms give a $\sim 1\%$ contribution.

In addition to the above, there are small corrections to the spin-orbit interaction (spin-spin and spin-other-orbit) which have

been included in recent work, Judd et al. (1968). Ab initio calculations of all of the corrections to the basic model have been
made, and the values of the computed terms are in good agreement
with those determined by fitting experimental data, Morrison and
Rajnak (1971), Balasubramanian et al. (1975). For the small corrections to ζ, ab initio results are often used directly and not
varied.

Significant improvements in the fits to experimental data by
incorporating the two-body terms in the model were first demonstrated by Rajnak (1965) in the analysis of PrIII ($4f^3$). The deviation in fitting 38 (of the 41) free-ion states dropped from
500 cm^{-1} using (3.2) to 149 cm^{-1} by adding the terms of (3.4).
Further improvement to 29 cm^{-1} was obtained by the addition of
three-particle terms. Computations have also been made using
data for $Nd^{3+}:LaF_3$ where 147 of the 182 crystal-field levels in
the $4f^3$-configuration representing 39 of the 41 free-ion states
have been assigned. Important new insights were gained when the
data were fit first with the free-ion parameters only, and then
with the addition of (3.3). Fits made to the pseudo free-ion
levels in LaF_3, i.e., the centers of gravity of the observed
crystal-field components of each state, showed deviations between
theory and experiment that were similar to those reported by
Rajnak for the same number of parameters. It should be emphasized
that the four-parameter model (3.2) did not provide useful correlation with the experimental data except at low energy, and values
of F^4 and F^6 particularly were distorted compared to those obtained
with the inclusion of (3.4) and the three-body corrections.

The data for PrIII ($4f^3$) fit by Rajnak (1965) were derived
from actual free-ion spectra, whereas the centers of gravity of
crystal-field components of states in $Nd^{3+}:LaF_3$ are, in fact,
perturbed from their true free-ion energies by the effects of the
environment. In addition, it is important to recognize that
there are interactions between neighboring crystal-field states
that can only be accommodated by ceasing to use J as a valid means
of describing a state and resorting to eigenvectors based on
crystal-field states (J-mixing). Use of this new description of
states requires the simultaneous diagonalization of the atomic
and crystal-field interactions. For the f^3-configuration the
computation is readily accommodated by available computers since
the largest matrix involved is 62 x 62 (for D_{3h}-symmetry).

The results of including the crystal-field interaction in
computing the energy level parameters for $Nd^{3+}:LaF_3$ starting with
the four free-ion parameters (3.2) and progressively adding the
two and three-body correction terms, parallel those for the free-
ion case. When the model was restricted to E_{CF} and the atomic
interactions in (3.2), the deviation was 350 cm^{-1}. Examination
of the data shows that the lowest-energy states are fit best and

that it is clearly the inability to account for the free-ion structure that produces very large deviations at higher energies. Thus the addition of the crystal-field parameters to the model does not automatically improve the fit to the experimental data. The new parameters are constrained by their angular dependence in the type of interactions that they can reproduce. The complete present model reproduced the 147 observed crystal-field transition energies with a deviation of 16 cm^{-1}, Carnall et al. (1977).

There are, of course, higher order interactions that could be introduced. However, with use of the present model including two and three-body electrostatic corrections and the indicated magnetic corrections there no longer appear to be a limited number of large "effects" that would further significantly reduce the error.

III.2.   Typical results for several organometallic lanthanides and actinides

Since the values of the radial integrals in the theoretical model are established by a fit to experimental data, the data base must be large enough to justify an extensive parametrization. In the early stages of this type of analysis, severe restrictions were placed on the bounds within which the parameters could vary and second order effects were not all included. Only the lower energy range of the spectra could be interpreted. As indicated in the previous section, a poor correlation with experiment results when the parametrization is limited. Analyses of data have now been carried out for lanthanide ions doped into single crystal $LaCl_3$ or LaF_3 which establish the systematic variation of parameters across the series in both cases, Crosswhite (1977). Similar results for actinides in $LaCl_3$ are presently being obtained. Thus even though the region in which the $LnCp_3$ and $AnCp_3$ optical spectra can be observed is limited, the groups can be compared in energy to those observed in the halide crystals. Where similar energy structures are indicated, it can confidently be assumed that the model parameters are similar. Detailed spectroscopic studies are necessarily carried out in host lattices. Dilution of the ion of interest in a suitable host minimizes ion-ion interactions and the intensities of the vibronically coupled bands which tend to broaden the spectral lines even at low temperatures. Corresponding host dilution studies of the cyclopentadienides have not been conducted. All the reported structure is of the pure compounds.

If we compare the reported band energies of several groups in $NdCp_3$, Pappalardo (1965a) or $ErCp_3$, Pappalardo (1968), with the corresponding data for Ln^{3+}:$LaCl_3$, Dieke (1968) or Ln^{3+}:LaF_3, Carnall et al. (1977), we see the clear parallel in the energy

level structures, Table 1. The absence of any large shifts in
the f^N-structure is also apparent in the solution spectra of
PrCp$_3$ and ErCp$_3$ in THF, Birmingham and Wilkinson (1956). While
the crystal-field splitting of the ground state in NdCp$_3$, for
example, is greater than that in Nd^{3+}:LaCl$_3$, it is quite comparable to that in the highly ionic lattice LaF$_3$. A similarity in
structure is also revealed in comparing data for the actinide
cyclopentadienides with the halides, Kanellakopulos et al. (1970),
Carnall et al. (1968a), Pappalardo et al. (1969), as shown in
Fig. 3, although, for the actinides, larger crystal-field splitting is observed. The interpretation of the data is therefore
more complex, particularly in the light half of the series where
the free-ion states are not as isolated in energy as they are at
lower energies in the heavier actinides.

We conclude that even though an extensive analysis of the
spectrum of an organometallic lanthanide or actinide may not be
possible, the energy level parameters may be approximated by reference to available extensive single crystal free-ion analyses.

IV. CALCULATION OF TRANSITION PROBABILITIES FROM THE ABSORPTION
     SPECTRA OF TRIVALENT f-ELEMENTS

Reference has already been made to the great differences in
transition probabilities (band intensities) observed when comparing the (electric dipole forbidden) f→f bands with those corresponding to f→d transitions or to charge transfer processes. We
now turn to the theory of f→f transition probabilities. Our purpose is to explore the extent to which the current model can reproduce observed absorption spectra, then use the model to compute the rates of relaxation of the longer-lived fluorescing
states.

We approach the problem of intensity correlation in the Ln^{3+}
and An^{3+} ions with the realization that for many systems of poten
tial interest we have no adequate model for the effect of the
environment on the central atom. The purely electrostatic model
which treats the nearest neighbor atoms as point charges in an
array defined by the crystal structure does not provide the basis
for a good approximation of crystal-field parameters. We therefore, discuss here only the free-ion approximation which relates
to the actual spectra observed at 25°C. Since $E_{CF} \ll E_F$, many distinct bands corresponding to the collective intensities of the
crystal-field components of individual free-ion states are observed. Based on the discussion in III, we bring to the intensity
calculation a reliable set of eigenvectors for these states.

TABLE 1. Comparison of Crystal-field Levels of Nd^{3+} Observed at 4°K in Different Environments

State	Nd^{3+}:$LaCl_3$[a] (cm^{-1})	Nd^{3+}:LaF_3[b] (cm^{-1})	$NdCp_3$[c] (cm^{-1})
$^4I_{9/2}$	0	0	0
	115	45	70
	123	136	310
	244	296	570
	249	500	670
$^4F_{7/2} + {}^4S_{3/2}$	13396	13514	13245
	400	590	394
	474	671	485
	488	676	555
	527	711	611
	531	715	721
$^4G_{5/2} + {}^4G_{7/2}$	17095	17306	16778
	(114)	316	818
	165	363	958
	228	511	17062
	(254)	518	197
	(294)	571	247
	297	605	370

a Dieke (1968).   b Carnall et al. (1977).   c Pappalardo (1965a).

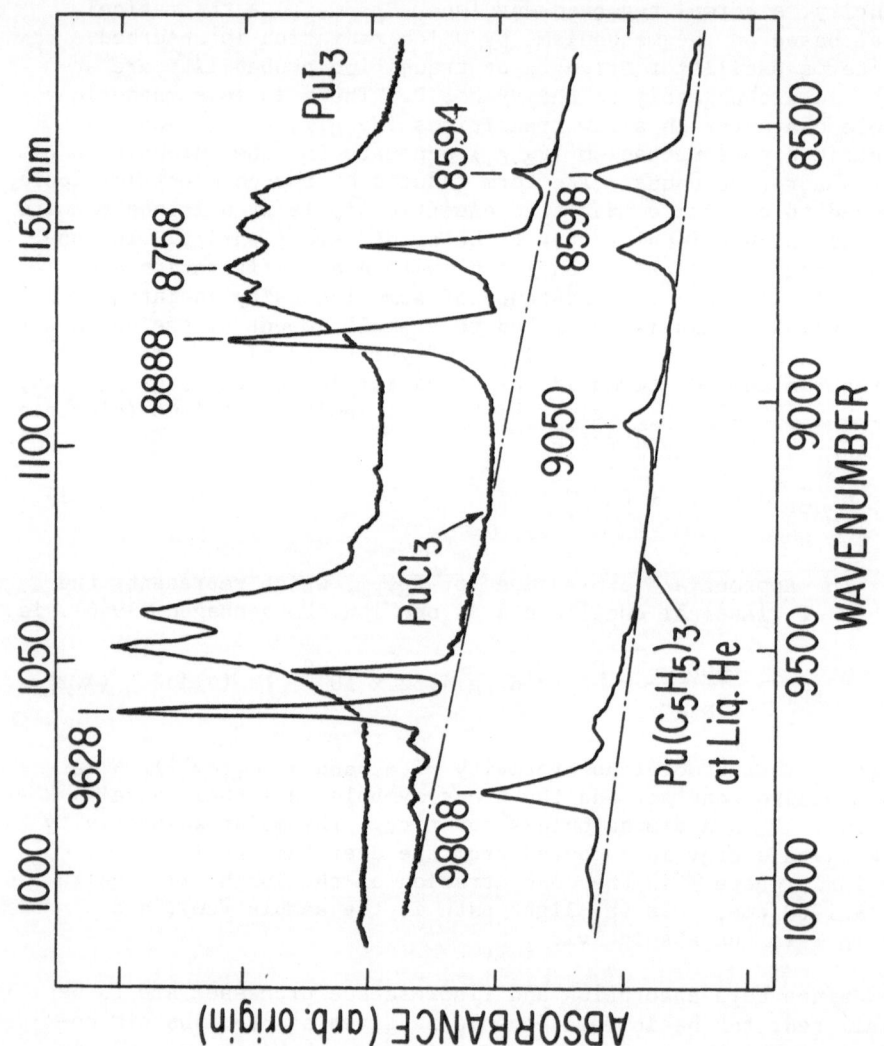

Fig. 3. Absorption Spectra of PuI₃, PuCl₃ and PuCp₃ at 4°K

IV.1.   Induced electric-dipole and magnetic-dipole transitions

The quantitative treatment of the intensities of trivalent
f-element absorption bands relates an experimentally determined
quantity, a normalized band envelope, P_{EXPT}, to a theoretical
model based on the mechanisms by which radiation is absorbed.
The terms oscillator strength or transition probability are applied interchangeably to the symbol P. There is some magnetic
dipole character in a few transitions ($P_{M.D.}$), but an induced
electric-dipole mechanism ($P_{E.D.}$) accounts for the intensities of
most absorption bands. The term induced or forced electric dipole
is used to emphasize that true electric dipole transitions require
the initial and final states to be of different parity. In contrast, magnetic dipole transitions within a configuration are
(parity) allowed. The existence of some intensity in intra-f^N
transitions is usually ascribed to a small amount of the character
of higher-lying opposite-parity configurations mixed into the f^N
states via the odd terms in the potential due to the ligand field,
Wybourne (1965). We neglect higher multipole mechanisms (electric
quadrupole, etc), and write

$$P_{EXPT} = P_{E.D.} + P_{M.D.} \qquad (4.1)$$

The appropriate expression for P_{EXPT}, which represents the
number of classical oscillators in one ion, Hoogschagen (1946), is

$$P_{EXPT} = \frac{2303 \ mc^2}{N\pi e^2} \int \varepsilon_i(\sigma)d = 4.32 \times 10^{-9} \int \varepsilon_i(\sigma)d\sigma \qquad (4.2)$$

where ε_i is the molar absorptivity of a band at $\sigma_i(cm^{-1})$, N is
the Avogadro constant and the other symbols have their usual
meaning. P is a dimensionless quantity. The molar absorptivity
at a given energy is computed from the Beer-Lambert law, $\varepsilon \mathscr{C} \ell =$
log I_0/I, where $\mathscr{C}$ is the concentration of the lanthanide ion in
moles/1000 cm^3, ℓ is the light path in the sample (cm), and
log I_0/I is the absorptivity.

Since both absorption and fluorescence processes are to be
considered, the basic relationship is given by the Einstein coefficient which expresses the transition probability due to dipole radiation

$$A(i,f) = \frac{64\pi^4 \sigma^3}{3h} \ |<i|D|f>|^2 \qquad (4.3)$$

where i and f signify the initial and final states, A is the
(spontaneous) transition probability per unit time, $\sigma(cm^{-1})$ is
the energy difference between the states, and $\underset{\sim}{D}$ is the dipole

operator, Condon and Shortley (1957). Following Broer et al.
(1945), we transform (4.3) via the substitution $P = mcA(i,f)/8\pi^2\sigma^2e^2$ to give the computed oscillator strength, P_{CALC}, of a
band

$$P_{CALC} = \frac{8\pi^2 mc\sigma}{3he^2(2J+1)} \left[\chi\overline{F}^2 + n\overline{M}^2 \right] \tag{4.4}$$

where $\overline{F}^2$ and $\overline{M}^2$ represent the matrix elements of the electric dipole and magnetic dipole operators, respectively, joining an
initial state J to the final state J', $\chi = (n^2+2)^2/9n$, and n is
the refractive index of the medium.

Judd (1962), and Ofelt (1962), independently derived expressions for the oscillator strength of induced electric dipole
transitions between free-ion states within the f^N-configuration.
However, Judd's expression (4.5), was cast in a form that could
be directly related to oscillator strengths derived from lanthanide absorption spectra

$$P_{E.D.} = \sum_{\lambda=2,4,6} T_\lambda \nu(\psi J \| U^{(\lambda)} \| \psi'J')^2 \tag{4.5}$$

where $\nu(sec^{-1})$ is the mean frequency of the transition $\psi J \rightarrow \psi'J'$,
$U^{(\lambda)}$ is a unit tensor operator of rank λ, the sum running over
the three values $\lambda = 2,4,6$, and the T_λ are three parameters which
can be evaluated from experimental data. These parameters involve
the radial parts of the $4f^N$ wave functions, the wave functions of
perturbing configurations such as $4f^{N-1}5d$, and the interaction between the central ion and the immediate environment. Axe (1963)
showed that (4.5) could be rewritten in the form

$$\overline{F}^2 = e^2 \sum_{\lambda=2,4,6} \Omega_\lambda (\psi J \| U^{(\lambda)} \| \psi'J')^2 \tag{4.6}$$

where Ω_λ is related by a constant to Judd's T_λ parameter. The
matrix elements of $U^{(\lambda)}$, (4.6), for transitions between various
excited states as well as between the ground and excited states
of the whole series of lanthanide ions using the energy level
parameters obtained from a systematic treatment of the $Ln^{3+}:LaF_3$
series have been tabulated, Carnall et al. (1977).

Following Condon and Shortley (1957), the magnetic dipole
operator is given as $M = -e(L + 2S)/2mc$. The matrix elements of
the operator $\overline{M}^2$ in (4.4) can then be written,

$$\overline{M}^2 = e^2(\psi J \| L + 2S \| \psi'J')^2/4m^2c^2 \tag{4.7}$$

TABLE 2. Oscillator Strengths for Er^{3+} (25°C)

Spectral Range[a] (cm^{-1})	S'L'J'	$P \times 10^6$			
		$Er^{3+}(HClO_4\text{-}DClO_4)$		$ErCp_3(MeTHF)$	
		Expt	Calc[b]	Expt	Calc[b]
6200–7000	$^4I_{13/2}$	2.19	$\left\{\begin{array}{l}0.41^c \\ 1.98\end{array}\right.$		
9400–10800	$^4I_{11/2}$	0.84	0.66	1.47	3.73
12000–12900	$^4I_{9/2}$	0.29	0.35	1.05	0.39
14800–15800	$^4F_{9/2}$	2.25	2.36	4.80	4.96
18000–18700	$^4S_{3/2}$	0.66	0.62	2.55	2.17
18700–19800	$^2H_{11/2}$	2.89	3.26	68.5	75.0
20000–21100	$^4F_{7/2}$	2.27	2.43	6.60	7.26
21800–23000	$\left.\begin{array}{l}^4F_{5/2} \\ ^4F_{3/2}\end{array}\right\}$	1.25	1.20	$\left\{\begin{array}{l}1.80 \\ 0.77\end{array}\right.$	$\left\{\begin{array}{l}2.62 \\ 1.52\end{array}\right.$
24100–25100	$(^2G,^4F,^2H)_{9/2}$	0.80	0.92	7.25	3.01
25800–27100	$^4G_{11/2}$	5.92	5.76	136	132
27100–27700	$^4G_{9/2}$	1.75	1.56		
27700–28400	$\left.\begin{array}{l}^2K_{15/2} \\ ^2G_{7/2}\end{array}\right\}$	0.91	$\left\{\begin{array}{l}0.05^c \\ 0.99\end{array}\right.$		
31300–31900	$(^2P,^2D,^4F)_{3/2}$	0.091	0.083		
32700–33600	$\left.\begin{array}{l}^2K_{13/2} \\ ^4G_{5/2} \\ ^2P_{1/2}\end{array}\right\}$	0.12	0.12		

a Range encompassing observed band(s).
b The parameters used to obtain these values were:

$Er^{3+}(HClO_4\text{-}DClO_4)$	$ErCp_3(MeTHF)$
$\Omega_2 = 1.59 \times 10^{-20}$ (cm^2)	$\Omega_2 = 64.1 \times 10^{-20}$ (cm^2)
$\Omega_4 = 1.95$	$\Omega_4 = 1.76$
$\Omega_6 = 1.90$	$\Omega_6 = 6.34$

c Calculated magnetic dipole oscillator strength.

The non-zero matrix elements will be those diagonal in the quantum numbers α, S, and L. The selection rule on J, $\Delta J = \pm 0, 1$, restricts consideration to three cases, J'=J, J'=J-1, and J'=J+1. In each instance (4.7) reduces to a quantity that can be computed exactly. The maxtrix elements of (4.7) have been tabulated, Carnall et al. (1968b).

IV.2.   Comparison with experiment

While intensity correlations in the actinides and lanthanides have been explored extensively in all phases, detailed analysis with organometallic compounds does not appear to have been pursued. Thus, to provide some basis for judging the extent to which correlation is possible, results for Er^{3+}(aquo), Carnall et al. (1968b) and $ErCp_3$ in 2-methyl tetrahydrofuran (MeTHF), Pappalardo (1968), are compared in Table 2. The data for Er^{3+} (aquo) were chosen because the numerous isolated absorption bands are both well characterized in terms of the transitions involved and occur over an extensive energy range. The excellent correlation between observed and calculated intensities consequently demonstrates the power of the theory.

The results for $ErCp_3$ in MeTHF were computed from data given by Pappalardo (1968). While it is apparent that the fit of the Er^{3+}(aquo) data is much better than that of the $ErCp_3$, it is the change in intensity pattern that is particularly striking. The interpretation of the increase in intensities of the $^4I_{15/2} \rightarrow$ $^4H_{11/2}(\sim 19000 \text{ cm}^{-1})$ and $^4G_{11/2}$ ($\sim 26000 \text{ cm}^{-1}$) transitions in $ErCp_3$ relative to Er^{3+}(aquo) is discussed in the next section.

IV.3.   Hypersensitive transitions

A comparison of the spectra of the 3+ lanthanides and actinides in many different environments reveals that the intensities associated with most absorption bands for a given f-element corrected to unit concentration are remarkably invariant. In contrast, the intensities of one or two transitions in each element can change significantly as a function of the environment. With the advent of the Judd-Ofelt Theory these "hypersensitive" transitions, Jørgensen and Judd (1964), were systematized. They were shown to follow electric quadrupole selection rules, $\Delta J \leq \pm 2$, and to be identified with large matrix elements of $U^{(2)}$. A particularly dramatic change in intensity was noted in (IV.2.) comparing data for Er^{3+}(aquo) to that for $ErCp_3$ (MeTHF).

Intense hypersensitive bands have also been identified in other $LnCp_3$ and clearly established in $AmCp_3$ as well, Pappalardo et al. (1969). Our interest for present purposes stems in part

from what these transitions reveal about bonding conditions, and
in part because they represent a possible avenue for pumping energy into the f^N-configurations in connection with potential lasing
experiments.

Since we have already recognized the relative insensitivity
of the energy of transitions in trivalent f-electron systems to
the immediate environment, considerable interest derives in possible correlations between bonding type and band intensity. Judd
(1962) showed that the values of Ω_λ (4.6) computed from an approximate model of Ln^{3+}(aquo) were consistent with those derived by a
fit to experimental data, and that, to a good approximation only
the value of Ω_2 was sensitive to changes in the environment.
Mechanisms for increased intensity that would uniquely influence
the value of Ω_2 were examined by Jørgensen and Judd (1964), but
subsequently Judd (1966b) pointed out that, without reference to
any detailed mechanism, symmetry arguments could be used as a
basis for classification. In effect, only a limited number of
point groups would permit Ω_2 to vary independently of Ω_4 and Ω_6
within the context of the original Judd theory. Most of these
groups, C_s, C_n(n=1-4,6), C_{nv}(n=2-4,6) represent low site symmetries, but a unique assignment to one of these symmetries would
not appear to be possible.

More recently, a dynamic coupling mechanism relating the
polarizability of the ligands to the intensity of the hypersensitive transitions has been advanced by Mason et al. (1975), Peacock
(1975). This represents a refinement to and extension of the Judd
theory, since it assumes that the intensities of most transitions
are correctly described by (4.4) and deals only with the increase
of intensity in the hypersensitive transitions. According to this
model, there is a Coulombic correlation between transient induced
electric dipoles in the ligands and the f-electron quadrupole
moments of Ln^{3+} (or An^{3+}). The transient charge distribution at
the metal site is induced by the f→f transition. The dynamically
induced oscillator strength for trivalent f-elements is given by

$$P_{DYN} = \frac{8\pi^2 mc\sigma}{3h(2J+1)} \times \Omega_{DYN}(f^N\psi J \| U^{(2)} \| f^N\psi'J')^2$$

$$\Omega_{DYN} = \frac{28}{5} <4f|r^2|4f>^2 \sum_{m=0}^{3} (2-\delta_{m,o}) \sum_{L} |R_L^{-4}\overline{\alpha}(L)C_m^{(3)}(L)|^2$$

$\overline{\alpha}(L)$ is the average polarizability of the ligand, R_L is the metal-
ligand bond distance, and $C_m^{(3)}(L)$ describes the orientation of
each ligand with respect to the central metal atom. Since R_L
occurs to a high power, R^{-12} and R^{-16} for Ω_4 and Ω_6, respectively,
it is obvious why Ω_4 and Ω_6 do not contribute to P_{DYN}.

The values of Ω_λ computed for Er^{3+} in a number of different environments are listed in Table 3. Only a limited number of calculations of P_{DYN} have been reported, but the results for $ErI_3(g)$, $ErBr_3(g)$, and Er^{3+}:LaF_3 are reported to be in agreement with the values of Ω_2 derived from experiment. This suggests that the large Ω_2 in $ErCp_3$ may also be reproduced with a judicious interpretation of α and R_L for the Cp ligands. This continues to be an active area of investigation, and new insights into bonding may be forthcoming.

V. LUMINESCENCE SPECTRA OF 3+ LANTHANIDES AND ACTINIDES

Interest in the luminescence properties of the 3+ lanthanides was at a peak in the mid 1960's as the laser potential of various systems including solutions of lanthanide chelates was explored, Lempicki and Samelson (1966), Crosby (1966), Lempicki (1971). The basis for the interest in chelates was the discovery by Weissman (1942), that they could absorb energy in the ultraviolet range and in certain cases transfer this energy "intramolecularly" to the lanthanide ion. The transfer of energy in Eu^{3+} resulted in emission characteristic of f→f transitions in the visible region. The process is indicated in Fig. 4. The potential efficiency of this mechanism in populating f-electron states is suggested by the much greater intensity of the bands characteristic of the ligands compared to those of the weak f→f transitions.

There are several reasons for continued basic interest in the luminescence properties of chelates and organometallic compounds of the f-elements. The Judd-Ofelt theory now makes it possible to compute the radiative component of the lifetimes of fluorescing states based on measurements of absorption spectra. This provides a means of developing new insights into the fundamental nature of relaxation processes and energy transfer. There have also been important advances in the ability to obtain relevant experimental data. Micro to nanosecond lifetimes are measured routinely, and tunable dye lasers provide a means, not available earlier, of pulsed excitation at selected energies throughout a wide range of the spectrum. Interestingly, little attention has been given to the study of the luminescent properties of f-elements in the vapor phase, yet the volatility of many organometallics is one of the unique properties of this class of compounds.

V.1. Mechanisms of relaxation of excited states and the computation of radiative emission rates

The forced-electric and magnetic dipole processes discussed relative to the absorption of radiation are also primarily responsible for the radiative decay of excited f^N-states. Non-radiative

TABLE 3. Values of Ω_λ for Er^{3+} in Various Systems

System	Phase	T °C	Ω_2	Ω_4	Ω_6	Ref.
				$\times\ 10^{20}\ cm^2$		
Er^{3+}(aquo)	Liquid	25	1.59	1.95	1.90	a
$ErCp_3$(MeTHF)	Liquid	25	64.1	1.76	6.34	b
Er^{3+}:LaF_3	Solid	25	0.74	0.19	0.44	c
Er^{3+}:$YAlO_3$	Solid	25	1.06	2.63	0.78	c
Er^{3+}:Li-KNO_3	Liquid	160	15.8	1.84	1.39	d
$Er(thd)_3$	Vapor	288	46	2.70	3.70	e
$ErCl_3(AlCl_3)_x$	Vapor	~400	25.8	2.70	2.01	f
ErI_3	Vapor	993	97.7	5.90	-1.47	e
$ErBr_3$	Vapor	1105	60	1.47	1.66	e

a Carnall et al. (1968b).
b Pappalardo (1968).
c Riseberg and Weber (1976).
d Carnall et al. (1978a).
e Gruen et al. (1967).
f Carnall et al. (1978b).

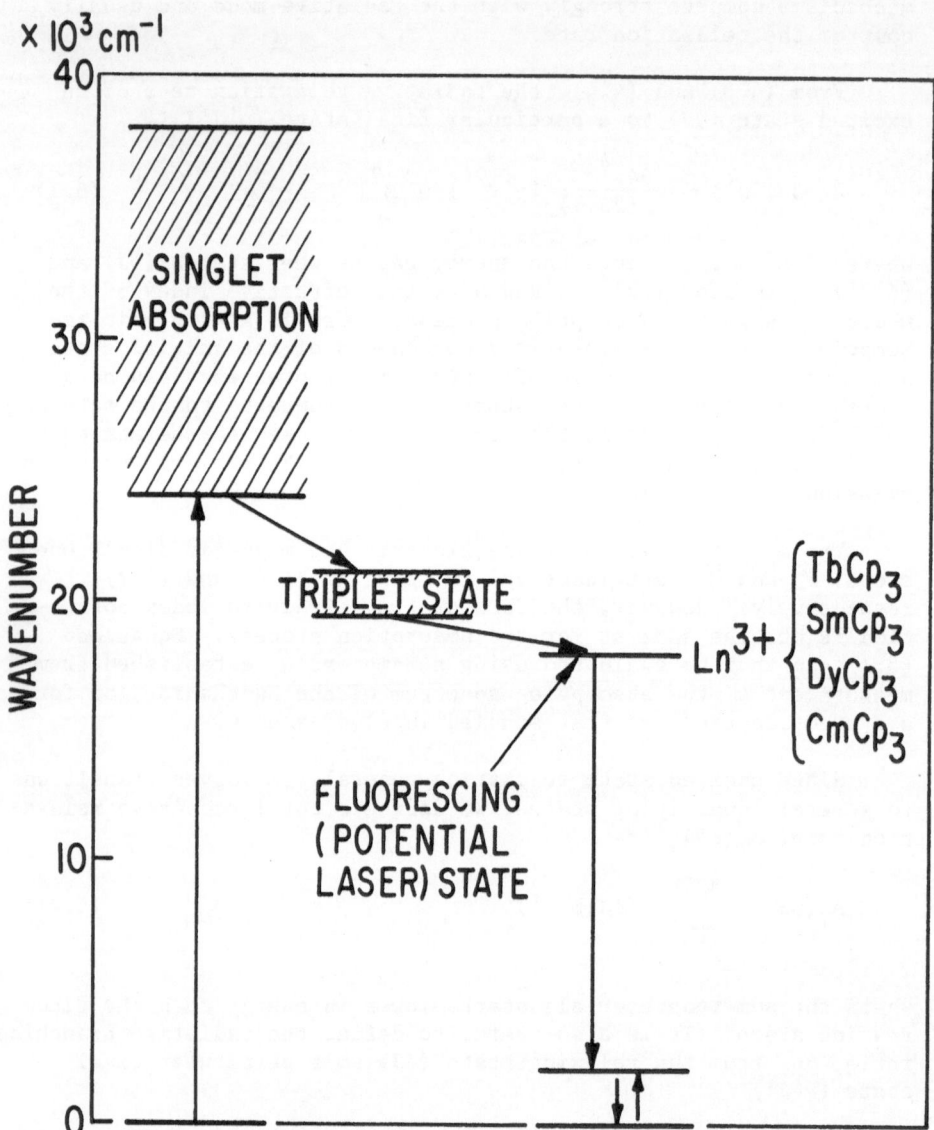

Fig. 4. Schematic Energy Transfer Cycle in Lanthanide–actinide Organometallic Compounds

mechanisms compete strongly with the radiative mode and usually control the relaxation rate.

From (4.3) and (4.4), the radiative relaxation rate of an excited state (ψJ) to a particular final state $(\psi'J')$ is

$$A(\psi J, \psi'J') = \frac{64\pi^4 \sigma^3}{3h(2J+1)} [\chi' \overline{F}^2 + n^3 \overline{M}^2] \tag{5.1}$$

where $\sigma(cm^{-1})$ represents the energy gap between states (ψJ) and $(\psi'J')$, $\chi' = n(n^2 + 2)^2/9$, and n is the refractive index of the medium. As in the absorption process, there is an implicit assumption that all crystal-field components of the initial state are equally populated. In principal, if fluorescence can be detected, the lifetime of the state is long compared to the rate at which it is populated in the excitation process, so thermal equilibrium at the temperature of the system can be achieved prior to emission.

The matrix elements of the electric and magnetic dipole operators, $\overline{F}^2$ and $\overline{M}^2$, are identical to those of (4.6) and (4.7), respectively. However, the form of the refractive index correction is not the same as for the absorption process. Equation (5.1) can thus be evaluated using parameters Ω_λ established from measurement of the absorption spectrum of the lanthanide ion for a system identical to that studied in fluorescence.

Since excited state relaxation generally involves transitions to several lower-lying states, we define a total radiative relaxation rate, $A_T(\psi J)$

$$A_T(\psi J) = \sum_{\psi'J'} A(\psi J, \psi'J') \tag{5.2}$$

where the sum runs over all states lower in energy than the fluorescing state. It is also useful to define the radiative branching ratio, β_R, from the relaxing state (ψJ) to a particular final state $(\psi'J')$

$$\beta_R(\psi J, \psi'J') = \frac{A(\psi J, \psi'J')}{A_T(\psi J)} \tag{5.3}$$

and the radiative lifetime of a state

$$\tau_R(\psi J) = [A_T(\psi J)]^{-1} \tag{5.4}$$

While the relaxation of an excited state may occur via a purely radiative process, this must be recognized as the exception

rather than the rule. Detailed studies of the fluorescence spectra of f-elements in various compounds reveal that only a few excited states can be observed to fluoresce. Most are rapidly re-
laxed by non-radiative processes even at low temperatures.
Typically, several different non-radiative mechanisms may operate
simultaneously.

Formally, we can express the total fluorescence lifetime of
a state as

$$(\tau_T)^{-1} = A_T(\psi J) + W(\psi J) \tag{5.5}$$

where A_T is the total radiative rate and $W(\psi J)$ is the sum of the
rates of the various non-radiative processes.

V.2. Comparison of experimental and computed excited state
lifetimes

While fluorescence from excited states of f-elements in
crystals, solution and in the vapor phase has been observed,
actual lifetime measurements have been made on relatively few
systems. A great deal of interest has been manifest in the
spectra of solid phases because of laser applications. As an
example of the use of the theory outlined, the following results
were obtained in single crystal $Er^{3+}:LaF_3$, Weber (1967):

Transition	Calculated τ_R(msec)	Observed τ_T(msec)
$^4I_{13/2} \rightarrow {}^4I_{15/2}$	10.9	13
$^4I_{11/2} \rightarrow {}^4I_{15/2}$	11.6	11
$^2P_{3/2} \rightarrow {}^4I_{15/2}$	0.43	0.29

It was recognized experimentally before any quantitative
theories were developed that the fluorescence lifetime of an excited level in a crystal host was related to the energy gap to
the next lower-lying state. For example, fluorescence in Ln^{3+}:
$LaCl_3$ was not observed at 25°C if the energy gap was <10000 cm^{-1},
Barasch and Dieke (1965). The principal mechanism identified in
this case was a multiphonon process involving transfer of energy
to the lattice, and it has been treated phenomenologically by
assuming that the appropriate phonon energy corresponds to the
cut-off in phonon states or $\sim$350 cm^{-1}, in LaF_3, Moos (1970),
Riseberg and Weber (1976). The corresponding expression for
$W(\psi J)$ as a function of temperature is:

$$W(\psi J) = Ce^{\alpha\Delta E}[1-e^{-\hbar w_i/kT}]^{-\Delta E/\hbar w_i} .$$

where $\hbar w_i$ is the maximum phonon energy, ΔE is the energy gap, α and C are parameters characteristic of the host. For example, the radiative lifetime of the $^4I_{9/2}$ state of $Er^{3+}:LaF_3$ at $\sim$12000 cm^{-1} is computed to be 20.7 msec but observed to be $\sim$0.15 msec, Weber (1967). Since ΔE is $\sim$2000 cm^{-1} we compute $W(\psi J) = \sim10^4$ sec^{-1} at 25°C. Thus the non-radiative lifetime of $\sim$0.1 msec is rate determining.

As indicated earlier, Pappalardo and Losi (1965) observed band emission in both $GdCp_3$ and $LaCp_3$ near 20000 cm^{-1} upon excitation further in the ultraviolet, while line fluorescence was reported in undiluted sublimates of $TbCp_3$, $SmCp_3$ and $DyCp_3$, Pappalardo (1965b). The f→f states in $GdCp_3$ are encompassed by the broad absorption characteristic of $C_5H_5^-$. In contrast, there exist f^N-states at lower energies for the ions where line fluorescence was observed. Thus intramolecular energy transfer in the latter appears to be well established. The band emission corresponds to a triplet-singlet transition characteristic of the Cp-ligand. Similar broad band fluorescence of lanthanide acetylacetonate chelates due to π→π* (T→S) relaxation has been reported by Crosby et al. (1971).

In $CmCp_3$, several sharp bands were observed near 15600 cm^{-1} upon excitation in the ultraviolet range, Laubereau and Burns (1970). These correspond well to expected $^5D_{7/2}$ → ^8S transitions in Cm^{3+}. Since the ground state is not split, the observed structure is due to fluorescence from several crystal-field levels of $^5D_{7/2}$ which are populated at the temperature of the measurement, 25°C. It should be emphasized that fluorescence in $LnCp_3$ and $AnCp_3$ has to date been studied in pure samples where concentration quenching can be an important relaxation mechanism. In the future studies using dilution techniques will need to be emphasized.

V.3. Energy transfer and emission in the vapor phase

One of the unique properties of organometallic f-elements as a class is their volatility. Just as renewed interest is being focussed on the fundamental mechanisms of energy transfer in solids and solutions as a result of increased experimental capabilities and theoretical insights, there is much to be learned about fluorescence properties in the gas phase. Indeed a number of limitations presently imposed on the generation of high power densities in solid state lasers would not exist in a circulating gas phase system.

Although the lifetimes associated with the fluorescing $^5D_4$ state of Tb^{3+} in vapor phase $Tb(thd)_3^*$ at elevated temperatures were predictably reduced compared to those observed in the liquid and solid at lower temperatures, pulsed excitation and fast detection techniques made it possible to investigate some aspects of the energy transfer process, Jacobs et al. (1975). The chelate ligands were pumped with the direct 29670 cm^{-1} pulse from a N_2-laser, and transient fluorescence from $Tb(thd)_3$ was observed at right angles to the exciting light. Emission characteristic of fluorescence from the $^5D_4$ state to several components of the ^7F ground multiplet was observed but no emission from the $^5D_3$ state was detected. This suggests that the lowest triplet state in the chelate may lie at an energy intermediate between the $^5D_3$ and $^5D_4$ states. The absorption spectra of a number of the $Ln(thd)_3$ compounds in the gas phase were reported by Gruen et al. (1967), and in each case intense hypersensitive transitions were observed.

The transfer of energy is typically rapid in the $Tb(thd)_3$ vapor since the observed risetime of the $^5D_4$ fluorescence was $\geq$ the pump pulse duration. This indicates a transfer time [singlet→ triplet→$^5D_4$] on the nanosecond time scale. Similar experiments with organometallic lanthanide and actinides could lead to valuable insights into the nature of the bonding and the types of relaxation mechanisms that are important.

---

*2,2,6,6-tetramethyl-3,5,heptanedione

REFERENCES

AXE, J. D. (1963), J. Chem. Phys. 39, 1154.
BALASUBRAMANIAN, G., M. M. ISLAM and D. J. NEWMAN (1975), J. Phys. B. 8, 2601.
BARASCH, G. E., and G. H. DIEKE (1965), J. Chem. Phys. 43, 988.
BIRMINGHAM, J. M., and G. WILKINSON (1956), J. Am. Chem. Soc. 78, 42.
BREWER, L. (1971), J. Opt. Soc. Am. 61, 1011, 1666.
BROER, L. L. F., C. J. GORTER, and J. HOOGSCHAGEN (1945), Physica 11, 231.
CARNALL, W. T., P. R. FIELDS, and R. G. PAPPALARDO (1968a), Proc. 11th Intern. Conf. Coord. Chem. Haifa.
CARNALL, W. T., P. R. FIELDS, and K. RAJNAK (1968b), J. Chem. Phys. 49, 4412.
CARNALL, W. T., H. M. CROSSWHITE and H. CROSSWHITE (1977), "Energy level structure and transition probabilities in spectra of the trivalent lanthanides in LaF$_3$," Chemistry Division Report, Argonne National Laboratory
CARNALL, W. T., JAN P. HESSLER, and F. WAGNER, JR. (1978a), J. Phys. Chem., in press.
CARNALL, W. T., JAN P. HESSLER, H. R. HOEKSTRA, and C. W. WILLIAMS (1978b), J. Chem. Phys. 68, 4304.

COHEN, E., and H. W. MOOS (1967), Phys. Rev. 161, 258, 268.

CONDON, E. U., and G. H. SHORTLEY (1957), "The Theory of Atomic Spectra", Cambridge University Press, London, pp. 91-109.

CROSBY, G. A. (1966), Molecular Crystals 1, 37.

CROSBY, G. A., R. J. WATTS, and S. J. WESTLAKE (1971), J. Chem. Phys. 55, 4663.

CROSSWHITE, H. M. (1977), Colloques Internationaux C.N.R.S., Paris, No. 255, p. 65.

DIEKE, G. H. (1968), "Spectra and Energy Levels of Rare Earth Ions in Crystals", H. M. Crosswhite and H. Crosswhite, eds., Wiley, New York.

GOFFART, J., and G. DUYCKAERTS (1978), Inorg. Nucl. Chem. Lett. 14, 15.

GRUEN, D. M., C. W. DE KOCK, and R. L. MC BETH (1967), Adv. Chem. Series 71, 102.

HAYES, R. G., and J. L. THOMAS (1971), Organomet. Chem. Rev. A, 7, 1.

HODGSON, K. D., F. MARES, D. F. STARKS, and A. STREITWEISER, JR. (1973), J. Am. Chem. Soc. 95, 8650.

HOOGSCHAGEN, J. (1946), Physica 11, 513.

JACOBS, R. R., M. J. WEBER, and R. K. PEARSON (1975), Chem. Phys. Lett. 34, 80.

JØRGENSEN, C. K., and B. R. JUDD (1964), Mol. Phys. 8, 218

JØRGENSEN, C. K. (1969), "Oxidation Numbers and Oxidation States," Springer Verlag, Berlin.

JUDD, B. R. (1962), Phys. Rev. 127, 750.

JUDD, B. R. (1963), "Operator Techniques in Atomic Spectroscopy," McGraw-Hill, New York

JUDD, B. R. (1966a), Phys. Rev. 141, 4.

JUDD, B. R. (1966b), J. Chem. Phys. 44, 839.

JUDD, B. R., H. M. CROSSWHITE and H. CROSSWHITE (1968), Phys. Rev. 169, 130.

KANELLAKOPULOS, B., E. O. FISCHER, E. DORNBERGER and F. BAUMGÄRTNER (1970), J. Organometall. Chem. 24, 507.

KANELLAKOPULOS, B., E. DORNBERGER and F. BAUMGÄRTNER (1974), Inorg. Nucl. Chem. Lett. 10, 155.

LAUBEREAU, P. G., and J. H. BURNS (1970), Inorg. Nucl. Chem. Lett. 6, 59.

LEMPICKI, A., and H. SAMELSON (1966), "Organic Laser Systems", in A. K. Levine, ed., Lasers 1, 181, Marcel Dekker, N. Y.

LEMPICKI, A. (1971), "Rare Earth Liquid Lasers," R. J. Pressley, ed., Handbook of Lasers, p. 355, Chemical Rubber Pub. Co.

MASON, S. F., R. D. PEACOCK and B. STEWART (1975), Mol. Phys. 30, 1829.

MOOS, H. W. (1970), J. Luminescence 1, 106.

MORRISON, J. C., and K. RAJNAK (1971), Phys. Rev. A4, 536.

NUGENT, L. J., P. G. LAUBEREAU, G. K. WERNER and G. K. VANDER SLUIS (1971), J. Organomet. Chem. 27, 365.

NUGENT, L. J. (1975), "Chemical Oxidation States of the Lantha-
    nides and Actinides", K. W. Bagnall, ed., International Re-
    view of Science, Inorganic Chemistry Series Two, 7, (Univer-
    sity Park Press, Baltimore), Chapter 6.
OFELT, G. S. (1962), J. Chem. Phys. 37, 511.
PAPPALARDO, R. (1965a), Helv. Phys. Acta 38, 178.
PAPPALARDO, R. (1965b), Proc. Fifth Rare Earth Conference, Ames,
    Iowa, p. 87.
PAPPALARDO, R., and S. LOSI (1965), J. Inorg. Nucl. Chem. 27, 733.
PAPPALARDO, R., and C. K. JØRGENSEN (1967), J. Chem. Phys. 46,
    632.
PAPPALARDO, R. (1968), J. Chem. Phys. 49, 1545.
PAPPALARDO, R. (1969), J. Mol. Spectr. 29, 13.
PAPPALARDO, R. G., W. T. CARNALL and P. R. FIELDS (1969), J. Chem.
    Phys. 51, 842.
PEACOCK, R. D. (1975), Struct. Bonding 22, 83.
RAJNAK, K., and B. G. WYBOURNE (1963), Phys. Rev. 132, 280.
RAJNAK, K. (1965), J. Opt. Soc. Am. 55, 126.
RISEBERG, L. A., and M. J. WEBER (1976), Prog. Optics 14, 89.
WEBER, M. J. (1967), Phys. Rev. 157, 262.
WEISSMAN, S. I. (1942), J. Chem. Phys. 10, 214.
WYBOURNE, B. G. (1965), "Spectroscopic Properties of Rare Earths,"
    Wiley, N. Y.

NUCLEAR MAGNETIC RESONANCE FOR RARE EARTH COMPLEXES: THEORY.

Bruce R. McGarvey

Department of Chemistry, University of Windsor,
Windsor, Ontario, Canada

1. INTRODUCTION - PARAMAGNETIC SHIFTS AND RARE EARTH IONS.

This chapter will deal with the theory of the Nuclear Magnetic Resonance (NMR) of ligand nuclei in lanthanide complexes. The theories will also be applicable to actinide complexes, but in the actinide complexes the larger crystal fields, the possibly larger covalent interactions and the greater importance of relativistic effects will make the theories more approximate.

The general nature of the paramagnetic shift and the conditions needed to observe it are best understood by first considering a simple system in which $S=\frac{1}{2}$ and $I=1$. If we have an isotropic electron spin interacting with an isotropic nuclear spin in a magnetic field H the energy levels can be represented by the Hamiltonian

$$H = g\beta_e S \cdot H - g_N \beta_N I \cdot H + a I \cdot S \qquad (1.1)$$

where β_e and β_N are the Bohr and nuclear magnetons ($\beta_e = 0.927120 \times 10^{-20}$ ergs/gauss and $\beta_N = 5.0494 \times 10^{-24}$ ergs/gauss), S and I are the electron and nuclear spin operators in units of $(h/2\pi)$ and a is the hyperfine interaction. For ligand nuclei in rare earth complexes studied by normal NMR spectrometers, the first term in Eq. (1.1) is much larger than the other two and therefore, to first order, the energy of the various spin levels can be written as

$$E(M_S, M_I) = g\beta_e H M_S - g_N \beta_N H M_I + a M_S M_I \qquad (1.2)$$

In Figure 1 the various energy levels for $S=\frac{1}{2}$, $I=1$ are shown. For the normal arrangement of an oscillating magnetic

T. J. Marks and R. D. Fischer (eds.), Organometallics of the f-Elements, 309–336.

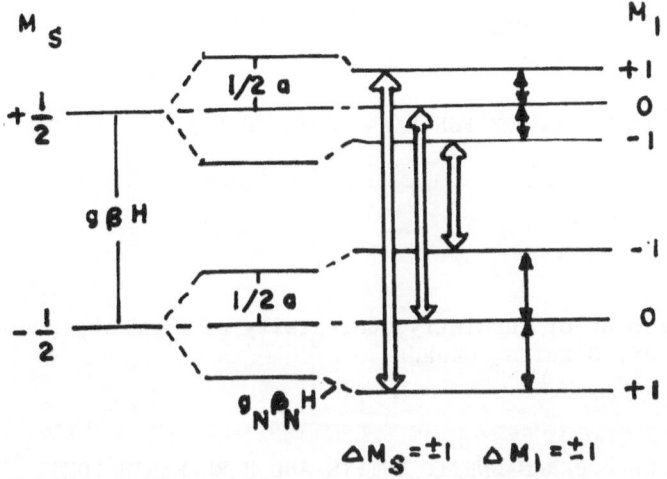

Figure 1. Energy levels for S=½, I=1 spin system.

field perpendicular to the main field, the allowed transitions are the Electron Spin Resonance (ESR) transitions of $\Delta M_S=\pm 1$, $\Delta M_I=0$ and the NMR transitions of $\Delta M_I=\pm 1$, $\Delta M_S=0$. These allowed transitions are designated in Figure 1. Examination of this figure reveals that for S=½ there are two NMR transitions of frequency

$$\Delta E = h\nu = |g_N\beta_N H \pm \tfrac{1}{2}a| \qquad (1.3)$$

The separate NMR transitions for each electron spin state are normally not detected directly because the lines are too broad in concentrated solutions or pure solids due to interactions with neighboring electron spins. At dilute concentrations the intensity is generally too low for detection although there are several recent reports of their detection (1-3) at very low temperatures. They are detected by the Electron Nuclear Double Resonance (ENDOR) method in which an ESR spectrometer which is saturating one of the ESR transitions is used to detect a $\Delta M_I=\pm 1$ transition caused by a second rf source.

When the relaxation time T_{1e} for electron spin states is very short ($T_{1e}<10^{-11}$ sec) the broadening effect of neighboring spins is effectively destroyed producing narrow NMR lines even in concentrated samples. In this situation we observe one NMR transition that has been averaged over all the electron spin states. If we assume a Boltzmann distribution among the two spin states of our example the averaged frequency for the NMR transition becomes

$$h\bar{\nu} = (g_N\beta_N H - \tfrac{1}{2}a)\rho_+ + (g_N\beta_N H + \tfrac{1}{2}a)\rho_- \tag{1.4}$$

$$\rho_\pm = e^{\mp g\beta_e H/2kT} \Big/ \Big[e^{-g\beta_e H/2kT} + e^{g\beta_e H/2kT} \Big] \cong \tfrac{1}{2} \mp \tfrac{1}{4}(g\beta_e H/kT) \tag{1.5}$$

$$h\bar{\nu} \cong g_N\beta_N H + (g\beta_e H/4kT)a \tag{1.6}$$

or if $\nu^o = g_N\beta_N H/h$

$$\Delta\nu/\nu^o = (\bar{\nu}-\nu^o)/\nu^o = (g\beta_e/g_N\beta_N)(a/4kT) \tag{1.7}$$

The approximate equation for the probability, $\rho_\pm$, of a given spin state is sufficient because $g\beta_e H \ll kT$ at normal temperatures. Since the magnetic susceptibility per atom of an $S=\tfrac{1}{2}$ system is $\chi=(g^2\beta_e^2/4kT)$, Eq. (1.7) can be written as

$$(\Delta\nu/\nu^o) = (\chi/g\beta_e g_N\beta_N)a \tag{1.8}$$

Thus if the magnetic field is kept constant, there is a shift in frequency proportional to a and χ. This is generally called the paramagnetic shift. If we keep frequency constant at ν_0 and vary the magnetic field, Eq. (1.6) yields

$$(H-H_o)/H_o \cong (H-H_o)/H = -(\chi/g\beta_e g_N\beta_N)a \tag{1.9}$$

$$H_o = h\nu_o/g_N\beta_N$$

From the preceding discussion we see that the paramagnetic shift is basically a measure of the hyperfine interaction. In general, it has been found that ions having short enough T_{1e}'s to give narrow NMR lines must have large spin-orbit interactions and small crystal field splittings (order kT) of the orbital ground states. This condition is met by all lanthanide and actinide ions except those having an S ground state (Eu^{+2} and Gd^{+3}).

2. LIGAND HYPERFINE INTERACTION WITH RARE EARTH IONS.

Since the paramagnetic shift is determined by the hyperfine interaction we will first consider the various types of interactions that lead to a hyperfine interaction and the theories that have been proposed to explain the experimental results.

2.1. Types of hyperfine interaction

The operator connecting the nuclear spin with the electron spin can be written as the sum of these terms

$$\mathbf{H}_N = \underset{\sim}{A}_N \cdot \underset{\sim}{I} = (\underset{\sim}{A}_F + \underset{\sim}{A}_D + \underset{\sim}{A}_L) \cdot \underset{\sim}{I} \tag{2.1}$$

$$\underset{\sim}{A}_F = (8\pi/3) g_e \beta_e g_N \beta_N \sum_j \delta(r_j) \underset{\sim}{s}_j \tag{2.2}$$

$$\underset{\sim}{A}_D = g_e \beta_e g_N \beta_N \sum_j [3(\underset{\sim}{s}_j \cdot \underset{\sim}{r}_j) \underset{\sim}{r}_j - r_j^2 \underset{\sim}{s}_j] r_j^{-5} \tag{2.3}$$

$$\underset{\sim}{A}_L = 2\beta_e g_N \beta_N \sum_j r_j^{-3} \underset{\sim}{\ell}_j \tag{2.4}$$

In Eq. (2.2) $\delta(r_j)$ is the Dirac delta function which, when integrated with the wave function, gives the value of the wave function at $r_j=0$. The electronic g factor $g_e=2.0023$ but is generally taken to be 2.0 exactly, while r_j and s_j are the distance from the nucleus and the electron spin for the jth electron. The orbital angular momentum operator, ℓ_j, in Eq. (2.4) is defined in a coordinate system centered at the nucleus. The first two terms in Eq. (2.1), A_F and A_D, are in reality two limiting forms of the same interaction. A_D is the ordinary dipole-dipole interaction for two dipoles that are not too close to each other. It is the proper form to be applied to p, d and f electrons. For s electrons, which have a finite probability of being at the nucleus, A_D is clearly inappropriate, since it gives a zero contribution at large values of r_j and does not hold for small values of r_j. From Dirac's relativistic theory of the electron, it is found (4) that A_F in Eq. (2.2) is the correct form when the electron is close to the nucleus. A_F is often called the <u>Fermi contact</u> term in the hyperfine interaction.

The dipolar term A_D gives a traceless tensor for the hyperfine interaction which averages to zero in solution while the Fermi contact term A_F gives an isotropic interaction term in the absence of any orbital momentum in the ground state. If the orbital momentum is quenched in the crystal field states, the third term A_L makes no contribution in first order to the hyperfine interaction. It is however, a significant term for the rare earth ions whose orbital angular momentum is not quenched by the crystal field interactions. A_L does not lead to a traceless tensor for the hyperfine interaction and will contribute to both the isotropic and anisotropic portions of the hyperfine interaction. Therefore for rare earth complexes one cannot readily extract the Fermi contact contribution from the total hyperfine interaction by averaging over all orientations as is done for most transition metal complexes.

An important contribution to A_D and A_L in the rare earths comes from electrons in orbitals centered on the rare earth ion at some distance from the ligand nucleus. In this case Marshall (5) has shown by making a multipole expansion that $(A_D + A_L)$ can be approximated by using the dipolar operators

$$(A_D + A_L)_x = g_N \beta_N \beta_e \{-(\tfrac{1}{2})(3\cos^2\Omega - 1) + (3/2)\sin^2\Omega\cos2\psi\}R^{-3}(L_x + 2S_x) \tag{2.5}$$

$$(A_D + A_L)_y = g_N \beta_N \beta_e \{-(\tfrac{1}{2})(3\cos^2\Omega - 1) - (3/2)\sin^2\Omega\cos 2\psi\} R^{-3}(L_y + 2S_y) \quad (2.6)$$

$$(A_D + A_L)_z = g_N \beta_N \beta_e (3\cos^2\Omega - 1) R^{-3}(L_z + 2S_z) \quad (2.7)$$

where R, Ω and ψ are the polar coordinates for the nucleus expressed in a coordinate system centered at the metal ion nucleus.
$\underset{\sim}{L}$ and $\underset{\sim}{S}$ are total orbital and spin momentum operators with $\underset{\sim}{L}$
expressed in coordinates of the metal ion, also. For rare earth
ions we can replace the $(\underset{\sim}{L} + 2\underset{\sim}{S})$ operator in the preceding
equations by

$$(\underset{\sim}{L} + 2\underset{\sim}{S}) = g_J \underset{\sim}{J} \quad (2.8)$$

$$g_J = [3J(J+1) + S(S+1) - L(L+1)]/2J(J+1) \quad (2.9)$$

for matrix elements involving functions all belonging to the same
JLS manifold. Eqs. (2.5-7) are an approximation in that smaller
terms involving R^{-5} and R^{-7} terms in the multipolar expansion are
ignored. Attention has focused recently (6-9) on calculating
R^{-5} and R^{-7} terms in the pseudocontact contribution to the NMR
shift. Golding (9) reports that calculations on a $3d^1$ transition metal ion show an error of 12.6% results from ignoring
higher-power terms when the nucleus is 3 Å from the transition
metal ion. It is unlikely that errors greater than one or two
percent would occur for the rare earths because the f orbitals
are much more tightly bound and are shielded from the ligand
nucleus by the outer valence electrons. Thus, Eqs. (2.5-7)
should be adequate for calculating any contributions from rare
earth f electrons to the hyperfine interaction of any ligand
nuclei, even the nearest neighbor nucleus.

2.2. Theories of rare-earth ligand hyperfine interactions.

Considerable evidence now exists that experimental hyperfine interactions for nearest neighbor nuclei cannot be accounted for entirely by assuming the unpaired electrons are only in
f orbitals of the rare earth ions. There must be a mechanism
to transfer some unpaired spin onto nearest neighbor atoms. The
two theories of this mechanism that have gained acceptance and
some experimental verification will be presented here.

2.2A. Polarization model. This mechanism of spin transfer
was first proposed by Watson and Freeman (10), who used their
earlier Hartree-Fock calculation (11) of spin density in Gd^{3+}
which showed that a negative spin density is produced in the
outer 5s and 5p electrons by a polarization interaction with the
inner unpaired 4f electrons. They proposed that this negative
spin density is transferred onto a neighboring atom by an

overlap or covalent interaction between the rare earth 5s and 5p electrons and the valence s and p electrons of the ligand atom. The distinguishing feature of this model is its prediction of hyperfine constants of opposite sign to that expected when the spin of the unpaired electron transferred to the ligand is the same as that of the unpaired f electron.

2.2B. Covalent model. Baker (12) first proposed this mechanism to explain ENDOR results on Yb^{3+} and Tm^{2+} in CaF_2. It was later reformulated (13) into a more useful form. This is identical to the covalent model used in transition metal complexes in that it is proposed that there is a covalent interaction between 4f electrons and the s and p valence electrons of the ligand. Since in this mechanism the unpaired electron occupies a molecular orbital made up of a 4f orbital and the ligand orbitals, the electron spin on the ligand will be the same as that on the rare earth ion.

It should be pointed out that both of these mechanisms transfer spin by a covalent or overlap interaction. They differ in that the covalent model transfers spin directly from 4f orbitals while the polarization model transfers spin from the valence 5s and 5p orbitals. Watson and Freeman (10) made a distinction between what they called the overlap interaction and the covalent interaction. To understand the distinction and why for our purposes they will be considered alike, consider the case of one electron on the metal ion in an atomic orbital ϕ and two electrons on the ligand in atomic orbitals χ. We assume in our case that when the two atoms are close to each other, the correct determinental wavefunction is of the form

$$(N\phi-a\chi)\alpha(N'\chi+a'\phi)\alpha(N'\chi+a'\phi)\beta \qquad (2.10)$$

where α and β are electron spin functions. We further assume for a correct wave function that $(N\phi-a\chi)$ is orthogonal to $(N'\chi+a'\phi)$. To preserve orthogonality the smallest allowed value of a in Eq. (2.10) is $S/(1-S^2)^{\frac{1}{2}}$ where S is the overlap integral of atomic orbitals ϕ and χ. In this case $a'=0$ and Eq. (2.10) becomes

$$(1-S^2)^{-\frac{1}{2}} (\phi-S\chi)\alpha\chi\alpha\chi\beta \qquad (2.11)$$

This is the overlap model of Watson and Freeman (10), in which some spin density is transferred to the ligand but no charge density is transferred. If a is greater than the overlap value, we will also have a net charge transfer from the ligand to the metal ion, and therefore a situation that is traditionally called covalency. In any case, the interaction always transfers electron spin onto the ligand atom with the same sign as the spin on the atomic orbital of the metal atom which is inter-

acting with the ligand orbitals.

3. GENERAL THEORY OF THE PARAMAGNETIC SHIFT.

3.1. Master equation.

In general the paramagnetic shift ($\Delta H = H - H_o$) for a fixed frequency is represented by the equation

$$\Delta H = \mathbf{h}_o \cdot (\Delta H / H_o) \cdot \mathbf{h}_o \tag{3.1}$$

where $\mathbf{h}_o$ is a unit vector parallel to the magnetic field $\mathbf{H}_o$. $(\Delta H / H_o)$ is a second rank tensor which we shall call the shift tensor. We assume that in the absence of a magnetic field our paramagnetic system has a set of energy states of energy ε_Γ and eigenfunctions $|\Gamma n\rangle$ where n designates the different degenerate states for each Γ. This degeneracy is removed by the electronic Zeeman interaction

$$\mathbf{H}_z = -\mathbf{\mu} \cdot \mathbf{H}_o = \beta_e (\mathbf{L} + g_e \mathbf{S}) \cdot \mathbf{H}_o \tag{3.2}$$

Kurland and McGarvey (14) have shown using the density matrix approach that the shift tensor components are given by the equation

$$(\Delta H / H_o)_{ij} = (kTq)^{-1} \{ \sum_{\Gamma n, \Gamma m} \langle \Gamma n | \mu_i | \Gamma m \rangle \langle \Gamma m | A_{Nj} / g_N \beta_N | \Gamma n \rangle e^{-\varepsilon_\Gamma / kT}$$

$$-kT \sum_{\Gamma n, \Gamma' m (\Gamma \neq \Gamma')} Q_{\Gamma \Gamma'} \langle \Gamma n | \mu_i | \Gamma' m \rangle \langle \Gamma' m | A_{Nj} / g_N \beta_N | \Gamma n \rangle \} \tag{3.3}$$

$$q = \sum_{\Gamma n} e^{-\varepsilon_\Gamma / kT} \tag{3.4}$$

$$Q_{\Gamma \Gamma'} = \{ e^{-\varepsilon_\Gamma / kT} - e^{-\varepsilon_{\Gamma'} / kT} \} / (\varepsilon_\Gamma - \varepsilon_{\Gamma'}) \tag{3.5}$$

where i and j are x, y and z.
This is the master equation from which we start to treat special cases. It was derived assuming T_{1e} was very short so we could average the hyperfine interaction over all electronic states. The derivation also assumed that ΔH is small compared to H_o for all orientations of the magnetic field.

Eq. (3.3) is applicable to the solid state. In solution, where rapid tumbling takes place, we will obtain the average shift

$$(\overline{\Delta H / H_o})_{Ave} = (1/3) [(\Delta H / H_o)_{xx} + (\Delta H / H_o)_{yy} + (\Delta H / H_o)_{zz}] \tag{3.6}$$

In earlier literature (15, 16) on the paramagnetic shift this

type of average was referred to as the solid state case and
other types of averaging were discussed for situations in which
the tumbling correlation times were shorter than the electronic
relaxation time, T_{1e}. Vega and Fiat (17, 18) have shown that
the relative magnitudes of the rotational correlation time and
T_{1e} have no effect on the averaging and that Eq. (3.6) is correct
in solution for all cases.

3.2. Shift equations for a simple spin state.

Let us consider first a system in which the lowest state
($\Gamma=0$) has $\varepsilon_0=0$ and has a degeneracy of $(2S+1)$. Further assume
all other Γ states have ε_Γ so large compared to kT to make
them unimportant in Eq. (3.3). The energy levels of such a state
can generally be considered to be the solution of an effective
spin Hamiltonian of the form

$$H_s = \beta_e \underset{\sim}{S} \cdot g \cdot \underset{\sim}{H} + \underset{\sim}{S} \cdot A \cdot \underset{\sim}{I} \tag{3.7}$$

for a spin of S. In this case Eq. (3.3) becomes

$$(\Delta H/H_o)_{ij} = [kT(2S+1)]^{-1} \underset{On,Om}{\Sigma} <On|\mu_i|Om><Om|A_{Nj}/g_N\beta_N|On> \tag{3.8}$$

Let i=j=z and recognize that the spin Hamiltonian terms in Eq.
(3.7) are given by (19)

$$g_{zz} = M_s^{-1} <SM_s|L_z+g_e s_z|SM_s> = -M_s^{-1}\beta_e^{-1}<SM_s|\mu_z|SM_s> \tag{3.9}$$

$$A_{zz} = M_s^{-1} <SM_s|A_{Nz}|SM_s> \tag{3.10}$$

Eq. (3.8) now becomes

$$(\Delta H/H_o)_{zz} = -[kT(2S+1)]^{-1} \beta_e g_{zz} A_{zz} \underset{M_j=-S}{\overset{S}{\Sigma}} M_S^2$$

$$= -g_{zz}\beta_e A_{zz} S(S+1)/3kT \tag{3.11}$$

Similar equations will be obtained for $(\Delta H/H_o)_{xx}$ and $(\Delta H/H_o)_{yy}$.
Note for $S=\frac{1}{2}$ that this is same as the Eq. (1.9) obtained earlier.

If we assume the hyperfine interaction to result from a
Fermi contact interaction in which we can represent $\underset{\sim}{A_F}$ by $a\underset{\sim}{S}$ and
the dipolar terms given by Eqs. (2.5-2.7) we obtain

$$(\Delta H/H_o)_{zz} = -g_{zz}\beta_e S(S+1)a/3kTg_N\beta_N$$

$$-g_{zz}^2\beta_e^2 S(S+1)(3\cos^2\Omega-1)/R^3kT \tag{3.12}$$

$$(\Delta H/H_o)_{xx} = -g_{xx}\beta_e S(S+1)a/3kTg_N\beta_N$$

$$-g_{xx}^2\beta_e^2 S(S+1)\{-\tfrac{1}{2}(\cos^2\Omega-1)+(3/2)\sin^2\Omega\cos2\psi\}/R^3kT \tag{3.13}$$

$$(\Delta H/H_o)_{yy} = - g_{yy}\beta_e S(S+1)a/3kTg_N\beta_N \tag{3.14}$$
$$- g_{yy}^2\beta_e^2 S(S+1)\{-(\tfrac{1}{2})(3\cos^2\Omega-1)-(3/2)\sin^2\Omega\cos2\psi\}/R^3kT$$

In solution this gives for the average shift

$$(\overline{\Delta H/H_o})_{Ave} = -\bar{g}\beta_e S(S+1)a/3kTg_N\beta_N \tag{3.15}$$
$$- [\beta_e^2 S(S+1)/9kT]\{[g_{zz}^2-(\tfrac{1}{2})(g_{xx}^2+g_{yy}^2)](3\cos^2\Omega-1)$$
$$+ (3/2)(g_{xx}^2-g_{yy}^2)\sin^2\Omega\cos2\psi\}R^{-3}$$

Eq. (3.15) was first derived by McConnell, et al (15, 19) who called the first term in Eq. (3.15) the Fermi contact shift and the second term the pseudocontact shift. Eq. (3.15) has been quoted and used a great deal in the literature on lanthanide and actinide complexes and shift reagents. It is totally incorrect, however, to apply it to these systems because the rare earth complexes do not satisfy the assumptions under which these equations are derived. Small crystal field splittings in these complexes produce excited states in which $\epsilon_{\Gamma'}$ is not large compared to kT and therefore we cannot ignore the second term in Eq. (3.3).

3.3. Shift equations for rare earth complexes (Second Order Zeeman Terms).

In the preceding section we have found that there is a direct connection between the NMR shift tensor and the hyperfine tensor when there are no excited states whose energies are comparable to kT. Thus, in these systems, the NMR shift can be used to measure the hyperfine tensor through the use of equations such as Eq. (3.11). When there are thermally accessible excited states the situation becomes more complex. The first term in Eq. (3.3) gives a thermal average of the hyperfine tensor for each electronic state and the second term in Eq. (3.3) gives rise to terms that are not related directly to the hyperfine tensor of any state. These terms are often referred to as the Second Order Zeeman (SOZ) terms. Calculations on both trans-ition metal complexes and rare earth complexes have shown these terms to be very significant when the value of $(\epsilon_{\Gamma'}-\epsilon_{\Gamma})$ is comparable to kT. The origin of these Second Order Zeeman terms is found in the presence of a magnetic field which mixes excited electronic states into the ground state through the matrix elements, $<\Gamma n|\mu_i|\Gamma'm>$, which are not all zero. The main point to be emphasized here, is that for rare earth complexes one cannot expect to find the simple connection between NMR shifts and the hyperfine tensor measured by ESR and ENDOR that is given by the equations found in Sect. 3.2.

Although one cannot obtain a simple general equation for rare earth complexes connecting NMR shifts to hyperfine measurements by other techniques, there is a simple relationship between the NMR shift and susceptibilities when the hyperfine interaction arises solely from the dipolar interaction between the nuclear spin and the electron spin on the rare earth ion. If Eq. (2.7) is substituted in Eq. (3.3) we obtain

$$
(\Delta H/H_o)_{zz} = -(\beta_e^2/kTq)(3\cos^2\Omega - 1)R^{-3}\{\textstyle\sum_{\Gamma n,\Gamma m}|<\Gamma n|(L_z + g_e S_z)|\Gamma m>|^2
$$
$$
-kT_{\Gamma n}\textstyle\sum_{\Gamma' m(\Gamma \neq \Gamma')}Q_{\Gamma\Gamma'}<\Gamma n|(L_z + g_e S_z)|\Gamma' m><\Gamma' m|(L_z + g_e S_z)|\Gamma n>\}
$$

(3.16)

Comparison of this equation with Van Vleck's theory (20) of magnetic susceptibilities reveals that

$$
(\Delta H/H_o)_{zz} = -\chi_{zz}(3\cos^2\Omega - 1)R^{-3}
$$

(3.17)

where χ_{zz} is the magnetic susceptibility tensor term for the rare earth ion. Similar expressions can be obtained for the other terms in the shift tensor. For the isotropic shift we have

$$
\overline{(\Delta H/H_o)}_{Ave} = -(1/3)[\chi_{zz} - (\tfrac{1}{2})(\chi_{xx} + \chi_{yy})](3\cos^2\Omega - 1)R^{-3}
$$
$$
-(1/2)[\chi_{xx} - \chi_{yy}]\sin^2\Omega\cos 2\psi R^{-3}
$$

(3.18)

Therefore the NMR shift due to the direct dipolar interaction can be related directly to the magnetic susceptibility tensor, which can be measured separately. Eq. (3.18) is the valid form for the pseudocontact shift in rare earth complexes rather than the form given in Eq. (3.15) that is often quoted in the literature.

If there is any transfer of spin onto the ligand nuclei by either mechanism discussed in Sect. 2.2 the equations become complex and depend strongly on the system in question. As will be discussed later, there is mounting evidence that such mechanisms are important only for nearest neighbor nuclei, (at least for the lanthanide complexes). Therefore it is anticipated that for nuclei further removed from the rare earth ion Eq. (3.18) will be adequate. Thus, if there is some means of measuring or estimating the components of the susceptibility tensor for the complex, the NMR shift can yield direct information on the geometry of the complex being studied.

3.4. Temperature expansion.

The splitting between the lowest J state of the free ion and the first excited J state is 2000 cm^{-1} or greater for all

rare earth ions except Sm^{3+} and Eu^{3+}. Since the crystal field interaction is a few hundred cm^{-1} at most, it is customary in theoretical treatments to start with the free ion J states and treat the crystal field interactions as a perturbation. Further since $kT \sim 200$ cm^{-1} at room temperature, it is not an unreasonable approximation, for all ions except Sm^{3+} and Eu^{3+}, to consider the summations in Eq. (3.3) to take place only over the states arising from the J ground state of the ion. If we further expand Eq. (3.3) into a power series of the form

$$(\Delta H/H_o)_{ij} = \sum_{k=1}^{\infty} (A_k)_{ij} T^{-k} \tag{3.19}$$

we shall find the first few terms have a very simple form.

If we take H_Q to be the crystal field operator, then

$$H_Q|\Gamma n> = \varepsilon_\Gamma |\Gamma n> \tag{3.20}$$

Further we shall take the conventional definition of zero for our energy scale to be that which satisfies the equation

$$\sum_{\Gamma n} \varepsilon_\Gamma = 0 \tag{3.21}$$

Using Eq. (3.21) $(A_1)_{ij}$ becomes

$$(A_1)_{ij} = k^{-1}(2J+1)^{-1} \sum_{\Gamma n \Gamma' m} <\Gamma n|\mu_i|\Gamma' m><m\Gamma'|A_{Ni}/g_N\beta_N|\Gamma n> \tag{3.22}$$

$$= k^{-1}(2J+1)^{-1} \sum_{\Gamma n} <\Gamma n|\mu_i A_{Nj}/g_N\beta_N|\Gamma n> \tag{3.23}$$

$$= k^{-1}(2J+1)^{-1} \sum_{M_J=-J}^{J} <JM_J|\mu_i A_{Nj}/g_N\beta_N|JM_J> \tag{3.24}$$

Eq. (3.23) comes from our assumption that the $|\Gamma n>$ are constructed from a complete set of linear functions and Eq. (3.24) comes from the invariance of the trace of a matrix to the representation. The $|JM_J>$ are the free ion J state functions. Since H_Q does not appear in Eq. (3.24) we see that A_1 requires no knowledge of the crystal field for its evaluation. In a similar fashion $(A_2)_{ij}$ can be shown to be

$$(A_2)_{ij} = -k^{-2}(2J+1)^{-1}(\tfrac{1}{2}) \sum_{\Gamma n, \Gamma' m}(\varepsilon_\Gamma + \varepsilon_{\Gamma'})<\Gamma n|\mu_i|\Gamma' m>$$
$$<\Gamma' m|A_{Nj}/g_N\beta_N|\Gamma n> \tag{3.25}$$

$$= -(\tfrac{1}{2})k^{-2}(2J+1)^{-1} \sum_{M_J=-J}^{J} <JM_J|(H_Q\mu_i+\mu_i H_Q)(A_{Nj}/g_N\beta_N)|JM_J> \tag{3.26}$$

The expression for $(A_3)_{ij}$ has also been reported (21). In evaluating these equations we use for μ_i the form

$$\mu_i = -g_J J_i \beta_e \tag{3.27}$$

where g_J is given by Eq. (2.9) and is also given in the Appendix for the lanthanide ion ground states.

The advantage of Eqs. (3.24) and (3.26) comes from the fact that H_Q and A_{Ni} can be expressed in terms of operator equivalents involving the operators J_i so that evaluation of the integrals in these equations can be done without knowing the exact form of the $|JM_J>$ wave functions. For the purpose of Eq. (3.26) the most useful form for H_Q is

$$H_Q = \sum_{L,M} k_{JL} a_{LM} <r^L> T_{LM} \qquad (3.28)$$

where T_{LM} are irreducible tensor operators, $<r^L>$ is the average value of r^L for an f electron, a_{LM} is a constant appropriate to a given crystal field and k_{JL} is a proportionality constant determined by L and the ground state of the particular ion in question. For crystal field calculations on f orbitals, only L=2,4 and 6 are needed in the summation in Eq. (3.28). The T_{LM} satisfy the equations

$$[J_z T_{LM} - T_{LM} J_z] = M T_{LM} \qquad (3.29)$$

$$[J_\pm T_{LM} - T_{LM} J_\pm] = [L(L+1) - M(M\pm1)]^{\frac{1}{2}} T_{L(M\pm1)} \qquad (3.30)$$

$$T_{L\pm L} = (\mp1)^L J_\pm^L = (\mp1)^L (J_x \pm i J_y)^L \qquad (3.31)$$

and since expressions for them are hard to find in the literature they are given for L=2,4 and 6 in the Appendix along with the values of k_{JL}. Most works on crystal field theory use the O_L^M equivalent set of operators which can be obtained from the T_{LM} set. A tabulation of O_L^M operators is given by Abragam and Bleaney (22), who also tabulate the values of k_{JL}. It is the relations in Eqs. (3.29-30) that make T_{LM} useful in evaluating the integrals found in Eq. (3.26).

A simple application of Eqs. (3.24, 3.26) is found in the dipolar shift. Combining Eqs. (2.7), (2.8), (3.24) and (3.27) gives

$$(A_1)_{zz} = -[g_J^2 \beta_e^2/k(2J+1)](3\cos^2\Omega - 1)R^{-3} \sum_{M_J=-J}^{J} <JM_J|J_z^2|JM_J> \qquad (3.32)$$

$$= -[g_J^2 \beta_e^2/k(2J+1)](3\cos^2\Omega - 1)R^{-3} \sum_{M_J=-J}^{J} M_J^2 \qquad (3.33)$$

$$= -[g_J^2 \beta_e^2 J(J+1)/3k](3\cos^2\Omega - 1)R^{-3} \qquad (3.34)$$

Since

$$<JM_J|J_x^2|JM_J> = 1/4 <JM_J|(J_+ + J_-)^2|JM_J> = 1/4 <JM_J|J_+J_- + J_-J_+|JM_J>$$

$$-(1/2)J(J+1)-(1/2)M_J^2 \qquad =<JM_J|J_y^2|JM_J> \qquad (3.35)$$

we find for $(A_1)_{xx}$ and $(A_1)_{yy}$

$$(A_1)_{xx}=-[g_J^2\beta_e^2J(J+1)/3k]\{-(\tfrac{1}{2})(3\cos^2\Omega-1)+(3/2)\sin^2\Omega\cos2\psi\}R^{-3} \qquad (3.36)$$

$$(A_1)_{yy}=-[g_J^2\beta_e^2J(J+1)/3k]\{-(\tfrac{1}{2})(3\cos^2\Omega-1)-(3/2)\sin^2\Omega\cos2\psi\}R^{-3} \qquad (3.37)$$

The average value of $\bar{A}_1$ in solution would be zero showing no T^{-1} term for the pseudocontact shift in rare earth complexes. Note that Eqs. (3.34-7) can be obtained from Eq. (3.18) if we assume that the magnetic susceptibility is isotropic and equal to that of the free ion in the absence of a crystal field.

$$X_{zz}=X_{xx}=X_{yy}=(g_J^2\beta_e^2J(J+1)/3kT) \qquad (3.38)$$

Thus the T^{-1} term is the high temperature limiting form in which the crystal field and any anisotropy it creates can be ignored.

For $(A_2)_{zz}$ we would calculate

$$(A_2)_{zz}=[g_J^2\beta_e^2/2k^2(2J+1)](3\cos^2\Omega-1)R^{-3}\textstyle\sum_{L,M}k_{JL}a_{LM}<r^L>$$

$$\textstyle\sum_{M_J=-J}^{J}<JM_J|T_{LM}J_z^2+J_zT_{LM}J_z|JM_J> \qquad (3.39)$$

$$=[g_J^2\beta_e^2/k^2(2J+1)](3\cos^2\Omega-1)R^{-3}\textstyle\sum_{L,M}k_{JL}a_{LM}<r^L>\sum_{M_J=-J}^{J}M_J^2$$

$$<JM_J|T_{LM}|JM_J> \qquad (3.40)$$

$<JM_J|T_{LM}|JM_J>$ is zero if $M\neq0$ so only a_{20}, a_{40} and a_{60} can contribute to $(A_2)_{zz}$. If we compute the average value of $\bar{A}_2$ in the usual manner we find that L=4 and 6 terms drop out yielding

$$\bar{A}_2=[g_J^2\beta_e^2J(J+1)(2J+3)(2J-1)/30k^2]$$

$$k_{J2}<r^2>[(2/3)^{\tfrac{1}{2}}a_{20}(3\cos^2\Omega-1)+(a_{22}+a_{2-2})\sin^2\Omega\cos2\psi]R^{-3}$$

which is an equation first derived by Bleaney (23). An expression for $\bar{A}_3$ has also been derived (24).

3.5. NMR shift and spin transfer mechanisms.

In the theory of spin transfer mechanisms it has been assumed that it is important for nearest neighbors only and that to a first approximation one can assume an axial symmetry about the axis connecting the rare earth and ligand atoms. If we adopt a coordinate system with the z axis parallel to the bond axis, the contribution to the NMR shift can be characterized by two parameters

$$(\Delta H/H_o) = (\Delta H/H_o)_{iso} + (\Delta H/H_o)_{anis} (3\cos^2\theta - 1) \qquad (3.42)$$

where θ is the angle between the magnetic field and the bond axis. In the coordinate system of the complex, the contribution to the shift tensor would be

$$(\Delta H/H_o)_{ij} = \delta_{ij}(\Delta H/H_o)_{iso} + (\Delta H/H_o)_{anis} (3\alpha_i\alpha_j - \delta_{ij}) \qquad (3.43)$$

where α_i is the direction cosine for the bond axis in the complex's coordinate system. In this approximation only $(\Delta H/H_o)_{iso}$ would contribute to the solution resonance.

3.5A. Polarization Model. Very few theoretical computations have been made. In general, it has been assumed that the hyperfine interaction was directly proportional to the spin angular momentum of the f orbitals causing the polarization of spin in the outer s and p orbitals, so that the hyperfine interaction could be represented by the operators

$$A_{Nz}^{P} = -K_{\parallel}S_z = -K_{\parallel}(L_z + 2S_z - J_z) = -(g_J - 1)K_{\parallel}J_z \qquad (3.44)$$

$$A_{Nx}^{P} = -K_{\perp}S_x = -(g_J - 1)K_{\perp}J_x \qquad (3.45)$$

$$A_{Ny}^{P} = -K_{\perp}S_y = -(g_J - 1)K_{\perp}J_y \qquad (3.46)$$

where the z axis is the bond axis. The parameters $K_{\parallel}$ and $K_{\perp}$ would be positive in this mechanism and must be evaluated from experiment.

Using Eqs. (3.44-6) plus (3.24) the T^{-1} terms for the polarization contribution to the shift are

$$(\Delta H/H_o)_{iso}^{P} = g_J(g_J - 1)\beta_e J(J+1)(K_{\parallel} + 2K_{\perp})/9kTg_N\beta_N \qquad (3.47)$$

$$(\Delta H/H_o)_{anis}^{P} = g_J(g_J - 1)\beta_e J(J+1)(K_{\parallel} - K_{\perp})/9kTg_N\beta_N \qquad (3.48)$$

In deriving these equations only the lowest J states were considered. Golding and Halton (25) attempted to include excited J states and, as expected, found them to be significant only for Sm^{3+} and Eu^{3+}. Note that g_J is less than one for all ions before Gd^{3+} in the Periodic Table and greater than one for those that come after Gd^{3+}. Thus Eq. (3.47) predicts for those liquid systems in which the polarization mechanism is dominant, that NMR shifts will be downfield for the first half of a rare earth series and upfield for the second half.

Until recently much of the literature on solution shifts assumed that any contribution from a spin transfer mechanism was via the polarization mechanism and the Fermi contact interaction and referred to this contribution as the Fermi contact shift. This literature uses Eq. (3.47) for interpretation. Because of this there are many statements in the literature that the "contact shift is pure T^{-1}" in its temperature behavior and therefore one could separate the contact and pseudocontact interactions by the temperature dependence, since as we saw in Sect. (3.4) the T^{-2} is the first non-zero term for the dipolar contribution in solution. Unfortunately, these researchers failed to realize that Eq. (3.47) is only the first term in the expansion and fails to account for crystal field splitting effects. Thus the polarization mechanism will also contribute a T^{-2} term which can easily be shown (26) from Eqs. (3.26) and (3.44-6) to be

$$-[g_J(g_J-1)\beta_e J(J+1)(2J-1)(2J+3)/45k^2 T^2 g_N\beta_N]$$
$$\times k_{J2}\{(2/3)^{\frac{1}{2}}a_{20}<r^2>(K_{\parallel}-K_{\perp})+(a_{22}+a_{2-2})<r^2>K_{\perp}\} \tag{3.49}$$

for the $(\Delta H/H_o)^P_{iso}$ term. Thus even if one would fit the NMR shift accurately to a power series in T^{-k} the coefficient of T^{-2} could not be assigned to the dipolar contribution because the polarization mechanism could make a significant contribution if $(K_{\parallel}-K_{\perp})$ were large enough.

3.5B. Covalent Model. Axe and Burns (27) and Baker (12) were the first to apply this model. They developed the MO's appropriate to a strong crystal field case and then applied the large spin-orbit interaction to obtain the wave functions appropriate to the experimental situation in which the spin-orbit interaction is much larger than the crystal field interaction. This is a complex calculation due mainly to the starting MO's being defined and quantized relative to the symmetry axes of the crystal field rather than the bond axis. In low symmetry cases it also leads to more parameters than can conveniently be handled. An alternative approach has been developed (13) in which the f orbitals are quantized relative to the bond axis being the z axis. Since the covalent interaction is small we can make the approximation that the local axial symmetry of the metal ion-ligand bond dominates, allowing admixture of ligand s and p orbitals only with the $4f_0$ and $4f_{\pm 1}$ orbitals of the metal ion. In this approach these orbitals are then replaced with the molecular orbitals.

$$4f_0 \rightarrow (N_0 4f_0 + a_\sigma s + b_\sigma p_0 + \text{ligand orbitals of other atoms}) \tag{3.50}$$

$4f_{\pm 1} \rightarrow (N_1 4f_{\pm 1} - b_\pi p_{\pm 1} + \text{ligand orbitals of other atoms})$ (3.51)

The coordinate system of the ligand atom has its z axis parallel to the bond axis and pointing away from the metal ion. The signs of a_σ, b_σ and b_π are chosen to make the MO's antibonding in nature. The $4f_{\pm 2}$ and $4f_{\pm 3}$ atomic orbitals do not mix with the ligand atom of interest along the z axis but do mix with that of other ligand atoms, but since we are only interested in the one atom, we need not concern ourselves with the form of these orbitals. These MO's are then put into the standard atomic wave function for the appropriate LSJ state of the rare earth ion. It has been shown in the case of Yb^{3+} in lower symmetry sites of CaF_2 (13) that this three parameter approach is adequate to explain the results even though the symmetry allows for more than three mixing coefficients.

This MO model has been applied to calculations of the F^- hyperfine interaction of Yb^{3+} (13), Tm^{2+} (13) and Ho^{2+} (28) in cubic sites of CaF_2 and Yb^{3+} (13) in tetragonal sites of CaF_2. It has also been used with the master equation, Eq. (3.3), to calculate the NMR shift for F^- in cubic sites of Yb^{3+} in CaF_2. All of these calculations have shown that the results are little influenced by the relative magnitudes of b_σ and b_π. This is due to the largest terms involving these two parameters being $b_\pi b_\sigma$ terms resulting from the mixing of p_0 and $p_{\pm 1}$ by the A_L (see Eq. (2.4)) term in the hyperfine interaction. The dominant term in the NMR equation was a Second Order Zeeman (SOZ) term of this type which produced a positive $(\Delta H/H_0)_{iso}^C$ term. Until this computation had been done, it was commonly assumed that the dominant term in the covalent mechanism would be a Fermi contact term which would produce a negative $(\Delta H/H_0)_{iso}^C$ term.

The complete NMR shift calculation on Yb^{3+} in CaF_2 showed that at room temperature a good estimate of the correct shift could be obtained from the T^{-1} term in the temperature expansion method (Sect. 3.4). Since calculation of this term can be done with no knowledge of the crystal fields, it has been done and applied to a variety of rare earth ions. The results can all be put into the following form

$$(\Delta H/H_0)_{iso}^C = (\beta_e/kTg_N\beta_N)\{C_{iso}^a a_\sigma^2 A_{2s} + C_{iso}^b b_\sigma b_\pi A_{2p}\}$$ (3.52)

$$(\Delta H/H_0)_{anis}^C = (\beta_e/kTg_N\beta_N)\{C_{anis}^a a_\sigma^2 A_{2s} + C_{anis}^b b_\sigma b_\pi A_{2p}\}$$ (3.53)

$$A_{2s} = (8\pi/3)g_e\beta_e g_N\beta_N |s(0)|^2$$ (3.54)

$$A_{2p} = (2/5)g_e\beta_e g_N\beta_N \langle r^{-3}\rangle_p$$ (3.55)

In obtaining these equations it was assumed that $b_\sigma^2 = b_\pi^2 = b_\sigma b_\pi$, because changing the relative values of b_σ and b_π had little effect. Values for the C coefficients have been calculated (21, 28, 29) for many rare earth ions and are tabulated in Table 1.

Table 1. MO Coefficients for T^{-1} Term in Covalent Contribution to NMR Shift

Ion	c^a_{iso}	c^b_{iso}	c^a_{anis}	c^b_{anis}
Pr^{3+}	0.108	0.735	-0.022	-0.880
Nd^{3+}	0.234	1.705	-0.095	-3.719
Tb^{3+}	-1.222	1.896	0.194	-2.080
Ho^{3+}	-1.0714	5.121	0.036	-4.456
Er^{3+}	-0.729	4.866	0.032	-4.298
Tm^{3+}	-0.389	6.385	0.065	-4.494
Yb^{3+}	-0.122	1.359	0.041	-1.232

From Table 1 we learn that the s and p orbital contributions are of opposite sign for the second half of the rare earth series but of the same sign in the first half. Therefore the covalent mechanism will always produce an upfield shift in solution for ions in the first half of the rare earth series but in the second half of the series the shift will depend on the relative importance of the s and p orbital contributions. Studies of solid state fluorides, so far, have shown the p orbital contribution to be dominant in the second half.

4. APPLICATION OF THEORY TO EXPERIMENTAL RESULTS.

4.1. Crystalline state.

Much of the best information about the applicability of these theories and the relative importance of spin transfer mechanisms versus the dipolar mechanism has come from ENDOR and NMR studies on single crystals of rare earth fluorides and hydrates.

4.1A. Rare Earth Fluorides. ENDOR studies (30-33) on alkaline earth fluorides doped with rare earth ions have shown that for fluorides that are not nearest neighbors the hyperfine interaction is entirely accounted for by the dipolar interaction alone with no contribution from spin transfer mechanisms. For nearest neighbor fluorides, however, this is not the case. For Gd^{3+} and Eu^{2+} the spin transfer contribution is negative and must be due to the polarization mechanism. For S state ions the covalent contribution would be zero in any case. Baker (12) found $K_{\parallel}$=3.60 MHz, $K_{\perp}$=0.93 MHz for Gd^{3+} in CaF_2 and $K_{\parallel}$= 3.49 MHz, $K_{\perp}$=1.60 MHz for Eu^{2+}. For Yb^{3+} and Tm^{2+} he found the spin transfer contribution to the hyperfine interaction to be positive and hence due to a covalent mechanism. Assuming the $K_{\parallel}$ and $K_{\perp}$ values of Gd^{3+} and Eu^{2+} to be true for Yb^{3+} and Tm^{2+}, respectively, the hyperfine results yield (13) the values of a_σ^2 and $b_\sigma b_\pi$ given in Table 2. A similar analysis (28) of ENDOR results (34) for Ho^{2+}

Table 2. MO Coefficients for Fluoride Ions Determined from
          ENDOR and NMR Measurements

System	Type of Measurement	$a_\sigma^2 \times 10^4$	$b_\sigma b_\pi \times 10^4$
Yb^{3+} in CaF_2	ENDOR	4.7	13.5
Tm^{2+} in CaF_2	ENDOR	3.8	6.4
Ho^{2+} in CaF_2	ENDOR	1.9	9.4
Tb^{3+} in Rb_2NaTbF_6	NMR	4.4	54.3
Ho^{3+} in Rb_2NaHoF_6	NMR	7.4	35.3
Er^{3+} in Rb_2NaErF_6	NMR	0.7	22.7
Pr^{3+} in Cs_2KPrF_6	NMR	~ 0	~ 0

in CaF_2 give the values also tabulated in Table 2. For Yb^{3+} and Tm^{2+} one could assume no polarization contribution and obtain reasonable values for the MO coefficients. In the case of Ho^{2+}, one can explain the magnitude and signs of the hyperfine parameters only by assuming a significant contribution from both spin transfer mechanisms.

The MO coefficients determined by ENDOR on doped CaF_2 systems should not be regarded as reliable because their contribution depended on estimating properly the dipolar contribution to the hyperfine interaction and this requires a precise knowledge of the extent of distortion in the CaF_2 lattice caused by replacing a Ca^{2+} ion with a rare earth +2 or +3 ion. The values serve to give an idea of the magnitude needed to explain experimental results. The ENDOR results do show conclusively that both the covalent and polarization mechanisms are needed to properly account for the results.

NMR studies have been done on CaF_2 (35) and CdF_2 (36) doped with Yb^{3+} and Er^{3+}. Both have a positive isotropic component in the shift tensor. The NMR shift has been calculated (21) for Yb^{3+} in CaF_2 using crystal fields determined from optical studies, wave functions determined by ESR and the crystal fields and MO parameters found by ENDOR. The calculation yielded $(\Delta H/H_o)_{iso} = 0.149 \times 10^{-3}$ and $(\Delta H/H_o)_{anis} = -1.17 \times 10^{-3}$. Experimental results give $(\Delta H/H_o)_{iso} = 0.32 \times 10^{-3}$, $(\Delta H/H_o)_{anis} = -1.17 \times 10^{-3}$ for CaF_2 and $(\Delta H/H_o)_{iso} = 0.12 \times 10^{-3}$, $(\Delta H/H_o)_{anis} = -1.06 \times 10^{-3}$ for CdF_2. The agreement in magnitude and sign between theory and experiment for $(\Delta H/H_o)_{iso}$ is striking evidence for the validity of the theoretical approach and the failure of the McConnell, et al

(15, 19) equations (Eq.(3.15)) to explain the shifts in rare earths. Since ENDOR shows the isotropic portion of the hyperfine interaction of the ground state to be positive, Eq. (3.15) would predict a negative value of $(\Delta H/H_o)_{iso}$ rather than the actual positive value.

NMR studies on pure single crystal rare earth fluorides have been done on rare earth trifluorides (28,37-41,59) and mixed rare earth alkali fluorides of the formulas $LiLnF_4$ (28, 42, 60) and A_2BLnF_6 (29), where A and B are different alkali metal ions. In all these studies a significant isotropic shift was found which must be due to a spin transfer mechanism with nearest neighbor rare earth ions. Values found, so far, are given in Table 3. In all cases the shift is positive for ions in the second half of the rare earth series and negative for those in the first half. There is considerable variability in magnitude between different sites indicating strong dependence on actual ionic distances. The number of nearest neighbor rare earth ions for a given fluoride ion is 3 for LnF_3, 2 for $LiLnF_4$ and one for A_2LnF_6, which accounts for the drop in average magnitude as we go from left to right in Table 3.

Table 3.   Values of $(\Delta H/H_o)_{iso} \times 10^3$ for ^{19}F in Rare Earth Fluorides[a].

Ion	LnF_3[b]	$LiLnF_4$	A_2BLnF_6
Pr^{3+}	–	–	-0.2
Nd^{3+}	(-0.5, -0.2, -0.8)	–	–
Tb^{3+}	(4.7, 2.8)	3.0	1.2
Dy^{3+}	(4.1, 2.4)	3.2	–
Ho^{3+}	(3.9, 2.9)	2.9	1.3
Er^{3+}	(2.1, 3.8)	2.4	2.0

[a]All at room temperature except Pr^{3+} at 137K and Nd^{3+} at 176K.
[b]NdF_3 has a trigonal structure with three nonequivalent sites for the fluoride ion and the rest have an orthorhombic structure with two sites in the unit cell.

The NMR studies have also shown that the anisotropic portion of the shift agrees with that calculated for the dipolar contribution to within 10%. In the case of LnF_3, it has been shown that uncertainties in our knowledge of the crystal structure and the susceptibility tensor is the major reason the experimental and the calculated dipolar values disagree. For the $LiLnF_4$ crystals, both the susceptibility tensor and structure are better known. Hansen and Nevald (42) claim the dipolar calculation agrees with experiment within experimental error for these crystals. However, the low site symmetry of the rare earth ion and the fluoride ion plus the presence of two nearest neighbor rare earth ions complicates the analysis. The cubic elpasolites (A_2BLnF_6) provide the best system for determining the presence of a spin transfer contribution to the anisotropic shift because all rare earth ions are in cubic sites giving an isotropic susceptibility and all fluorides are in sites of axial symmetry and have only one nearest neighbor rare earth ion. NMR studies (29) on these elpasolites found a definite negative contribution to the anisotropic shift from a spin transfer mechanism for Tb^{3+}, Ho^{3+}, Er^{3+} and Pr^{3+}. The isotropic and anisotropic shift terms for Pr^{3+} are close to what one would estimate using the polarization model and the $K_{\parallel}$ and $K_{\perp}$ values of Gd^{3+}. However, for Tb^{3+}, Ho^{3+} and Er^{3+} the polarization mechanism predicts a positive contribution to both the isotropic and anisotropic terms while the covalent mechanism predicts a positive isotropic term and a negative anisotropic term. Thus the covalent mechanism must be important for Tb^{3+}, Ho^{3+} and Er^{3+}. Analysis of these results using Eqs. (3.52-5) gave the MO coefficients in Table 2. This analysis showed that one could not completely ignore the polarization contribution for the ions Tb^{3+}, Ho^{3+} and Er^{3+}.

The ENDOR and NMR data obtained, so far, shows that the polarization contribution to the spin transfer in rare earth fluorides is present and fairly constant throughout the rare earth series but the covalent mechanism is also important and becomes the dominant mechanism in the second half of the rare earth series.

4.1B. <u>Rare Earth Hydrates</u>. There are two recent reports on proton ENDOR studies of H_2O complexed to Nd^{3+} (43) and Gd^{3+} (44). These studies find the hyperfine interaction to be almost pure dipolar with isotropic components no greater than ~0.04 MHz. In fact both of these studies have used the hyperfine interaction to determine the position of the H atoms to a high accuracy. One set of authors (43) claims to have determined the rare earth – H atom distance to ±.003Å and to have demonstrated that no significant transfer of electron spin to the O atoms has occurred.

A very low temperature (<4.2K) NMR study of protons in yttrium ethylsulfate doped with Yb^{3+}, Tb^{3+}, Dy^{3+} and Nd^{3+} has been reported (3). This work found, also, that the interaction was purely point-dipolar in nature with no evidence of any spin transfer. Room temperature NMR studies of several rare earth ethylsulfate crystals has been initiated (45) but analysis of the data was not complete at the time of this writing.

4.2. Solution spectra

Due to rotational averaging we can only determine the isotropic portion of the NMR shift for complexes in solution. This coupled with the possible lack of structural integrity of rare earth complexes in solution makes a theoretical analysis less exact or reliable. The much higher resolution of solution spectra, however, does allow for the detection of much smaller shifts and effects. Also solution studies can be applied to many more systems than could be studied in the solid state.

The sensitivity can be better illustrated by making an order of magnitude calculation for a proton in a ligand bonded to Yb^{3+}. Since the minimum paramagnetic shift one could expect to detect is 0.1 ppm, we might ask what is the largest value of R in the dipolar interaction that could produce such a shift. Optical experiments on Yb^{3+} in several crystalline environments give values of $(2/3)^{\frac{1}{2}}a_{20}<r^2>\sim 100$ cm^{-1}. Using this value in Eq. (3.41) gives a value $R\sim 30$ Å for a 0.1 ppm shift. Thus measureable shifts via the dipolar or pseudocontact mechanism can be expected for ligand atoms far removed from the rare earth ion. A similar calculation for the polarization mechanism using Eq. (3.47) gives a value of $(K_{\parallel}+2K_{\perp})/3 \sim 0.002$ MHz, which is two orders of magnitude less than the isotropic hyperfine term found for protons by ENDOR on rare earth hydrate complexes. Therefore small contributions via the spin transfer mechanisms can also be detected. However, spin transfer mechanisms, via σ bonds, tend to fall off an order of magnitude for every bond so that we might expect the dipolar interaction to dominate the shift for atoms that are separated by more than two bonds from the rare earth ion.

Only a few studies have been done on the NMR shift of nuclei adjacent to the rare earth ion. ^{17}O studies have been reported (46, 47) for rare earth hydrates in water solutions. ^{35}Cl studies (48) and ^{14}N studies (49) have also been reported. In general these give negative shifts for ions in the first part of the rare earth series and positive shifts for the second half. The experimental shifts satisfy Eq. (3.47) assuming a nearly constant and positive value for $(K_{\parallel}+2K_{\perp})/3$ over the whole rare earth series and for this reason it has been generally agreed that the dominant term in the shift for nearest neighbor atoms is due to a spin transfer mechanism and that this mechanism is the polarization mechanism. The covalent mechanism would give a positive shift for the whole series (See Sect. 3.5B.) and thus cannot be dominant in the first half. As in the case of the ionic fluorides (See Sect. 4.1A.) the covalent mechanism could also make an important contribution in O and N cases, in the second half of the series. Until solid state studies are done to determine the anisotropic portion of the shift there is no way to determine the relative importance of the covalent mechanism for the complexes studied only in solution.

The sign of dipolar shifts changes more often across the rare earth series. It has been shown (24) by calculating the T^{-3} term in the dipolar shift, that at room temperatures the dipolar shift should be given by Eq. (3.41) to an accuracy of 10 to 20%. Since the crystal field terms $a_{2M}<r^2>$ should vary only slowly for the same complex of various lanthanide ions across the series, the sign of the shift should be determined only by the sign of k_{J2} and the magnitude of the shift should be proportional to $g_J^2 J(J+1)(2J+3)(2J-1)k_{J2}$. A plot of this function is given in Figure 2. Thus the theory of the dipolar shift predicts three alternations in sign across the series rather than only the one change predicted by the polarization mechanism. These additional changes occur between Nd^{3+} and Pm^{3+} and between Ho^{3+} and Er^{3+}. Further, maximum or minimum shifts occur at Pr^{3+}, Dy^{3+} and Tm^{3+}. An extension of the theory by Bleaney (23) to include excited J levels for Sm^{3+} and Eu^{3+} indicates another maximum for Eu^{3+}. These general features have been found for protons in water molecules complexed to rare earth ions (47) and for protons in other ligands (50–55) complexed to rare earth ions. Thus the evidence for protons indicates that the dipolar shift is dominant for protons which are two or more bond lengths from the rare earth ion. A similar behavior has also been observed for ^{13}C shifts (56).

Many methods have been tried to separate the dipolar shift

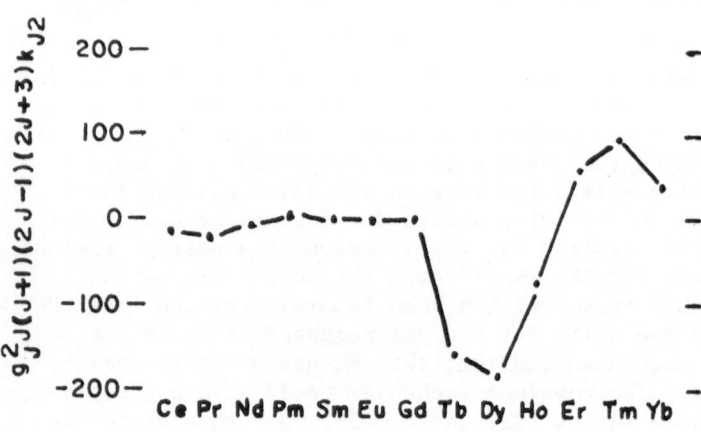

Figure 2. Plot of $g_J^2 J(J+1)(2J+3)(2J-1)k_{J2}$ for Ln^{3+} across lanthanide series.

from the shift due to spin transfer. These methods have been
reviewed in a recent article by deBoer, et al (57). All of
these methods have assumed the dipolar shift is given by Eq.
(3.18) and the spin transfer mechanism to be given by Eq. (3.47)
for the polarization mechanism. No consideration has been given
to the possibility that the covalent mechanism could make a
significant contribution and that Eq. (3.47) is not appropriate.
Most of these methods attempt to estimate either the dipolar or
spin transfer contribution from other data. The special method
using the temperature dependence has been discussed earlier (See
Sect. 3.5A.) and shown to be of little value.

    One method of separation is to measure the shift for Gd^{3+}
complexes (56) assuming that S state ions produce no dipolar
shifts so that any shift measured is a measure of $(K_N + 2K_L)/3$
for the polarization shift measurement. The spin transfer contribution for other rare earth complexes of the same ligand is then
calculated from Eq. (3.47) assuming a constant value of $(K_N + 2K_L)/3$
and this is subtracted from the experimental shift to give the
dipolar shift. This method has limited usefulness because the
Gd^{3+} resonances are difficult to detect and measure accurately
due to the large width of the resonances. The assumption that
S state ions have no dipolar shift contributions is not strictly
true as has been pointed out earlier (14, 23). The spin
Hamiltonian for such ions can be written

$$H = \beta_e S \cdot g \cdot H + \sum_M b_{SM} T_{SM} \tag{4.1}$$

where T_{SM} are irreducible tensor operators using spin operators
S. Following procedures outlined in Sect. 3.4 we can obtain
for $\bar{A}_2$ the equation

$$\bar{A}_2 = [4\beta_e^2 S(S+1)(2S+3)(2S-1)/30k^2][(2/3)^{1/2} b_{20}(3\cos^2\Omega - 1)$$

$$+ (b_{22} + b_{2-2})\sin^2\Omega\cos 2\psi]R^{-3} \tag{4.2}$$

ESR measurements on Gd ethylsulfate (58) give $(2/3)^{1/2} b_{20} \approx 0.0068$
cm^{-1} for $Gd(H_2O)_9^{3+}$ which will give a shift of ~ 0.6 ppm for
$R = 3\text{Å}$ and $\Omega = 0$. ESR measurements on more distorted systems give
b_{20} values 5 to 10 times larger so that dipolar contributions of
a few ppm are possible in Gd complexes. Observed shifts in ^{13}C
(56) complexes for carbon atoms two bonds removed are ~ 100 ppm
so that the dipolar contribution is probably within the experimental error and can be safely neglected. Shifts for water
protons (47), however, are only ~ 18 ppm so that the dipolar
contribution cannot be entirely neglected in this case. The
covalent mechanism makes no contribution to the shift for Gd^{3+}
but can contribute for the other ions, so that the method may not
obtain the dipolar shift to the accuracy claimed for it. Since
the dipolar shift is considerably larger than that due to spin

transfer, this error is probably not serious.

    Another method of separating the dipolar shift from the total
is to estimate the dipolar shift from Eq. (3.18) and experimental
measurements from single crystals of the anisotropy in the
magnetic susceptibility. This method requires knowledge also of
the geometry of the complex. This method is, of course,
difficult to apply in general because it requires information
not easy to acquire for most systems. The most serious drawback
to the method is found in the requirement that the location of
the principal axes for the susceptibility tensor of a given
complex ion must be accurately known. This cannot be determined
for most systems that have been studied from susceptibility
measurements due to low symmetry and the presence of more than
one complex in the unit cell. Therefore most attempts to apply
this method have resorted to making educated guesses as to the
location of the principal axes. Thus most applications of this
method have not obtained spin transfer shifts of any great
reliability. Another problem of the method is that it assumes
no geometrical changes in the complex between the liquid and
solid state and further must assume no spin exchange contributions
to the susceptibility are present in the solid state.

    Most other methods start from Eqs. (3.18) and (3.47) and
study the same complex for different rare earth ions and assume
$(K_{\parallel}+2K_{\perp})/3$ is constant for a given nucleus for all rare earth
ions. If m is the number of rare earth ions and n the number of
nonequivalent nuclei measured in the ligand, then m × n is the
number of shifts measured. There are n values of $(K_{\parallel}+2K_{\perp})/3$
values to be found, 2m values of $[\chi_{zz}-(\frac{1}{2})(\chi_{xx}+\chi_{yy})]$ and
$[\chi_{xx}-\chi_{yy}]$ plus 2n values of geometric parameters. Therefore, if
m × n > 2m+3n we have more equations than unknowns. In general
the number of parameters is lessened by making various geometrical assumptions. These assumptions are sometimes questionable,
such as assuming axial symmetry. A general application of this
method in which the geometrical assumptions were kept to a
minimum has been done by de Boer, et al (57). They found in
their system that reasonable agreement between theory and
experiment occurred by assuming no spin transfer for ^{13}C or ^{1}H.
Slightly better agreement came by including spin-transfer but
the values were small. Since reasonable agreement was obtained
without spin transfer, we cannot regard any reported values of
spin-transfer shifts by this method to be very reliable.

5. APPENDIX

5.1. Irreducible tensor operators for L=2, 4 and 6.

$$T_{20} = (2/3)^{\frac{1}{2}}[3J_z^2 - J(J+1)]$$

$$T_{2\pm1} = \mp[J_zJ_\pm + J_\pm J_z]$$

$$T_{2\pm2} = J_\pm^2$$

$$T_{40} = (2/35)^{\frac{1}{2}}\{35J_z^4 - [30J(J+1)-25]J_z^2 - 6J(J+1)+3J^2(J+1)^2\}$$

$$T_{4\pm1} = \mp(2/7)^{\frac{1}{2}}\{[7J_z^2 - 3J(J+1)-1]J_zJ_\pm + J_\pm J_z[7J_z^2 - 3J(J+1)-1]\}$$

$$T_{4\pm2} = (7)^{-\frac{1}{2}}\{[7J_z^2 - 2J(J+1)-10]J_\pm^2 + J_\pm^2[7J_z^2 - 2J(J+1)-10]\}$$

$$T_{4\pm3} = \mp(2)^{\frac{1}{2}}[J_zJ_\pm^3 + J_\pm^3 J_z]$$

$$T_{4\pm4} = J_\pm^4$$

$$T_{60} = 2/(231)^{\frac{1}{2}}\{231J_z^6 - [315J(J+1)-735]J_z^4 + [105J^2(J+1)^2 - 525J(J+1)$$
$$+294]J_z^2 - 5J^3(J+1)^3 + 40J^2(J+1)^2 - 60J(J+1)\}$$

$$T_{6\pm1} = \mp(2/11)^{\frac{1}{2}}\{33J_z^4 - [30J(J+1)-15]J_z^2 + 5J^2(J+1)^2 - 10J(J+1)+12\}J_zJ_\pm$$
$$\mp (2/11)^{\frac{1}{2}}J_\pm J_z\{33J_z^4 - [30J(J+1)-15]J_z^2 + 5J^2(J+1)^2 - 10J(J+1)+12\}$$

$$T_{6\pm2} = [(5/11)^{\frac{1}{2}}/2]\{33J_z^4 - [18J(J+1)+123]J_z^2 + J^2(J+1)^2 + 10J(J+1)+102\}J_\pm^2$$
$$+[(5/11)^{\frac{1}{2}}/2]J_\pm^2\{33J_z^4 - [18J(J+1)+123]J_z^2 + J^2(J+1)^2 + 10J(J+1)+102\}$$

$$T_{6\pm3} = \mp(5/11)^{\frac{1}{2}}\{11J_z^2 - [3J(J+1)+59]\}J_zJ_\pm^3$$
$$\mp(5/11)^{\frac{1}{2}}J_\pm^3 J_z\{11J_z^2 - [3J(J+1)+59]\}$$

$$T_{6\pm4} = [(6/11)^{\frac{1}{2}}/2]\{[11J_z^2 - J(J+1)-38]J_\pm^4 + J_\pm^4[11J_z^2 - J(J+1)-38]\}$$

$$T_{6\pm5} = \mp3^{\frac{1}{2}}[J_zJ_\pm^5 + J_\pm^5 J_z]$$

$$T_{6\pm6} = J_\pm^6$$

5.2 Matrix proportionality constants for rare earth ions.

Ion	Ground State	g_J	k_{J2}	k_{J4}	k_{J6}
Ce^{3+}	$4f^1$	$^2F_{5/2}$	6/7	$(-2/5\cdot7)$	$(2/3^2\cdot5\cdot7)$	0
Pr^{3+}	$4f^2$	$^3H_4$	4/5	$(-2^2\cdot13/3^2\cdot5^2\cdot11)$	$(-2^2/3^2\cdot5\cdot11^2)$	$(2^4\cdot17/3^4\cdot5\cdot7\cdot11^2\cdot13)$
Nd^{3+}	$4f^3$	$^4I_{9/2}$	8/11	$(-7/3^2\cdot11^2)$	$(-2^3\cdot17/3^3\cdot11^3\cdot13)$	$(-5\cdot17\cdot19/3^3\cdot7\cdot11^3\cdot13^2)$
Pm^{3+}	$4f^4$	$^5I_4$	3/5	$(2\cdot7/3\cdot5\cdot11^2)$	$(2^3\cdot7\cdot17/3^3\cdot5\cdot11^3\cdot13)$	$(2^3\cdot17\cdot19/3^3\cdot7\cdot11^3\cdot13^2)$
Sm^{3+}	$4f^5$	$^6H_{5/2}$	2/7	$(13/3^2\cdot5\cdot7)$	$(2\cdot13/3^3\cdot5\cdot7\cdot11)$	0
Eu^{3+}	$4f^6$	$^7F_0$	0	-	-	-
Gd^{3+}	$4f^7$	$^8S_{7/2}$	2	-	-	-
Tb^{3+}	$4f^8$	$^7F_6$	3/2	$(-1/3^2\cdot11)$	$(2/3^2\cdot5\cdot11^2)$	$(-1/3^4\cdot7\cdot11^2\cdot13)$
Dy^{3+}	$4f^9$	$^6H_{15/2}$	4/3	$(-2/3^2\cdot5\cdot7)$	$(-2^3/3^3\cdot5\cdot7\cdot11\cdot13)$	$(2^2/3^3\cdot7\cdot11^2\cdot13^2)$
Ho^{3+}	$4f^{10}$	$^5I_8$	5/4	$(-1/2\cdot3^2\cdot5^2)$	$(-1/2\cdot3\cdot5\cdot7\cdot11\cdot13)$	$(-5/3^3\cdot7\cdot11^2\cdot13^2)$
Er^{3+}	$4f^{11}$	$^4I_{15/2}$	6/5	$(2^2/3^2\cdot5^2\cdot7)$	$(2/3^2\cdot5\cdot7\cdot11\cdot13)$	$(2^3/3^3\cdot7\cdot11^2\cdot13^2)$
Tm^{3+}	$4f^{12}$	$^3H_6$	7/6	$(1/3^2\cdot11)$	$(2^3/3^4\cdot5\cdot11^2)$	$(-5/3^4\cdot7\cdot11^2\cdot13)$
Yb^{3+}	$4f^{13}$	$^2F_{7/2}$	8/7	$(2/3^2\cdot7)$	$(-2/3\cdot5\cdot7\cdot11)$	$(2^2/3^3\cdot7\cdot11\cdot13)$

6. REFERENCES

1. J.P. Wolfe and R.S. Markiewicz, Phys. Rev. Lett. 30, 1105 (1973).
2. A.R. King, J.P. Wolfe and R.L. Ballard, Phys. Rev. Lett. 28, 1099 (1972).
3. J.P. Wolfe, Phys. Rev. B 16, 128 (1977).
4. C.P. Slichter, Principles of Magnetic Resonance, (Harper, New York, 1971).
5. W. Marshall, in Paramagnetic Resonance, edited by W. Low (Academic Press, New York, 1963), Vol. 1, p.347.
6. A.D. Buckingham and P.J. Stiles, Mol. Phys. 24, 99 (1972).
7. P.J. Stiles, Mol. Phys. 27, 501 (1974); 29, 1271 (1975).
8. R.M. Golding, R.O. Pascual and L.C. Stubbs, Mol. Phys. 31, 1933 (1976).
9. R.M. Golding, in Magnetic Resonance and Related Phenomena, Proc XIX Congress Ampere (H. Brunner, K.H. Hausser and D. Schweitzer, Eds., Groupement Ampere, Heidelberg/Geneva, 1976).
10. R.E. Watson and A.J. Freeman, Phys. Rev. 156, 251 (1967).
11. R.E. Watson and A.J. Freeman, Phys. Rev. Lett. 6, 277 (1961).
12. J.M. Baker, Proc. Phys. Soc. London 1, 1670 (1968).
13. B.R. McGarvey, J. Chem. Phys. 65, 955 (1976).
14. R.J. Kurland and B.R. McGarvey, J. Magn. Resonance 2, 286 (1970).
15. H.M. McConnell and R.E. Robertson, J. Chem. Phys. 29, 1361 (1958).
16. J.P. Jesson, J. Chem. Phys. 47, 579 (1967).
17. A.J. Vega and D. Fiat, Pure Appl. Chem. 32, 307 (1972).
18. A.J. Vega and D. Fiat, J. Chem. Phys. 60, 579 (1974).
19. H.M. McConnell and D.B. Chestnut, J. Chem. Phys. 28, 107 (1958).
20. J.H. Van Vleck, Theory of Electric and Magnetic Susceptibilities, (Oxford Univ. Press, London/New York, 1933).
21. B.R. McGarvey, J. Chem. Phys. 65, 962 (1976).
22. A. Abragam and B. Bleaney, Electron Paramagnetic Resonance of Transition Ions, (Clarendon Press, Oxford, 1970).
23. B. Bleaney, J. Magn. Resonance 8, 91 (1972).
24. B.R. McGarvey, J. Magn. Resonance, In Press.
25. R.M. Golding and M.P. Halton, Aust. J. Chem. 25, 2577 (1972).
26. This result is not yet in the literature, having been obtained during the writing of this review.
27. J.D. Axe and G. Burns, Phys. Rev. 152, 331 (1966).
28. M.R. Mustafa, B.R. McGarvey and E. Banks, J. Magn. Resonance 25, 341 (1977).
29. A. Reuveni and B.R. McGarvey, J. Magn. Resonance, In Press.
30. J.M. Baker, E.R. Davies and J.P. Hurrell, Proc. Roy Soc. Ser. A 308, 403 (1968).
31. D. Kiro, W. Low and E. Secemski, J. Phys. (Paris) Colloq. C1, 950 (1971).
32. J.M. Baker and J.P. Hurrell, Proc. Phys. Soc. 82, 742

(1963).

33. R.G. Bessent and W. Hayes, Proc. Roy. Soc. A285, 430 (1965).

34. E. Secemski and W. Low, Phys. Rev. B9, 4954 (1974).

35. R.J. Booth, M.R. Mustafa and B.R. McGarvey, Phys. Rev. B17, 4150 (1978).

36. M.R. Mustafa, W.E. Jones, B.R. McGarvey, M. Greenblatt and E. Banks, J. Chem. Phys. 62, 2700 (1975); E. Banks, M. Greenblatt and B.R. McGarvey, ibid. 58, 4787 (1973).

37. A. Reuveni and B.R. McGarvey, J. Magn. Resonance, 29, 21 (1978).

38. K. Lee, Solid State Comm. 7, 367 (1969).

39. A.G. Lundin and S.P. Gabuda, Bull. Acad. Sci. USSR, 1, 90 (1968).

40. M.L. Afanas'ev, S.P. Gabuda and A.G. Lundin, ZhETF Pis'ma 7, 451 (1968).

41. M.L. Afanasiev, S.P. Habuda and A.G. Lundin, Acta Cryst. B28, 2903 (1972).

42. P.E. Hansen and R. Nevald, Phys. Rev. B16, 146 (1977).

43. C.A. Hutchison,Jr. and D.B. McKay, J. Chem. Phys. 66, 3311 (1977).

44. R. de Beer, F. Biesboer and D. van Ormondt, Physica, 83B, 314 (1976).

45. S. Sato, A. Reuveni and B.R. McGarvey, private communication.

46. W.B. Lewis, J.A. Jackson, J.E. Lemon and H. Taube, J. Chem. Phys. 36, 694 (1962).

47. J. Reuben and D. Fiat, J. Chem. Phys. 51, 4909 (1969).

48. F. Barbalat-Rey, Helv. Phys. Acta 42, 516 (1969).

49. M. Witanowski, L. Stefaniak, H. Januszowski and Z.W. Wolkowski, Chem. Commun. 1573 (1971).

50. E.L. Muetterties and C.M. Wright, J. Amer. Chem. Soc. 87, 4706 (1965).

51. T.H. Siddall,III, W.E. Stewart and D.B. Karraker, J. Inorg. Nucl. Chem. 3, 479 (1967).

52. J.M. Briggs, G.H. Frost, F.A. Hart, G.P. Moss and M.L. Staniforth, Chem. Commun. 749 (1970).

53. F.A. Hart, J.E. Newbery and D. Shaw, J. Inorg. Nucl. Chem. 32, 3585 (1970).

54. D.R. Crump, J.K.M. Sanders, and D.H. Williams, Tetrahedron Lett. 4419 (1970).

55. W.D. Horrocks,Jr. and J.P. Sipe,III, J. Amer. Chem. Soc. 93, 6800 (1971).

56. K. Ajisaka and M. Kainosho, J. Amer. Chem. Soc. 97, 330 (1975).

57. J.W.M. de Boer, P.J.D. Sakkers, C.W. Hibbers and E. de Boer, J. Magn. Resonance 25, 455 (1977).

58. B. Bleaney, H.E.D. Scovil and R.S. Trenam, Proc. Roy. Soc. A223, 15 (1954).

59. P.E. Hansen and R. Nevald, Phys. Rev. B, 17, 2866 (1978).

60. R. Nevald and P.E. Hansen, Phys. Rev. B, 19,   (1978).

NMR-SPECTROSCOPY OF ORGANOMETALLIC COMPOUNDS OF THE f-ELEMENTS: PRACTICAL APPLICATIONS

R. Dieter Fischer

Institut für Anorganische und Angewandte Chemie der Universität Hamburg, D-2000 Hamburg 13, Germany

INTRODUCTION

The purpose of this contribution is not to present a complete compilation of all nmr data and corresponding experimental results that have been obtained with organometallic nf^q-systems until now; it is rather intended to indicate the specific fields of activity in correlating data with distinct problems confined to the molecular and/or electronic structure of a paramagnetic complex. From the point of view of the reviewer, it appears particularly worthwhile to trace the various routes that have been demonstrated to be practicable so far, and to point out what still could, and should, be done to make optimal use of the fairly well developed theory of the nmr-spectra of paramagnetic samples.

Historically, the very first nmr experiments with paramagnetic organometallic lanthanide and actinide complexes in solution date back to the years 1965/66. When the magnetic susceptibility of various paramagnetic triscyclopentadienyl complexes was determined by means of Evans' method[1], it was realized that in various cases also specific pmr lines of the dissolved complexes were surprisingly narrow[2]. Since then, an appreciably large number of articles dealing more or less intensely with the nmr spectra of organoactinide and -lanthanide systems have been published, giving rise very soon to a first culmination point around 1973. In 1973 the first monograph on "NMR-Spectroscopy of Paramagnetic Systems" also appeared[3] including already a brief chapter on the specific properties of systems with an electronic f^q-ground configuration. The very rapid development of, and interest in, paramagnetic "nmr-shift

· 337

T. J. Marks and R. D. Fischer (eds.), Organometallics of the f-Elements, 337–377.
All Rights Reserved. Copyright © 1979 by D. Reidel Publishing Company, Dordrecht, Holland.

reagents"(4) has, since 1969, drawn much attention towards non-organometallic lanthanide chelate complexes. Likewise, in the field of organometallic chemistry, structural considerations involving mainly the dipolar term (vide infra) have clearly dominated over concise discussions of the contact term. Owing to the good solubility of most organometallics in organic solvents, practically all published data originate from solution spectra.

It appears to the reviewer that since 1973 until now a lot more of experimental material was collected by various research groups than has ever been published in the open literature. Hopefully, the two contributions on the nmr spectroscopy of paramagnetic (organometallic) compounds will also stimulate a better understanding, and further assessment, of as many "data in the drawers" as possible.

PRACTICAL ASPECTS

Basic Terminology and Sign Conventions:

The quantity of paramount interest in high resolution nmr spectroscopy of diamagnetic samples, the chemical shift δ^{dia}, is, by convention, always referred to a suitable standard signal such that δ^{dia} is positive when the standard signal lies at a higher magnetic field strength than the signal in question. The essential quantity in case of a paramagnetic nmr probe is called the "paramagnetic shift", or (for solutions) "isotropic shift", and results as the difference of the nmr shift of a distinct nucleus in a paramagnetic molecule, δ^{para}, and the chemical shift δ^{dia} for the same nucleus in a "chemically equivalent" diamagnetic molecule:

$$\Delta H/H_o \equiv \Delta H^{iso} = \delta^{para} - \delta^{dia} \tag{1}$$

Eq.1 accounts for the fact that most commercial nmr spectrometers are designed for magnetic field modulation at a constant radio frequency. Although no general sign convention has ever been established(5), the majority of authors consider ΔH^{iso} to be positive if δ^{para} corresponds to a higher magnetic field than δ^{dia}. Thus, if $\Delta H^{iso} > 0$ the internal magnetic field at the position of the nucleus in question counteracts the external field of the instrument (and vice versa).

The internal magnetic field is generally composed of two terms,

$$\Delta H^{iso} = \Delta H^{dip} + \Delta H^{con} \tag{2}$$

accounting, respectively, for all magnetic dipolar interactions
(ΔH^{dip}) and for a net free spin density on the nucleus under
consideration (ΔH^{con}). It is important to keep in mind that the
most frequently used expression for ΔH^{dip} in case of an axially
symmetric molecule (or ligand field acting on the paramagnetic
metal ion),

$$\Delta H_i^{dip} = -\frac{1}{3N} (\chi_\| - \chi_\perp) \frac{3\cos^2\theta_i - 1}{r_i^3} \tag{3}$$

holds only when the sign of ΔH_i^{iso} is properly chosen as outlined above. The term $G_i = (3\cos^2\theta_i - 1)/r_i^3$ is called the geometric
factor of the i-th nucleus, θ_i being the angle between the radius
vector pointing from the paramagnetic centre (in general the
paramagnetic metal ion) to the nucleus and the main rotational
axis. N is Avogadro's number, and r_i the distance from the metal
ion to the nucleus. For magnetically anisotropic systems
without axial symmetry, the expression for ΔH^{dip} becomes somewhat more complicated:

$$\Delta H_i^{dip} = -\frac{1}{3N} (\chi_{zz} - \frac{1}{2}\chi_{yy} - \frac{1}{2}\chi_{xx}) \frac{3\cos^2\theta_i - 1}{r_i^3}$$

$$-\frac{1}{2N} (\chi_{xx} - \chi_{yy}) \frac{\sin^2\theta_i \cos2\omega_i}{r_i^3}$$

ΔH^{dip} is now a function of the three internal coordinates r,
θ and ω of the nucleus (fig.1) as well as of the three principal
components of the magnetic susceptibility tensor, χ_{xx}, χ_{yy},
χ_{zz} (6).

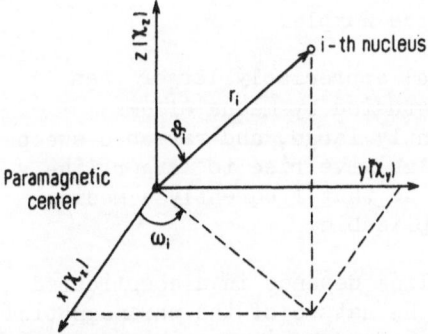

Fig. 1. Internal coordinates of a nucleus relative to the
        paramagnetic center. (The "geometric" and "magnetic"
        coordinates are not necessarily coincident(6)).

Following the above sign convention, the "contact" or "spin delocalization" term ΔH^{con} relates the (super)hyperfine coupling parameter A_i with the actual spin moment $<S_z>$ of the f^q-system as follows:

$$\Delta H^{con} = \frac{2\pi A_i}{h\gamma_N} \cdot \frac{<S_z>}{H} \qquad (5)$$

A_i is proportional to the spin density ρ_i generated at the site of the i-th nucleus:

$$A_i \sim \gamma_N \Sigma_m |\Psi_{mN}(0)|^2 \sim \rho_i/2S \qquad (6)$$

Hence, e.g. by delocalization of "α-spins" from the paramagnetic centre ($<S_z>$ negative) onto the i-th nucleus, ΔH^{con} will become negative, and A_i positive. Conversely, positive ΔH^{con} would, under the same conditions, indicate a different spin transfer mechanism (i.e. "spin reversal", according to $A_i < 0$). As, owing to eq. 6, the gyromagnetic ratio γ_N occurs both in the nominator and in the denominator of eq. 5, the sign of ΔH^{con} is, for a given spin delocalization mechanism, uniquely determined by the sign of $<S_z>$. For positive A_i/γ_N holds:

"α-spins": $\rho > 0$; $<S_z> < 0$
"β-spins": $\rho < 0$; $<S_z> > 0$ (7)

The quantity $<S_z>$ is proportional to the product $g_J(1-g_J)$ where g_J is the Landé factor of a distinct Russell Saunders state $^{2S+1}L_J$. Further details concerning the theory of ΔH^{con} and ΔH^{dip} are found in the preceding chapter and in Ref.3.

Instrumentation and Quality of the Sample:

As $|\Delta H^{iso}|$ is in general appreciably larger than $|\delta^{dia}|$, paramagnetic samples should be studied by means of nmr spectrometers allowing sufficiently large, and variable sweep widths. Also, paramagnetic samples give rise to larger line widths than diamagnetic probes, so that frequently a medium resolution spectrometer will applicable.

The actual width of a line depends in a complicated way on various factors such as the nature of the paramagnetic ion, the specific crystal field caused by its environment, the position of the nucleus in question relative to the paramagnetic centre, various exchange phenomena, the temperature etc.(7). For organometallic compounds of the trivalent lanthanides, the relative variation of the line width is very similar to that in typical nmr shift reagents and their various adducts with

substrates, respectively(8,9,10). Unlike for d-element organometallics where the half widths frequently are of the order of
10^3 Hz, the half widths of most f-element complexes range from
a few Hz to several 10^2 Hz, the ions U^{4+}, Pr^{3+}, Eu^{3+}, Sm^{3+} and
Ce^{3+} causing usually the weakest line broadening.

Contrary to many diamagnetic nmr probes, it is highly
desirable to carefully study the temperature dependence of ΔH^{iso}
in almost every case. Therefore, the nmr spectrometer ought to
be equipped with an efficient variable-temperature device, and
the sample should be dissolved in appropriate solvents with low
freezing points (e.g. toluene-d_8 or, if more polar solvents are
needed, tetrahydrofuran-d_8). To arrive at higher temperatures,
perdeuterated naphthalene or decalin have turned out to be very
useful.

Some attention should also be paid to the appropriate
choice of the diamagnetic standard. This demand becomes particularly imperative for nmr studies of nuclei whose δ^{dia} values
show large variations themselves (e.g. ^{13}C and ^{31}P) where even
the question for the correct sign of ΔH^{iso} might require the
full synthesis of a new compound.

STUDIES OF MOLECULAR STRUCTURE AND REACTIVITY

Neglect of ΔH^{dip}-ΔH^{con}-Separation

In many instances, the paramagnetism of the central ion
of an organometallic species gives rise to large variations of
ΔH^{iso} even when the chemical composition undergoes minor changes,
and may thus be used for different analytical purposes. E.g.,
all Yb(III)-complexes of the type $(\eta^5\text{-}C_5H_5)_3YbL$ exhibit at room
temperature slightly different isotropic C_5H_5 ring proton shifts
(table 1). However, for complexes of the type $[(\eta^5\text{-}C_5H_5)_2YbX]_2$
where X is an anionic ligand, the ring proton signals appear in
a fairly different region and depend, moreover, strongly on the
nature of X (11). While the dimeric character of the latter species appears to be preferred in most nonpolar solvents, monomeric adducts $(C_5H_5)_2YbX\cdot Ln$ are formed in polar solvents like
THF, pyridine, ect. This process is again reflected in the change
of ΔH^{iso}. In the case of Yb(III)-complexes, changes of ΔH^{iso} are
frequently accompanied by considerable changes of the line
width (table 1).

A similar situation is found for uranium(IV)-cyclopenta-
dienyl complexes $(C_5H_5)_nU^{IV}X_m$ where the C_5H_5 ring proton signal
displays ΔH^{iso} values which are strongly governed by both the
number of C_5H_5 ligands and the nature of X (table 2). Very recently, it was shown that by permethylation of the C_5H_5 rings

Table 1. ΔH^{iso} of C_5H_5 ring protons in the complexes
$(\eta^5-C_5H_5)_3YbL$, $(\eta^5-C_5H_5)_2YbX$ and $(\eta^5-C_5H_5)_2YbXL$

L or X	solvent	ΔH^{iso} (ppm)	$\Delta\nu_{1/2}$ (Hz)
-	C_6D_{12}	59	300
-	C_6D_6	56	290
THF	THF	54	280
pyridine	C_6D_6	52	270
CNC_6H_{11}	C_6D_6	51	290
NH_2	$THF-d_8$	21,4	50
NH_2	C_6D_6	24,0	45
Cl	$THF-d_8$	47,8	245
Cl	C_6D_6	74,0	260

rather stable complexes of the type $\left[C_5(CH_3)_5\right]_2MXX'$ (M = Th or U)
are accessible(15). The data in table 3 indicate that the variation of X and/or X' is again very clearly reflected in changes of

Table 2. Isotropic ring proton shifts of various $(C_5H_5)_nU^{IV}X_m-$
systems (references in parentheses)

$(C_5H_5)_3UX$		$(C_5H_5)_2UX_2$		$C_5H_5UX_3(dme)^{xxx}$	
X	ΔH^{iso}	X	ΔH^{iso}	X	ΔH^{iso}
OC_2H_5	24,4(48)	$(tdt)^+$	ca. 0 (14)		
Cl	9,6(11)	BH_4	ca. 0,9(13)	Cl	ca. 38(12a)
		$(cat)^{++}$	ca. 7 (14)		
BH_4	12,6(35)	$N(C_2H_5)_2$	ca. 21 (14)		

$^+$ toluene-3,4-dithiol, $^{++}$ catechol, $^{+++}$ 1,2-dimethoxyethane

the methyl proton shift if M = U while the δ-values of the
diamagnetic Th-homologues undergo only minor changes. The
marked increase of the line width for X = X' = Cl might be due
to an equilibrium of the monomer with a small amount of the dimer.

     More significant evidence for the presence of a
distinct complex, or class of complexes, is often obtained from
the characteristic temperature dependence of ΔH^{iso}. Thus the
majority of $(C_5H_5)_3UX$-systems give rise to rather similar

Table 3. Methyl proton shifts of various $M[C_5(CH_3)_5]_2XX'$-systems

X, X'	$\delta(Th)$	$\delta(U)$	$\Delta H^{iso}(U)$
Cl_2	1,98	13,16[+]	$-$ 11,18
Cl, CH_3	2,01	8,96	$-$ 6,95
$(CH_3)_2$	1,92	5,03	$-$ 3,11
$C_4(C_6H_5)_4$	2,31	6,02	$-$ 3,71
H_2	1,98	$-$ 2,15	$+$ 4,13

[+] line width 190 Hz, otherwise 4 to 7 Hz

$(\Delta H^{iso})^{-1}$-vs-T-curves when T is varied between ca. 200 and 340 K. While the curves of the three halide complexes with X = Cl, Br and I are in fact of extremely similar shape, the corresponding fluoride system shows a markedly different temperature dependence which has been explained by pronounced association in non-polar solvents. This view matches nicely with a drastic increase of the line width below 230 K (17), and is supported by molecular weight studies in benzene solution(16).

In the presence of tetrahydrofuran (THF) or other Lewis bases (such as pyridine etc.), however, the formation of an adduct with the respective Lewis base is favoured. The new $(\Delta H^{iso})^{-1}$-vs-T-curves tend to be S-shaped suggesting here a rapid dynamic equilibrium of the homo- and the hetero-associated forms. Below ca. 250 K, the latter species is expected to dominate.

The unusual nmr and optical spectra of $(C_5H_5)_3UF$ (16) have initiated a single crystal X-ray study which has revealed that even in the solid state some weak self-association is accomplished. Thus, unlike for $(C_5H_5)_3UCl$ and other $(C_5H_5)_3UX$-systems, distinct linear ··U-F··U-F·· chains with strongly alternating UF-distances are formed(18). This observation has, in turn, stimulated further studies on the capability of other $(C_5H_5)_3UX$-systems to act as Lewis-acids. There is, meanwhile, no doubt that a large number of complexes of the general type $\{(C_5H_5)_3UXY\}^{-n}$ or $\{(C_5H_5)_3UX_2\}^{-n}$ exists, the most prominent ones being the adduct $(C_5H_5)_3UNCS \cdot CH_3CN$(for which a strict trans-position of NCS and CH_3CN has recently been confirmed by an X-ray study(19)),the oligomeric system $[(C_5H_5)_3UC(CN)_3]_\infty$ (in which each $(C_5H_5)_3U$-moiety most probably coordinates two CN-groups of different $C(CN)_3$-ligands(20) and the aquo-cation $[(C_5H_5)_3U(H_2O)_2]^+$ which was observed as early as in 1956(21).

All the complex systems that have so far been suspected to contain two trans-coordinated ligands in excess of the three η^5-C_5H_5 rings exhibit not only optical spectra different from those of trigonal-pyramidal $(C_5H_5)_3UX$-systems, but likewise very characteristic ΔH^{iso}-vs-T-curves. Thus, one soluble precursor of the completely insoluble $[(C_5H_5)_3UC(CN)_3]_\infty$ which has been assumed to involve the anion $[(C_5H_5)_3U(C(CN)_3)_2]^-$ displays surprisingly small ring proton ΔH^{iso}-values which even tend to give rise to an inverse temperature dependence(20). Very similar observations have been made on the aquo complex(21,22) and on other sufficiently soluble species, but in many instances measurements over wide temperature ranges were hampered by unfavourable properties of the solvent. The observation of rather low ΔH^{iso}-values, and their small temperature dependence, are in good accordance with the results of static magnetic suscepti-bility measurements on some $(C_5H_5)_3UXY$-systems(23), where χ_{exp} usually is found to be lower than for genuine $(C_5H_5)_3UX$ complexes.

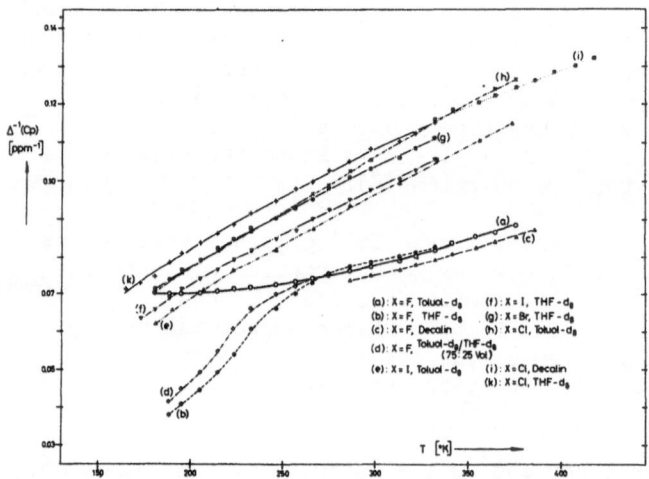

Fig. 2. $(\Delta H^{iso})^{-1}$-vs-T-diagrams of $(C_5H_5)_3U^{IV}$-halide systems in solution(16)

An interesting example of the direct monitoring of chemical reactions by means of nmr (and optical spectroscopy) is the attempted reduction of the complex $(C_5H_5)_3UCl$ by means of $LiBH(C_2H_5)_3$ and $LiAlH_4$, respectively(24). From the successive disappearance of signals, and the appearance of new ones at different positions, it was deduced that in both cases the same intermediate U(IV) complex was formed. While in the first experi-ment this intermediate is, in THF, transformed into the well-identified complex $(C_5H_5)_3UOC_2H_5(25)$, genuine reduction appears to finally take place only in the second experiment, yielding

presumably $(C_5H_5)_3U(III)\cdot THF(26)$. While further confirmation on
a preparative basis is still lacking, the common intermediate
was suspected to be the hydride system $(C_5H_5)_3UH$.

In situ pmr studies have also been carried out on complexes of the type $(C_5H_5)_3UR$ (R = alkyl). It was claimed that
exchange of the alkyl group is possible in the presence of
aluminium alkyls, $AlR'_3(27)$. Based on this observation, a very
unstable intermediate of the type $(C_5H_5)_3URR'AlR'_2$ with two
different bridging alkyl groups was postulated.

From some characteristic and unexpected changes in the
pmr-spectra of the paramagnetic Ce(III)-complex
$C_8H_8Ce(O-i-C_3H_7)_2Al(C_2H_5)_2\cdot CH_3CN$ with temperature (-15 to 105°C),
and in presence of a strong Lewis base B, it was suggested that
structural rearrangements owing to the following equilibria might
take place(28):

Crease and Legzdins have studied the interaction of various
organometallic Lewis bases with the organometallic Lewis acids
Cp_3Ln (Cp = C_5H_5 or $C_5H_4CH_3$)both by infrared and pmr spectroscopy
(29a). While, occasionally, rather strong changes in the
vibrational spectra were observed, the reported pmr shifts induced by $(C_5H_5CH_3)_3Nd$ were, however, in no instance larger than
ca. 2 ppm. The stronger adducts are apparently not sufficiently
soluble for nmr experiments. Considerable line shifts (and
broadening effects) have been obtained with some soft bases as
phosphides and sulfides in the presence of the "shift reagent"
$(C_5H_5)_3Yb$ (29b).

Nmr spectroscopy has also provided a valuable piece of
information to support the formation of the thermally very
unstable tetraalkyluranium(IV)-systems, UR_4 (30). Thus solutions
suspected to involve the species "$U(CH_3)_4$" exhibited at -30°C a
broad high-field proton resonance ($\tau \sim 32,7$, $lw_{1/2} \sim 50$ Hz) which
must be attributed to a paramagnetic species that carries methyl
groups and is soluble in toluene.

Nmr spectroscopy has also proved very useful for the

determination of some important structural details of the two
tetra(allyl)uranium complexes $(C_3H_5)_4U$ (31) and $(C_3H_4CH_3)_4U$ (32a)
which are only stable below −20°C. The role of the paramagnetic
uranium ion as an "inherent shift reagent" converts the otherwise
more complicated pmr spectra into simple spectra of first order
which unequivocally reveal that all four allyl groups are equivalent and trihapto-bonded(table 4). The surprisingly large
separation of the resonances of the syn- and anti-protons
precludes any kind of ligand mobility except rotations about
axes between the central metal and a point close to the centre
of gravity of the C_3-skeleton.

Table 4. Isotropic shifts of protons on the allyl ligand in
$U(\eta^3-C_3H_5)_4$, $U(\eta^3-2-CH_3C_3H_4)_4$ (32a) and
$[(\eta^3-C_3H_5)_2U(O-i-C_3H_7)_2]_2$ (32c)

H_{syn}	H_{anti}	$H_{methine}$ /	CH_3	Temp.
18,72	53,8	28,74		(248 K)
J_{vic} = 8 Hz	J_{vic} = 14 Hz	(unres.)		
18,24	53,28		18,08	(241 K)
131	146	70		(∼300 K)

Similar pmr spectroscopic features are also displayed
by some lanthanide systems of the type $LiLn(C_3H_5)_4$ (Ln = Ce, Nd,
Sm, Gd and Dy) which involve four virtually equivalent $\eta^3-C_3H_5$
ligands at least below room temperature(32b), and by the complexes
$[(\eta^3-C_3H_5)_2U(OR)_2]_2$ as long as a cleavage into monomeric adducts
with basic solvent molecules is avoided(32c).

While it has not been possible to slow down the rotation
of an unsubstituted $\eta^5-C_5H_5$ ligand such that the ring proton
singlet would be split into several components, the amplification
effect of the paramagnetic $(C_5H_5)_3U^{IV}$-moiety has been sufficient
to successfully study the dynamics of the pseudorotation of the
BH_4 ligand bonded in $(C_5H_5)_3UBH_4$(33,34). Under the conditions of
broad band decoupling of ^{11}B(35), the BH_4 pmr singlet was found
to completely collapse at −120°C, and by a very careful argu-
mentation Marks et al. have deduced that the most reasonable
explanation should be a slow-down of the otherwise very rapid
chemical exchange of BH_4-protons at different sites. The vibrational spectra indicate a η^3-coordination of the BH_4 ligand(34,
36), giving at any moment rise to one terminal and three bridging
H-atoms, so that in the static state two distinct singlets with
the relative intensities 3:1 should appear. The low solubility

and great reactivity of $(C_5H_5)_3UBH_4$ have, however, prevented successful nmr studies at lower temperatures than $-121°C$, so that the situation of very slow exchange could not be reached as yet.

No spectral collapse is, over the same temperature range, observed for the modified complexes $(C_5H_5)_3UH_3R$ which are hence no longer fluxional (as far as the borohydride ligand is concerned), the position of the terminal BH_4-proton being now occupied by an organic group. Under rapid exchange (e.g. at 28°C), the calculated ΔH^{iso}-values of the BH_4 ligand are ca. 69 ppm (bridging H-atoms) and 44 ppm (terminal H-atom), corresponding to a virtual line separation of ca. 25 ppm or 2250 Hz. This distance increases considerably with decreasing temperature, and was extrapolated to lie between 15000 and 36000 Hz at the estimated coalescence temperature, $-140±20°C$ (34). From these data, an activation barrier for the hydride exchange of $5,0±0,6$ kcal/mol was obtained. The lowest barrier so far evaluated for a fluxional diamagnetic tetrahydroborate complex, $(C_5H_5)_2VBH_4$, amounts to $7,6±0,3$ kcal/mol(37), corresponding to a line separation of ca. 2500 Hz and $T_c = -87,7°C$.

While it has been demonstrated that the mechanism of hydrogen exchange in the complexes $(C_5H_5)_2M(BH_4)_2$ (M = Zr, Hf) also involves a scrambling of BH_4- and C_5H_5-hydrogen atoms(38), details of the exchange process in $(C_5H_5)_3MBH_4$ (M = Th, U) are still lacking. It is known, however, that the BH_4-pmr spectra of the methylated system $(C_5H_4CH_3)_3UBH_4$ is, between +30 and -90°C, almost identical to the spectra of the parent compounds (39). No evidence for an exchange of ring and/or methyl proton is provided by the $C_5H_4CH_3$ proton spectra up to +80°C.

The rather new complex $(C_5H_5)_2U(BH_4)_2$ (13) has not yet been subjected to variable-temperature nmr studies. In view of the lack of axial symmetry, also the bridging BH_4 H-atoms should have different geometric factors, a circumstance which might introduce further interesting facets into the spectral behaviour.

The fluxional allyl complexes $(C_5H_5)_3MC_3H_5$ (M = U or Th) exemplify systems for which the essential structural and dynamic features can be derived from the variable-temperature pmr spectra both in case of para- and diamagnetism(40,41). Rather unexpectedly, the allyl group turns out to be η^1-bonded in both complexes, the pmr spectra of the C_3H_5-group being typical for a dynamic A_4X species at room temperature, but for a static A_2BCD system at fairly low temperatures (-43°C for M = U, and below ca. -100°C for M = Th).

For the paramagnetic uranium complex, the A_4X-pattern is only approached in that the four A-protons give rise to one broad singlet with ΔH^{iso} = ca. +126 ppm(30). However, the X-proton which

lies not so close to the paramagnetic centre undergoes much less
line broadening and exhibits as well-resolved quintet at
+38,2 ppm (relative to internal benzene). In the low-temperature
pmr spectrum of the allyl group all proton coupling is apparently
quenched. While the signal with highest ΔH^{iso} (-90°C : ca.
+344 ppm(!), rel. int. 2) is easily assigned to the two equivalent
α-protons, an unambiguous assignment of the three other, fairly
sharp, singlets (-90°C, ΔH^{iso} = +60, +57,5 and -41,0 ppm; rel.
int. 1) is less straightforward.

The diamagnetic thorium complex exhibits a high-
temperature spectrum completely consistent with the expectation
for other diamagnetic systems with corresponding fluxionality
involving pronounced multiplet splitting. Spectra of the non-
fluxional σ-allyl complex were, however, not obtained ($T_c \simeq$
-100°C). It is very interesting that within experimental error
both complexes have identical activation energies for the sigma-
tropic rearrangement, i.e. ca. 8 kcal/mol(40,41). If steric
congestion about the central metal ion were the main reason for
the preference of the η^1-coordination of the allyl ligand, $\Delta G^{\ddagger}$
(Th) should be slightly smaller than $\Delta G^{\ddagger}$(U).

The data of table 5 illustrate the influence of the large
line separations $\Delta \nu$ on the coalescence temperature, T_c.

Table 5. Characteristic data of some fluxional tetrahydroborate
         and allyl complexes

Complex	T_c/K	$\Delta \nu$/kHz	$\Delta G^{\ddagger}$/kJmol^{-1}
$(C_5H_5)_3UBH_4$	133±20	15–36	20,9±2,5
$(C_5H_5)_2VBH_4$	182,2	ca. 2,5	31,8±1,3
$(C_5H_5)_3ThBH_4$	< 120	≤ 0,5	ca. 20(assumed)
$(C_5H_5)_3UC_3H_5$	230	25±3	34,9*
$(C_5H_5)_3ThC_3H_5$	170±10	0,11±0,02	33,0**

*   recalculated on the basis of the published data and eq.9
** estimated on the basis of eq.9

Studies Involving the Separation of ΔH^{con} and ΔH^{dip}

Obviously, more sophisticated nmr studies of structural
details, and, in particular, insights into the bonding, will be
possible as soon as ΔH^{dip} has been separated from ΔH^{con}. Owing
to the very simple form of eq.2, highly preferred objects of such
studies are axially symmetric.

Fortunately, an axially symmetric crystal field (CF), and thus the situation:

$$\chi_{xx} = \chi_{yy} = \chi_{\perp} \; ; \; \chi_{zz} = \chi_{\parallel}$$

can be expected for most $(\eta^5-C_5H_5)_3M$-systems. A rather simple check may be a look at the temperature dependence of the fractional spectrum of the C_5H_5-ring protons. In the majority of instances, only one singlet appears down to $-80°C$ (the lower limit of solvents like toluene-d_8 and THF-d_8). Evidently, the three C_5H_5 ligands are rotating very rapidly about their five-fold axes, and at least during the period of a single uptake of $h\nu$ (ν being the radio frequency of nmr spectrometer) the three C_5H_5 ligands may be considered equal.

Marks et al. have found one interesting case where steric hindrance precludes the three C_5H_5 ligands from being completely equal at temperatures below ca. 190 K(40). Thus, the complex $(C_5H_5)_3UCH(CH_3)_2$ exhibits only one ring proton resonance above 190 K, but two distinct lines with relative intensities 1:1, and line separation of less than 2 ppm, below 180 K. Throughout the temperature range considered, the lineshape of the isopropyl proton lines remains unchanged, and no significant influence of the solvent was observed. The interpretation of the phenomenon is straightforward in terms of appropriate Newman projections (fig. 3) indicating that at sufficiently low temperatures the rotation about the U-C-σ bond must be restricted, while the rotation of the C_5H_5-rings about their five-fold axes (as well as the rotations of the isopropyl CH_3 groups) remain sufficiently

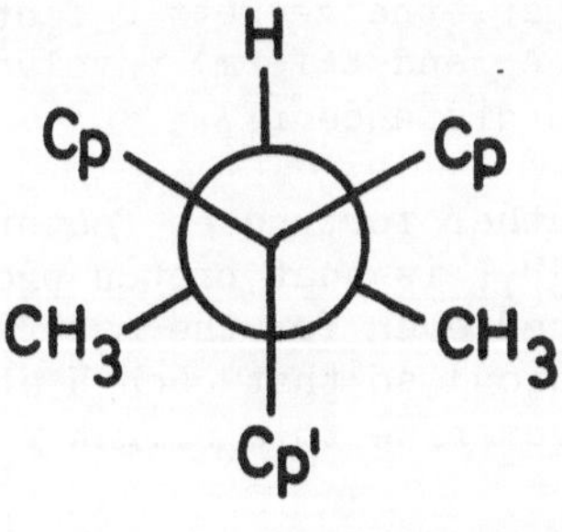

Fig. 3.

Newman projection of one of the three equivalent, staggered conformations of the complex $(C_5H_5)_3UCH(CH_3)_3$ (observed along the U-C bond)

rapid. For the restricted rotation, an activation barrier of $10,5\pm0,5$ kcalmol^{-1} was estimated(40).

The most trivial way to separate ΔH^{dip} from ΔH^{con} is either to study nuclei too far away from the paramagnetic centre to carry any spin density themselves (i.e. $\Delta H^{iso} = \Delta H^{dip}$), or to choose magnetically isotropic molecules where the observed ΔH^{iso} must be identical to ΔH^{con}. The former way is the most common one applied in the shift reagent technique(4), but likewise in organometallic chemistry. As a rule of thumb, in lanthanide complexes the attenuation of ΔH^{con} is practically complete when the nucleus in question is separated by at least two (σ-)bonds from the magnetic centre, while for actinide complexes, even for considerably more remote atoms, individual considerations appear advisable. Magnetically isotropic organometallics are, on the other hand, very rare. One example is the complex $(\eta^5-C_5H_5)_4U$ (39) (vide infra).

Adducts of the type $(\eta^5-C_5H_5)_3Ln(III)B$, where B is an uncharged Lewis base molecule, have so far been the best candidates for systematic studies mostly of the coordinated ligand B, for the following reasons: (a) The entire series of lanthanide complexes with Ln = La-Lu can be, and has been, prepared with the only exception of Pm. (b) All complexes of the type $(C_5H_5)_3LnB$ have the same molecular structure with strict axial symmetry. (c) With Lewis base molecules B which themselves have a good solubility in non-polar organic solvents, sufficiently high concentrations of the organometallic adduct, too, can be obtained.

The best studied ligand B so far has been cyclohexylisocyanide, CNC_6H_{11}, which after coordination to a $(C_5H_5)_3Ln$-moiety, can be considered as just another monosubstituted cyclohexane. It was shown in a series of studies(43,45) that the interconversion of the equatorially substituted cyclohexane (E-form) into the axially substituted conformer (A-form) which is rapid at room temperature, is sufficiently slowed down at -70°C so that then the individual pmr spectra of the A- and E-form are readily observable. For all cyclohexyl protons, except for the α-proton, the assumption $\Delta H^{iso} = \Delta H^{dip}$ will be fulfilled, and as long as the LnCNC-arrangement remains linear, the geometric factor G_i of none of the C_6H_{11}-protons (of the A- and E-form) involves any averaging over the angle θ_i or the distance r_i.

A very surprising, and rather fortunate, feature of all paramagnetic adducts $(C_5H_5)_3LnCNC_6H_{11}$ is that proton-proton coupling remains completely quenched even for the most remote H-atoms (i.e. those in the δ-position) so that each H-atom at a distinct spatial position only gives rise to a slightly broadened singlet (cf. fig. 4).

The best complex to study in great detail is the Pr-system(43) as here the signals turn out to be sharpest while the various singlets are nevertheless spread over a fairly wide spectral range. Accounting for the fact that either conformer (A and E) has, apart from one α-proton, two equatorial (e) and axial (a) protons, respectively, in the β- and γ-positions, and just one δe- and δa-proton, a minimum of 12, and a maximum of 14, lines is expected. From the appearance of separate lines even for the βAe- and βEe-protons it has been deduced that somewhere

in the atomic chain -LnCNC- slight deviations from linearity
occur. This conclusion was later confirmed by an X-ray study
on single crystals of $(C_5H_5)_3PrCNC_6H_{11}$(46) where a PrCN-angle
of 174,1° was found.

Fig. 5 demonstrates that at -70°C the ΔH^{iso}-values of
all C_6H_{11}-protons of the complexes with Ln = Ce, Pr and Nd are

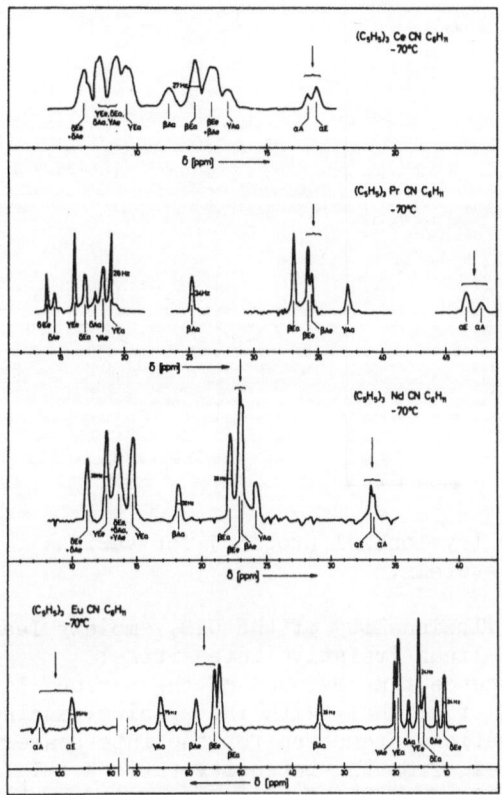

Fig. 4. Low-temperature pmr spectra of the complexes
$(C_5H_5)_3LnCNC_6H_{11}$ (Ln = Ce, Pr, Nd, Eu; region
of cyclohexyl protons).

practically identical with the corresponding ΔH^{dip}-values. For
Ln = Eu, however, where the largest shifts of all systems with
Ln = Ce-Gd are observed, at least the isotropic shifts of the
α-protons appear to involve non-negligible ΔH^{con} contributions.
The appearance of one Aα- and one Eα-signal throughout the series
seems to indicate a different spin transfer in the two conformers.
Furthermore, the separation $\Delta(\Delta H^{iso}_{A\alpha} - \Delta H^{iso}_{E\alpha})$ tends to increase
with increasing deviations of the respective isotropic shifts

from the straight line in fig. 5, giving rise to the sequence:
Sm < Ce < Nd < Pr < Eu.

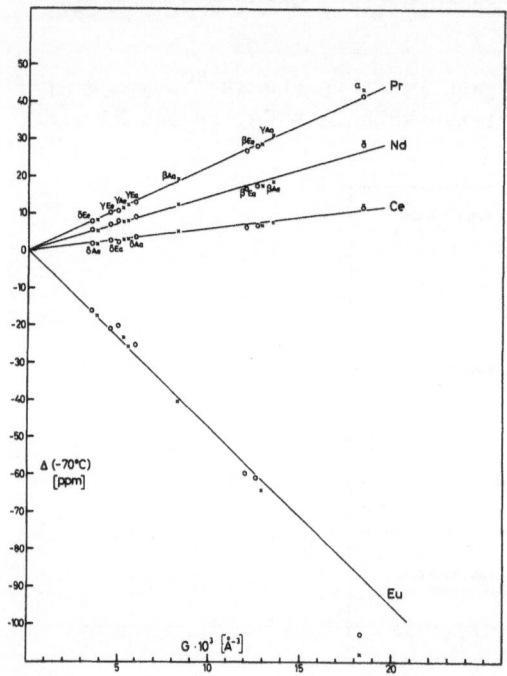

Fig. 5. ΔH_i^{iso}-vs-G_i-plots (cyclohexyl protons) for various $(C_5H_5)_3LnCNC_6H_{11}$-systems.

       Above $-50^\circ C$, the fluxionality of the C_6H_{11}-moiety leads
to the appearance of but 7 lines (relative intensities:
1:2:2:2:2:1:1). The coalescence temperatures of the various low-
temperature line pairs differ slightly with the spacing, and no
attempt to determine activation parameters for the interconversion
$A \rightleftharpoons E$ has been made. However, from the intensity ratio $K = I_E/I_A$
of corresponding pairs of lines of the A- and E-conformers at
$-70^\circ C$, the differential free energy $\Delta G = RTlnK$ was determined in
a number of cases(43,44,47). It is seen from the data in table 6
that the conversion of the free ligand CNC_6H_{11} into the adduct
$(C_5H_5)_3LnCNC_6H_{11}$ is always accompanied by a decrease of ΔG
indicating that the A-form becomes less stable relative to the
E-form. This result matches quite well with conclusions to be
drawn from a steric model of the adduct (fig. 6) as at least one
ring H-atom of the rapidly rotating C_5H_5-ligands can interfere
with the two γAa-protons. Surprisingly, ΔG is even slightly more
negative for adducts with paramagnetic, than with diamagnetic,
Ln-ions(44). When one H-atom of each ring ligand is replaced by
a methyl group, ΔG decreases again significantly(47). This observation supports the above suggestion of some steric interaction

Table 6. $\Delta G(A\text{-}E)$ values (in $Jmol^{-1}$, mean error ≤ 100 $Jmol^{-1}$)
of CNC_6H_{11} and various $Cp_3LnCNC_6H_{11}$-systems (43,44,47).

Cp = $\eta^5\text{-}C_5H_5$		Cp = $\eta^5\text{-}CH_3C_5H_4$
free CNC_6H_{11}	879	
Y	837	
La	975	
Ce	1101	1318
Pr	1197	1670
Nd	1164	1549
Eu	1247	
Lu	1038	

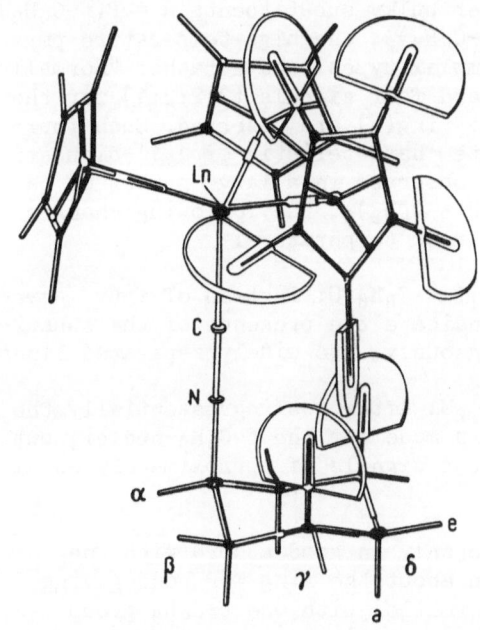

Fig. 6. Spatial arrangement of ligand atoms in the A-
conformation of $(C_5H_5)_3LnCNC_6H_{11}$-complexes.

between the cyclopentadienyl ligands and the γAa-protons and
raises the question if after partial ring methylation also the
ring rotation at least of the A-conformer could be frozen out.
Experimental evidence is so far only provided by the low-temperature spectra of $(CH_3C_5H_4)_3PrCNC_6H_{11}$ (47).

As soon as the atom(s) linking a $(C_5H_5)_3M$-moiety and an organic residue R prefer a non-linear arrangement MXY··R, the determination of reliable geometric factors of nuclei within R becomes difficult. A typical situation is met in the alkoxides $(C_5H_5)_3UOR(48,49)$. In one exceptional case, namely where R is the very rigid and bulky cholesteryl group(49), a correlation of at least ten ΔH^{iso}-values of R with their geometric factors was achieved. The best numerical fit affords the assumption of a non-linear UOC-linkage. There is no evidence for rotation about the OC-bond, and no low-temperature splitting of the C_5H_5 proton signal. However, rapid oxygen inversion cannot be fully excluded. An interesting phenomenon is that with R = cyclohexyl some signals of the C_6H_{11}-protons are detectably broadened over a limited temperature range(49).

Structural considerations on uranocene derivatives of the types $(\eta^8-C_8H_7R)_2U$ and $(\eta^8-C_8H_7R)(\eta^8-C_8H_8)U$ have recently been challenged in case of the rather bulky substituents R = $P(t-C_4H_9)_2$ and $P(C_6H_5)_2(50)$. With R = $P(t-C_4H_9)_2$, the high-temperature pmr spectra (T $\geq$ 340 K) of both complex types behave rather "normally" in that just one signal for R and four signals (2:2:2:1) for the substituted cyclooctatetraenide ring(s) are observed. Such four-line spectra are otherwise quite characteristic of 1,1'-disubstituted uranocenes and have been observed in a large number of cases before(51). However, when R = $P(t-C_4H_9)_2$, the following changes occur in the pmr spectra below room temperature:

(a) Complex $\left[\eta^8-C_8H_7P(t-C_4H_9)_2\right](\eta^8-C_8H_8)U$: Instead of four, seven lines of equal intensity indicate the presence of the substituted ring, while, simultaneously, two widely separated lines of the $t-C_4H_9$-protons occur.
(b) Complex $\left[\eta^8-C_8H_7P(t-C_4H_9)_2\right]_2U$: After cooling essentially the same observation as above is made for the $t-C_4H_9$-nuclei, but now a total of 14 ring proton signals of approximately equal intensity is found.

These observations are only in good accord with the assumption of hindered rotation about the ring C-P bond giving rise to a low-temperature conformation with one $t-C_4H_9$ group somewhere above the ring plane (exo position) and one $t-C_4H_9$ group somewhere below this plane (endo position). Any conformation of this type would preclude the existence of a mirror plane passing simultaneously through the ring carbon atoms C_1 and C_5, the U-atom and the P-atom, and must in fact give rise to the observation of seven separate ring proton signals.

The only aspect where the consideration of ΔH^{dip}, and not of ΔH^{iso}, becomes essential, is dealing with the observation of different signs for the two $t-C_4H_9$ proton signals indicating that if $\Delta H^{iso} = \Delta H^{dip}$ the two geometric factors, G_{endo} and G_{exo},

should also be different in sign. Simple calculations based upon
reasonable structural data have in fact confirmed that the average
value of G_{exo} always corresponds to a positive $(3\cos^2\theta-1)$ while
the reverse is true for G_{endo}. A corresponding relationship holds
for the methyl carbon atoms of the two $t-C_4H_9$ groups which matches
nicely the observed sign difference of the two corresponding
low-temperature ^{13}C nmr signals of the complex $[C_8H_7p(t-C_4H_9)_2]_2U$.

    A rather interesting observation is that, with decreasing
temperature, the line width of the signal of the endo protons increases much more rapidly than the width of the exo-signal, so
that below $-80°C$ the former signal becomes too broad to be detected while the latter still has a width of ca. 180 Hz (fig.7).
In view of the larger steric crowding around the endo-$t-C_4H_9$
group, one might argue that the rotation about the P-C(butyl)
bond in endo position will be frozen out at a higher temperature
than in exo position. However, the two corresponding ^{13}C nmr
signals of the methyl C-atoms exhibit no different line broadening
whatsoever. Thus, as an alternative explanation for the behaviour
of the proton lines, one might speculate about a kind of direct
transfer of free spin density ("through space") from the uranium
atom onto the closest-lying methyl proton(s) of the endo $t-C_4H_9$
groups(50).

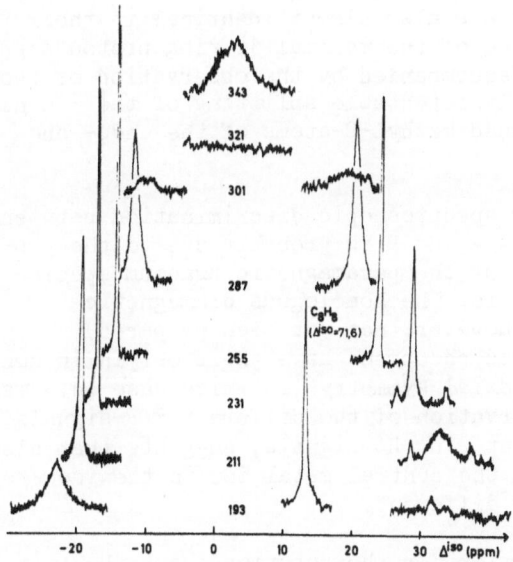

Fig. 7. Variation of lineshape and position of the $t-C_4H_9$ pmr
        signals in $C_8H_7P(t-C_4H_9)_2(C_8H_8)U$ with temperature.

    Immediate consequences of the restricted rotation about
the (ring-)C-P bond are for the complex of type (b) that here,

in view of the rather high barrier to inversion of the tertiary
phosphine, two different isomers A and B must be expected(fig.8).

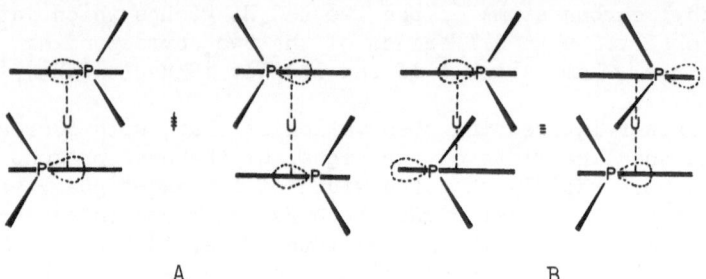

                    A   B

Fig. 8. Schematic description of the two low-temperature isomers
        of the complex $|C_8H_7P(t-C_4H_9)_2|_2U$.

The interconversion A $\rightleftharpoons$ C is only determined by the activation
barrier of the rotation about the C-P bond, regardless of whether
in addition, the mutual rotation of the two eight-membered rings
is also restricted. Thus, out of the 14 ring proton signals observed in case (b), one half must be ascribed to isomer A and the
other half to isomer B. The almost equal intensities of all lines
indicate that the two forms are also almost identical in their
free energies. The occurrence of the maximal 14 ring proton signals
in this particular case is accompanied by the observation of two
^{31}P-nmr signals and an easily detectable splitting of the ^{13}C nmr
signals of the above-mentioned methyl-C-atoms of the endo- and
exo-t-C_4H_9 groups.

        The successful nmr spectroscopic discrimination between
the two very similar isomers A and B is probably due to the pronounced magnetic anisotropy of the paramagnetic uranium species
and will hardly be possible for the homologous diamagnetic
thorium system (which has, however, not yet been prepared).
Although some of the observed phenomena have their origin in non-
negligible deviations from axial symmetry (in which case eq.3 is
replaced by eq.4), the observation of two different ^{31}P-signals,
and likewise of two different ring-Hδ-signals, suggests that also
the crystal field acting on the central metal ion in the isomers
A and B has become slightly different.

        The activation barrier for the rotation about the (ring)
C-P bond has been estimated to amount in case (a) and (b) to
56 - 59 kJmol^{-1}(50). It is noteworthy that adopting the approxi-
mation(52):

$$\Delta G^{\ddagger}_c = 0,01913\ T_c\ (9,972 + \log T_c - \log\Delta\nu_o) \qquad (9)$$

very similar $\Delta G^{\ddagger}$-values are obtained even if the coalescence of quite different pairs of nuclei is considered(table 7).

Table 7. Determination of $\Delta G_c^{\ddagger}$ of the rotation about the (ring) P-C bond from different coalescence phenomena of the complex (C_8H_8) $C_8H_7P(t-C_4H_9)_2U$.

nucleus	T_c/K	$\Delta\nu/Hz$	$\Delta G_{T_c}^{\ddagger}/kJmol^{-1}$
$C(\underline{C}H_3)_3$	313 ± 10	1165 ± 50	56,3 ± 2
$C(\underline{C}H_3)_3$	328 ± 5	1490 ± 50	58,4 ± 1
$\underline{P}(t-C_4H_9)_2$	278 ± 8	52 ± 4	57,4 ± 1,5
	303 ± 9	252 ± 12	58,3 ± 2
$C_8\underline{H}_7$	313 ± 10	360 ± 25	59,2 ± 2

While PR_2-substituted uranocenes with $R = C_6H_5$ apparently give rise to such high rotational barriers about the (ring-)C-P (and likewise the P-C(phenyl) bonds that even above room temperature all attempts to interpret the corresponding pmr spectra (both in the C_8H_7- and in the C_6H_5-region) have been unsuccessful(50), the transition to $R = C_2H_5$ gives rise to a system (of type (a)) where down to -80°C, the rotation about all σ-bonds remains rapid. A comparatively simple pmr spectrum (exhibiting but four lines for the substituted ring, and one pseudoquintet for the methyl protons (J_{PH} = 11 Hz has, quite fortuitously, twice the value of J_{HH} = 5,5 Hz) is observed. One peculiarity of this system is, however, that the methylene protons give rise to two spin-quartets (relative intensity: 1:2:2:1) with an unexpectedly large separation (J_{PH} is expected to be no larger than ca. 1,5 Hz)which is, moreover, markedly temperature dependent. The most satisfactory explanation of this observation is based upon the diastereotopic nature of the two H-atoms in each methylene group, as may be visualized from appropriate Newman projections(fig.9).

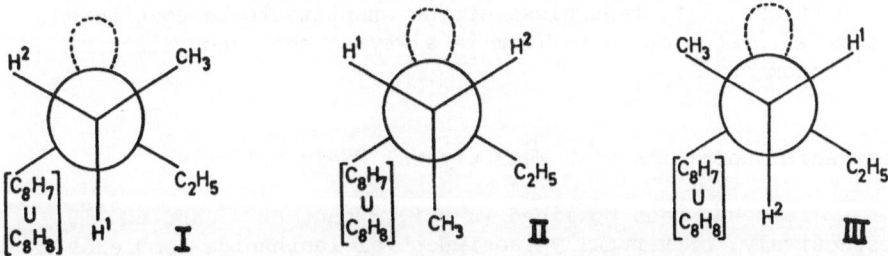

Fig. 9. Different staggered conformations of $(C_8H_8)C_8H_7P(C_2H_5)_2U$ with respect to the (methylene)C-P bond.

While numerous cases involving pairs of diastereotopic
nuclei in diamagnetic systems are known(53), so far only one
other example has been reported where the diastereotopic splitting
is substantially magnified by the strong inherent paramagnetism
of the probe(54). It has been shown that from the specific temperature dependence of the two corresponding ΔH^{iso}-values some
conclusions about the relative statistic weights of competing
conformations can be arrived at(54). With the same arguments and
assuming in our case also that $\Delta H^{iso} = \Delta H^{dip}$, the relative weights
of the three possible staggered conformations(fig.9) should
correspond to I >> III > II. The same qualitative sequence results
from simple arguing in view of the different spatial demands of
the groups $C_8H_8UC_8H_7$, C_2H_5, CH_3, the free electron pair of the
P-atom and H, respectively. As, owing to the rapid mutual rotation
of the two eight membered ring ligands, axial symmetry might be
anticipated for the calculation of the average geometric factors
of the diastereotopic nuclei in different conformations, even more
quantitative conformational studies might become accessible with
such paramagnetic organometallic samples.

CONSIDERATIONS OF THE ELECTRONIC STRUCTURE

General

Different aspects related to the electronic structure of
a compound can in principle be studied either via the dipolar term,
ΔH^{dip}, or via the so-called contact term, ΔH^{con}. While from ΔH^{dip}
(of axially, or at least pseudo-axially, symmetric systems) the
magnetic anisotropy $(\chi_\parallel - \chi_\perp)$ and its temperature dependence can
be evaluated, ΔH^{con} provides a measure of the net density of free
electron spins that has been generated on a particular nucleus
in question. It appears that at least for complexes of the f-
elements the knowledge of $(\chi_\parallel - \chi_\perp)$ offers a more immediate clue
for correlations with other data related to the electronic struc-
ture (i.e. first of all, the crystal field splitting pattern) of
a complex. Almost any closer interpretation of ΔH^{con} requires,
on the other hand, the aid of rather sophisticated quantum chemical
calculations. Yet, regardless of the quantity to be considered,
the very first step to perform is always a neat separation of
ΔH^{dip} from ΔH^{con}.

Lanthanide Complexes with Weak Crystal Field Splitting

It has been outlined in ref. 3 that ΔH^{con} and ΔH^{dip},
respectively, of virtually isostructural lanthanide complexes
whose lowest-lying J-manifolds are split by ≤ 200 cm^{-1} depend
essentially on the electronic ground configuration $4f^q$. Thus,
ΔH^{con} is proportional to the expection value of the free spin

density on the central metal ion:

$$<S_z>_{J,av} = - g_J(g_J-1)J(J+1)\beta H/kT \qquad (10)$$

ΔH^{iso}-values whose variation with q follows in fact very closely
that of the quantity $-g_J(g_J-1)J(J+1)$ (where g_J is the Landé-factor
of the respective ground manifold) have been observed for the ^{17}O-
resonances of the aquo complexes of all Ln^{3+}-ions (except Pm^{3+})
present in ^{17}O-enriched aqueous solution(55). While in case of
predominant ΔH^{con} the sign of ΔH^{iso} changes abruptly (but only
once) between q = 5 and 6, the pure dipolar term ΔH^{dip} varies
in quite a different way in that between q = 1 and 13 three sign
reversals occur (fig.10).

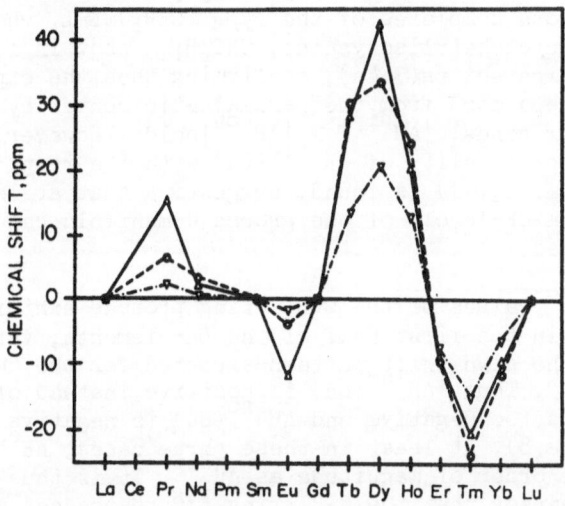

Fig. 10. Variation of ΔH^{iso} of various typical lanthanide shift
reagent-substrate adducts (4) with the ionic configu-
ration, f^q (cf. refs. 4a-b, 57).

Bleaney has shown first(56) that, in case of full thermal
population of the respective lowest J-manifold, ΔH^{dip} of corresponding nuclei in isostructural f^q-systems should be proportional
to the product

$$A_2^{0} \cdot <r^2>g_J^2 J(J+1)(2J-1)(2J+3) \cdot <J|\alpha|J> \qquad (11)$$

It was demonstrated subsequently that actually a large number of
lanthanide shift reagent-substrate adducts nicely fulfill this
condition as long as nuclei not adjacent to the paramagnetic center
are considered(57). It is noteworthy that inspection of the rather
large amount of experimental material available suggests that
Bleaney's correlation might also serve as a useful criterion for

the retention of one particular geometric structure throughout
a series of homologous f^q-systems. Conversely, predictions of the
approximate ΔH^{iso}-values for missing links within a series of
homologues might be possible.

Returning to purely organometallic lanthanide complexes,
it may be recalled that presently only the complex series
$(C_5H_5)_3LnL$ (L being a neutral Lewis base), $M[(C_8H_8)_2Ln]$ (58),
$C_8H_8LnC_5H_5(60)$ and probably also $Ln[P(CH_2)_2(CH_3)_2]_3$ (78) cover
all Ln elements while, e.g. the series $(C_5H_5)_2LnX$ is restricted
to the heavier lanthanides beyond Ln = Nd (59). Very recently,
a new promising series, $Li[Ln(C_3H_5)_4]$ has been found(32b).

Most ΔH^{iso} values of corresponding protons of the Lewis
base L in homologous complexes of the type $(C_5H_5)_3LnL$, where L =
most frequently, cyclohexylisocyanide, CNC_6H_{11}, show in fact the
predicted three sign shifts(3,4a), confirming thus the expectation
that for nuclei separated from the paramagnetic center by more
than three to four bonds, $|\Delta H^{dip}| > |\Delta H^{con}|$ holds. However, no
satisfactory proportionality of $H_i^{dip}(Ln)$ with the respective
Bleaney-factor (cf. eq.11) is found, suggesting that at least in
some cases not all CF-levels of the ground J-manifold can be fully
thermally populated.

The ΔH^{iso} values of the C_5H_5 -ring protons exhibit,
particularly within the first half of the Ln-elements, various
deviations from the sign shift pattern expected for the "high
temperature limit". Thus, $\Delta H^{iso}(Nd)$ is positive instead of negative, $\Delta H^{iso}(Sm)$ is too negative and $\Delta H^{iso}(Eu)$ is negative instead
of positive (table 8). At least in these three cases, ΔH^{con} seems
to be of the same order of magnitude as ΔH^{dip}. It is thus not
unreasonable to expect that the C_5H_5 ring ^{13}C resonances will be
strongly dominated by the contact term so that in this case ΔH^{iso}
should display but one sign reversal (at Ln = Sm).

Table 8. Various ΔH-values (in ppm) of different protons in
$Cp_3LnCNC_6H_{11}$-systems (Cp = η^5-C_5H_5, temperature: -70°C)

Ln	Ce(44,47)	Pr(43,47)	Nd(44,47)	Eu(44,45)
$\Delta H^{iso}(Cp)$	-0,9	-16,7	+4,3	ca. -12
$\Delta H^{con}(Cp)$	0,75 to 0,94	-3,8 to -5,3	11,4 to 12,0	ca. -26
$\Delta H^{dip}(Cp)$	< 0	< 0	< 0	> 0
$\Delta H^{con}(H\alpha)$	3,4	1,25 to 2,5	2,4 to 2,6	ca. -20
$\Delta H^{iso}(\beta Ae)$	11,77	34,66	24,7	-70
$\approx \Delta H^{dip}(\beta Ae)$				

Simple Attempts to Separate ΔH^{dip} and ΔH^{con}

The simplest way to separate ΔH^{dip} from ΔH^{con} is, at least for axially symmetric molecules of known structure and sufficient rigidity, to pick out protons with precisely cal-culable geometric factors which lie too remote from the para-magnetic center to accomodate any substantial free spin density. The quantity $(\chi_{\parallel} - \chi_{\perp})$ which is then easily obtained by means of eq. 3 is subsequently used to calculate ΔH^{dip} for other nuclei whose positions are also sufficiently well known. This procedure has so far been successfully adopted for some lanthanide com-plexes of the type $(C_5H_5)_3LnL(45,47)$, the uranium(IV) systems $(C_5H_5)_3UX$ with X = alkoxide, OR, (49) and borohydrides, BH_3R (34), respectively, and two uranocene derivatives, $(C_8H_7R)_2U$, with R = $Si(CH_3)_3$ and $P(t-C_4H_9)_2$ (50).

In case of the lanthanide complexes $(C_5H_5)_3LnCNC_6H_{11}$ it has been confirmed that the α-protons of the cyclohexyl group are undoubtedly furnished with weak free spin density (table 8). This result suggests that spin density of a comparable order of magni-tude is also transferred to the equivalent atoms C_2 and C_6 of the C_6H_{11} ring and, thence probably, onto the β-protons, too. This result might, in turn, serve to explain the observation of diffe-rent ΔH^{iso} values for the α-protons in the A- and E-form (viz. fig. 11). While the ΔH^{con} values of the α-protons indicate (for Ln = Ce-Eu) only one sign reversal, the corresponding ΔH^{con} values of the C_5H_5 ligands reveal an exceptional behaviour of the Pr-complex where ΔH^{con} is not, as expected, positive(44). This fea-ture suggests, along with the surprisingly small (positive) ΔH^{con} values for Ln = Ce and Nd, that ΔH^{con} itself might be composed of two contributions with different signs.

While for most complexes systematic analyses of ΔH^{dip} and ΔH^{con} at variable temperature (and for the complexes with Ln = Tb-Yb, even room temperature data) are still lacking, a rather detailed study has been carried out in case of the europium com-plex $(C_5H_5)_3EuCNC_6H_{11}$ (45). Although the variable-temperature pmr spectra of the C_6H_{11} group suffer, as usual, from occasional line broadening, and coalescence, it has been possible to determine $(\chi_{\parallel} - \chi_{\perp})$, and thence ΔH^{con}, over the temperature range T = 183 - 363 K. While ΔH^{iso} of the C_5H_5 ring protons behaves highly unusually in that the almost linear ΔH-vs-T^{-1}-curve passes ΔH^{iso} = 0 at -55°C (fig. 12), the corresponding quasi-linear curve of ΔH^{con} exhibits a temperature dependence which is no longer in accordance with the Curie-Weiss law. In view of the rather weak slope, there is some reminiscence of a temperature independent ΔH^{con}. As a matter of fact, the low-temperature susceptibility of Eu(III) compounds is known to be practically temperature independ-ent, and most likely similar conditions are also met here for the spin transfer.

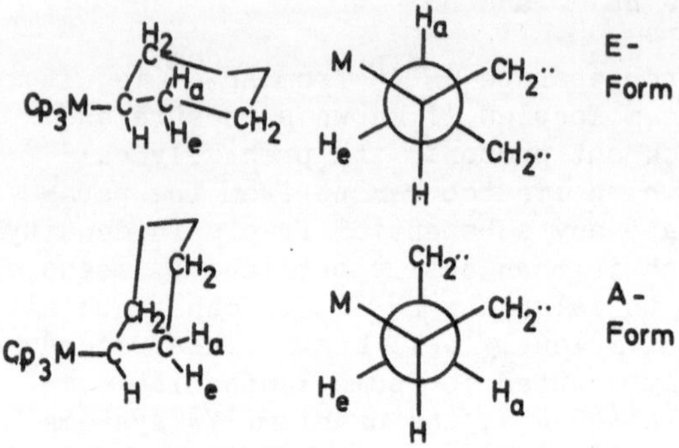

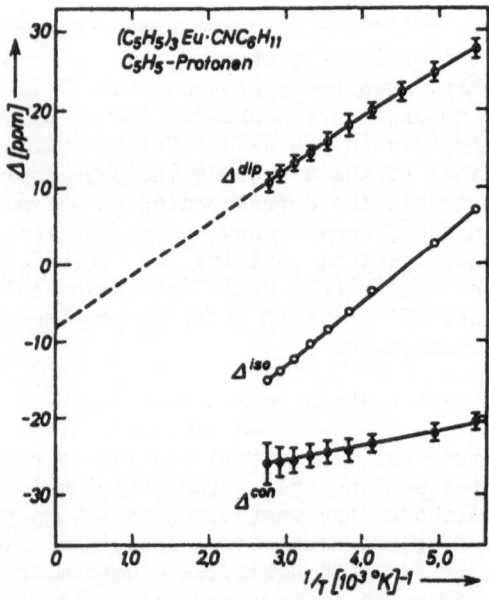

Fig. 11. Newman projections of the A- and E-conformation of
$(C_5H_5)_3LnCNC_6H_{11}$ (view along the C_1-C_2-bond of the
cyclohexyl group).

Fig. 12. Temperature dependence of the C_5H_5 ring proton isotropic,
dipolar, and contact shifts, respectively.

For the Eu complex, also ΔH^{iso} of the C_5H_5 ring ^{13}C
atoms was deduced (T = 316 K, with the corresponding Y complex
as a standard). From the resulting data, ΔH^{iso} = 187 ppm, ΔH^{dip} =
61 ppm and ΔH^{con} = 126 ppm, respectively, and from an estimate
of the spin polarization <S>/H on the Eu^{3+} ion at room temperature,

a hyperfine coupling parameter $A(^{13}C$ ring) of -1,15 MHz was de-
duced(45). Similarly, $A(^{1}H$ ring) was calculated as 0,97 MHz,
which value would, on the basis of the well known McConnell
equation(61),

$$A_H = Q_{CH}\, \rho_C\; /\; 2S \tag{12}$$

correspond to formally 1,38 free β-spins on the 15 C_5H_5 ring
carbon atoms. This unexpectedly large spin delocalization casts,
however, along with the relation $|A_H| \leq |A_C|$, considerable doubt
on the applicability of McConnell's equation on metal complexes
of π-systems.

In view of the fact that Bleaney's relation (eq. 10)
allows an estimate of the quadratic crystal field splitting
parameter $<r^2>A_2{}^0$ from ΔH^{dip}, provided that the axially symmetric
system in question fulfills the conditions for the "high temperature limit", it has been tempting to adopt eq. 10 to complexes
with particularly small $\Delta H^{iso}(H_{ring})$ and $\Delta H^{con}(H_{ring})$ values,
respectively. The resulting estimates of $<r^2>A_2{}^0$ are of the expected order of magnitude (100 - 300 cm^{-1}), but surprisingly
sensitive to the nature of Ln as well as to cyclopentadienyl
ring methylation (e.g. $(C_5H_5)_3CeCNC_6H_{11}$: -115 cm^{-1} and
$(CH_3C_5H_4)_3CeCnC_6H_{11}$: -80 cm^{-1})(47,62). This casts some doubt even
on the assumption that the $^2F_{5/2}$ ground manifold of Ce(III) is
completely thermally populated at room temperature.

Concerning the two related uranium complexes $(C_5H_5)_3UX$
with X = OR and BH_3R, respectively, it has turned out that the
magnetic anisotropy term, $(\chi_\| - \chi_\perp)$, appears to be positive in
the former case, but negative in the latter. The conclusions
referring to the class of alkoxide complexes are essentially
based on observations on the cholesteryl derivative in which
case only one favourite molecular conformation with respect to
the rotation about the OC-bond was assumed(49). Accounting for
the known structure of the bulky cholesteryl group, the resonances of most β- and γ-protons of R could be assigned, and
confirmed to be mainly due to magnetic dipolar interactions.
As hardly any negative geometric factor can be attributed to
these protons, the exclusively negative ΔH^{dip} values signal
that $(\chi_\| - \chi_\perp)$ should be positive. While ΔH^{con} of the α-proton
of R was calculated to amount to -30,5 ppm (10°C), ΔH^{con} of the
C_5H_5 ring protons turns out to be positive (17,6 ppm at room
temperature). The temperature dependence of all calculated ΔH^{dip}-
and ΔH^{con}-values comes out as expected (i.e. Curie-Weiss-behaviour
is usually approached).

The evaluation of pmr data of the borohydride complexes
with X = BH_3R is based on the firm assumption of three equivalent
UHB bridges and was started with the sufficiently rigid complex

where R = phenyl(34). As the observed phenyl proton shifts could
be correlated very well by means of eq. 3, it seems unlikely
that much unpaired spin density will get as far as the phenyl
protons. However, unlike for X = OR, the quantity $(\chi_\parallel - \chi_\perp)$ turns
out negative. From a further, very careful analysis of the room
temperature pmr data of the three compounds with R = C_6H_5, C_2H_5
and H it was suggested that the three bridging H-atoms most
probably exhibit strongly negative contact shifts(table 9).

Table 9. Observed and calculated isotropic, dipolar, and contact
shifts, respectively, of various nuclei in $(C_5H_5)_3UBH_4$
(room temperature, av=average, b=bridging, t=terminal)

	ΔH^{iso}	ΔH^{dip}	ΔH^{con}
$C_5H_5(^1H)$	12,5	−7,1 to −10,3	19,6 to 22,8
$C_5H_5(^{13}C)$	−125,0	−31,0*	−94,0*
$BH_4(^1H_{av})$	63,2	109,9	−46,7
$BH_4(^1H_b)$	69,6	130,6	≥ −61,0
$BH_4(^1H_t)$	44,0	47,8	−3,8
$BH_4(^{11}B)**$	−79,3(35)	161,6*	−240,9*

  * simple estimates on the basis of eq.3;
 ** diamagnetic standard: $(C_5H_5)_2Zr(BH_4)_2$

Although sufficient information is still lacking to determine extremely accurate values of ΔH^{con} (and ΔH^{iso}) of the bridging and
the terminal BH_4-atom(s), Marks et al. have estimated that the
contact shift of the H-atoms directly bound to uranium fall in
the range −36 to −61 ppm from the anticipated diamagnetic po-
sitions(34).

          The strongly negative ΔH^{con} value resulting from the
observed room temperature isotropic ^{11}B-nmr shift(35) indicates
an appreciable spin density transfer onto the boron atom(79),
suggesting direct U-B-bonding. Actually, the three bridging H-
atoms and the B-atom lie almost on one sphere around the U-atom.

          Considerable attention has been drawn by Marks and
collaborators towards the evaluation of the (room temperature)
pmr data of a number of complexes of the type $(C_5H_5)_3UR$, R being
various alkyl, alkenyl, alkinyl and aryl groups, respectively(40).
Striking evidence for appreciable contact shift contributions is
offered by the vinylic compounds (R = CH=CHR) where some protons
of R exhibit downfield shifts which cannot be explained in terms
of dipolar interactions only. Therefore, it was attempted to
assess the signs and magnitudes of ΔH^{con} in these complexes on
the basis of unrestricted Hartree-Fock INDO/2 MO calculations

for various alkyl radicals(63). This calculational procedure
appears to be quite useful for the determination of reliable
hyperfine coupling parameters a_i of a large number of organic
radicals. Assuming throughout that the free spin densities, and
thus the ΔH^{con} values, at different sites of a radical attached
to the paramagnetic metal ion are proportional to the corresponding a_i-values of the free radical, Marks et al. succeeded
in constructing, and solving, a number of simultaneous equations
for some selected protons. Using the results of this initial
step to determine the ΔH^{con} values of other protons, all H-atoms
of most members of the $(C_5H_5)_3UR$-series could finally be treated
(40).

The procedure appears to work well for all protons of R
except the ones directly attached to the α-carbon atom, where
e.g. hyperconjugative effects could be operative and compete with
the usual mechanism of spin transfer. It is noteworthy that here
the anisotropy term $(\chi_\parallel - \chi_\perp)$is negative like for the borohydride
complexes.

Very roughly, the following sequence of ΔH^{dip} of the
C_5H_5 ring protons can be figured out(table 10), convincingly
indicating that the nature of the ligand X in complexes of the
type $(C_5H_5)_3UX$ governs the magnetochemistry and the underlying
crystal field splitting patterns in a subtle way. ΔH^{con} appears
to show considerably less variation with X.

Table 10. Comparison of calculated C_5H_5 ring proton dipolar and
contact shifts for various $(C_5H_5)_3UX$-systems.

X	ΔH^{dip}	ΔH^{con}
OR*	+ 6,4	+17,6
$\eta^5-C_5H_5$	± 0	+21,5
η^3-H_3BR*	-7 to -10	+19 to +23
R*	-19	+28
Cl**	-1,64 to -2,83	+10,8 to +11,6
OR***	neg.	pos.

* R = alkyl or aryl; ** calculated from assumed crystal field
data(67); *** R = $CCo_3(CO)_9$, ΔH^{iso} = +9,2 ppm(80).

It is noteworthy that conventional alkoxide systems have positive
$\chi_\parallel - \chi_\perp$, while a different situation seem to appear when X is the
oxygen-bridged cluster $CCo_3(CO)_9$ (80).

Unrestricted Hartree-Fock INDO/2 SCF calculations of the

a_i-values have also been carried out for the methoxyl radical and for $H_3BC_2H_5$ (34). The results of these calculations agree in fact with essential conclusions drawn from other observations in that for X = OCH_3 and $H_3BC_2H_5$, respectively, ΔH^{con} both of the methyl protons and of the triply bridging H-atoms is strongly negative (-20 ± 10 ppm for OCH_3 and -35 to -50 ppm for $H_3BC_2H_5$, respectively). While these deductions nicely support the important conclusion that, for X = OR, $(\chi_{\parallel} - \chi_{\perp})$ is positive, the treatment of X = $H_3BC_2H_5$ also indicates the limits of the procedure. Thus, the INDO/2 calculations also predict some spin density on the ethyl group which is not confirmed by direct observation.

Although the class of complexes of the type $(C_5H_5)_3UX$ is particularly rich in species with different X, it becomes very hard to arrive at resonable ΔH^{dip} : ΔH^{con} separations as soon as the ligand X is devoid of nuclei that could be easily subjected to nmr spectroscopy. This argument applies particularly to complexes with X = halide or pseudohalide (e.g. NCS, OCS, N_3 etc.). In many cases also low solubilities prevent extensive nmr studies at variable temperature. For the weakly soluble complex $(C_5H_5)_3UNCBH_3$ (34)(where X is N-bonded to the metal ion) at least a reasonable guess for the sign of $(\chi_{\parallel} - \chi_{\perp})$ can be made: As ΔH^{iso} for the three equivalent BH_3 protons is clearly positive like the corresponding geometric factor, it follows immediately that $(\chi_{\parallel} - \chi_{\perp}) < 0$ as long as we may assume that $\Delta H^{dip} \gg \Delta H^{con}$.

To overcome solubility problems, and in the hope of obtaining further information through the case where a single C_5H_5 ring proton (and likewise ^{13}C) signal is no longer observed, we have also studied the variable-temperature ^1H and ^{13}C nmr spectra of a large number of systems of the type $(CH_3C_5H_4)_3UX$ where each ring ligand carries one methyl group(64). As expected for three equivalent, η^5-bonded $CH_3C_5H_4$ ligands, three pmr signals with the relative intensities 2:2:3 are now apparent throughout the series. Frequently, very widely separated signals of the two non-equivalent ring protons with ΔH^{iso} of opposed signs are observed, suggesting that the otherwise free rotation of the ring ligands about their five-fold axes is strongly restricted in favor of only one conformer (still with C_{3v}-symmetry). However, in view of the fact that now at least three unknown quantities have to be considered, i.e. $(\chi_{\parallel} - \chi_{\perp})$ and two different ΔH^{con} values for the ring-protons), a complete quantitative treatment of the ΔH^{iso} data has not yet been achieved(81).

The easiest $(C_5H_5)_3UX$-system to examine is, on the other hand, the one with X = η^5-C_5H_5, since X-ray crystallographic studies have proved that here all four C_5H_5 ligands are equivalent (65). According to the appearance of only one pmr signal (down to -90°C) (42), very rapid ring rotation should render the time-

averaged crystal field completely tetrahedral, and thus isotropic,
so that here the assumption $\Delta H^{iso} = \Delta H^{con}$ should be valid(82).
Again, ΔH^{con} turns out to be very similar to the ring proton
ΔH^{con} values of anisotropic $(C_5H_5)_3UX$-systems (table 10).

Although the electronic structure of the complex
$(\eta^5-C_5H_5)_4U$ has been the subject of a detailed theoretical study
(66), a clear-cut correlation of e.g. the well-known magneto-
chemistry(66) of this complex and the temperature dependence of
its ΔH^{con} is still lacking. On the other hand, it was attempted
to correlate the temperature dependence of the bulk magnetic
susceptibility $\bar{\chi}$ of solid $(C_5H_5)_3UCl$ with the electronic absorption spectrum of this complex, anticipating at the same time that
the crystal field splitting pattern of this anisotropic (C_{3v})
system results from a moderate perturbation of the CF-splitting
pattern of the T_d-system $(C_5H_5)_4U$ (67). The resulting ΔH^{dip}
values, and thus $(\chi_\parallel - \chi_\perp)$, too, are again negative and decrease
when the temperature is lowered; it is, however, not completely
stringent to use the excellent agreement of $\bar{\chi}_{calc}$ and $\bar{\chi}_{exp}$ for
$(C_5H_5)_3UCl$ also as a criterion for optimal $\chi_\parallel$- and $\chi_\perp$-values. It
is at least suspicious that the calculated $(\chi_\parallel - \chi_\perp)$ values fail
to interpret the pmr spectra of the methylated homologue,
$(CH_3C_5H_4)_3UCl$ (68). A rather rigorous treatment of the magnetic
properties, and the isotropic nmr shifts, of the ferricenium
cation(72) might serve as a good model also for future work on
f-element organometallics.

Uranocene and Uranocene Derivatives

The problem of separating ΔH^{con} from ΔH^{dip} in the case
of uranocene, $(\eta^8-C_8H_8)_2U$, has so far been tackled in two different
ways. While one group of authors has tried to arrive at a reliable
(room temperature) magnetic anisotropy $(\chi_\parallel - \chi_\perp)$ from susceptibility measurements and specific a priori assumptions about the
underlying CF-splitting pattern(s) (68,69), Dr. A. Spiegl in our
Laboratory has attempted to arrive at reasonable $(\chi_\parallel - \chi_\perp)$ values
simply from the dipolar shifts of suitable protons located in
various ring substituents R of 1,1'-disubstituted uranocenes(50).

The most essential assumption of the former approach has
been that the quadratic CF-parameter $<r^2>A_2^0$ of all complexes
$(C_8H_8)_2M$ with a f^q ground configuration be positive and so dominant as compared with the other two CF-parameters (i.e. B_4^0 and
B_6^0) that for M = U and Np at room temperature only the respective
CF doublet with highest J_z will be thermally populated (i.e. for
U(IV) with the ground manifold $^3H_4$, $J_z = \pm4$, and for Np(IV) with
$^4I_{9/2}$, $J_z = \pm9/2$) (68,69). An attractive aspect of this model is
that for M = Pu(IV) with the ground manifold $^5I_4$ the non-magnetic
CF singlet with $J_z = 0$ should lie lowest, and, in fact, "plutono-

cene" has been found to be diamagnetic(70). The reason is that
the so-called Stevens-parameter $<J|\alpha|J>$, is negative for both
U(IV) and Np(IV), but positive for Pu(IV). Further arguments
supporting this CF-approach have been that also the observed
bulk susceptibilities of the U and Np complexes are in fair
agreement with the corresponding data expected for the two
isolated CF-ground states mentioned.

Neglecting all second-order Zeeman contributions
(affecting particularly $\chi_\perp$), the simple model finally infers
that for, M = U and Np, $\chi_\perp$ = 0, and $(\chi_\| - \chi_\perp) = 3\bar{\chi}\exp(J_{z,\max})$
(68). Thence, the quantity ΔH^{dip} can be calculated provided
that also reasonable geometric factors are chosen. Although very
accurate X-ray structure analyses both of uranocene ($(COT)_2U$)
and of bis(1,3,5,7-tetramethylcyclooctatetraen)uranium ($(TMCOT)_2U$)
are available(71), the proximity of the quantity $(3\cos^2\theta-1)$ to
zero makes the entire geometric factor of the ring protons very
sensitive to small uncertainties about the ring proton position.
In particular, it is unknown whether these hydrogen atoms lie
exactly in the C_8 ring plane, or are slightly bent towards the
metal ion. It should also be recalled that the ring-H G-factors
of 3d-element metallocenes, $(\eta^5\text{-}C_5H_5)_2M$, may adopt slightly
positive or negative values according to the nature of M (72). By
appropriate variation of θ it was shown that for the ring protons
of $(COT)_2U$ the resulting ΔH^{iso} and ΔH^{con} values remain positive,
the ratio $\Delta H^{con}/\Delta H^{iso}$ decreasing from ca. 4 to ca. 2 if the (negative) G-factor is decreased between two extreme situations(69).

Table 11. Characteristic data, derived from isotropic room temperature pmr shifts (in ppm) of uranocene and neptuno-
cene(69).

| | $(TMCOT)_2U$ | | $(COT)_2U$ | $(TMCOT)_2Np$ | |
	Ring-H	CH_3-H	Ring-H	Ring-H	CH_3-H
ΔH^{iso}	+ 41,3	+ 6,0	+ 42,6	+ 41,5	- 9,9
ΔH^{dip}	+ 7,9	+ 23,6	+ 7,9	+ 5,2	+ 13,2
ΔH^{con}	+ 33,4	- 17,6	+ 34,7	+ 36,3	- 23,1
A/MHz	+ 0,98	- 0,52	+ 1,02	+ 0,95	- 0,61

Table 11 presents a survey of the total of pmr shift
data as observed, and calculated by means of appropriate geometric
factors, for the complexes $(COT)_2U$ and $(TMCOT)_2M$ (M = U and Np),
along with the corresponding hyperfine coupling parameters A. It
is worth noting that the contact shifts of the ring and methyl
protons are always opposite in sign, a feature which is typical
of metal-free aromatic radicals $C_nH_n^\cdot$, where the unpaired free
spins reside mainly on carbon $2p_z$-orbitals.

Rather fortuitously, the two $(TMCOT)_2M$-complexes show similar spectra in that the corresponding resonances exhibit quite similar shifts and shift variations with temperature. Essential differences are, however, displayed by the much broader lines of the $(TMCOT)_2Np$-spectrum (methyl group: ca. 300 Hz, ring proton: ca. 2000 Hz). Apart from the room temperature C_5H_5 proton resonance of the complex $(C_5H_5)_3NpCl$ (ΔH^{iso} = 27,4 ppm, line width: 30 Hz (11)) the report on $(TMCOT)_2Np$ appears to be the only successful nmr study of a Np(IV) system. Various attempts with $(C_5H_5)_4Np$ have failed(73), owing presumably to even larger line broadening effects than observed for $(TMCOT)_2Np$.

The alternative approach to $(\chi_\| - \chi_\perp)$ for uranocenes(50) starts with the complex $(C_8H_7Si(CH_3)_3)_2U$ anticipating that here ΔH^{con} of the methyl protons be negligible and that for this, strictly speaking non axially symmetric system, χ_{xx} is still practically identical to χ_{yy}. Experimental evidence for the absence of significant changes of the electronic structure after monosubstitution is provided by the fairly constant electronic spectra in the near infrared range and by the observation of $(\Delta H^{iso})^{-1}$-vs-T-curves identical with the corresponding diagrams of $(COT)_2U$ and $(TMCOT)_2U$, respectively, when only one ring ligand is substituted (e.g. in the cases of $(C_8H_8)(C_8H_7R)U$ with R = $P(C_2H_5)$ and $P(t-C_4H_9)_2)$.

Table 12. Magnetic anisotropy data of $(C_8H_7Si(CH_3)_3)_2U$ as calculated from the observed ΔH^{iso} and $\bar{\chi}$-values.

T/K	ΔH^{iso}	$\chi_\| - \chi_\perp$ $\cdot 10^{-6}$	$\mu_\|^2$ (B.M.)	$\mu_\perp^2$ (B.M.)	$\bar{\chi}_{exp}$ (23) $\cdot 10^{-6}$
367	7,1	2670	13,1	5,21	2666
337	8,1	3050	12,9	5,06	2892
315	9,3	3500	13,7	4,83	3085
273	11,5	4330	14,0	4,56	3533
251	12,9	4860	14,2	4,43	3824
222	16,0	6020	14,8	4,05	4290
202	18,8	7080	15,2	3,76	4683
182	23,3	8770	16,0	3,25	5156

With an appropriate geometric factor accounting for the two rotational degrees of freedom within the C_{ring}-$Si(CH_3)_3$ section of the complex, the data in the first column of table 12 are obtained. Combining the $(\chi_\| - \chi_\perp)$ values with the likewise known magnetic susceptibilities $\bar{\chi}$ = 1/3 $(\chi_\| + 2\chi_\perp)$ of a polycrystalline sample(23), also the quantities $\chi_\|$, $\chi_\perp$, $\mu_\|^2$ and $\mu_\perp^2$ are easily arrived at. Very similar data have been obtained for the complexes $[C_8H_7P(t-C_4H_9)_2]_2U$ and $(C_8H_8)[C_8H_7P(t-C_4H_9)_2]U$ at temperatures where the rotation about the C_{ring}-P-axis is restricted. In these two cases, $(\chi_\| - \chi_\perp)$ has been derived from the ΔH^{iso} ($= \Delta H^{dip}$) values of the tert-butyl protons after optimizing the dihedral angle between the C_8H_7P-ring plane and the α-carbon atom of the $t\text{-}C_4H_9$ group in exo position(50). This latter result strongly supports our initial assumption about the exclusive dipolar origin of ΔH^{iso} (CH_3) in the trimethylsilyl derivative.

It is immediately apparent from table 12 that the condition $\chi_\| > \chi_\perp > 0$ is met throughout, and that $\mu_\|^2$ and $\mu_\perp^2$ exhibit an inverse temperature dependence. Thus, all CF-treatments of uranocene involving negative(66) or zero(74) $\chi_\| - \chi_\perp$ can be ruled out, but likewise the approach by Edelstein et al.(68) must be discarded as $\chi_\perp$ does not vanish, nor do the calculated $\chi_\|$ values match the values required for $J_z = \pm4$. It can be shown by a series of CF model calculations based on a "simultaneous diagonalization" of the $5f^2$ ground configuration that rather different choices of CF parameter sets $\{B_2^{\,0}, B_4^{\,0} \text{ and } B_6^{\,0}\}$ give rise to a CF ground state with $J_z = \pm3$ which is closely followed by the state with $J_z = \pm2$, but well-separated from all other CF levels(50). It is also possible to simulate the observed temperature dependence of χ of uranocene and of its derivatives satisfactorily assuming simply $E(J_z = \pm3) = 0$ and $E(J_z = \pm2) = 300\pm50 \text{ cm}^{-1}$ with a zero field splitting of $35\pm5 \text{ cm}^{-1}$, all other states lying much higher than $E(J_z = \pm2)$.

Table 13. Observed and calculated nmr shifts of various nuclei in substituted uranocenes (calculations based on the data of table 12).

Ring substituent	nucleus studied	$G_i \cdot 10^{22}$ cm^{-3}	ΔH^{iso}	ΔH^{dip} ppm	ΔH^{con}
$Si(CH_3)_3$	^1H ($C\underline{H}_3$)	$-0,48$	$9,9$	$9,9$	0
	^{13}C ($\underline{C}H_3$)	$-0,50$	$2,4$	$10,6$	$-8,2$
$Sn(CH_3)_3$	^1H ($C\underline{H}_3$)	$-0,36$	$9,6$	$7,4$	$2,2$
	^{13}C ($\underline{C}H_3$)	$-0,40$	$8,9$	$8,5$	$0,4$
CH_3	^1H ($C\underline{H}_3$)	$-0,59$	$9,0$	$12,5$	$-3,5$
	^{13}C ($\underline{C}H_3$)	$-0,44$	$71,0$	$9,4$	$61,6$
$P(t\text{-}C_4H_9)_2$	^{31}P	$-0,50$	$-13,9$	$10,6$	$-24,5$
H	^1H ($C_8\underline{H}_8$)	$-0,21$	$42,6$	$4,5$	$38,1$
	^{13}C ($\underline{C}_8H_8$)	$3,15$	$-207,9$	$-67,1$	$-140,8$

Accounting for the somewhat modified magnetic anisotropy of this latter approach, the (room temperature) data presented in table 13 for various nuclei in unsubstituted and substituted uranocenes, respectively, are arrived at. Again, ΔH^{con} of the methyl protons (of methyl groups adjacent to ring carbon atoms) and of ring protons have opposite signs. The table also contains some results of ^{13}C- and ^{31}P-nmr studies. The sign of ΔH^{iso} of the ^{31}P-resonance is, however, not completely firm as the underlying diamagnetic standard system (the phosphino-cyclooctatetraene) might not be as good as the dianion. It should also be kept in mind that for all other nuclei than ^{1}H considerable errors may be introduced owing to the neglect of dipolar coupling via spin density in non-s orbitals. While most of the assessments as summarized in table 13 are not dramatically different from those for comparable uranium complexes in table 11, a somewhat different situation might be met in the case of the neptunium homologues as soon as the bordering condition: $\chi_\perp = 0$ will be dropped[83].

A particularly interesting approach was adopted for the separation of ΔH^{iso} and ΔH^{con} of the two complexes tetra(allyl)-uranium(IV) and tetra(2-methylallyl)uranium(IV) in that here for the first time the specific temperature dependence of ΔH^{iso} was considered[32]. The pmr spectra of both complexes show patterns unambiguously typical of four equivalent, trihapto-coordinated allyl ligands. In spite of the absence of any dynamical process (except the rotation of the η^3-allyl ligands about their axes perpendicular to the C_3-planes), Curie-like behaviour is not displayed by any of the ΔH^{iso} values. ΔH^{iso}-vs-T^{-1}-curves with fairly different, and unexpected, intercepts are found. Assuming a "quasi-tetrahedral" CF with a well-separated T_2-ground state which undergoes slight zero-field splitting, appropriate conditions for an application of the Kurland-McGarvey theory[75] were provided. The experimental shifts were fitted to functions of the type

$$\Delta H^{iso} = a/T + (b/T)^2 \qquad (13)$$

With some further assumptions, it was finally possible to arrive at (by order of magnitude) reasonable estimates for the hyperfine coupling parameters of the various ligand protons[32].

It should be noted in this context that, since 1970, the original "Kurland-McGarvey theory" has been considerably improved[76], and likewise, that also the temperature dependence of ΔH^{iso} for a large number of systems is now known. In very many instances, ΔH^{iso} does not follow the Curie law. Most frequently, the Curie-Weiss law is at least approached, and unusually large intercepts become apparent. For all future work it may, however, be worth noting that even excellent fits of experimental data to expansions in terms of $(T)^{-n}$ are not unambiguous.

Spin Density Distribution

One final goal of most advanced studies of paramagnetic shifts is, of course, the examination of ΔH^{con} with respect to the transfer of free spin density from the metal ion onto the ligands. As ΔH_i^{con} is, according to eq. 5, proportional to the average spin moment $<S_z>/H$ of the paramagnetic metal ion, ΔH^{con} itself, or even better, the ratio $\Delta H^{con}/\Delta H^{con}(max)$ may be used for a mapping of the relative spin density propagation over the ligand where $\Delta H^{con}(max)$ are tabulated values(84) corresponding to the contact shift in case of the location of one unpaired electron in an orbital of the nucleus in question. If also $<S_z>/H$ is known, it will be more appropriate to choose the hyperfine coupling parameter A_i for all further considerations. For a system with strong spin-orbit coupling, and various thermally accessible crystal field levels, it is frequently very hard to determine reliable values of $<S_z>/H$ and, consequently, of A_i. Likewise, all separations of ΔH^{con} and ΔH^{dip} for other nuclei than ^1H have only very approximative character.

In principle, the quantity A_i involves two different components, which account for two different spin transfer mechanisms. The so-called direct spin delocalization (SD) is directly related to the synergic metal-ligand electron transfer and is thus rather directly connected with the covalency of the metal-ligand bond. The other, so-called spin polarization (SP) term, is due to the exchange energy, or exchange integral, which plays a predominant role in basic quantum chemistry, but is, nevertheless, fairly often neglected in MO-calculations.

It seems quite illustrative to introduce a sideview on the present state of the interpretation of experimental A-values of paramagnetic 3d-metallocene systems. Although various theoretical approaches have been carried out, most of them have neglected the interelectronic repulsion (i.e. the exchange integrals), and thus the spin polarization term, too. Only recently, Fantucci et al.(77) have published the results of a systematic INDO treatment of various 3d element sandwich complexes with the essential conclusion that the spin polarization mechanism in metal complexes is not necessarily similar to that operating in the free ligand. Thus, it is not justified to correlate the SP-term on the ring hydrogen atoms with the SD term on the ring carbon $2p_z$ orbital via the McConnell equation(61) by transferring the factor Q simply from the organic radical to the paramagnetic complex.

In particular, the detailed results obtained for the two high-spin systems $(C_5H_5)_2V(II)$ $(3d^3)$ and $(C_5H_5)_2Cr(II)$ $(3d^4)$ (table 14) might serve as a guiding example also for uranocene and its homologues. According to the approach of Fantucci et al., the SP mechanism is dominant, but involves in the complexes all orbitals (e.g. also $2p_{x,y}C$ and $2sC$) without any restriction of symmetry. The regular alternation of sign of the SP-contribution,

$$metal(+) \cdot\cdot carbon(-) \cdot\cdot hydrogen(+) \cdot\cdot$$

is in fact reminiscent of the sign alternation found for uranocene (see table 11), tris(cyclopentadienyl)uranium borohydride (see table 9) and tris(cyclopentadienyl)europium(cyclohexylisocyanide) (see fig. 12). $\langle S_z \rangle / H$ is expected to vary as follows:

U(IV):   uranium(-) $\cdots$ carbon(+) $\cdot\cdot$ hydrogen(-)

Eu(III):   europium(+)$\cdots$ carbon(-) $\cdot\cdot$ hydrogen(-)

Table 14. Spin delocalization (SD) and spin polarization (SP) in two d-element metallocenes

| | Observed ΔH^{iso} values* | | Calculated Spin Densities | | | |
| | | | $2p_z C$ | | 1sH | |
	C	H	SD	SP	SD	SP
$(C_5H_5)_2V$	+588	-307	0,0054	-0,0095	0,0021	0,0015
$(C_5H_5)_2Cr$	+325	-314	0,0027	-0,0080	0,0025	0,0012

* in ppm, assumed to be equal to ΔH^{con}

It is well-known that the spin density on the $5f^2$-system U^{4+} is negative, but positive for the $4f^6$-system Eu^{3+} where free spins are generated by second-order Zeeman interactions(55).

The order of magnitude of the calculated $\Delta H^{con}(^{13}C)$ value of uranocene (-140,8 ppm) is almost of the same order of magnitude as the corresponding ΔH^{con} values of the two 3d-metallocenes. Although no serious objection to the term "f-orbital effect" appears adequate, it is still up to rather sophisticated quantum chemical calculations to provide a final, and more satisfactory, interpretation of the experimental data.

REFERENCES AND FOOTNOTES

1. (a) D.F. Evans, J. Chem. Soc. (London), 1959, p. 2003
   (b) concerning the adaptation of Evan's method on organo-
       metallic complexes, see H.P. Fritz and K.E. Schwarzhans,
       J. Organomet. Chem. 1, 208(1964)
2. H. Fischer, Dissertation, Technische Hochschule München, 1965
3. NMR of Paramagnetic Molecules, G.N. LaMar, W.deW. Horrocks,
   Jr. and R.H. Holm, Eds., Academic Press, New York and London,
   1973
4. (a) R. von Ammon and R.D. Fischer, Angew. Chem., Int. Ed. 11,
       675(1972);
   (b) W.deW. Horrocks, Jr., Loc. cit.3, Chap. 12, p 479;
   (c) "Nuclear Magnetic Resonance Shift Reagents", R.E. Sievers,
       Ed., Academic Press, New York, 1973;
   (d) K.A. Kime and R.E. Sievers, Aldrichimica Acta 10, No. 4,
       p. 54(1977), and further references therein
5. See footnote on p. 137 of ref. 3
6. A very comprehensive description of the internal magnetic
   field, and of the derivation of eqns. 3 and 4 is presented
   in ref. 56
7. See ref. 3, pp. 72 and 128
8. W.deW. Horrocks, Jr. and J.P. Sipe, III, J. Amer. Chem. Soc.
   93, 6800(1971)
9. L. Tomić, Z. Majerski, M. Tomić and D.E. Sunko, Croat. Chem.
   Acta 43, 267(1971)
10. See also ref. 3, p. 529 (table 13-II)
11. R. von Ammon, B. Kanellakopulos, R.D. Fischer and P. Laubereau,
    Inorg. Nucl. Chem. Lett. 5, 315(1969); R.D. Fischer and G.
    Bielang, unpublished data
12. (a) R.D. Ernst, W.J. Kennelly, C.S. Day, V.W. Day and T.J.
        Marks, J. Amer. Chem. Soc., in press;
    (b) K.W. Bagnall and J. Edwards, J. Organomet. Chem. 80,
        C14(1974)
13. P. Zanella, G. de Paoli and G. Bombieri, J. Organomet. Chem.
    142, C21(1977)
14. J.D. Jamerson and J. Takats, J. Organomet. Chem. 78, C23(1974)
15. J.M. Manriquez, P.J. Fagan and T.J. Marks, J. Amer. Chem. Soc.
    100, 3939(1978)
16. R.D. Fischer, R. von Ammon and B. Kanellakopulos, J. Organo-
    met. Chem. 25, 123(1970)
17. R. von Ammon, B. Kanellakopulos and R.D. Fischer, Abstr. Int.
    Symp. on NMR Spectroscopy, Birmingham, 1969
18. R.R. Ryan, R.A. Penneman and B. Kanellakopulos, J. Amer. Chem.
    Soc. 97, 4258(1975)
19. R.D. Fischer, E. Klähne and J. Kopf, Z. Naturf. 33b (1978),
    in press
20. R.D. Fischer and G.R. Sienel, Z. anorg. allg. Chem. 419, 126
    (1976)
21. L.T. Reynolds and G. Wilkinson, J. Inorg. Nucl. Chem. 2, 246
    (1956)

22. R. von Ammon, personal communication
23. B. Kanellakopulos, personal communication
24. H. Marquet-Ellis and G. Folcher, J. Organomet. Chem. <u>131</u>, 257(1977)
25. It has been known for long that also direct reaction of UCl_3 with NaC_5H_5 in THF may yield $(C_5H_5)_3UO(n-C_4H_9)$. G.L. Ter Haar and M. Dubeck, Inorg. Chem. <u>3</u>, 1648(1964)
26. B. Kanellakopulos, E.O. Fischer, E. Dornberger and F. Baumgärtner, J. Organomet. Chem. <u>24</u>, 507(1970); ^1H-nmr studies of $(C_5H_5)_3U(III)$ complexes have been carried out by R. von Ammon(22)
27. V.K. Vasil'ev, V.N. Sokolov and G.P. Kondratenkov, J. Organomet. Chem. <u>142</u>, C7 (1977)
28. A. Greco, B. Bertolini and S. Cesca, Inorg. Chim. Acta, <u>21</u>, 245(1977)
29. A.E. Crease and P. Legzdins, J. Chem. Soc.(Dalton) <u>1973</u>, p. 1501
30. T.J. Marks and A.M. Seyam, J. Organomet. Chem. <u>67</u>, 61(1974)
31. N. Palladino, G. Lugli, U. Pedretti, M. Brunelli and G. Giacometti, Chem. Phys. Lett. <u>5</u>, 15 (1970)
32. (a) M. Brunelli, G. Lugli and G. Giacometti, J. Mag. Res. <u>9</u>, 247(1973)
    (b) S. Poggio, M. Brunelli, U. Pedretti and G. Lugli, Abstr. Contrib. Sem., NATO ASI on "Organometallics of the f-Elements", Sogesta/Urbino, Sept. 11-22, 1978;
    (c) M. Brunelli, G. Perego, G. Lugli and A. Mazzei, J. Chem. Soc., Dalton Trans., in press
33. T.J. Marks and J.R. Kolb, Chem. Communic. <u>1972</u>, p. 1019
34. T.J. Marks and J. R. Kolb, J. Amer. Chem. Soc. <u>97</u>, 27(1975)
35. Otherwise, the study of the hydrogen exchange would be complicated by the "thermal decoupling" of the nuclei ^1H and ^{11}B via ^{11}B-quadrupolar relaxation. See, e.g. R. von Ammon, B. Kanellakopulos, G. Schmid and R.D. Fischer, J. Organomet. Chem. <u>25</u>, C1 (1970)
36. T.J. Marks, W.J. Kennelly, J.R. Kolb and L.A. Shimp, Inorg. Chem. <u>11</u>, 2540(1972), and references therein
37. T.J. Marks and W.J. Kennelly, J. Amer. Chem. Soc. <u>97</u>, 1439 (1975)
38. T.J. Marks and J.R. Kolb, J. Amer. Chem. Soc. <u>97</u>, 3397(1975); see also: T.J. Marks and J.R. Kolb, Chem. Rev. <u>77</u>, 263(1977)
39. R.D. Fischer and E.J. Mayer, unpublished results
40. T.J. Marks, A.M. Seyam and J.R. Kolb, J. Amer. Chem. Soc. <u>95</u>, 5529(1973)
41. T.J. Marks and W.A. Wachter, J. Amer. Chem. Soc. <u>98</u>, 703 (1976)
42. R. von Ammon, B. Kanellakopulos and R.D. Fischer, Chem. Phys. Lett. <u>2</u>, 513(1968); ibid. <u>4</u>, 553 (1970)
43. R. von Ammon, R.D. Fischer and B. Kanellakopulos, Chem. Ber. <u>104</u>, 1072 (1971)
44. R. von Ammon and B. Kanellakopulos, Ber. Bunsenges. Phys. Chem. <u>76</u>, 995 (1972)

45. R. von Ammon, B. Kanellakopulos, R.D. Fischer and V. Formacek, Z. Naturf. 28b, 200 (1973)

46. J. H. Burns and W.H. Baldwin, J. Organomet. Chem. 120, 361 (1976)

47. W. Wagner, Dissertation, Universität Heidelberg, 1974

48. R. von Ammon, B. Kanellakopulos and R.D. Fischer, Radiochim. Acta 11, 162 (1969)

49. R. von Ammon, R.D. Fischer and B. Kanellakopulos, Chem. Ber. 105, 45 (1972)

50. A.W. Spiegl, Dissertation, Universität Erlangen-Nürnberg, 1978; R.D. Fischer and A.W. Spiegl, to be published

51. C.A. Harmon, D.P. Bauer, S.R. Berryhill, K. Hagiwara and A. Streitwieser, Jr., Inorg. Chem. 16, 2143(1977), and further references therein

52. J. Sandström and J. Seita, Acta Chem. Scand. B31, 86(1977)

53. W.B. Jennings, Chem. Rev. 75, 307(1975)

54. D.F. Evans and G.C. de Villardi, J. Chem. Soc. (Dalton), 1977, p. 2256

55. W.B. Lewis, J.A. Jackson, J.F. Lemons and H. Taube, J. Chem. Phys. 36, 694(1962)

56. B. Bleaney, J. Mag. Resonance 8, 91(1972)

57. B. Bleaney, C.M. Dobson, B.A. Levine, R.B. Martin, R.J.P. Williams and A.V. Xavier, Chem. Comm. 1972, p. 791

58. K.O. Hodgson, F. Mares, D.F. Starks and A. Streitwieser, Jr., J. Amer. Chem. Soc. 95, 8650 (1973);

59. R.E. Maginn, S. Manastyrskyj and M. Dubeck, J. Amer. Chem. Soc. 85, 672 (1963)

60. J.D. Jamerson, A.P. Masino and J. Takats, J. Organomet. Chem. 65, C33 (1974); A.P. Masino, Ph.D. Thesis, University of Alberta (Edmonton, Canada), 1978

61. H.M. McConnell, J. Chem. Phys. 24, 764 (1956); H.M. McConnell and D.B. Chesnut, J. Chem. Phys. 28, 107 (1958)

62. Note that in Bleaney's paper(56) the sign of $\Delta H/H_0$ is opposed to that of ΔH^{iso} as applied throughout this Chapter

63. J.A. Pople and D.L. Beveridge, "Approximate Molecular Orbital Theory", McGraw Hill, New York, 1970, p. 80; J.A. Pople, D.L. Beveridge and P.A. Dobosh, J. Chem. Phys. 47, 2026(1967)

64. G.R. Sienel, Dissertation, University of Erlangen-Nürnberg (Germany), 1976; R.D. Fischer and G.R. Sienel, to be published

65. J.H. Burns, J. Organomet. Chem. 69, 225 (1974)

66. H.-D. Amberger, R.D. Fischer and B. Kanellakopulos, Z. Naturf. 31b, 12(1976); H.-D. Amberger, J. Organomet. Chem. 110, 59 (1976)

67. H.-D. Amberger, J. Organomet. Chem. 116, 219(1976)

68. N. Edelstein, G.N. LaMar, F. Mares and A. Streitwieser, Jr., Chem. Phys. Letters 8, 399(1971)

69. A. Streitwieser, Jr., D. Dempf, G.N. LaMar, D.F. Karraker and N. Edelstein, J. Amer. Chem. Soc. 93, 7343(1971)

70. D.G. Karraker, J.A. Stone, E.R. Jones and N. Edelstein, J. Amer. Chem. Soc. 92, 4841(1970)

71. A. Avdeef, K. N. Raymond, K.O. Hodgson and A. Zalkin, Inorg. Chem. $\underline{11}$, 1083(1972); K.O. Hodgson and K.N. Raymond, Inorg. Chem. $\underline{12}$, 458 (1973)

72. S.E. Anderson and R. Rai, Chem. Phys. $\underline{2}$, 216 (1973)

73. R. von Ammon, B. Kanellakopulos, E. Dornberger and R.D. Fischer, unpublished results

74. C. Aderhold, Dissertation, University of Heidelberg, 1975

75. R.J. Kurland and B.R. McGarvey, J. Magn. Reson. $\underline{2}$, 286(1970)

76. B.R. McGarvey, J. Chem. Phys. $\underline{65}$, 955 (1976); see also the Chapter by B.R. McGarvey in this Book

77. P. Fantucci, P. Balzarini and V. Valenti, Inorg. Chim. Acta, $\underline{25}$, 113 (1977)

78. H. Schumann and S. Hohmann, Chem. Z. $\underline{100}$, 336 (1976)

79. For further systems displaying unusual ^{11}B nmr shifts see e.g. G.R. Eaton and W.N. Lipscomb, NMR-Studies of Boron Hydrides and Related Compounds, W.A. Benjamin Inc. New York and Amsterdam, 1969

80. B. Stutte and G. Schmid, J. Organomet. Chem. $\underline{155}$, 203 (1978)

81. H.-D. Amberger, private communication

82. According to more recent experience with η^n-coordinated cyclic π-electron systems C_nH_n (see, e.g., ref. 77), the evaluation of the free spin density on C_5H_5 as given in ref. 42 definitely deserves a revised view

83. Most recent calculations (simultaneous diagonalization of the total f^q-configurations) carried out at Argonne National Laboratory have confirmed the J_z-sequence: $\pm3 < \pm2 < 0 \ll \pm1$, ±4 for uranocene, and suggest strongly that J_z of the ground state of neptunocene is either $\pm5/2$ or $\pm7/2$, but not $\pm9/2$ (W.T. Carnall, personal communication)

84. Cf. B.R. McGarvey and R.J. Kurland in Ref. 3, p. 559 (table 14-I)

CATALYSIS AND OTHER APPLICATIONS OF f-ELEMENT ORGANOMETALLICS.

Alessandro Mazzei

Snamprogetti S.p.A., Direzione Ricerca e Sviluppo,
20097 S.Donato, Milan, Italy

I.   AVAILABILITY OF RAW MATERIALS.

Among actinides, depleted uranium is already a potentially
cheap material as a by-product of nuclear fuel cycles. Table 1
gives a summary of the estimated cumulative generation of depleted uranium in Europe. The consumption of uranium for commercial
non-nuclear applications is expected to grow as the properties
of depleted uranium are recognized and exploited.

Table 1. Estimated cumulative Eurodif capacity of depleted ura-
         nium

	1975	1980	1985	1990
Eurodif	3900	12100	78000	148000
Italy	–	1500	10500	22100

By admitting a production of U enriched to 2.6%, expressed as
U metal (tons)

Further education of potential users is needed to remove
the fear created by the word "uranium" as synonymous with "highy
radioactive material". It has been demonstrated that uranium can
be safely produced and used with very few special precautions;
it is demonstrably safer to handle than some other materials
that carry less legal restriction such as lead or arsenic.

Also the term "rare earth" is somewhat misleading: actually
they are metals and not earths, and most of them are not rare.
For example Table 2 shows the relative abundance of lanthanides

T. J. Marks and R. D. Fischer (eds.), Organometallics of the f-Elements, 379–393.

in the earth's crust, compared with a number of other metals.
Cerium, the most abundant rare earth, is at the same level as
cobalt, molybdenum, tin and lead, and thulium (the rarest one)
is more plentiful than bismuth, gold, silver and platinum. All
lanthanides can now be obtained in very pure form using the new
solvent-extraction techniques.

Table 2.   Relative abundance of rare earths (in the earth's
           crust, ppm)

Rare earths		Other metals	
Ce	46	Ni	80
Y	28	Cu	70
Nd	24	W	69
La	18	Sn	40
Sm	6,5	Co	23
Gd	6,4	Pb	16
Pr	5,5	Mo	15
Yb	2,7	U	4
Er	2,5	Hg	0,8
Lu	0,8	Ag	0,1
Tm	0,3	Au,Pt	0,005

     In the light of the above considerations and also of the
unique electronic and stereochemical characteristics of these
elements, many organometallic and polymer researchers are devo-
ting more attention to actinides and lanthanides as new promising
initiators for catalytic reactions.

II.   LIMITS AND SCOPE OF THE REVIEW

     It is beyond the scope of this chapter to cover all the
catalytic applications of these elements. Most references concern
the use of oxides, especially supported oxides, in typical hete-
rogeneous catalytic reactions. As far as uranium is concerned,
multiple and variable valence as well as non-stoichiometry of
uranium in some compounds (especially oxides) indicates major
interest in its use as a catalyst for the oxidation of organic
compounds: for example oxidative dehydrogenation of ethylbenzene to styrene (1); oxidation of propylene to acrolein (2); oxidation of monoolefins to diolefins (3); oxidation of propylene
to acetaldehyde and propylene oxide (4) or simply to propylene
oxide (5); oxidation of acrolein to acrylic acid (6). As known,
improved catalysts for the ammonoxidation of propylene to acrylonitrile have been claimed (7) and used in the past several
years. Some references are related also to a commercial interest

in uranium reforming catalysts, particularly in form of mixed
oxides containing, U, Ni, Al and K or nickel uranate.

The largest application of rare earths (RE) was found in
the preparation of catalysts for the cracking of petroleum (8).
Tests of a commercial catalyst based on RE showed that it was more than 100 times active than a conventional silicaalumina catalyst. The catalytic properties of rare earth oxides in oxidation
reactions, in the hydrogenation/dehydrogenation of hydrocarbons,
olefin isomerization, dehydration of alcohols, conversion of nitrogen oxides, cracking of butane, synthesis of ketones from alcohols and acids, have been recently reviewed by Rosynek (9). Ceric ion redox systems are usually employed in the emulsion polymerization of acrylonitrile (10); polyvinylalcohol is oxidative-
ly degraded by ceric ion (11); transparent nylon-6 is claimed to
have been obtained following injection molding of the polymer
in the presence of lanthanum chloride (12).

According to the purpose of this chapter, it seems to me
appropriate to focus the attention on the catalytic aspects of
actinides and lanthanides in homogeneous or pseudohomogeneous
reactions, where well identified organometallic compounds are involved or, at least, where the precursor of the catalyst system
has been identified with certainty. Stereospecific polymerization has received more attention with respect to other catalytic applications and therefore will be discussed here in greater detail. However, other applications of the f-elements in homogeneous catalysis will also be reviewed briefly, although they
may presently be said to represent hopes rather than established
possibilities. The few publications in this field will be integrated with some experimental data obtained in our laboratory.

III. STEREOSPECIFIC POLYMERIZATION OF DIOLEFINS

1.   Uranium based catalysts.

Generic uranium catalysts have been claimed, already 20
years ago, for polymerization reactions. Uranium salts such as
nitrate, chloride, acetate etc. and especially uranyl derivatives are claimed to be of interest for the polymerization of olefins, photopolymerization of vinyl chloride and acrylonitrile,
polyester interchange reactions, polyurethane manufacture, and
graft reactions (13).

As far as the diolefin polymerization is concerned, of paramount importance is the possibility of controlling the type of
addition of the monomeric units. It is well known that diolefins
can give macromolecules made up almost exclusively of 1,2 (3,4)
or 1,4 units. Geometrical isomers arise from cis and trans struc-

ture of 1,4-units: in general trans-polymers exhibit a high
melting temperature, whereas cis polymers are rubber-like materials.

Controlled polymerization of diolefins by various kinds of
transition metal based catalysts is a well known reaction whose
industrial application for the production of cis, trans, and
1,2 polymers is now well established. However, the mechanism of
polymerization has not yet been completely clarified in spite
of the large amount of published data.

The more widely employed processes for the synthesis of
high 1,4-cis polybutadiene are based on catalytic systems formed by reaction between an organic salt of cobalt or nickel and
aluminum alkyl halides. A third component (Lewis acids or bases)
is sometime added to improve the activity and/or the stereospecificity (14).

The more generally accepted theory is that the growing polymer chain is bonded to the transition metal by a π-allyl
bond and that the incoming monomer is inserted into the transition metal-allyl bond. Some authors (15) suggest that the insertion occurs in two steps: coordination of the monomer to the
transition metal with formation of a short-lived σ-allyl species (rapidly reversible) and transfer of the σ-allyl to the
coordinated monomer, with reformation of a π-butenyl bond
(scheme 1):

(1)

Experimental findings obtained with pure π-allyl metal complexes support such a hypothesis:in particular π-C_3H_5NiCl gives
1,4-cis polybutadiene, π-C_3H_5NiBr gives a mixture of cis and
trans polymer and π-C_3H_5NiI pure trans polymer (16). In this
case a mechanism frequently proposed in the past was associated
essentially with the isomerism of the growing chain end in the
form of a π-allyl complex: trans-1,4 units can derive only from
a syn π-butenyl group, whereas cis-1,4 units can be obtained
when the last polymerized unit is in the anti form (scheme 2):

"SYN" FORM   TRANS-1,4   "ANTI" FORM   CIS-1,4

$$(2)$$

However, NMR data obtained in solution seem to indicate that only the syn form exists as a stable complex in both π-allylnickel chloride and iodide, which yield high cis-1,4 and high trans-1,4 polybutadiene, respectively (17). Nevertheless, an unstable transition state, i.e., an anti π-allyl complex that undergoes polymerization prior to rapid isomerization to the stable syn form, should not be ruled out a priori.

Of great importance, therefore, is the study of the intimate behaviour of transition metal π complexes, because most of these complexes, being relatively stable and well defined species, are considered as good models (or precursors) of the active site structure in the stereospecific polymerization of diolefins. In view of this, new uranium derivatives of the π-allyl type have been prepared in our laboratories in order to investigate their catalytic behaviour (18).

Tetraallyluranium itself is not active in the polymerization of diolefins, giving in fact only traces of dimerization and trimerization products. However by reacting U(π-allyl)$_4$ with HCl(HBr,HI), tris-allyl U halides are obtained which are fairly active homogeneous catalysts for butadiene polymerization, giving cis-polymers with an extremely high cis content, up to 99% (Table 3). When a Lewis acid is added to allyl uranium halides or to tetraallyluranium, the catalyst activity is very much improved (Table 4). A peculiar characteristic of these uranium catalysts, as opposed to those of transition metals, is that they give high cis content independently of both the nature of the halogen bonded to uranium and the nature of the metal of the Lewis acid. The high cis-tacticity of uranium catalyzed polybutadiene influences its crystallization and affects considerably the processing and mechanical properties of the obtained rubber. Other diene monomers such as isoprene, piperylene and 2,3-dimethylbutadiene, give homopolymers with substantially cis structure, as well as copolymers of various composition, always in the cis configuration.

In order to avoid the synthesis of allyl uranium complexes

Table 3. Polymerization of butadiene with (π-allyl)$_3$UX.

		Microstructure		$[\eta]$	Yield
X	cis %	trans %	1.2 %	dl/g	%
Cl	98.4	1.3	0.3	2.26	74
Br	98.5	1.0	0.5	2.7	25
J	98.5	1.2	0.3	2.7	52
(π-allyl)	–	–	–	–	traces

Experimental conditions: solvent, n-hexane; butadiene, 3 mole/
/liter; U(C$_3$H$_5$)$_3$X, 3.10^{-3} mole/liter;
polymerization time,15 hr; temperatu-
re,+5°C.

Table 4. Polymerization of butadiene with (π-allyl)$_3$UX and
Lewis acids

X	L.A. Type	L.A./U molar. ratio	cis %	trans %	1.2 %	Yield %
Cl	AlEtCl$_2$	0.3	98	1	1	50
	BCl$_3$	0.3	96	3.5	0.5	50
	TiCl$_4$	0.2	97	2	1	40
	AlCl$_3$	0.2	99	0.7	0.3	30
Br	AlEtCl$_2$	0.25	98.2	0.8	1	45
	TiCl$_4$	0.3	98.5	0.5	1	40
I	AlEtCl$_2$	0.3	98	1.2	1	40
	AlI$_3$	0.15	98	1.5	0.5	35
(C$_3$H$_5$)	AlEtCl$_2$	0.8	99	0.7	0.3	70

Experimental conditions: solvent, n-hexane; butadiene, 2 mole/
/liter; U compound, 0.1 x 10^{-3} mole/liter;
polymerization time, 1 hr; temperature,
+20°C; catalyst aged 4 hr at room temp.

and their storage at low temperature, we have extended our research to include more available and stable uranium compounds
suitable to give "in situ" the required π-allyl intermediate
by reaction with aluminum alkyls in the presence of the monomer.
Simple uranium derivatives such as UCl$_4$, UF$_4$ etc. are not suitable because of their insolubility in hydrocarbon solvents and
their inertness toward AlR$_3$. Soluble derivatives such as
U(acetylacetonate)$_4$, U(NR$_2$)$_4$, U(OOCNR$_2$)$_4$ give active catalysts,

but the reaction rate is somewhat low (Table 5). Very good results have instead been obtained with uranium alkoxides, in particular $U(OCH_3)_4$, employed in combination with AlR_3 and a Lewis
acid (19), Table 6. No activity was shown by other recently reported uranium organometallics, namely (π-COT)$_2$U, Cp_3UR (R=me-
thyl, phenyl, n-butyl) and (dipy)$_4$U, also in the presence of a
Lewis acid. The higher stability of these complexes and their
coordinative saturation may prevent the possibility of coordination of the monomer with both double bonds in the cis conformation.

Table 5.   Polymerization of butadiene with some uranium derivatives

Compound	$Al(C_2H_5)_3$ mmole	Lewis Acid mmole	Yield %	cis %	trans %	1,2 %
U(Acac)$_4$	1.8	$AlBr_3$(0.04)	60	93	5.5	1.5
U[N(C$_2$H$_5$)$_2$]$_4$	1.8	$AlBr_3$(0.1)	64	97.7	1.8	0.5
UO$_2$(Acac)$_2$	1.5	AlEt$_2$Br(0.12)	34	89.5	9.1	1.4
UCl$_4$·THF	2.0	AlRCl$_2$(0.05)	30	85.5	13.5	1.0

Experimental conditions: U derivative 0.1 mmole; solvent, 80 ml;
                         temperature, 50°C; time, variable;
                         Acac = acetylacetonate.

    As far as the mechanism of polymerization is concerned, we
have up to now been unable to obtain direct evidence of the nature of the active sites by applying the NMR technique on uranium π-allyl systems, both in presence and in absence of butadiene. The monometallic catalysts, $U(C_3H_5)_3X$, although well characterized in the solid state by chemical analyses, show solution NMR spectra not clearly explainable. Probably a disproportionation reaction occurs with formation of tetraallyl uranium
(clearly recognizable) and more halogenated π-allyl species.
This behaviour shows up also at very low temperature, e.g. -80°C.
On the other hand, all attempts to synthesize π-2-butenyl uranium derivatives did not succeed, thus preventing some conclusions on the syn-anti isomerism to be reached (20). We believe
that also in the case of the catalyst system $U(OCH_3)_4$+AlR_3+Lewis
acids, a π-allyl intermediate is formed "in situ" when the catalyst is prepared in the presence of the monomer. This view is
supported by the similarity of results obtained either with uranium alkoxides or with preformed (π-allyl)$_3$UX (i.e., the same
stereospecificity, activity, polymer properties, etc.). In order to verify this hypothesis, the synthesis of mixed allyl-al-
koxo complexes of uranium was attempted. New complexes of the
general formula U(π-C$_3$H$_5$)$_2$(OR)$_2$ (R = ethyl, isopropyl, tertbutyl) have been prepared by reaction of tetraallyluranium and
alcohol (21).

Table 6. Polymerization of butadiene with $U(OCH_3)_4$, Lewis acids and AlR_3.

Lewis acid type	$\dfrac{\text{Lewis acid}}{U(OCH_3)_4}$ molar ratio	$\dfrac{Al(C_2H_5)_3}{U(OCH_3)_4}$ molar ratio	Polymeriz. Temperature °C	$[\eta]$ dl/g	cis %	trans %	1,2 %	Yield %
$AlBr_3$	0.5	15	20	5.28	98.5	0.8	0.7	30
$AlBr_3$	0.5	15	30	4.63	98.3	0.9	0.8	45
$AlBr_3$	0.5	15	40	3.99	98.4	0.9	0.7	60
$AlC_2H_5Cl_2$	0.8	15	20	5.52	98.2	0.7	1.1	30

Experimental conditions: solvent, n-hexane; butadiene, 2 mole/liter; $U(OCH_3)_4$, $0.1 \cdot 10^{-3}$ mole/liter; polymerization time, 1 hr; catalyst aged at room temperature for 4 hr.

$$(C_3H_5)_4U + 2ROH \longrightarrow U(C_3H_5)_2(OR)_2 + 2C_3H_6$$

Infrared and NMR data suggest a trihapto geometry for the allyl moiety as well as the presence of a monomeric structure in THF and of a dimeric one in toluene.

The crystal and molecular structure of the bis(η^3-allyl)- -bis(isopropoxo)-U(IV) dimer has been determined from single crystal X-ray diffraction data (22). Alkoxo-allyl uranium derivatives alone do not show any catalytic activity, as expected, although, in the presence of a Lewis acid they give high cis-po- lybutadiene in good yield, Table 7.

Table 7. Polymerization of butadiene with (π-C_3H_5)$_2$U(OC_2H_5)$_2$ and Lewis acids

Lewis acid	L.A./U molar ratio	Yield %	cis %	trans %	1,2 %
None	–	0	–	–	–
AlEtCl$_2$	1	50	96.0	3.0	1.0
AlEt$_2$Cl	2	95	98.4	1.1	0.5

Experimental conditions: U concentration, 0.2×10^{-3} mole; ben- zene, 80 ml; C_4H_6, 10 g; Temperature, 50°C; time 10 hrs.

2.   Lanthanides based catalysts.

Since 1964 a series of catalysts based on lanthanide halides have been reported by Chinese (23), American (24) and Russian (25) scientists for the stereospecific polymerization of diole- fins. More active catalyst systems are now employed to obtain both high 1,4-cis polybutadiene ($\geqslant$ 97%) and high 1,4-cis polyisopre- ne ($\geqslant$ 95%), as well as the stereospecific copolymer of the two monomers (26). The polymerization activity depends on the nature of the metal employed, being higher for Ce, Pr and particularly Nd, and lower for Sm and Eu. The nature of the halogen in the Lewis acid affects activity in the decreasing order Br > Cl > I > F.

A more extensive research has been made (27) of the system cerium octanoate-AlR$_3$-halide. As in the case of the preceding catalysts, the stereospecificity of polybutadiene is not affec- ted by the nature of the halogen (ranging from F to I), even though the activity is higher for Br. Cerium is believed to be the primary stereoregulating factor. It is unknown whether it can form π-allyl or π-crotyl complexes with butadiene, but this seems to be a likely possibility in view of cerium's vacant inner shell. The polymer has high tack, good processability and relatively good physical and dynamic properties. A drawback of this elastomer is that the catalytic residues must be completely removed because of the high activity of cerium in promoting rubber oxidation.

In our laboratories we have also devoted some attention to the catalytic behaviour of lanthanide elements. Our interest, as in the case of uranium catalysts, has been firstly directed to the synthesis and characterization of π-allyl derivatives of some lanthanides, not yet discovered. All attempts to prepare pure tris-allyl derivatives of lanthanides failed, independently of the route employed. Instead we succeeded in obtaining new allyl complexes of the type $LiLn(allyl)_4$ by the following reaction (28):

$$LnCl_3 + 4LiR + Sn(allyl)_4 \longrightarrow LiLn(allyl)_4 + SnR_4 + 3LiCl$$

The reaction solvent (THF) is evaporated and the residue washed with n-hexane to remove SnR_4 and then extracted with diethylether. The new complexes can finally be obtained by drying the ether solution under vacuum. They can also be precipitated with dioxane from the ether solution obtaining, in this case, the $LiLn(allyl)_4$ diox. complex. IR and NMR evidence shows they are typical π-allyl systems, fluxional with temperature. Complexes isolated until recently include Ce, Nd, Sm, Gd and Dy. Their catalytic activity in diolefin polymerization is completely different with respect to the previously mentioned lanthanide catalysts (29). Some results are reported in Table 8.

Table 8. Polymerization of butadiene with $LiLn(allyl)_4$-dioxane

	Yield %	cis %	trans %	1,2 %
$LiCeA_4$.D	85	3.8	91.6	4.6
$LiNdA_4$.D	90	2.6	92.8	4.6
$LiSmA_4$.D	82	10.5	79.4	10.1
$LiGdA_4$.D	78	15.5	67.1	17.4
$LiDyA_4$.D	87	11.7	74.6	13.7

Experimental conditions: Ln derivative, 0.1×10^{-3} mole; toluene, 80 ml; C_4H_6, 10 g; temperature 50°C; time 15 hrs.

The microstructure of the obtained polymers is predominantly 1,4-trans, with very low cis content. In addition, the activity is pratically the same for all tested elements, in opposition to the previous findings. On the other hand the polymeric structure is different from those obtained by pure anionic catalysis based on lithium alkyls alone. The addition of a Lewis acid enhances the activity and causes a small change of microstructure of the polymers (unlike anionic catalysis), whereas the addition of a Lewis base changes progressively the stereospecificity toward an increasingly large vinyl content (like lithium alkyl catalysts) (Table 9). Further research is necessary to explain the catalytic properties of π-allyl derivatives of lanthanides.

Table 9. Catalytic activity of $LiLn(allyl)_4$ with Lewis acids or bases

	Lewis acid or base	LA or LB/Ln molar ratio	Yield %	cis %	trans %	1,2 %
$LiCeA_4 \cdot D$	$AlBr_3$	0.3	85	8	88	4
$LiNdA_4 \cdot D$	$AlBr_3$	0.3	58	5	90	5
$LiSmA_4 \cdot D$	$AlBr_3$	0.3	78	6	88	6
$LiGdA_4 \cdot D$	$AlBr_3$	0.3	70	4	90	6
$LiCeA_4$	TMEDA	5	86	5	15	80
$LiGdA_4 \cdot D$	THF	6	89	13	42	45
$LiGdA_4 \cdot D$	THF	200	94	6	15	79

Experimental conditions: as in Table 8; polymerization time, 2 hrs for Lewis acid tests and 24 hrs for Lewis base tests.

In conclusion, the debated question of whether π-allyl intermediates, rather than σ-type ones, are involved in the stereospecific polymerization of diolefins and the knowledge of the driving factors in the stereoregulation is, in my opinion, still unresolved.

IV. OLIGO- AND POLYMERIZATION OF OLEFINS

The polymerization of α-olefins with actinide and lanthanide catalysts has received minor attention and generally, the results are fair. The use of rare earth halides and aluminum alkyls (30), of RE halides or oxides and lithium alkyls (31), of RE metals used as promotors of $Mo_2O_3-Al_2O_3$ catalysts (32), of RE oxides with metal hydrides (33) all have been claimed useful for ethylene polymerization.

Definite cerium organometallic derivatives have been recently prepared in our laboratories (34), namely $Ce(COT)_2$, $Ce_2(COT)_3$ and the bimetallic (COT) $Ce(O-iC_3H_7)_2Al(C_2H_5)_2$. All these complexes, especially the last one, which resembles in some way a bime-

tallic complex thought to be involved in Ziegler-Natta catalysis,
as well as anhydrous cerium halides, were tested for ethylene polymerization under various conditions (35). It has thus been
found that, even though these compounds show some catalytic activity, they are by far inferior to the "high yield" catalysts currently employed industrially. Olefin oligomerization is also
catalyzed by lanthanidé compounds. For example, lubricating oils
with high viscosity indexes are prepared by oligomerization of
C_{10}-C_{13} normal α-olefins with aluminosilicate catalysts containing 25% of rare earth oxides (36). The synthesis of higher,
straight chain olefins, used for biodegradable detergents, from
ethylene with $CeCl_3$ and butyl-lithium or Grignard catalysts has
also been reported (37). Finally, propylene dimerization was obtained with rare earth acetylacetonate and aluminum alkyl halides in the presence of triphenylphosphine (38).

V.   MISCELLANEOUS REACTIONS

The use of f-organometallics in homogeneous catalytic reactions, other than polymerization, has been scarcely investigated.
Often the few attempts that were carried out have not been subjected
to further studies leaving little room for comment: therefore my
remarks are necessarily limited to some hypotheses of application
based mainly on our own experience in this field.

By reaction of $(C_3H_5)_4U$ with an excess of a nitrile (i.e.,
CH_3CH_2CN, $CH_2 = CH$-CH_2-CN, CH_3-$CH = CH$-CN) an insertion reaction
takes place, with formation of a new, unidentified uranium com-
plex and an unsaturated primary amine:

$$R-C \equiv N + (C_3H_5)_4U \rightarrow [\text{new U complex}] + R-\underset{\underset{C_3H_5}{|}}{\overset{\overset{C_3H_5}{|}}{C}}-NH_2$$

In the case of 3-butenenitrile the excess of monomer reacts with
the intermediate complex (which contains an atomic ratio U:N=1:8)
giving almost exclusively 2-butenenitrile (90%) and a small amount
of dimer (10%). Such isomerization reaction is also operative with
other uranium complexes, for ex. $U(dipy)_4$. Employing phtalonitri-
le with a uranyl ion template, Marks (39) obtained an expanded
"superphthalocyanine" thus recognising the possibility of creating
new macrocycles by carrying out oligomerizations in the presence
of large, coordinatively specific actinide templates.

Alkylation reactions should also be catalyzed by strongly cationic uranium complexes. In fact, the reported (40) arene derivative $UCl_3 \cdot 3AlCl_3 \cdot C_6H_6$ reacts with olefins in benzene giving al-
kylbenzene. The behavior of the uranium complex in Friedel-Crafts

reactions is similar to that of the well known $AlCl_3$, when higher olefins are employed. In contrast, with ethylene the reaction product is a mixture of ethylbenzene and butylbenzene with molar ratio 1:4, instead of the typical ratio 4:1, perhaps because of ethylene dimerization preceding the alkylation reaction (41). The same arene complex reacts with diphenylacetylene in benzene giving tetraphenyl ethane (41).

When $UCl_3 \cdot 3AlCl_3 \cdot C_6H_6$ is reacted with aluminum alkyls, new complexes are formed, such as $2UCl_3 \cdot AlRCl_2$ and $UCl_3 \cdot 3AlRCl_2$, with loss of the complexed arene. Both these complexes, as well as the starting one, are catalytically active in the cationic polymerization of vinyl monomers (isobutylene, vinylethers, etc.)(41).

VI. OTHER POSSIBLE APPLICATIONS

Since 1971 we have been carrying out some research on the synthesis of divalent complexes of ytterbium. This metal, as well as europium, is soluble in liquid ammonia, yielding the characteristic blue solution, like alkali metals. By reacting the ammonia solution of Yb with organic substrates containing reactive protons, such as acetylacetone, salicylaldehyde, dibenzoylmethane, dipivaloylmethane, acetic acid and others, compounds which show a considerable solubility in hydrocarbon solvents have been isolated (41). The evaluation of these complexes as antiknock agents of automotive fuels gave rather good results, although less satisfactory than alkyllead.

Other β-ketoenolates of lanthanides in their higher valence state, like Pr, Nd, Yb and Ce 2,2,6,6 tetramethyl-3,5-heptanedionates, have been recently claimed as knock-inhibiting agents (42). All of the rare earth ketoenolates gave results superior to those obtainable with tetraethyllead, especially $Ce(THD)_4$, Table 10.

Table 10. Antiknock properties of lanthanides derivatives

Additive	gr/gal	mmols of metal / gal	Octane number (Res. Met.)
$Pr(THD)_3$	0.447	0.65	102.0
$Nd(THD)_3$	0.451	0.65	102.0
$Yb(THD)_3$	0.470	0.65	102.3
$Ce(THD)_3$	0.568	0.65	103.7
$Pb(C_2H_5)_4$	–	0.65	101.6
Isooctane	–	–	100.0

THD: 2,2,6,6-tetramethyl-3,5-heptanedione.

From the economic point of view, it is interesting to note that
a mixture of RE can be used, for ex. that derived from naturally
occurring ore rich in cerium, without any expensive separation
of the elements in their pure form.

VII. CONCLUSIONS.

I have tried to select, among more than one thousand references, the catalytic aspects which are more strictly related
to organometallic derivatives of the f-elements. It is inevita-
ble that some of the topics selected have not been covered with
the proper degree of thoroughness because of either the lack of
published data or their insufficient rationalization. Increasing
knowledge of the organometallic chemistry of these elements,
their availability on a commercial scale at a reasonable prices
and some example of industrial interest to their utilization,
reported here,are all deemed to be important "catalytic" factors
for the development of new applications.

VIII. REFERENCES

1. N.Fumio, and I. Fumio, Nippon Kagaku Kaishi, 254 (1973).
2. Ibid, 842 (1972).
3. Jap. P. 7,403,495 (1974) to Ube Ind.
4. Jap. P. 7,327,281 (1973) to Mitsui Chem.
5. Ger. Offen 2,312,282 (1974) to Metalgesellschaft.
6. Ger. Offen 2,459,092 (1976) to Standard Oil.
7. Ger. Offen 2,450,886 (1976) to Monsanto; 2,641,846 (1977)
    and 2,640,099 (1977) to Gulf; Neth. Appl. 74 11.326 (1975)
    to BP; Ger. Offen 2,264,403 (1973) to Snamprogetti; USP
    4,018,712 (1977) to Monsanto; B.J.Evans, J.Catal., 41, 271
    (1976).
8. Chem. Eng. News, May 10, 78 (1965); Nesmejanova et al., Kinet.
    y Cataliz. 16, 551 (1975); Brit. P. 1,392,312 (1971) to Mobil
    Oil.
9. M.P.Rosynek, Catal. Rev.-Sci. Eng. 16, 111 (1977).
10. A.Rout, S.P.Rout, B.C.Singh, and M. Santappa, Europ. Polym. J.
    13, 497 (1977); Makromol. Chem., 178, 639 and 1971 (1977).
11. Y.Ikada, Y.Nishizaki, H.Iwata, and I.Sakurada, J.Polymer Sci.,
    Chem. Ed., 15, 451 (1977).
12. Ger. Offen 2,529,691 (1977) to BASF.
13. USP. 3,065,217 (1962) to Dal Mon Research; Brit. P. 861,712
    (1961) to Goodyear Tire; C.A. 61, 4488e (1964); C.A. 61,
    9628g (1964).
14. W.M.Saltman, "The Stereo Rubbers", Wiley,Intersci. N.Y. 1977.
15. See for ex.L.Porri, Chimia, 541 (1974).
16. L.Porri, G.Natta, and M.C.Gallazzi, J.Polym. Sci., C16, 2525
    (1967); V.A.Kormer, B.D.Babitski, and M.I.Lobach, ibid, C16

4351 (1969); B.A.Dolgoplosk, E.I.Tinyakova, P.A.Vinogradov, and O.Parenago, ibid, C16, 3685 (1968).

17. J.P.Durand, F.Dawans, and Ph. Teyssie, J.Polym. Sci., A8, 979 (1970); T.Matsumoto, and J.Furukawa, ibid, B5, 935 (1967); R.Warin,Ph. Teyssie, and F.Dawans, ibid B11, 177 (1973); V.A.Kormer, and M.I.Lobach, Macromolecules, 10, 572 (1977).

18. G.Lugli, W.Marconi, A.Mazzei, N.Palladino, Inorg. Chim.Acta, 3, 253 (1969); G.Lugli, A.Mazzei, S.Poggio, Makromol. Chem., 175, 2021 (1974); N.Palladino,G.Lugli, U.Pedretti, and M.Brunelli, Chem. Phys. Letters, 5, 15 (1970); M.Brunelli, G.Lugli, and G.Giacometti, J.Magn. Reson., 9, 247 (1973).

19. M.Bruzzone, A.Mazzei, and G.Giuliani, Rubber Chem. & Technol. 47, 1175 (1974).

20. T.J.Marks, and A.M.Seyam, J.Organometal. Chem. 67, 61 (1974); and our unpublished results.

21. G.Lugli, M.Brunelli, and A.Mazzei, Proceed. 7th Intern. Conf. Organometal. Chem., Venice, Sept. 1975.

22. M.Brunelli, G.Perego, G.Lugli and A.Mazzei, J.Chem.Soc., Dalton, in press.

23. Sheng et al., Kexue Tongbao, 335 (1964); Sheng et al. Scientia Sinica, 13, 1339 (1964).

24. Belg. P. 644,291 (1964) and USP. 3,297, 667 (1967) to Union Carbide.

25. Y.B.Monakov, Dokl. Akad. Nauk. SSSR, 234, 1125 (1977).

26. Scientia Sinica, 17, 656 (1974); Sun-Shu-Ji, private communication (1975).

27. M.C.Throckmorton, Kautschuk Gummi Kunst., 22, 293 (1969).

28. Ital. P. Appl. 20,140 A/77 (1977) to Snamprogetti.

29. Ital. P. Appl. 24,137 A/78 (1978) to Snamprogetti.

30. Brit. P. 840,327 (1960) to ICI, USP 3,111,511 (1963) to Grace.

31. USP. 3,429,864 (1969) to Phillips.

32. USP. 2,921,058 (1960) to Standard Oil.

33. C.P.Brown, and J.Saunders, J.Polym. Sci., 43, 579 (1960).

34. A.Greco, G.Bertolini and S.Cesca, J.Organometal. Chem., 113, 321 (1976); Inorg. Chim. Acta, 21, 245 (1977).

35. Ital. P. Appl. 26,220 A/75 (1975) to Snamprogetti).

36. USP. 3,322,848 (1967) to Mobil Oil.

37. USP. 3,366,704 (1968) to Phillips.

38. USP. 3,641,185; 3,803,053 (1968) to Atlantic Richfield.

39. V.W.Day, T.J.Marks, and W.A.Wachter, J.Am.Chem.Soc., 97, 4519 (1975).

40. M.Cesari,U.Pedretti, A.Zazzetta, G.Lugli and W.Marconi, Inorg. Chim. Acta, 5, 439 (1971).

41. Unpublished data.

42. USP. 3,794,473 (1974) to K.J. Eisentrant.

MÖSSBAUER SPECTROSCOPY OF ^{237}Np ORGANOMETALLIC COMPOUNDS

D. G. Karraker

Savannah River Laboratory
E. I. du Pont de Nemours and Company
Aiken, South Carolina   29801

I.   INTRODUCTION

Mössbauer Effect (ME) spectra are an important tool in the investigation of the properties of solids. The electron charge density, electric field gradient and magnetic hyperfine field at the actinide nucleus are influenced by covalent contributions to bonding, site symmetry of the actinide ion, and magnetic interactions between the ion and the nucleus, respectively.

Isotopes of all the actinides from thorium to americium have been shown to give Mössbauer spectra, but only the ME of ^{237}Np has been extensively exploited for chemical purposes. The ^{237}Np ME has excellent resolution and is comparatively easy in experimentation; Np(IV) and Np(III) ions form most of the organometallic compounds that can be synthesized in the actinides. This paper will deal exclusively with ^{237}Np ME on Np organometallics.

II. ^{237}Np MÖSSBAUER EFFECT

The ME on ^{237}Np was discovered by Stone and Pillinger[1] at the Savannah River Laboratory and developed further by the group at Argonne National Laboratory, then directed by G. M. Kalvius. The fundamentals of the ^{237}Np ME have been discussed in two excellent reviews[2,3] and will only be briefly reviewed here. The useful gamma ray is the 59.5 keV transition from ^{237m}Np ($t_{\frac{1}{2}}$, 63 nsec) $5/2^-$ to the ^{237}Np, $5/2^+$ ground state. The 59.5 keV - ^{237m}Np results from the α-decay of ^{241}Am ($t_{\frac{1}{2}}$, 433 yr), the β-decay of ^{237}U ($t_{\frac{1}{2}}$, 6.75 da) or the electron capture decay in ^{237}Pu ($t_{\frac{1}{2}}$, 45.6 da) (Figure 1). The most convenient of these

395

T. J. Marks and R. D. Fischer (eds.), Organometallics of the f-Elements, 395–420.
All Rights Reserved. Copyright © 1979 by D. Reidel Publishing Company, Dordrecht, Holland.

for a Mössbauer source is the ^{241}Am; normally ^{241}Am is diluted
in a matrix of thorium metal to give a gamma source with a narrow
line width.

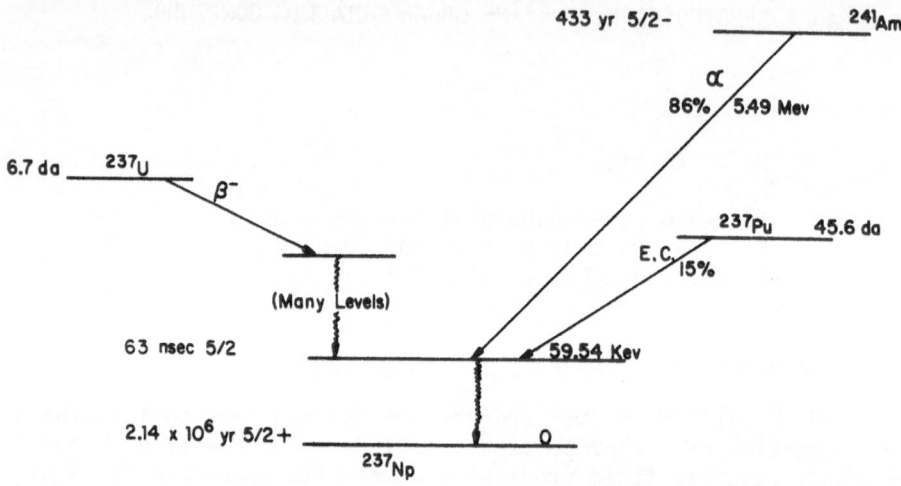

FIGURE 1. Decay Schemes to the 59.5 keV ^{237}Np Level (Simplified)

If the emitting and absorbing ^{237}Np nuclei are not in
identical environments, an energy increment equal to the recoil
energy of the emitting matrix must be added mechanically to
achieve a resonance. The energy increment required for resonance
is affected by ^{237}Np valence, site symmetry, and magnetic interactions. The Hamiltonian for the system may be written

$$H_{Total} = H_{IS} + H_Q + H_M$$

where H_{IS} represents the interactions of the central field with
the nucleus (valence); H_Q represents the interaction of the
quadrupole moment with the electric field gradient (site
symmetry); and H_M represents the interaction of the nuclear
dipole moment with the magnetic field, internal or external.
The transitions induced between the ^{237m}Np and ^{237}Np levels are
shown in Figure 2, for a single level, a quadrupole-split level,
magnetically split levels, and a magnetic-quadrupole combined
splitting.

The central field interactions that lead to the isomer shift
in ME are spherically symmetric. Only S electrons have an appreciable density at the nucleus, so the isomer shift reflects the
density of electrons in the S orbitals. In the case of ^{237}Np,

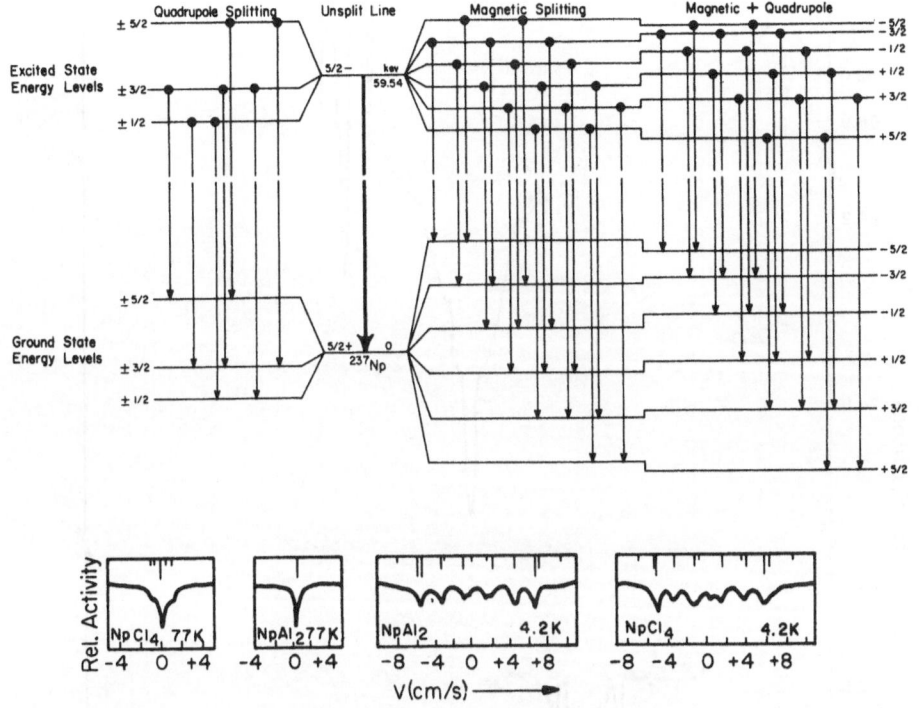

FIGURE 2. Splitting of the Ground State and 59.5 keV Level of
[237]Np in Magnetic and Electric Fields

the 6s orbitals are shielded by the inner 5f orbitals: as the 5f
shell is filled, progressing from Np(VII) to Np(III), the shielding
of the 6s orbitals is increased, and the isomer shift becomes progressively more positive. Thus, the decrease in S-electron
density is actually measured by the ME isomer shift of [237]Np
compound. Figure 3 shows the effect of valence on isomer shift.[3]

The isomer shift in [237]Np is very large - from -6.9 cm/sec
to +3.5 cm/sec, compared with the normal 2 to 4 cm/sec shift
found for most Mössbauer isotopes. This large isomer shift
allows excellent precision in determinations of the isomer shift,
and can also allow a quantitative measure of covalent effects in
bonding. Covalency involves electronic contribution from a ligand
to the metal ion orbitals, which in the [237]Np ME results in an
isomer shift toward a lower valence. An example (Figure 4) is
the comparison of Np fluorides with oxygen-bonded Np in four
valences.[3,4]

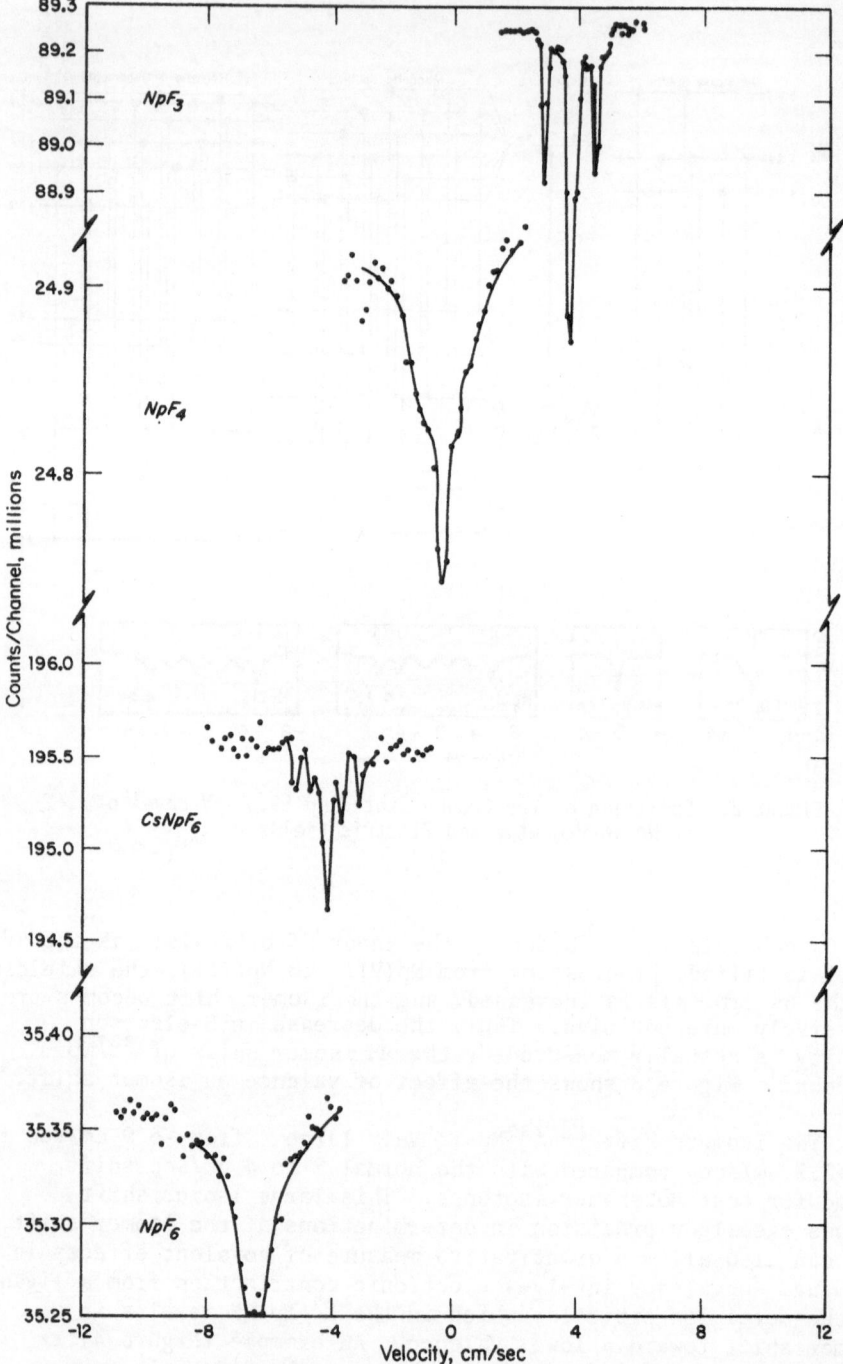

FIGURE 3.   Mössbauer Spectra of Various Neptunium Fluorides

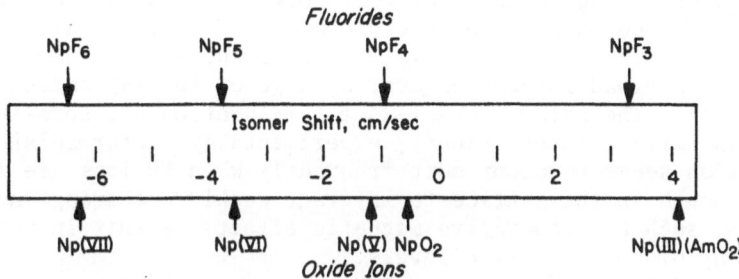

FIGURE 4.   Isomer Shifts of Np Fluorides and Np Oxide Ions

Quadrupole splitting arises from the interaction between the
nuclear quadrupole tensor and the electric field gradient tensor
due to the distribution of charges around the ion. The nuclear
quadrupole tensor is assumed to be rotating rapidly compared with
the lifetime of the emitting nucleus and can be considered a
scalar, the nuclear quadrupole moment Q. The electric field
gradient is a more-complicated function, but in general leads to
quadrupole splitting only when the field on the absorbing ^{237}Np
nucleus has an n-fold axis, where n>2. If the interaction is
axially symmetric, the quadrupole splitting of the resonance will
give a 5-line equally spaced pattern for perfect n>2 axial symmetry; if the charge symmetry is nonaxial, the pattern will no
longer have equal spacing and may be a three-line pattern with
high asymmetry.

Magnetic splitting results from the interaction of the
nuclear energy levels of the absorbing ^{237}Np nucleus with either
an internal or an external magnetic field. The unpaired electrons
of a paramagnetic ion produce a magnetic field on the nucleus;
but at room temperature, the direction of the field changes so
rapidly that the nuclear magnetic moment is unable to respond and
thus no effect results. At low temperatures, the rate of relaxation will be slower, and in cases where the absorber becomes
magnetically ordered - ferro or antiferromagnetic - the relaxation time becomes essentially infinite, and magnetic splitting
is observed. Quadrupole splitting may occur simultaneously with
magnetic splitting, but since normally E_m>>E_q, quadrupole splitting only perturbs the hyperfine splitting due to the magnetic
field (Figure 2) of ME spectra.

So far only the time-independent effects of ME spectra have
been discussed. In these effects, the relaxation time is either
much faster (single line or pure quadrupole splitting) or much
slower (magnetic splitting) than the lifetime of the emitting

^{237m}Np. When the relaxation time is of the same order as the
lifetime of ^{237m}Np, a ME spectrum results that is not easily
predictable and is often uninterpretable. ME spectra affected
by intermediate relaxation have little definition and usually
appear as a broad absorption over a range of several cm/sec
(Figure 5). The major mechanisms for relaxation are spin-spin
and spin-lattice interactions;[5] experimentally, intermediate
relaxation seems to occur most frequently when Np ions are located
close enough in the lattice to interact weakly. (Strong interactions, such as cooperative magnetic effects, result in magnetic
splitting of the hyperfine levels.)

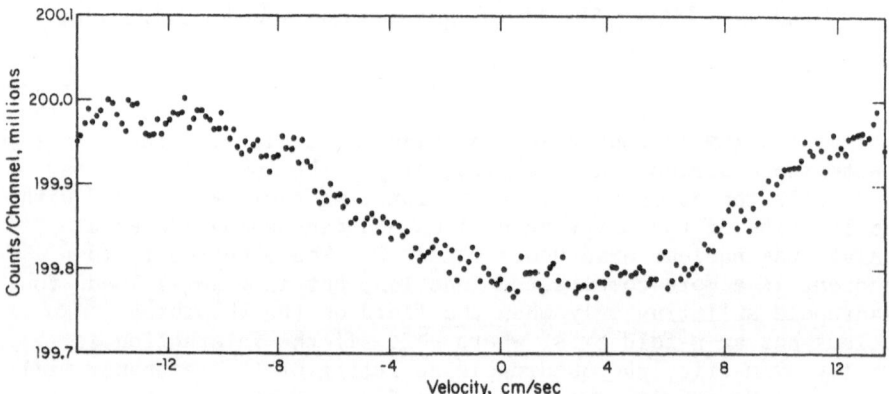

FIGURE 5. Mössbauer Spectrum of Cp$_3$NpCl

     In principle, intermediate relaxation effects can be avoided
by measuring the spectrum under conditions that increase the
relaxation time (higher temperatures) or decrease the relaxation
time (external magnetic field, low temperature; dilution of the
sample in an inert matrix). In practice, some difficulties
prevent these measures from being generally effective. The recoilless fraction is severely reduced at higher temperatures, so the
running time to obtain a spectrum may be prohibitive. Applying
an external magnetic field requires very expensive equipment, and
dilution of the sample may not be effective within any reasonable
sample size. As will be discussed later, sometimes intermediate
relaxation effects can be avoided by synthesizing a compound that
retains the essential of structure under investigation. For
example $[(C_6H_5)_4As]_2NpCl_6$ may be studied instead of the Cs_2NpCl_6
since the ME spectrum of Cs_2NpCl_6 is dominated by uninterpretable
intermediate relaxation effects. Relaxation effects are particularly important for triscyclopentadienyl neptunium(IV) compounds.

III. ME OF NEPTUNIUM - CYCLOOCTATETRAENYL COMPOUNDS

A. Np(IV) Compounds

The first compound between a quadrivalent actinide ion and the cyclooctatetraenyl dianion, $U(COT)_2$ ($COT = C_8H_8^{-2}$) was reported by Streitwieser and Müller-Westerhoff in 1968.[6] The properties of $U(COT)_2$ - solubility in aromatic solvents, an unusual optical absorption spectra, stability to water, sublimation in vacuum at $160°C$ - suggested strongly that its bonding involved some covalent contribution. Preparation and study of $Np(COT)_2$[7] and substituted analogues[8] confirmed that a strong covalent contribution was present in the bonding of $Np(COT)_2$, and by inference, in the isostructural $U(COT)_2$ and $Pu(COT)_2$ as well.

The Mössbauer spectrum of $Np(COT)_2$ (Figure 6) has combined magnetic and quadrupole splitting. The Mössbauer parameters for the three Np(IV) - cyclooctatetraenyl compounds studied are shown in Table 1.

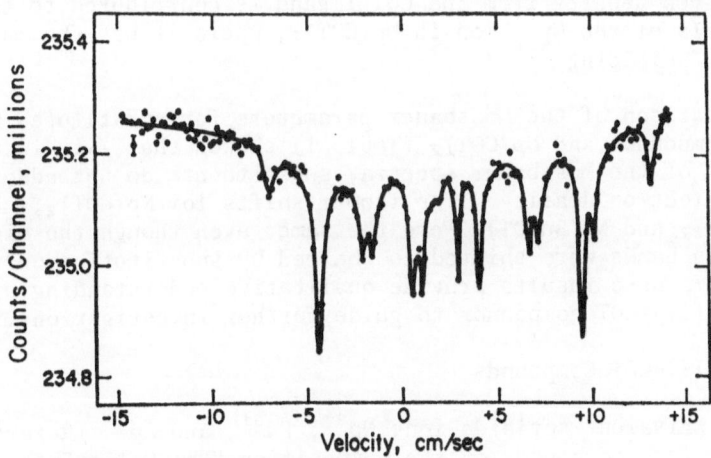

FIGURE 6. Mössbauer Spectrum of $Np(COT)_2$ at 4.2 K

Table 1.   Isomer Shifts and Hyperfine Parameters

Compound	Isomer Shift,[a] cm/sec	Magnetic Hyperfine Constant $g_0\mu_N H_{eff}$, cm/sec	Quadrupole Coupling Constant, eqQ/4, cm/sec
Np(COT)$_2$	+1.94 ±0.05	6.12 ±0.05	-0.46 ±0.05
Np(EtCOT)$_2$	+1.90 ±0.10	6.15 ±0.10	-0.5 ±0.02
Np(BuCOT)$_2$	+1.94 ±0.05	6.10 ±0.05	-0.46 ±0.05

[a]Relative to NpAl$_2$ = 0.

The most striking result from the Mössbauer spectrum of Np(COT)$_2$ is the isomer shift of +1.94 cm/sec. The isomer shift for a normal Np(IV) ion is about -0.4 cm/sec (compared with NpO$_2$, -0.56 cm/sec, or NpCl$_4$, -0.34 cm/sec) and for a Np(III) ion, about +3.5 cm/sec (NpCl$_3$, +3.54 cm/sec; NpBr$_3$, +3.64 cm/sec). Thus, the Np(COT)$_2$ isomer shift of +1.94 cm/sec shows a strong contribution to the shielding of the 6s orbitals. Presumably, the electron density from the COT ligand is contributed to the 5f orbitals of the Np^{4+} ion in Np(COT)$_2$, where it has the maximum effect in shielding.[3,4]

Comparison of the Mössbauer parameters for substituted COT-Np(IV) compounds and Np(COT)$_2$ (Table 1) showed that, within the precision of the Mössbauer spectra, substituents do not add further electron density. The isomer shifts for Np(COT)$_2$, Np(EtCOT)$_2$, and Np(BuCOT)$_2$ were the same, even though the visible absorption bands were shifted to the red by substitution on the COT ring. These results provide qualitative understanding of the actinide (IV)-COT compounds to guide further investigations.

B.   Np(III)-COT Compounds

The trivalent actinide ions Np^{3+}, Pu^{3+}, and Am^{3+} (References 9 and 10) form compounds of the composition KNp(COT)$_2 \cdot$xS (x = 1 or 2; S = solvent). These compounds are isostructural with their lanthanide analogues, which were previously discovered by Mares, Hodgson, and Streitwieser.[11] Structurally, the actinide (III) and lanthanide (III) ions are in a sandwich between two planar COT rings.[12] The Mössbauer spectra of KNp(COT)$_2 \cdot$2THF shows five-line quadrupole splitting (Figure 7) with an isomer shift of +3.92 cm/sec, a quadrupole coupling constant eqQ/4 of +0.75 cm/sec, and an asymmetry parameter of zero. This zero asymmetry parameter indicates that the Np^{3+} ion is in a site with a 3-fold or higher axis and is consistent with the D$_{8d}$ symmetry found by x-ray studies on lanthanide analogues.

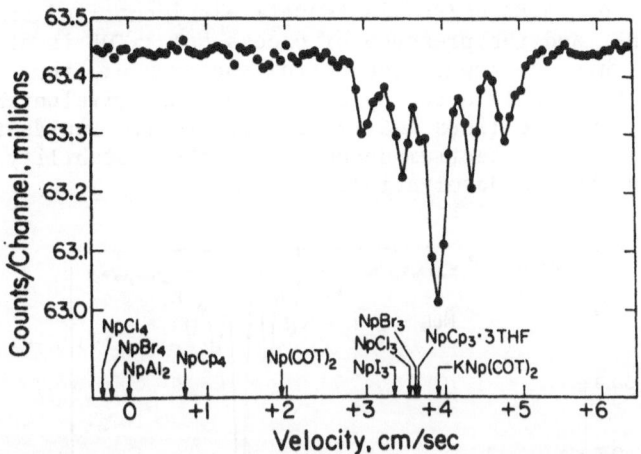

FIGURE 7.   Mössbauer Spectrum of KNp(COT)₂·2THF

The isomer shift of +3.92 cm/sec is the largest positive
shift yet observed for a Np(III) ion, but the shift is only
slightly more positive than NpCl₃ (+3.54 cm/sec) or NpCp₃
(+3.64 cm/sec). The greater isomer shift for Np^{3+} in KNp(COT)₂
indicates a covalent contribution to the bonding but not a major
contribution. A study of the absorption spectra of KAm(COT)₂
also shows that covalency in these compounds is low and is consistent with their chemical properties.[13]

The ME spectra were also used to study the reaction between
NpI₃ and K₂COT in the THF solution and to identify, tentatively,
the mixed COT-halide compounds Np(COT)I·xTHF.[14] This extremely
unstable species is crudely analogous to lanthanide-COT compounds
of the general composition Ln(COT)Cl·2THF discovered by Mares,
Hodgson, and Streitwieser.[15] ME spectra of the starting
materials, intermediates, and products were used to follow the
reaction NpI₃+K₂COT $\xrightarrow{\text{THF}}$ Np(COT)I·xTHF since no useful information could be obtained by normal techniques.

Figure 8 shows the ME spectra for a) NpI₃, b) NpI₃·6THF,
c) NpI₃·6THF and Np(COT)I·xTHF (damp-dry), and d) the decomposition products after thorough drying of c) in vacuum. The NpI₃
spectrum shows quadrupole splitting with an asymmetry parameter,
η ≈ 0 (Reference 4), which indicates the Np^{3+} ion lies on an
axis of 3-fold or higher symmetry. Solvation of NpI₃ with THF
does not change the isomer shift, but solvation does change the
quadrupole splitting pattern to give an asymmetry parameter
approaching unity; thus, solvation creates a loss of symmetry
for the Np^{3+} ion. The third spectrum shows the product of the
reaction between NpI₃·xTHF and K₂COT; the major species with

isomer shift of +3.83 cm/sec is tentatively identified as
Np(COT)I·xTHF, and the presence of excess NpI$_3$·xTHF is also
observed in this spectrum. The bottom spectrum of Figure 8 is
that of the thoroughly dried Np(COT)I·xTHF; no trivalent Np^{3+}
is observed, and the lines near +1 cm/sec are from unidentified
Np^{4+} species. This spectrum demonstrates the instability of
this compound toward desolvation.

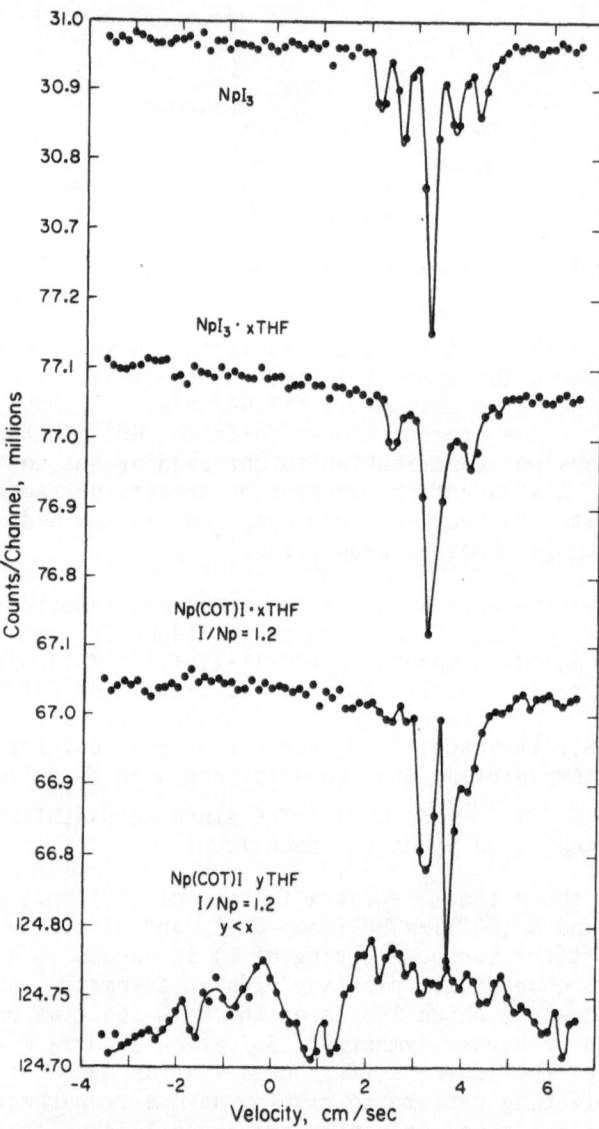

FIGURE 8. Mössbauer Spectra of NpI$_3$, NpI$_3$·xTHF, Np(COT)I·xTHF, and
Np(COT)I·yTHF

The new resonance line in spectrum c of Figure 8 is considered evidence for Np(COT)I•xTHF. The isomer shift of +3.83 cm/sec is intermediate between NpI₃ (+3.33 cm/sec) and KNp(COT)₂•2THF (+3.92 cm/sec), as would be expected for a Np(COT)I•xTHF species. The Mössbauer parameters for these compounds are listed in Table 2.

Table 2. Mössbauer Parameters of Np(III) Compounds

Compound	^{237}Np Isomer[a] Shift, cm/sec	eqQ/4, cm/sec	Asymmetry Parameter, n←eta, η
NpI₃	+3.33	0.81	∿0
NpI₃•6THF	+3.32	0.74	∿1
Np(COT)I•xTHF(?)	+3.83	-	-
K₂Np(COT)₂•2THF	+3.92	0.73	∿0

[a]Relative to NpAl₂

IV. ME OF NEPTUNIUM - CYCLOPENTADIENE COMPOUNDS

A. Np-Cp Compounds

The U(IV) compounds with Cp, UCp₃Cl (Reference 16) and UCp₄ (Reference 17) were among the first actinide organometallic compounds discovered. The analogous neptunium compounds NpCp₃Cl (Reference 18) and NpCp₄ (Reference 19) were synthesized soon after and were shown to have identical properties. The chemical properties of these compounds - solubility in aromatic solvents, sublimation in vacuum, slow reaction with water and ferrous chloride - indicated a substantial degree of covalency. The Mössbauer spectrum[20] of NpCp₄ (Figure 9) is quadrupole-split (eqQ/4 = 1.66 ±0.02 cm/sec) and has an isomer shift of +0.72 cm/sec. This isomer shift can be compared with the isomer shift of Np^{4+} in NpO₂ at -0.56 cm/sec and shows a strong covalent contribution to the bonding in NpCp₄. From the difference in the isomer shifts of +3 and +4 ionic compounds of Np, an isomer shift of +0.72 cm/sec corresponds to an effective valence of 3.7 for the Np^{4+} ion in NpCp₄. The quadrupole splitting has an asymmetry parameter η ≈ 0 which indicates a 3-fold or higher axis through the Np^{4+} site. A subsequent single-crystal x-ray study of UCp₄ (Reference 21) showed its molecular structure to be a distorted tetrahedron.[22]

The Mössbauer spectra of NpCp$_3 \cdot$3THF (Figure 9)[20] is a single sharp line with an isomer shift of +3.64 cm/sec, not appreciably different from NpCl$_3$ whose isomer shift is +3.54 cm/sec. The ME spectrum of In$_3$Np·xTHF

$$\left(\text{In = indenyl,} \quad \text{—} \bigcirc\!\!\!| \; | \right)$$

(Reference 23), is quadrupole-split (Figure 10) and has the Mössbauer parameters δ = +3.55 ±0.15 cm/sec and eqQ/4 = 1.37 ±0.07 cm/sec. The Mössbauer results for both NpCp$_3 \cdot$3THF and NpIn$_3 \cdot$xTHF indicate that both are ionic compounds, analogous to the lanthanide-Cp compounds.

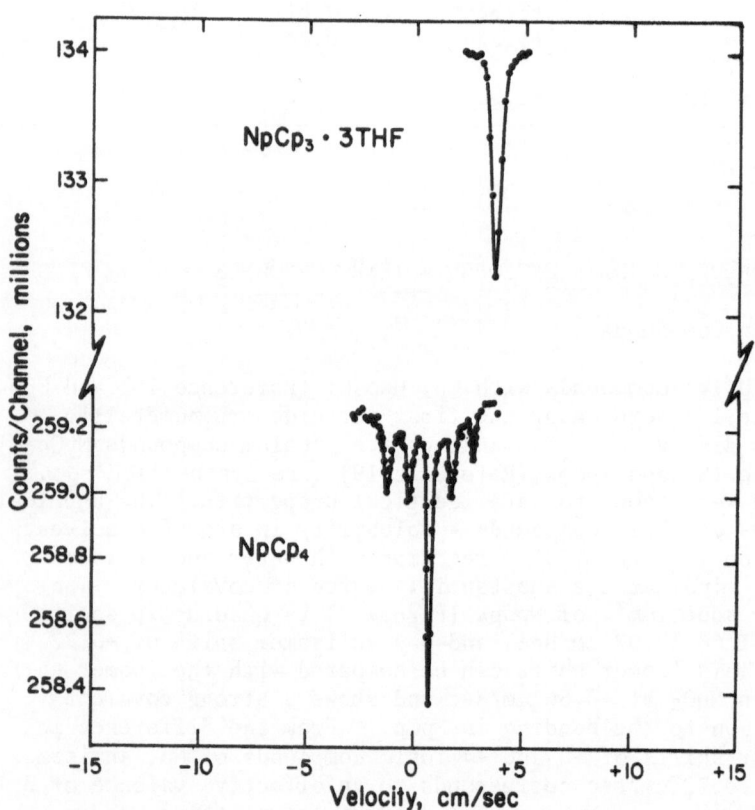

FIGURE 9. Mössbauer Spectra of NpCp$_4$ and NpCp$_3 \cdot$3THF at 4.2 K

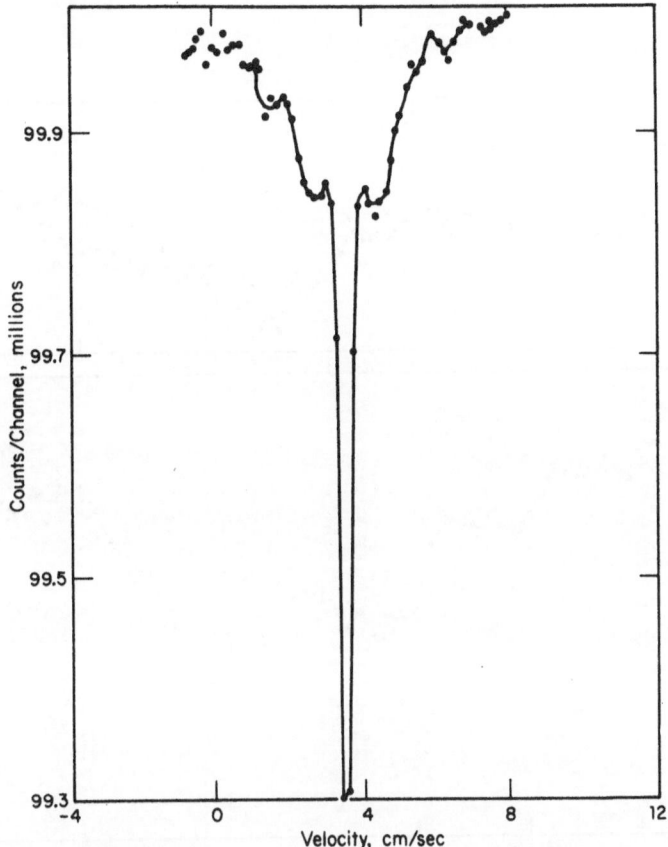

FIGURE 10. Mössbauer Spectrum of NpIn₃·xTHF

B. Cp₃NpX Compounds

The Mössbauer spectra of Cp₃NpX compounds (X = Cl, Br, BH₄) are dominated by intermediate relaxation, but offer some useful information. Cp₃NpCl (Figure 5) and (MeCp)₃NpCl (Figure 11a) have broad, featureless spectra. Assuming that the relaxation time of Cp₃NpCl is near the fast relaxation limit, an isomer shift of +1.4 ±1.0 cm/sec can be derived. Substitution of a bulkier Cp ligand, indenyl, was attempted to reduce the inter-actions responsible for the relaxation effects; the spectrum of In₃NpCl (Figure 11b)[23] still lacks sufficient definition for analysis, but appears to be approaching a magnetically split spectrum. The spectrum of Cp₃NpBr (Figure 11c)[23] showed no more resolution than Cp₃NpCl.

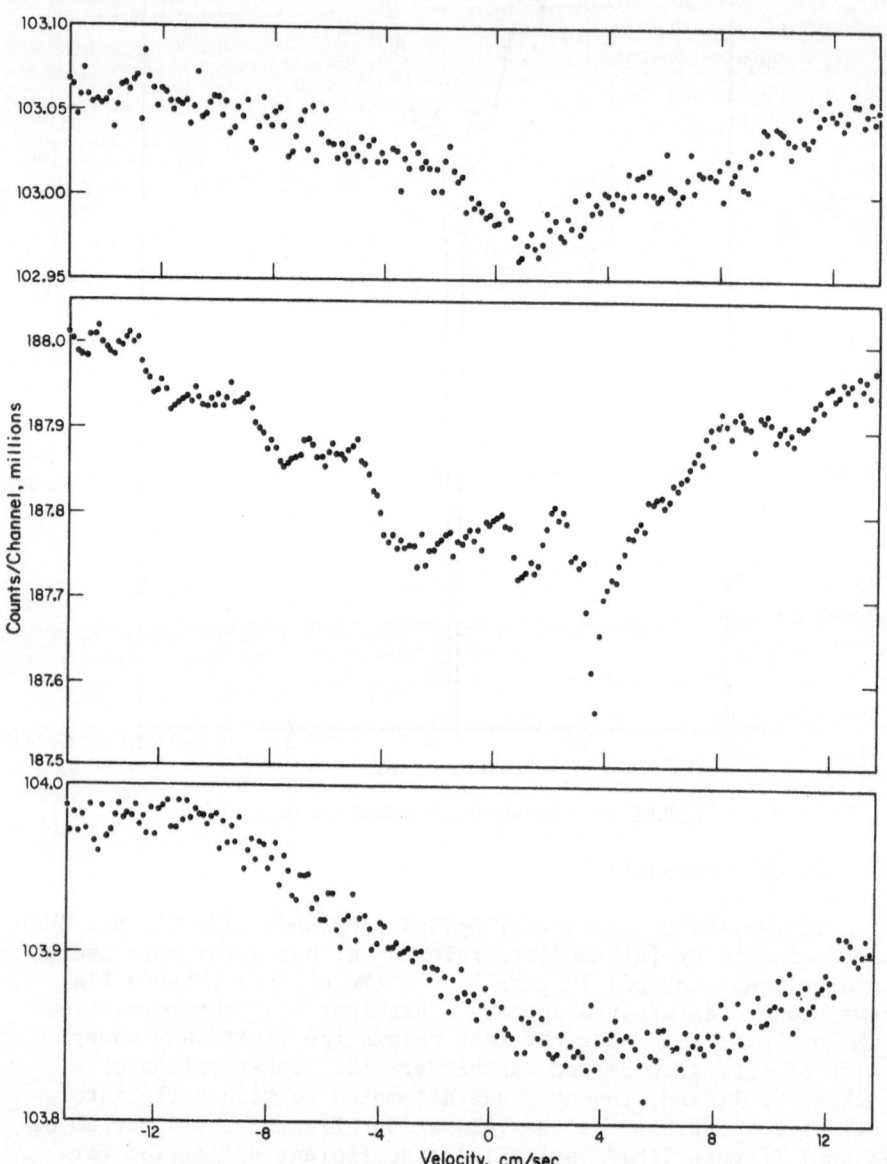

FIGURE 11. Mössbauer Spectra of a) MeCp₃NpCl, b) In₃NpCl (the Sharp Peak
at 3.5 cm/sec is In₃Np), and c) Cp₃NpBr

The Mössbauer spectrum of $(MeCp)_3NpBH_4$ (Figure 12),[23] though influenced by relaxation broadening, shows a broad, wide resonance with an isomer shift of 1.45 ±0.4 cm/sec. This large isomer shift (compared with $NpCp_4$, δ = 0.72 cm/sec) indicates a greater covalent contribution in the bonding of $(MeCp)_3NpBH_4$ than for $NpCp_4$.

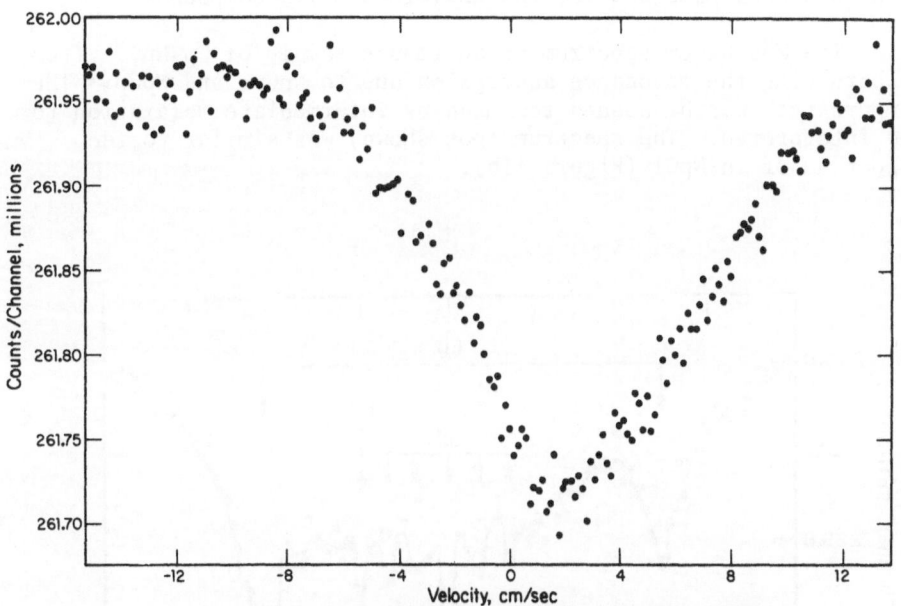

FIGURE 12. Mössbauer Spectrum of $(MeCp)_3NpBH_4$

C. $CpNp^nBu$, $Cp_3Np\phi$

Cp_3UR (where R is alkyl or aryl) compounds have been extensively studied since their discovery.[24,25,26] Since Np(IV) organometallics are less stable toward reduction than U(IV) organometallics, Cp_3Np^nBu and $Cp_3Np\phi$ could not be prepared in high purity. Gas chromatographic analyses of the products of the decomposition of Cp_3NpR compounds by ethanol showed that Cp_3Np^nBu preparations were 80 to 90% pure. The best preparation of $Cp_3Np\phi$ was 40% pure, with $NpCp_4$ and $NpCp_3Cl$ major impurities.

The Mössbauer spectrum of Cp_3Np^nBu (Figure 13)[23] showed
three species in the spectrum — the five-line spectrum of $NpCp_4$,
a single line probably due to $NpCp_3 \cdot 3THF$, and two strong
resonances broadened by relaxation effects that are interpreted
as magnetically-split resonance due to Cp_3Np^nBu. The Mössbauer
parameters are $\delta = 0.27$ cm/sec, and $g_0\mu_NH_{eff} = 5.8$ cm/sec. The
isomer shift for Cp_3Np^nBu is more negative than the shift for
$NpCp_4$ (0.72 cm/sec) and is consistent with an assignment of a
sigma bond to the Cp_3Np-^nBu link. Sigma bonding of the Cp_3U-^nBu
link has been proposed for the analogous U(IV) compounds.[24-26]

The Mössbauer spectrum of an impure sample of $Cp_3Np\phi$, after
subtracting the resonance absorption due to $NpCp_4$ and $NpCp_3 \cdot 3THF$
impurities, was broadened too much by intermediate relaxation to
be interpreted. The spectrum (not shown) was similar to the
spectrum of In_3NpCl (Figure 11b).

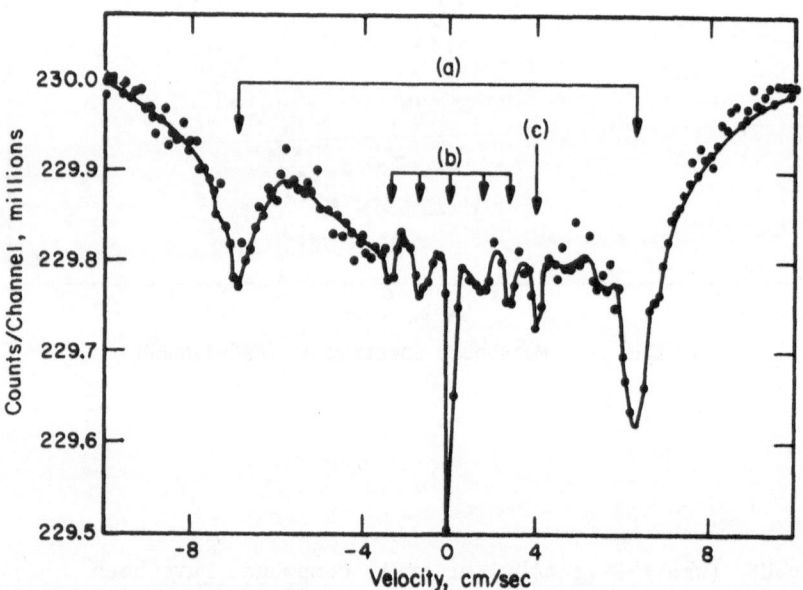

FIGURE 13. Mössbauer Spectra of a) $(C_5H_5)_3Np^nC_4H_9$, b) $(C_5H_5)_4Np$, and
c) $(C_5H_5)_3Np$

D. Cp_3NpOR

The Mössbauer spectra of $Cp_3NpOCH(CF_3)_2$ and $(MeCp)_3NpO^iC_3H_7$ have broadened resonances because of relaxation effects but are better resolved than Cp_3NpX or Cp_3NpR compounds and provide some data of interest. These compounds were synthesized by treating Cp_3NpCl with the corresponding potassium alkoxide.[27] The spectrum of $(MeCp)_3NpO^iC_3H_7$ (Figure 14) is magnetically split, with an isomer shift $\delta = 0.93 \pm 0.07$ cm/sec, and $g_0\mu_N H_{eff} = 5.72 \pm 0.40$ cm/sec. The spectrum of $Cp_3NpOCH(CF_3)_2$ (Figure 15) is more seriously affected by intermediate relaxation, but if magnetic splitting is assumed, $\delta = 0.79 \pm 0.3$ cm/sec and $g_0\mu_N H_{eff} = 5.7 \pm 1$ cm/sec. The isomer shifts of the two alkoxide derivatives are about the same, within the resolution of the spectra, and slightly greater than the isomer shift (0.72 cm/sec) of Cp_4Np.

These isomer shifts indicate that a σ-bonding alkoxy ion may be substituted for a π-bonding Cp ligand without an appreciable change in covalency of the Np^{4+} ion. This surprising result may be explained by reference to the structure of the analogous Cp_3U^+ compounds.

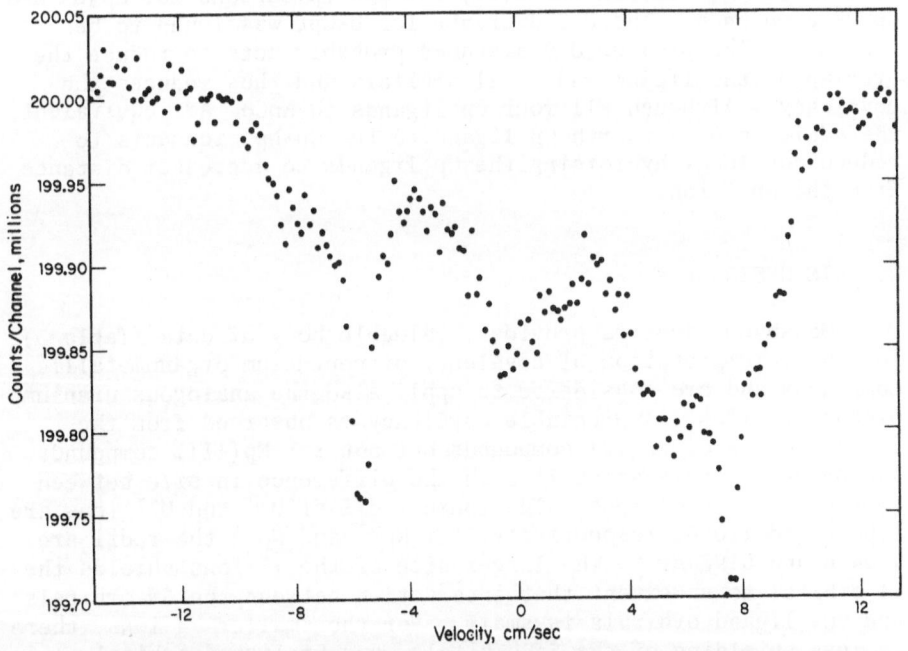

FIGURE 14. Mössbauer Spectrum of $(MeCp)_3NpOCH(CH_3)_2$

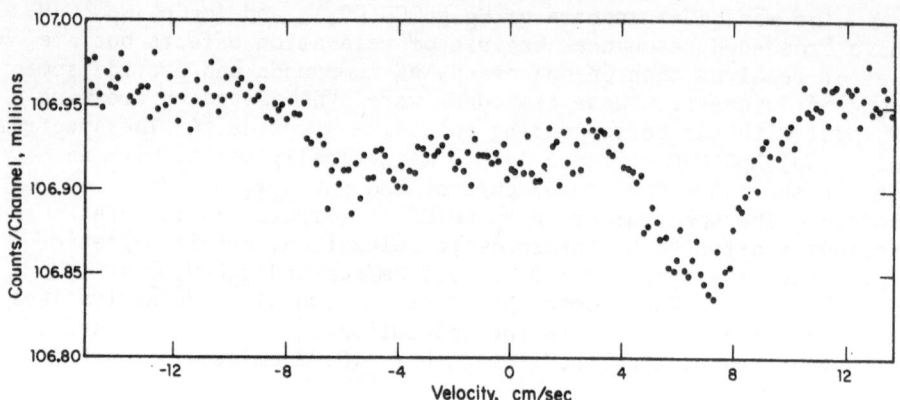

FIGURE 15. Mössbauer Spectrum of Cp₃NpOCH(CF₃)₂

Structure studies[22] on U(IV)-Cp compounds find a distance of 2.72 to 2.74 Å between the U(IV) ion and Cp-carbons for Cp₃UX and Cp₃UR compounds. The U-C distance for U-Cp₄ was found to be 2.84 Å.[21] The greater U-C distance probably acts to reduce the overlap of the ligand and metal orbitals and thus reduces the covalency. Although all four Cp ligands in NpCp₄ are equivalent, the addition of a fourth Cp ligand to the Cp₃Np⁺ ion acts to reduce covalency by forcing the Cp ligands to a greater distance from the Np⁴⁺ ion.

V.   DISCUSSION

Mössbauer spectra provide a valuable body of data (Table 3) for the interpretation of covalency of neptunium organometallic compounds and are considered to apply also the analogous uranium organometallics. Appreciable covalency is observed from the isomer shifts of Np(IV) compounds but not for Np(III) compounds. The probable explanation lies in the difference in size between +3 and +4 actinide ions. The ionic radii of U^{3+} and U^{4+} ions are 1.06 Å and 1.0 Å, respectively; for Np^{3+} and Np^{4+} the radii are 1.04 Å and 0.98 Å.[28] The larger size of the +3 ions shields the 5f orbital to an extent that interaction between the 5f orbitals and the ligand orbitals is small. For the smaller +4 ions, there is less shielding of the 5f orbitals, and 5f-ligand orbital interactions can occur.

Table 3. Mössbauer Parameters for Np-Cp and Np-COT Compounds

Compound	Isomer Shift,[a] δ, cm/sec	Quadrupole Coupling Constant eqQ/4, cm/sec	Magnetic Hyperfine Constant $g_0\mu_N H_{eff}$, cm/sec
$NpCp_4$	0.72 ±0.02	1.66 ±0.02	
$NpCp_3 \cdot 3THF$	3.65 ±0.10		
$NpIn_3 \cdot xTHF$	3.55 ±0.15	1.37 ±0.07	
$NpCp_3Cl$	1.4 ±1.0		
$Np(MeCp)_3BH_4$	1.45 ±0.4		
$NpCp_3{}^nBu$	0.27 ±0.07		5.8 ±0.2
$Np(MeCp)_3O^iC_3H_7$	0.93 ±0.07	5.0 ±1.0	5.7 ±0.2
$NpCp_3OCH(CF_3)_2$	0.79 ±0.3		5.7 ±0.5
$NpCp_3C_6H_4C_2H_5$	0.42 ±0.28		5.5 ±0.4
$NpCp_3O^iC_3H_7$	0.86 ±0.2		5.4 ±0.5
$NpCp_3O^tBu$	0.86 ±0.3		5.2 ±0.4
$Np(MeCp)Cl_3 \cdot 2THF$	-0.31 ±0.07		5.15 ±0.06
$Np(COT)_2$	1.94 ±0.05	0.46 ±0.05	6.12 ±0.05
$KNp(COT)_2$	3.92 ±0.10	0.75 ±0.10	
$Np(COT)I \cdot xTHF(?)$	3.83 ±0.10		
$NpCl_4$	-0.34) Included for		
$NpCl_3$	3.54) comparison		

[a]Relative to $\delta = 0$ for $NpAl_2$

The isomer shift of $Np(COT)_2$ indicates a greater degree of covalency than for $NpCp_4$, although if 10π electrons per COT ligand and 5π electrons per Cp ligand are considered, their covalencies could be equal. However, the U-C bond length is 2.65 Å for $U(COT)_2$, and the U-C bond is 2.81 Å for UCp_4; the crowding[22] of the Cp ligands forces longer bonds in the UCp_4. The lower covalency of $NpCp_4$ indicated by the isomer shifts in the Mössbauer spectra of the isostructural $NpCp_4$ and $Np(COT)_2$ probably arises from the differences in bond distances; the longer Np-C bond in $NpCp_4$ diminishes the overlap of 5f and Cp orbitals.

For $Np(MeCp)_3BH_4$ and $NpCp_3Cl$, the isomer shifts are 1.4 cm/sec, and the difference between 1.4 cm/sec and -0.4 cm/sec is considered to represent the covalent contribution of the three Cp ligands. BH_4^- and Cl^- ions are assumed to have little or no covalency in their bonding to Np(IV). The four $NpCp_3OR$ compounds all have isomer shifts in the range 0.8 to 0.93 cm/sec; compared with the 1.4 cm/sec shift of $NpCp_3Cl$, this demonstrates that the sigma bonding -OR group is withdrawing some of the electron density contributed by the three Cp ligands from the

Np(IV) ion. The isomer shifts of $NpCp_3^nBu$ and $NpCp_3C_6H_4C_2H_5$, 0.27 and 0.4 cm/sec, respectively, show a much stronger electron-withdrawing tendency, and verify the strong sigma bonding reported for the $-^nBu$ and $-C_6H_4C_2H_5$ ligands.[10,11] The isomer shift of $NpCp_3^nBu$ represents a withdrawal of electron density of more than 1 cm/sec with respect to $NpCp_3^+$, equivalent to about 1/4 the isomer shift difference between Np(IV) and Np(III). In view of the nature of the $Np(IV)-^nBu$ bond, it is hardly surprising that no stable Np(IV) or U(IV) tetraalkyl compounds are known.

The Mössbauer spectra of $Np(MeCp)Cl_3 \cdot 2THF$ has an isomer shift of -0.31 cm/sec, about the same as the isomer shift of $NpCl_4$ (-0.35 cm/sec). Comparing these isomer shift values, the MeCp ligand and the chloride ion are essentially equivalent in donation of the electron density to the Np^{4+} ion. This equivalence indicates that the MeCp ligand is σ-bonded in $NpMeCpCl_3 \cdot 2THF$ and infers that the Cp ligand in analogous compounds ($UCpCl_3 \cdot 2THF$, $UCpCl_3$ DME, etc.) is probably σ-bonded also. The infrared data of Bagnall, et al[29] show an average ν(U-Cp) of 262 cm^{-1} for seven compounds of the general formula $UCpX_3 \cdot xS$ (X = Cl$^-$ or Br$^-$, x = 1 or 2, S = ligand) compared with an average ν(U-Cp) of 243 cm^{-1} for UCp_3Cl and UCp_3Br. The difference suggests a difference between the U-Cp bond in UCp_3X and the U-Cp bond in UCp_3X compounds, consistent with the Mössbauer results.

ACKNOWLEDGMENT

The author is deeply indebted to John A. Stone for experimental assistance, theoretical advice, and, over all, many illuminating discussions on "what to do next."

REFERENCES

1. J. A. Stone and W. L. Pillinger. *Phys. Rev. Letters 13*
    200 (1964).

2. W. L. Pillinger and J. A. Stone. *Mössbauer Effect Methodology,*
    Vol. 4, Ed. I. J. Gruverman, Plenum Press, NY (1968) p 217-236.

3. G. M. Kalvius in "Plutonium 1970 and Other Actinides,"
    Proceedings of the 4th International Conference on Plutonium
    and Other Actinides, Santa Fe, NM, Oct. 5-9, 1970, Ed.
    W. N. Miner (Metallurgical Society of American Institute of
    Mining, Metallurgical and Petroleum Engineers, Inc. 345 E.
    47th St. NY 10017) p 296-330.

4. J. A. Stone, W. L. Pillinger and D. G. Karraker. *Inorg. Chem. 8*
    2519 (1969).

5. H. H. Wickman. *Mössbauer Effect Methodology,* Vol. 2, Ed. I. J.
    Gruverman, Plenum Press, NY (1966) p 39-66.

6. A. Streitwieser, Jr. and U. Müller-Westerhoff. *J. Am. Chem.*
    Soc. 90 7364 (1964).

7. D. G. Karraker, J. A. Stone, E. R. Jones, Jr. and N. Edelstein.
    J. Am. Chem. Soc. 92 4841 (1970).

8. D. G. Karraker. *Inorg. Chem. 12* 1105 (1973).

9. D. G. Karraker and J. A. Stone. *J. Am. Chem. Soc. 96*
    6885 (1974).

10. D. G. Karraker. *Transplutonium Elements.* Ed. W. Müller and
    R. Lindner. North-Holland, Amsterdam (1976) p 131.

11. F. Mares, K. Hodgson and A. Streitwieser, Jr.

     J. Organometallic Chem. 24 C68 (1970).

12. K. O. Hodgson and K. N. Raymond. *Inorg. Chem.* 11 3030 (1972).

13. D. G. Karraker. *J. Inorg. Nucl. Chem.* 39 87 (1977).

14. D. G. Karraker and J. A. Stone. *J. Inorg. Nucl. Chem.* 39

     2215 (1977).

15. F. Mares, K. Hodgson and A. Streitwieser, Jr.

     J. Organometallic Chem. 28 C24 (1971).

16. L. T. Reynolds and G. Wilkinson. *J. Inorg. Nucl. Chem.* 2

     246 (1956).

17. E. O. Fischer and Y. Hristidu. *Z. Naturforsch B17* 275 (1962).

18. E. O. Fischer, P. Laubereau, F. Baumgärtner and B. Kanellakopulos.

     J. Organometallic Chem. 5 583 (1966).

19. F. Baumgärtner, E. O. Fischer, B. Kanellokopulos, and P.

     Laubereau. *Angew. Chem.* 7 634 (1968).

20. D. G. Karraker and J. A. Stone. *Inorg. Chem.* 11 1742 (1972).

21. J. H. Burns. *J. Am. Chem. Soc.* 95 3815 (1973).

22. E. C. Baker, G. W. Halsted and K. N. Raymond. *Structure and*

     Bonding 25 23 (1976).

23. D. G. Karraker and J. A. Stone. (unpublished 1978).

24. G. Brandi, M. Brunelli, G. Lugli, and A. Mazzei, *Inorg. Chem.*

     Acta. 7 319 (1973).

25. M. Tsutsui, N. Ely and R. Dubois. *Accts. Chem. Res.* 9

     219 (1976).

26. T. J. Marks. *Accts. Chem. Res.* 9 223 (1976).

27. R. von Ammon, R. B. Kanellakopulos and R. D. Fischer,
     Radiochem. Acta. 11 162 (1969).

28. R. D. Shannon and C. T. Prewitt. *Acta Cryst B25* 925 (1969).

29. K. W. Bagnall, J. Edwards and A. C. Tempest. *J. Chem. Soc.*
     Dalton 295 (1978).

APPENDIX

Other Actinide and Lanthanide Mössbauer Isotopes

Recoilless resonant absorption or emission of nuclear gamma rays, now known as the Mössbauer effect, was discovered by R. L. Mössbauer in 1957.[1] Since then, over 73 isotopes of 43 elements have been shown to have a detectable Mössbauer effect, among them isotopes of all the lanthanides except cerium and lutecium, and isotopes of the first six actinides.[2,3] For a variety of reasons, only a few of these isotopes can be used to explore the chemical bonding, and none has the potential of the ^{237}Np Mössbauer effect. Actinide isotopes, other than ^{237}Np, either have very broad resonances (2 to 5 cm/sec); or the half-life of the source isotope is inconveniently short. The ^{243}Am Mössbauer effect has a very large isomer shift, 5.3 cm/sec between Am^{3+} and Am^{4+} compounds; but the disadvantages of a short source half-life (^{243}Pu, 5 hr) source and the self-damage to absorbers from the ^{243}Am radiation have prevented application to chemical problems.

Isotopes of eight lanthanides have useable Mössbauer resonances and have been exploited in studies of the magnetic properties. A potential for the study of chemical bonding is best for Sm, Eu, and Yb compounds, where compounds with both dipositive and tripositive valances can be studied. The ^{151}Eu Mössbauer effect has a well-resolved spectrum and an isomer shift of about 1.5 cm/sec between the +2 and +3 valances. The ^{151}Eu Mössbauer effect is also experimentally convenient, with either of two long-lived sources possible, and a large recoilless fraction at room temperature. Consequently, almost all of the chemical studies on lanthanide chemistry have been with europium compounds. A recent review by Barton and Greenwood[4] compiles the data for studies with ^{151}Eu Mössbauer effect.

The isomer shifts found in ^{151}Eu Mössbauer spectra fall in three ranges: Eu(II) -1.2 ±0.2 cm/sec, Eu(III) 0 ±0.1 cm/sec and metallic alloys and band systems -0.7 to -1.1 cm/sec with the EuF$_3$ isomer shift taken as zero. The isomer shift of a number of representative Eu(II) and Eu(III) compounds are compiled in the following table:

Isomer Shifts of Eu Compounds[a]

Compound	Isomer Shift[b] cm/sec
EuF_2	-1.36
$EuCl_2$	-1.37
$EuBr_2$	-1.34
$EuSO_4$	-1.40
$Eu(C_5H_5)_2$	-1.22
$Eu_2(SO_4)_3$	0.052
$EuCl_3 \cdot 6H_2O$	0.050
$Eu(C_5H_5)Cl_2 \cdot 3THF$	0.041
$Eu(C_9H_7)_3 \cdot 3THF$[c]	0.058

[a]Reference 4

[b]EuF_3 isomer shift = 0

[c]C_9H_7 = indenyl

In general, the isomer shifts are fairly constant for compounds and coordination complexes of both Eu(II) and Eu(III). A small effect in Eu(III) chelates has been attributed to a change in coordination number[5] and also to a small contribution to the 6s orbitals from the ligands.[6] No real evidence for ligand contribution to the 4f orbitals is apparent from these studies. These results conform to a model that assumes the 4f orbitals of the lanthanides are so well shielded from inter-action with ligand that covalent effects will be very small for trivalent ions, and even less for divalent ions. Appreciable covalent effects are observed in ^{237}Np Mössbauer only for Np(IV) and higher valences.

The isomer shifts for both Sm and Yb isotopes are quite small, 0.16 cm/sec for the ^{152}Sm Mössbauer effect between Sm(II) and Sm(III) and 0.063 cm/sec between Yb(II) and Yb(III) for the ^{171}Yb effect. Chemical studies of isomer shifts effects on Sm and Yb would be quite difficult to perform successfully.

REFERENCES FOR APPENDIX

1. R. L. Mössbauer. *Z. Physik 151* 124 (1958).

2. N. N. Greenwood and T. C. Gibb. *Mössbauer Spectroscopy*.
   Chapman and Hall, Ltd, London, 1971 (an excellent reference
   book on chemical studies by Mössbauer spectroscopy).

3. G. M. Kalvius, D. Cohen, B. D. Dunlap, and G. K. Shenoy.
   Phys Rev B (November 1978).

4. C. M. P. Barton and N. N. Greenwood.   *Mössbauer Effect
   Data Index*. J. G. and V. E. Stevens, Ed., Plenum Press,
   New York, NY 395-446   (1973).

5. S. Z. Ali, S. Chandra and M. L. Good.   *Proceedings of the
   9th Rare Earth Research Conference*. P. E. Field, Ed.,
   Virginia Polytechnic Institute and State University,
   Blackburg, VA p. 164-75 (1972).

6. A. H. Zakeer, I. Bliss, N. B. Keck, W. G. Bos, and P. J.
   Ouseph. *JINC 36* 2515 (1974).

PHOTOELECTRON SPECTROSCOPY

Ignazio L.Fragalà

Istituto Dipartimentale di Chimica
Università di Catania
V.le A.Doria 6   95125 Catania   Italy

1. GENERAL INTRODUCTION

When an electromagnetic radiation of sufficiently high energy
impinges on some material (in solid,liquid or gas phase) emission
of electrons occurs. This phenomenon,originally described by
Einstein as the photoelectric effect,nowadays is referred to as
photoionization or (by scientists having physical background) as
photoemission. The energetics of the process are simply defined
through the relationship:

$$E_k = h\nu - I_k$$

where E_k is the total energy shared between the emitted electron
and the recoiling ion, I_k is the ionization energy of the kth
electron in the material and,finally, $h\nu$ is the energy associated
with the ionizing monochromatic radiation. For a free molecule
it is a straightforward matter to show that the E_k term can be
equated with electron kinetic energy since the relationship of
conservation of momenta:

$$P_k \text{ (ion)} = P_e \text{ (electron)}$$

implies that the ratio of kinetic energies of photoemitted
electron to that of recoiling ion:

$$\frac{P_e^2}{2m_e} \Big/ \frac{P_k^2}{2m_k} = \frac{m_k}{m_e}$$

is very large even for the lightest ions (for instance this ratio
is ∿3600 for the H_2^+ ion). Photoionization is a threshold process:
it requires a minimal energy of irradiating photons ($h\nu_o$). This
minimal energy is,obviously,related to the first ionization energy

T. J. Marks and R. D. Fischer (eds.), Organometallics of the f-Elements, 421–466.

(or commonly first ionization potential) I_1.

If the photon possesses sufficient exciting energy (with respect to $h\nu_o$) it may ionize several electrons (in distinct one electron-one photon processes) having different ionization energies (IE). As a consequence, the photoionization process although due to monochromatic photons, will result in a polychromatic emission of electrons. Since electronic shells in molecules are quantized, the distribution of photoelectron kinetic energies will consist of a series of discrete bands. The experimental analysis of photoelectron kinetic energy is the basis of the photoelectron spectroscopy while a photoelectron (PE) spectrum consists of the plot of the number of photoelectrons connected with a particular kinetic energy. Noteworty, the most compelling feature of PE spectroscopy is its capability of singling out individual molecular subshells.

Strictly speaking, a PE spectrum should be handled in terms of many-electron states of molecular ion rather than those of one-electron molecular orbitals (MO) of the neutral system. In other words, along with the photoionization process:

$$M + h\nu \rightarrow M^+ + e$$

the ionized molecule (M^+) may be produced in various electronic states (M_i^+) having different electronic energies (E_i^+). The neutral system M, on the contrary, will be always in its ground state (E_1). So, the IEs measured in a PE spectrum:

$$I_i = E_i^+ - E_1$$

will correspond to various excited electronic states of M^+ whereas only the threshold IE can be correlated with both ground states of molecular ion and neutral system:

$$I_1 = E_1^+ - E_1$$

Accordingly, the PE spectral pattern can be considered as a picture of conceivable transitions in the molecular ion. As a matter of curiosity, the lowest IE thus far measured in organic molecules corresponds to first band in the PE spectrum of (3,3,3)Cyclazine (1). For metal derivatives, one of the lowest values is connected with $5f^{-1}$ ionization in bis(tertbutyl,cyclo-octatetraenylphosphina)uranium(IV) (2).

Photoelectrons are generally excited by soft X-ray or by UV vacuum emission of rare gases. These latter exciting sources allow studies of valence molecular subshells while the former are usually connected with investigation of 'core' orbitals. The application of UV PE technique to gases has been due to the work of Turner et al., Villesov et al., Price and coworkers and to McDowell's group. The use of this technique to investigate inorganic systems has been extensively introduced sometime later by A.F.Orchard at Oxford.

1.2 Photoionization processes:the one electron transition.

As a starting basis to understand the photoionization processes,attention must be paid to the ionization process itself. Photoionization is an electronic transition due to interaction with electric dipole radiation. Generally the only selection rule which must be fulfilled in the ionization processes is that they must be one-electron transitions. That is removal of a single electron from a neutral molecule without changing the quantum number of any electron more. For a free atom,the variation in the total angular momentum between atom plus photon and ion plus electron are restricted by the usual electric dipole selection rules:

$$\Delta L = \pm 1$$

Nevertheless this rule does not introduce any restriction in possibilities of photoionization because the ejected electron can carry whatever units of angular momentum (ranging from zero to the value needed) to fulfil the above selection rule.

1.3 Forbidden processes: shake-up and shake-off.

The only forbidden processes in PE spectroscopy are those involving two electrons simultaneously i.e. ionization of one electron along with excitation of another or simultaneous ejection of two electrons. The usual selection rules introduce these restrictions;they depend upon the peculiar characteristics of the one-electron orbital model used to describe the photoionization process. The fundamental rule implicit with this model lies in the indipendent motion of any electrons in a free molecule. Thus, any transition induced by electromagnetic radiation can only change the quantum numbers of the electron directly involved in the transition and must leave unchanged the quantum numbers of all the other electrons in the molecule.

Actually,two electron transitions which are sometimes observed in PE spectra (generally as low intensity bands) depend on correlation effects which must be introduced to describe properly the motion of electrons. Obviously the intensity of such bands is a measure of the probability of the corresponding two electron transition and,in turn,of electron correlation in molecules. For instance,the very weak band in the PE spectrum of mercury atom in the gas phase (3) corresponding to the process:

$$Hg\ (5d^{10}\ 6s^2 \ldots\ ^1S\ ) + h\nu \rightarrow Hg^+(5d^{10}\ 6p\ldots^2P) + e$$

is one example of such forbidden transitions. Another interesting example is provided by the PE spectrum of the Cs^- anion in the gas phase (4). In addition to the 'primary' band corresponding to the transition:

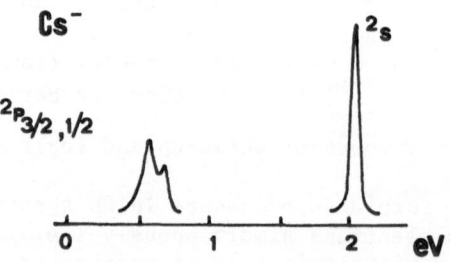

Fig.1.   UV PE spectrum of Cs⁻ under the excitation of
2.54 eV photons (argon ion laser).
Note that the abscissa scale refers to <u>electron kinetic
energy</u>. The doublet structure in the low kinetic energy
region (^{2}P 'forbidden' ion states) is due to operation
of spin-orbit coupling mechanism.   The ordering of the
two (J=3/2,J=1/2) spin-orbit components is the reverse
to that observed in,for instance,the PE spectra of noble gases because the ^{2}P ion state of Cs⁻ represents
a p^1 (one electron) rather than a p^5 (one hole) configuration (ref.4).

$$Cs^- \ (6s^2) \ + \ h\nu \rightarrow \ Cs \ (6s^1) \ + \ e \qquad ^2S$$

another quite intense doublet (the doublet structure depends on
spin-orbit coupling, <u>vide infra</u>) is observed (Fig.1). It is
related to the forbidden transition:

$$Cs^- \ (6s^2) \ + \ h\nu \rightarrow \ Cs \ (6p^1) \ + \ e \qquad ^2P$$

and it is connected with electron correlation in the unionized Cs^-
system. This sort of simultaneous co-excitation is often called
shake-up and the associated PE feature is referred to as a shake-
up satellite of the main allowed band. When the result of the
coexcitation is a final state of the doubly-ionized system, the
related process:

$$M \ + \ h\nu \rightarrow \ M^{2+} \ + 2e$$

is referred to as shake-off. In any case, it must be noticed that
forbidden processes seldom contribute to detectable structure
especially in the case of UV PE spectra. On the other hand, they
give rise to PE bands of considerably high intensity as satellites
of 'core' ionizations in X-ray PE spectra.

1.4 Autoionization.

    This kind of 'indirect' ionization process occurs when
the interaction of free molecules with electromagnetic radiation
produces a neutral excited state (with energy above the ionization
limit) which, in turn, emits electrons spontaneously:

$$M \ + \ h\nu \rightarrow \ (M^\dagger) \rightarrow \ M^+ \ + \ e$$

The first step is, obviously a resonance process governed by the
usual optical selection rules. The second one, because no radiation
is involved, is subject to the monopole selection rule; that is the
symmetries of the autoionizing state and of the final state (ion
plus free electron) must be preserved. If the energy of radiation
used to excite PE spectra coincides with that of some autoionizing
state the spectrum can be seriously distorted. Autoionizing states
are Rydberg states of the neutral molecule. They generally have
energies a few electron volts below the convergence limit of the
series to which such states belong.
    Actually, valence molecular subshells possess IEs considerably
below the energy of ionizing radiation mostly used in PE spectroscopy (He-I, 21.21 eV). As a consequence, autoionization process
is rare in He-I PE spectroscopy and even rarer in the He-II one.
In any case, a very convenient method to check wheter observed He-I
structure could be connected with some autoionizing state is
simply to switch to He-II radiation in order to excite the spectrum;
this structure must disappear. Note, however, that <u>all</u> autoionizing

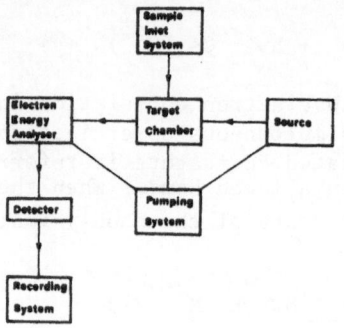

Fig.2.   The 45° parallel plate analyser.

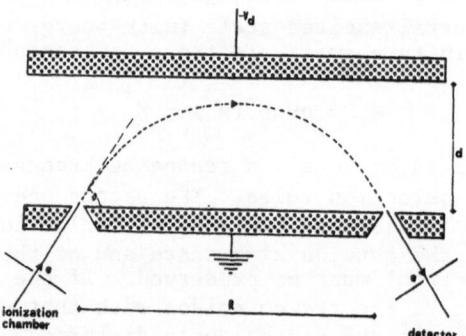

Fig.3.   Block diagram of a PE spectrometer.

states are encountered in techniques using continously varying
ionizing sources (for instance the syncrotron radiation) while
they sometimes can be encountered with rare gas resonance lines
below 21 eV (Ne, 16.67 eV or Ar, 11.62 eV).

1.5   Relation between PE spectroscopy and other electronic spectroscopy techniques.

Emission and absorption spectra of neutral systems. Even
though no direct relation exists between PE spectroscopy and
optical spectra (only the former are connected with resonance
processes) one could feel confident in obtaining a complete
mapping of both the 'occupied' and 'virtual' molecular subshells
through a combined use of the above two techniques. In fact,
within the 'orbital approximation'(see section 3.1),PE spectroscopy should provide the energies of occupied MOs whilst from
optical spectra quantities related to the energy differences
between filled (i) and virtual (j) MOs should be obtained.
However the energies of optical transitions between two i and j
levels do not afford the real energy difference between them.
In fact:

$$\Delta E \ (i \rightarrow j) \ = E_j - E_i - J_{ij} + 2K_{ij}$$

where J_{ij} and K_{ij} are respectively coulomb and exchange integrals.

So,optical spectroscopy can provide information on the energies of
virtual levels only if both J_{ij} and K_{ij} are well known. This is

not generally the case,since such evaluation requires accurate
knowledge of orbital wave-functions which,in turn, are available
only occasionally and,in any case, only for simple molecules.
As a consequence,one must be prepared to find differences in MO
ordering suggested by PE spectra with respect to that provided
by optical spectra.

Rydberg spectra of molecules. Far UV spectra of some molecules show a series of correlated bands analogous to Rydberg
series observed in spectra of atoms. The origins of these bands
converge to a continuum limit corresponding to ionization. The
related electronic transitions are due (in absorption) to promotion of one electron from some filled MO to a Rydberg orbital.
The main characteristics of such orbitals are their highly
diffuse nature which allows the excited electron to be considered
as seeing the rest of the molecule as a point positive charge.
The energy measured at the convergence limit of each series
obviously corresponds to the threshold IE of the particular electron being excited. Thus, each such Rydberg series should find a
counterpart in the PE bands related to ionizations from the corresponding molecular subshells. With respect to this particular
kind of optical spectroscopy,close identity exists between the

Table 1.

Sources generally used in X-ray and UV PE spectroscopy.

Noble gas discharge sources.

Gas	Line	eV	Intensity [†]
Helium	He-Iα	21.217	100
	He-Iβ	23.086	2
	He-Iγ	27.741	0.5
	He-IIα	40.813	<1
	He-IIβ	48.370	
Neon	Ne-I	16.670	15
		16.847	100
	Ne-II	26.813	100
		26.910	100

X-ray sources.

Anode material	Line	eV	Line halfwidth
Yttrium	YMζ	132.3	∿ 0.4
Magnesium	MgKα	1253.6	∿ 0.9
Aluminium	AlKα	1486.6	∿ 0.9

[†]

Referred to typical conditions for optimum He-I output.

PE spectra of molecules and their optical spectra.

2.   EXPERIMENTAL ASPECTS

   Basic requirements. The main components of any PE spectrometer are represented in Figure 2. It is important,at this point,
to stress the importance of vacuum conditions inside the PE
spectrometer. Photoelectrons have a very short mean free path
within a gas. Thus,it is fundamental to maintain high vacuum
conditions ($10^{-5} \sim 10^{-6}$ mm Hg) in the region including source,
target chamber ,electron energy analyser and detector.

2.2 Sources.

   The sources widely employed to excite valence electrons are
those using discharge of noble gases in an appropriate capillary.
The radiation is excited by discharge of high voltage direct
current or by microwave discharge,the former type being the one
used in the most of commercial instruments. The discharge in
pure helium (using relatively high helium pressure and low voltage)
produces the 584 Å (21.21 eV) resonance line:

$$\text{He (1s 1p, } ^1\text{P)} \rightarrow \text{He (1s}^2, \, ^1\text{S)}$$

Obviously higher members of this series (1s, np $\rightarrow$ 1s^2) are
excited as well,but their intensities are substantially lower
than that of the main line. Table 1 collects the main lines due
to the discharge of some noble gases,the notation used commonly,
their energies and their relative intensities. With the adoption
of particular conditions in helium discharge sources,it is possible to produce high photon flux at wavelength shorter than 584 Å.
This is a series of lines,with the main line at 303 Å (40.81 eV),
from excitation of ionized helium. To maximize the flux of such
photons,reduced differential helium pressure and higher voltage
(8-10 kV) must be used. Actually,other problems arise in production of high yields of these He-II photons because of the higher
sensitivity (than in He-I conditions) of gas discharge to foreign
contaminants. For this reason the best grade of boron nitride
and very pure tantalum or tungsten metals are used to build the
lamp body and the electrodes,respectively. The basic interest
in He-II radiation for PE spectroscopy is connected with its
capability of deeply probing inside the molecular subshells;
practically it makes all the valence shells accesible. Nonetheless
it must be noted that He-II emission is always accompained by
simultaneous emission of various parasitic radiations (mainly
He-I). This introduces some limitation in the accessible spectral
range of ionized photoelectrons (practically 35-18 eV kinetic
energy). Pure He-II radiation can be obtained either using a
monochromator or by total absorption of parasitic radiation

through a gas cell or through a very tin aluminium foils.

2.3 Other sources.

The sources (other than those using noble gas discharges)
most commonly used to excite PE spectra are X-ray and namely the
Al K_α (1487 eV) and Mg K_α (1254 eV) lines.[†] These higher energy
photons,even though they allow exploration of all the more tightly
bound 'core' levels,possess larger (than rare gas photons)
inherent linewidths ($\simeq 1$ eV). This fact always results in poorer
information obtained from X-ray PE spectra. Quite recently,new
sources have been introduced in ultrasoft X-ray region. They
possess photon energies in the 100-200 eV region of electromagnetic
radiation and allow one to cover the gap in photon energy between
the UV sources (up to 50 eV) and conventional X-rays (5). The
need in ultrasoft X-rays has been otherwise met by the use of
syncrotron radiation.

2.4 Electron energy analysers.

These components are,of course,the crucial part of any spectrometer. They allow discrimination of photoelectrons according
to their kinetic energy. The key in the design of such devices
is the achievment of high resolution together with high sensitivity. Even though various types of analysers have been described
up to date,we restrict our comments to dispersive analysis of
photoelectrons using electrostatic deflection analysers[††].
They,in fact,are the most used in commercial instruments (for a
more detailed discussion of analyser systems see J.H.D.Eland in
ref.6). Figure 3 illustrates the principles of dispersive analysis by reference to the 45° degree parallel plate analyser.
Photoelectrons entering the analyser through the inlet slit are
deflected by an electric field (V_d) applied between the two plates
of the analyser . For a 45° value of the incidence angle θ and
for a given deflection potential,the distance R that the electrons
travel before emerging from the outlet slit is only a function of
the electron energy (eV) :

$$R = \frac{2eV}{eV_d}$$

[†]

K_α characterizes emission processes $2p \rightarrow 1s$ after creation of
a 1s or K hole by electron bombardment.

[††]

Magnetic fields are not suitable for analysis of low energy
photoelectrons like those obtained using UV sources. In fact in
this case,weak deflecting fields are required. This enhances the
sensitivity of the analyser system to stray fields.

Since the field between the plates is V_d/d, the following relation holds:

$$\frac{V_d}{V} = \frac{2d}{R}$$

Thus, if $d = R/2$, the potential V_d to be applied to focus electrons corresponds numerically to their energy expressed in electron volts. The resolution $\Delta V/V$ of any PE instrument obviously depends both on the inherent resolution of the analyser used and on various experimental factors that, always, tend to diminish the resolution. Among the latter factors must be mentioned: i) the broadening of band widths (by Doppler effect) due to the thermal distribution of velocities of target molecules, ii) the broadening of photoionizing radiation lines, again by Doppler effect, due to the velocities of hot gases in the discharge lamp and iii) finally the contamination of slits, analyser and/or target chamber. In any case, $\Delta V/V$ being constant, the resolution (quoted as the half width of the argon line at 15.75 eV) will be greater the smaller the kinetic energy of electrons. This implies that better resolution can be achieved by using low energy ionizing sources. The usual resolution values quoted in UV PE spectroscopy are 15-25 meV using the He-I radiation and 40-70 meV using the He-II one. Better values can be achieved but only at expenses of sensitivity and simplicity of detection system.

2.5 Calibration problems.

Absolute calibration of the energy scale in PE spectroscopy is precluted by the effects of contact potentials and by the energy shift which depends on the vapor pressure of target molecules. As a consequence, PE spectra are usually calibrated by reference to lines of inert gasses simultaneously admixed into the target chamber. Table 2 lists the more useful lines used for this purpose.

2.6 Intensity measurements.

Intensities of PE bands, obviously, relate to cross-sections of MOs from which photoelectrons originate. The knowledge of such quantities can be very important in PE spectroscopy because variation of band intensities on changing the energy of the ionizing source can provide useful assignment criteria. Serious difficulties in the attainment of reliable quantities can arise because of variation of analyser sensitivity. For instance, deflection analysers used in some commercial instrument (127° sector deflection analyser) suffer a marked fall-off of sensitivity below 5-6 eV kinetic energy. Beyond these values the analyser trasmission function depends linearly on electron kinetic energy. Thus, reliable intensity data can be obtained by normalizing band intensities to unit kinetic energy.

Table 2.

Calibration lines obtained under He-Iα excitation.

Substance	Ion state	IE/eV [†]
Ar	$^2P_{3/2}$	15.76
	$^2P_{1/2}$	15.94
Xe	$^2P_{3/2}$	12.13
	$^2P_{1/2}$	13.43
He	^2S	4.99 [††]
N_2	$^2\Sigma_g^+$	15.57
	$^2\Pi_u$	16.69
CH_3I	$^2E_{3/2}$	9.53
	$^2E_{1/2}$	10.16

[†] All the values are referred to the apparent He-I IE scale.
[††] Under He-II excitation.

3. ELECTRONIC ENERGIES AND IONIC STATES

Basically a PE spectrum furnishes values of IEs which,in turn,can be related to particular states of the molecular ion. In some case both adiabatic and vertical IEs can be measured. This possibility is peculiar of PE spectroscopy. The adiabatic IE can be defined as the energy difference between a particular electronic state of unipositive ion (in its lowest vibrational and rotational levels) and the ground state (electronic,vibrational and rotational) of the neutral molecule (0-0 transition). Adiabatic transitions are sometimes very weak (substantial changes in molecular geometry being involved on ionization) but can be detected in PE spectra of very simple molecules. Even in the case of triatomic molecules the adiabatic IEs sometimes cannot be identified. Unless the PE band shape allows one to identify the 0-0 transition ,it is usual to connect the IE corresponding to the onset of any broad band with the adiabatic value.

The vertical IE can be defined as the energy difference between some particular electronic state of molecular ion and the ground state of the molecule. It corresponds to the maximum of the PE band.

3.2 Orbital approximation:Koopmans theorem.

Although it has been stressed that PE spectra should be interpreted in terms of many-electron states of molecular ion, quite often details of PE spectra are explained more simply in terms of the molecular orbital structure of the neutral system. This procedure which originates from the long familiarity with such concepts,assumes that the IE of the ith electron can be equated to the negative value of its self-consistent field (SCF) orbital energy :

$$I_i = - \varepsilon_i^{SCF}$$

This is Koopmans' approximation.

There are two main errors in the above relation. First, like all one-electron SCF models,it ignores correlation in electron motion. Techniques of configuration interaction may allow one to overcome this limitation but with the final result that individual MOs have lost any meaning. Secondly,Koopmans' approximation involves the neglect of orbital rescaling and relaxation on ionization. In other words,all the orbitals are assumed frozen upon ionization. This is a very drastic assumption which ignores the fact that simultaneous with the ionization process,all the orbitals tend to relax (expand or contract) in order to achieve a more stable state. As a consequence,the IEs evaluated through the above relation will always be larger than the actual values including relaxation. On the balance,correlation energies are dominated by two-electron contributions so that correlation effects tend to make $-\varepsilon_i$ smaller than I_i.

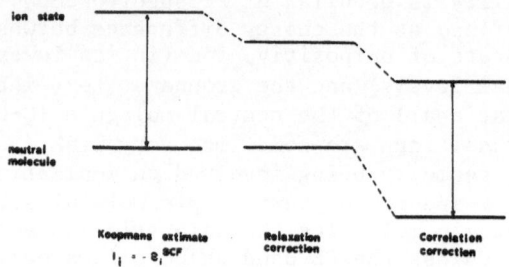

Fig.4.   The corrections to Koopmans'approximation.

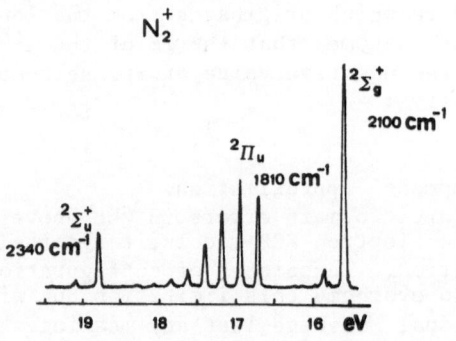

Fig.5.   The He-I PE spectrum of molecular nitrogen.

Overall correlation and relaxation correction to Koopmans'esti-
mate tend to cancel each other so that the application of
Koopmans' approximation in PE spectroscopy is not a bad practice
(Fig.4).

Deviation of experimental IEs from SCF eigenvalues are,
generally,comparable for all valence electrons in organic molecules. Thus,even if the numerical results of an SCF calculation
cannot be taken too seriously,the ordering of MOs inferred from
such a calculation is the same as the experimental sequence of
PE bands. Only in some cases deviations are large enough to
upset the expected sequence of ionizations. This is true of
molecular nitrogen. The first two bands in its PE spectrum
(Fig.5) have been unequivocally attributed,respectively, to
ionization from σ_g^+ (2p) and π_u (2p) MOs (7). However,according
to SCF theory,the expected sequence is just the reverse ($\pi_u > \sigma_g^+$).
This inversion is certainly the result of electron correlation
in the molecular ion N_2^+.

Relaxation effects can play an important role in determining
the final ordering of ionization in molecules containing transition metals (say organometallic or classical coordination compounds). This because large relative relaxation energies are
connected with nd^{-1} ionizations (8). These problems,however,
will be treated in more detail further.

Finally we consider the open-shell molecules. In these
cases Koopmans'theorem does not apply. The reasons are connected
to the possibility that,in open-shell systems,various ionic states
can be obtained upon ionization of one electron from the same MO.
In this respect,the PE spectrum of NO (Fig.6) provides an interesting example of such effects. NO possesses the valence electronic configuration :

$$(\sigma)^2 (\pi)^4 (\pi^*)^1$$

Ionization of the unpaired antibonding electron results in
the production of only the $^1\Sigma^+$,$(\sigma)^2 (\pi)^4$,ground state of the
unipositive ion;the first band accounts for this ionization. The
rather complicated structure in the 15.5-20 eV spectral region
arises from the variety of final states produced upon removal of
one electron from the remaining π or σ MOs. Ionization of the π
subshell produces six different states of NO^+. Namely,the states

$$3,1_\Delta \qquad 3,1_{\Sigma^+} \qquad 3,1_{\Sigma^-}$$

result from the electrostatic coupling between the π hole and the
unpaired electron ($\pi^*)^1$. Similarly,ionization of the σ subshell
produces two different states:

$$^3\Pi \quad \text{and} \quad ^1\Pi$$

The energy ordering of all these eight states has been set-up as
pictured in Figure 6 for various reasons (7).

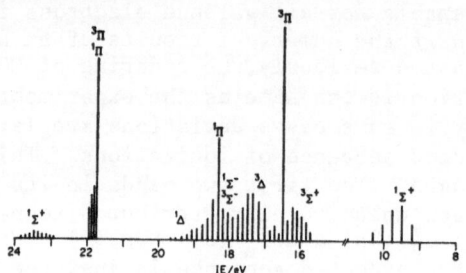

Fig.6.   The He-I PE spectrum of NO.

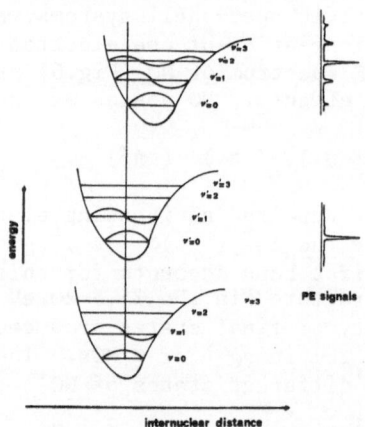

Fig.7.   Potential energy curves for a neutral molecule and some of its ionized states (see text).

Basically this example indicates two facts: (i) when dealing with open-shell systems, even in the case of simple molecules, rather complicated PE spectra must be expected; (ii) the one-to-one correspondence between PE spectral bands and MOs of the neutral molecule is not always found.

3.3 Vibrational fine structure.

PE bands, at least in the case of very simple molecules, may show structures due to vibrational excitations. Because in light molecules only the ground vibrational state has significant population at room temperature, details of PE bands vibrational in origin must relate to energy levels of molecular ion. Light diatomic molecules are the best examples of well resolved vibrational structures (see for example Fig.5). In the case of polyatomic molecules, vibrational details are seldom resolved. Various factors contribute to this fact : (i) small values of vibrational quanta, (ii) simultaneous co-excitation of various vibrational quanta and (iii) thermal population of excited vibrational states in neutral molecule. Resolved vibrational structures provide direct information on the nature of ionic state produced. In fact, ionizations are very fast processes and can be treated in terms of the Franck-Condon principle. Consider Figure 7 representing potential energy curves (a) for a neutral diatomic molecule A-B in its ground vibrational level, (b) for the corresponding molecular ion A-B$^+$ having the same internuclear distance and (c) for a state of A-B$^+$ where the bond strength has been weakened. Probabilities of vibrational transitions $0 \rightarrow v$ (v is the vibrational quantum number in the ion) are:

$$P\ (0,v\ ) = \left| \int \psi_0\ \psi_v^+\ dx \right|^{\,2}$$

That is the square of overlap integrals between initial and final state vibrational wave functions.

In the case of ionizations of non bonding electrons (the internuclear distance is preserved , curves a and b in the Figure) the 0-0' overlap will be substantial and the corresponding PE band will consist of a very short vibrational progression , probabilities for transitions other than 0→0' being very small. The first and the third bands in PE spectrum of N_2 are representative of such cases. If a bonding electron is ionized, the internuclear distance will increase in the molecular ion (curve c). Several 0→ v' transitions have significant probabilities and an extensive vibrational progression will be attached to the corresponding PE band. The same is true for antibonding electrons with the only difference that, in this case, shortening of the A-B$^+$ distance occurs. It transpires that the simple comparison of vibrational quanta of the molecular ion (the spacing in the vibrational progression) with those in the neutral molecule (as deduced from IR or Raman spectra) allows one to distinguish bonding or anti-

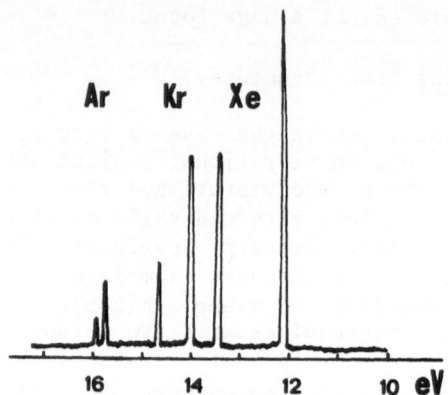

Fig.8.   The He-I PE spectra of noble gases.

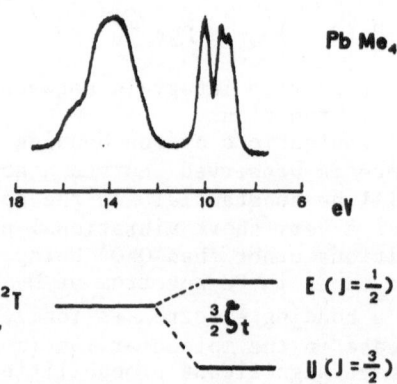

Fig.9.   The UV PE spectrum of tetramethyl lead under
He-I excitation. The main doublet structure in the
low IE region refers to the ^2T components split by
spin-orbit coupling. The incipient splitting of the
first component is due to Jahn-Teller effect.

bonding electrons. Actually, a reduction in frequency will be observed for bonding electrons;the contrary for antibonding ones.

3.4 Photoelectron bands related to degenerate ionic states.

Photoionization from fully occupied degenerate MOs results in orbitally degenerate electronic states of molecular ions. Their degeneracy can be lifted either by magnetic coupling between spin and orbital angular momenta (spin-orbit coupling) or because of the unstability of degenerate electronic state toward distortion (Jahn-Teller effect). If the effects are strong, the number of PE bands can be as high as the number of electron pairs in the original degenerate MO. Orbitally degenerate ionic states can arise also from ionization of one closed shell in an open shell system but,in any case,the mechanisms which lift the degeneracy of such states are always the same.

3.5 Spin-orbit coupling.

The mechanism of spin-orbit coupling can be easily illustrated by considering simple atoms. For instance,ionization of one electron from np^6 ground configuration of noble gases only yields the ^2P ion state. Its degeneracy is resolved according to the two possible values of total electronic angular momentum:

$$J = L + S = 1 \mp 1/2$$

which result from vector combination of both spin and angular momenta. As a consequence,PE spectra of noble gases (Fig.8) consist of well defined doublets whose separation increases on descending the group in the periodic table.

In molecules,the mechanism of spin-orbit coupling is not so obvious as in atoms. Here,the electronic angular momentum does not behave so simply as in the atomic case. For instance in highly symmetric cubic molecules (tetrahedral or octahedral) the orbital degeneracy can display all the values 1,2 and 3 while combined spin-orbital degeneracies span values up to 4. In such cases the number and characteristics of ion states resolved by spin-orbit coupling can be determined only through the use of extented point group.

Effects of spin-orbit coupling are evident in the PE spectra pictured in Figure 9. The first band in PE spectra of IVB group tetramethyl derivatives is resolved in two main components † (A,A') only in $(CH_3)_4Pb$. These bands relate to ^2T ion state produced by removing one electron from the t_2 MO (σ_{Pb-C}). The degeneracy

---

† The incipient splitting of band A is due to Jahn-Teller effect (vide infra).

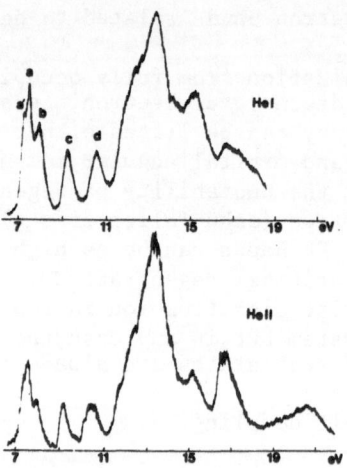

Fig.10.   The He-I and He-II PE spectra of $(C_7H_8)W(CO)_3$.

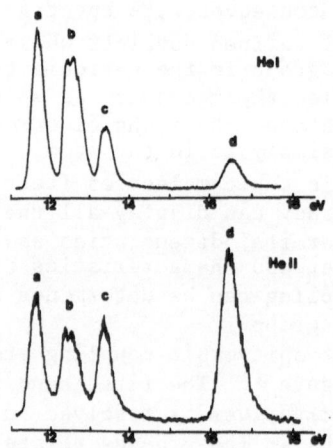

Fig.11.   The He-I and He-II PE spectra of CCl_4.

of this state is resolved by spin-orbit coupling into the U' and E" multiplet components. It transpires that the relative intensities of bands A and A' (2:1) reproduce well the relative degeneracies of U' and E" components (4:2) while the multiplet separation (≈ 0.9 eV) is well within the limit $3/2\ \zeta_{6p}$ determined by the spin-orbit coupling constant of 6p AOs in a free lead atom (9).

Similarly the well resolved splitting of first band in PE spectrum of $W(CO)_6$ has been explained (10) in terms of multiplet splitting of the $^2T_{2g}$ ion state into U' and E" terms (double group O*).

Molecules not having cubic symmetry can possess a twofold orbital degeneracy only if their symmetry allows a proper rotational axis of order three or higher. In this case the ^2E state produced by ionizing one electron from the degenerate orbital (symmetry label e) can be split,by spin-orbit coupling,into two components each having a total degeneracy of two.

Finally,it is worth mentioning that in some cases details of PE spectra can be understood in terms of spin-orbit coupling even though molecular symmetry precludes any orbital degeneracy. For instance,the splitting of first band in PE spectrum of low symmetry (C_s) (C_7H_8) $W(CO)_3$ (Fig.10) is certainly due to spin-orbit coupling effect not completely quenched by the low symmetry. Such effects are mediated by off-diagonal terms in the spin-orbit matrix (11).

3.6 Vibronic coupling: Jahn-Teller effect.

Interaction between molecular vibrations and orbital motion of electrons allows excitation of non-totally symmetric molecular vibrations during an electronic transition (breakdown of the Born-Oppenheimer approximation). As an extreme manifestation of such an interaction,it happens that degenerate electronic states in non-linear molecules become unstable toward distortion (Jahn-Teller theorem). In fact distortion due to non-totally symmetric vibrations will result in a lower molecular symmetry,with consequent lifting of orbital degeneracy. The influences of the Jahn-Teller effect are apparent in PE spectrum of $(CH_3)_4Pb$ (Fig.9) where the first band A shows an incipient splitting because of removal of the degeneracy of the correspondent ion state by Jahn-Teller effect.

3.7 Photoelectron cross-sections.

In a PE spectrum the parameters which can afford information on the electronic properties of molecules under study are (i) the IEs and (ii) the intensities of various bands. In the previous sections PE band intensities were ignored,but in practice they can be crucial guides in the interpretation of PE spectra. The intensity of any PE band obviously relates to the probability of transition between electronic molecular ground state and a par-

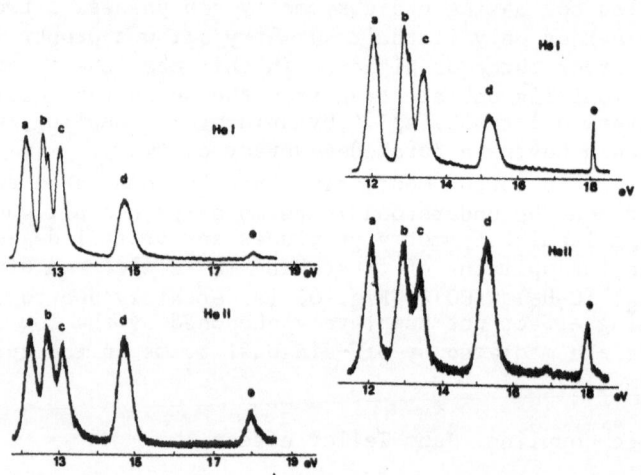

Fig.12.   The He-I and He-II PE spectra of GeCl$_4$
(left side) and of SiCl$_4$ (right side).

ticular final ion state. Accordingly,it transpires,the partial
photoelectron cross-section should be proportional to the statistical weight of the ion state produced. More simply,a linear
relation should exist between the occupancy of the ionized MO
and the intensity of the corresponding PE band,at least for MOs
having the same atomic orbital composition and IE. However such
a linear relations cannot always be verified.

Consider,for example,the He-I and He-II spectra of CCl_4
reported in Figure 11. In the He-I spectrum (12) the ratios
between intensities of the first three bands are 1.8:2.7:1.0.
These values,even sufficient to indicate an energy ordering of
the corresponding MOs $t_1 > t_2 >$ e,are far from those expected on
occupancy ground (3:3:2). In the He-II spectrum different ratios
can be measured (1:1:1) but,again,not coincident with statistical
ones (13).

A rigorous theory of PE cross-sections is beyond the scope
of this chapter,so,we shall be solely concerned with a simple
approach to the problem:that is the Gelius model (14). Within
the framework of Gelius'model,the partial PE cross-section (σ_i)
of the ith MO is given by :

$$\sigma_i = \Sigma_{ja} N(ja) \cdot \sigma(ja)$$

This means that σ_i results from the contribution of each atomic
orbital (j) on atomic centers (a) weighted by its effective occupancy in the MO in question. Application of Gelius'model allows
one,in some cases,to extimate the contributions of various AOs to
the MO responsible for a particular PE band. In this context the
following evaluation criteria can be used:

Ligand cross-sections. (i) Ionization from 'organic'ligands
present in metal complexes give rise to PE bands in both the 7-
12 eV (π system) and 12-16 eV (σ system) spectral region†.
The trend in relative cross-sections is such that PE bands related to π systems increase in intensity on passing from He-I to
He-II radiation when compared to PE band of σ systems. This is
an indirect effect due to the contribution,in σ systems,of H 1s
AO whose PE cross-section is reduced at He-II wavelengths. Only
in He-II spectra are ionizations from MOs mostly C 2s based as
detectable in the 18-20 eV IE region. This depends in part on
instrumental factors but must be attributed mainly to the substantially higher C 2s cross-sections under He-II excitation.

(ii) Ionizations from halogen np based MOs give PE bands
which have reduced He-II intensities (with respect to He-I). The
trends generally observed suggest the following sequence of He-II

---

† IEs referred to the apparent He-I scale.

cross-sections:

$$Br > Cl > I$$

Metal cross-sections. (i) ns cross-sections. Figure 12 pictures He-I and He-II PE spectra of IVB group tetrahalides. The band e has been related to the a_1 MO which is mainly ns in character. The trend in relative intensities suggests the following ordering for He-II metal ns cross-sections:

$$4s > 3s > 5s$$

(ii) np cross-sections. Reference again to the same spectra of Figure 12 allows one to gauge the relative magnitude of He-II cross-sections of metal np AOs. The intensity variation of band d (t_2 MO mainly np in nature) indicates the following sequence of He-II cross-sections:

$$4p > 3p > 5p$$

(iii) nd cross-sections. PE bands related to nd subshells generally have higher (with respect to He-I) He-II intensities. Furthermore, information obtained from PE spectra of complexes of metallic elements of various groups of the Periodic Table clearly indicate that such effects are strongly reduced on going toward heavier elements in the group (heavy metal effect) (10).

(iv) 5f cross-sections. PE spectra of actinide complexes thus far recorded (vide infra) indicate that the 5f cross-sections increase dramatically upon switching to He-II radiation. At the He-II wavelength the cross-sections exceed those of MOs which are C 2p, O 2p or Cl 2p based.

4.   PHOTOELECTRON SPECTRA OF TRANSITION METAL COMPLEXES

A.   Closed shell systems

Interpretation and comparison with theoretical calculations. PE spectroscopy has been applied to probe, often with great detail, on the electronic structure of inorganic substances. The following section, without any pretention of a complete survey, will be devoted to illustrating both the methods used to interpret PE spectra and, in turn, the information which can be drawn from them. The sequence of MOs in molecules can be delineated either through MO calculations (at various levels of sophistication) or through a qualitative 'localized bonding model'. This latter method assumes that details of the bonding in molecules can be reproduced by the interaction of appropriate sets of basis orbitals which correspond to well known conventional localized bonds. The ordering of various final MOs is dictated both by the relative

energies of the starting basis orbitals and by the magnitude of
interaction between them.   Sometimes the use of SCF-MO calculation in PE spectroscopy is precluded by large relaxation effects
upon ionization.   One of the most celebrated examples of breakdown of Koopmans theorem are the π-allyl complexes.   PE spectrum
of bis(π-allyl)nickel has been assigned (15) assuming that first
three bands correspond to ionization from nearly pure Ni 3d MOs.
In the framework of Koopmans'approximation,this implies that the
same Ni 3d MOs are the upper filled ones.   This interpretation
has been questioned by Veillard et al.(16) whose ab initio calculations indicated that the upper filled MOs in $(C_3H_5)_2Ni$ are
mainly ligand in character whilst those having Ni 3d contribution
are rather internal.   Nevertheless ionizations from these latter
MOs still remain those at the lowest IEs,relaxation energies
(evaluated through ΔSCF calculation †) associated with these ionizations being very large. However,despite the accuracy of the
above theoretical calculations,there remain ambiguities in the
assignment; quite recent He-I and He-II measurements (17) on
$(C_3H_5)_2Ni$ and related complexes have suggested that the ^2A ionic
state ,originating from a MO ligand in character $(7a_u)$,contributes
to the second band in the spectra of these complexes.

4.2 Carbonyl systems. ·

     Let us consider first carbonyl systems which provide a good
example of application,but also of limitation,of sophisticated
MO calculation in PE spectroscopy.

     Tetrahedral systems (T_d point group):Ni(CO)$_4$. The main fea-
tures of a 'localized bond' scheme applied to this system are the
splitting of Ni 3d AOs (ligand field effect) and the bonding interaction involving symmetry adapted linear combinations(SALCs)
from ligand 5σ AOs. A more detailed scheme would take into
account other interactions involving SALCs from ligand $4\sigma,1\pi$,and
2π MOs. However in such a complex system it is difficult to
gauge the extent of various interactions.   Unfortunately,even
sophisticated MO calculations,though providing quite the same
ordering of eigenvalues,do not improve the picture of the bonding.
For instance,a multiple scattering X_α (MSX$_\alpha$) calculation (18)
emphasizes only σ interactions while discrete variational X_α
(DVX$_\alpha$) (19) and Hartree-Fock calculations indicate appreciable
π back-bonding. But,according to all the various calculations
including ΔSCF ones (20),the first two bands of the PE spectrum
of Ni(CO)$_4$ should be associated with ionizations from 9t$_2$ and 2e
MOs (Ni 3d in nature) respectively;the amorphous nature of the

---

†

     ΔSCF IEs are computed as the energy differences between
total energies of various ionic states and that of ground state.

higher IE spectral region and discrepacies in MO calculations
leave the rest of the spectrum unassigned(20).

Octahedral systems (O_h point group):VIA group hexacarbonyls.

The same problems underlined in the case of $Ni(CO)_4$ again limit
the assignment of PE spectra of $M(CO)_6$ systems. According to
X_α or Hartree-Fock calculations (21),there is certainty only for
the assignment of first band in $M(CO)_6$ PE spectra (10) to ionization of 'metal' t_{2g} MOs. This assignment is in tune with two
experimental observations: (i) the peculiar trend,along the
series,of intensity variation relative to first bands on passing
from He-I to He-II radiation (heavy metal effect) and (ii) the
evident splitting of first band in PE spectrum of $W(CO)_6$ due to
spin-orbit coupling. The assignment of bands present in the
higher IE region (>13 eV) can be made only in terms of 'spectral
zones corresponding to a certain group of ionizations. This is
partly because the ordering of MOs depends upon the method of
calculation and partly because details of band structures are
not always the same in the published spectra.

Mixed π-olefin-carbonyl complexes. The following examples
illustrate the capability of combined He-I/He-II technique in
elucidating assignment of PE spectra of complex systems.
Figure 10 displays He-I and He-II spectra of $(C_7H_8)W(CO)_3$
molecule. The most notable differences with respect to corresponding spectra of simpler hexacarbonyls (10) are two new bands
labelled c and d in the figure,and the splitting of the first band
into two main components. These latter band clearly relate to
ionizations from MOs mainly nd in character . They correspond
to the MO $9t_{2g}$ of hexacarbonyls but are split by the effect of
lower symmetry. Such an assignment is corroborated by the
'heavy metal effect' observed along the series on passing from
He-I to He-II radiation (22) and by the splitting of the first
band due to spin-orbit coupling (see Sec.3.5). Furthermore the
bands c and d have been assigned to ionizations of π MOs localized on the C_7H_8 ligand. However the different variations of intensity of these bands on passing from He-I to He-II excitation
suggest an assignment (22) at variance to that proposed previously.
In particular the band c must be assigned to ionization of a
π system (only C 2p in nature) whilst the band d,which is of
lower intensity (with respect to band c) in the He-II spectrum,
correlates with some σ system having significant contributions
both from H 1s and C 2p AOs.
PE spectra of $(C_5H_5)_2Ti(CO)_2$ -(C_5H_5)=cyclopentadienyl anion-
(Fig.13) do not pose serious assignment problems (23). Band a,
whose relative intensity strongly increases in the He-II spectrum,
correlates with a MO mainly Ti 3d in nature; band b is due to
electron ejection from SALCs originating from cyclopentadienyl e_1
(D_{5h}) MOs :corresponding bands appear in the same spectral region

of PE spectra of several parent systems. Finally the band envelope in the 12-15 eV region relates to σ systems localized on the ligands. However the most interesting aspects of the case in question are some details of PE spectra which afford a very definite picture of the metal-ligand bonding: (i) The marked vibrational progression ($h\nu=1900$ cm^{-1}) associated with band a correlates with C-O stretching mode in the neutral molecule. This indicates significant mixing of Ti 3d AOs and CO MOs and, consequently, a strong titanium-carbonyl bonding. (ii) The band b which does not show evidence of any splitting (the corresponding band in PE spectrum of $(C_5H_5)_2TiCl_2$ (24) is split in four components) indicates that the titanium atom acts as a small perturbation on the cyclopentadienyl ligand.

4.3 Cyclopentadienyl systems.

Dealing with bonding in cyclopentadienyl systems, it is well known (25) that the cyclopentadienyl (cp) ring MOs which must be considered are the symmetry orbitals (SOs) $a_2''<e_1''<e_2''(D_{5h})$. The ordering is as suggested by the simple Hückel theory; among them the a_2'' and e_1'' are filled while, based on PE spectra, only the second seems to be generally involved in metal-ligand bonding.

Half-sandwich systems (C_{5v} point group) : Tlcp. Tlcp is a rare example of a half-sandwich molecule. The quite simple PE spectrum (Fig.14) has been interpreted (26) assigning the band a to ionization of e_1 (π) cp MO. Band b certainly relates to a MO of a_1 symmetry but it is difficult to decide whether dominant contribution comes from a_1 (cp π MO) or from thallium $6s^2$ (a_1) inert lone pair. However the changing intensity pattern between He-I and He-II spectra argues for a majority of metal 6s character in this MO: the non statistical branching ratio between bands a and b in the He-II spectrum would be particularly difficult to understand were they to be associated with MOs of similar composition. The bands f, f' in the higher IE region correspond to the ^2D multiplet states expected for a $(5d)^9$ ion configuration. No fine structure, due to ligand field effects, is associated with these bands, thus, arguing for a weak electrostatic field experienced by the inner 5d subshells ($5d^{-1}$ ionizations in Tl(I) halides (27) display such fine structure).

Sandwich systems (D_{5h} point group): main group and transition metal MCp_2 complexes. In the above symmetry, the two MOs of type e_1 localized on the cyclopentadienyl rings transform as e_{1g} and e_{1u}. These levels are expected to be split due to non-bonding interligand interactions. Furthermore the e_{1u} MO in main group Mcp_2 complexes, can be stabilized via interaction with metal p AOs. The proposed assignment of the PE spectrum of $Mgcp_2$ (25) follows from the above arguments. The first two bands have been associated with the ^2E ion states produced upon removal of electrons

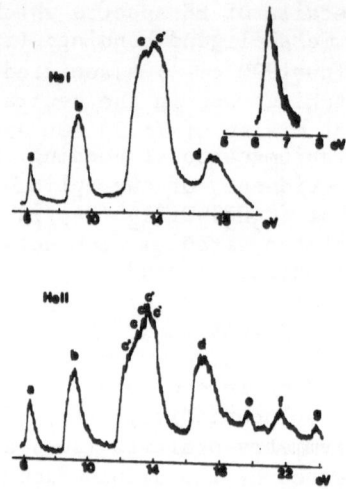

Fig.13.   The He-I and He-II PE spectra of mixed olefin-carbonyl complex $(C_5H_5)_2Ti(CO)_2$.

Note the vibrational structure of the $3d^{-1}$ band.

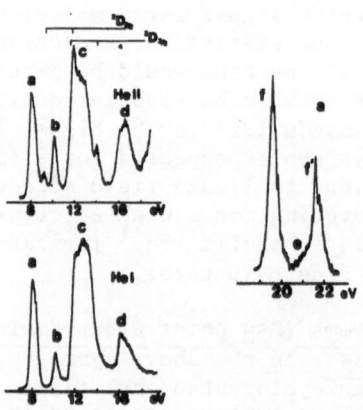

Fig.14.   The He-I and He-II PE spectra of $(C_5H_5)Tl$.

from e_{1g} and e_{1u} MOs respectively.

In transition metal sandwich complexes the most important metal-ligand interaction involves metal nd AOs and the e_{1g} cp MO. So, the reversed ordering of ligand e_{1u} and e_{1g} MOs to that found in Mgcp$_2$ ($e_{1g} < e_{1u}$) must be expected. In addition some additional bands connected with nd^{-1} ionizations should be present at the onset of PE spectra. Take the case of ferrocene (25) for example. In this diamagnetic d^6 species the six 3d electrons fill the outer subshells :

$$(a_{1g})^2 \quad \text{and} \quad (e_{2g})^4$$

The relative intensities of first two bands (2:1) in Fecp$_2$ PE spectrum argue clearly for assignment of first band to ionization of e_{2g} (d$_{x^2-y^2}$,d$_{xy}$) and of the second one to a_{1g} (d$_{z^2}$) electrons. Within Koopmans'approximation, this implies the orbital sequence in the Fecp$_2$ electronic ground state : $e_{2g} > a_{1g}$. This ordering is just the reverse of that inferred from optical absorption spectra. However if electron repulsion within the perturbed d subshells is allowed for, the energy difference between ^2E and ^2A ion states is:

$$E (^2A) - E (^2E) = 20 B - (e_2 , a_1)$$

(B is the Racah parameter and Δ the crystal field splitting). Now, a proper choice of the B parameter leads to a negative value of the above energy difference. Thus, the ground state of Fecp$_2^+$ is $^2E_{2g}$ and, as a consequence, the e_{2g} ionization must occur at the lowest IE. Finally it must be noted that accurate ab initio calculation on ferrocene indicate that the filled Fe 3d subshells are rather internal (28). These discrepacies between MO calculation and the PE spectrum are certainly due to relaxation effect.

B. Open shell systems

As pointed out in a previous section, PE spectra of metal complexes having open shell ground states are expected to be more complex than those of closed shell complexes. Such complexity arises because ionization of each molecular subshell will give rise to various states of molecular ion. Useful criteria as regard to assignment can be gained from Cox's theory of relative photoionization cross-sections (29).

4.4 Transition metal cyclopentadienyl complexes.

Consider, for example, the case of cobaltocene. Cocp$_2$ is a d^7 open shell complex having a $^2E_{1g}$ ($a_{1g}^2, e_{2g}^4, e_{1g}^1$) electronic ground state.
Ionizations from outer d orbitals generate the following states of Cocp$_2$ unipositive ion:

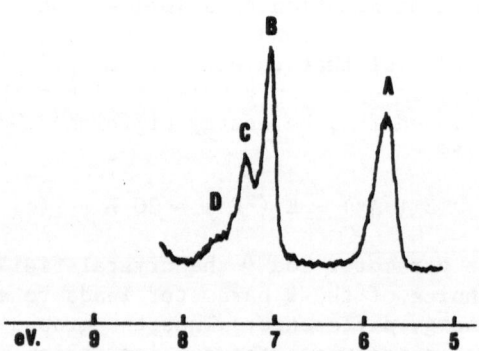

Fig.15.   The He-I PE spectrum of chromocene.

ion configuration	ion state
$a_{1g}^2 \quad e_{2g}^4$	$^1A_{1g}$
$a_{1g}^1 \quad e_{2g}^4 \quad e_{1g}^1$	$^{3,1}E_{1g}$
$a_{1g}^2 \quad e_{2g}^3 \quad e_{1g}^1$	$^{3,1}E_{1g} + {}^{3,1}E_{2g}$

These seven ion states must be correlated with the four bands present in the region below 8 eV of PE spectrum of cobaltocene. The proposed interpretation (30) attributes the first band to production of $^1A_{1g}$ ion state. The second band (based on ligand field analysis and on the band intensities) is likely to represent transitions to both $^3E_{2g}$ and $^3E_{1g}$ (the latter at lower energy) states. Assignment of the third and fourth bands remains unclear.

Another point which must be stressed when dealing with open systems,is the possibility of using PE spectroscopy to gain information about ground state configurations of molecules. Consider the well known case of chromocene. There was doubt whether

$^3A_{2g}$ $(a_{1g}^2 \ e_{2g}^2)$ or $^3E_{2g}$ $(a_{1g}^1 \ e_{2g}^3)$ configurations would describe the electronic ground state of $Crcp_2$. The possibility of a $^3A_{2g}$ ground state can be dismissed because,in this case,only three bands corresponding to $^{4,2}A_{2g}$ and $^2E_{2g}$ ion states should be observed in the low IE region of the spectrum. In fact the spectrum of $Crcp_2$ in the low IE region consists of four bands,consistent only with a $^3E_{2g}$ configuration: on ionizing one d electron from a $^3E_{2g}$ configuration five ion states can be reached (30)(Fig.15).

5. PHOTOELECTRON SPECTRA OF f BLOCK ELEMENT COMPLEXES

The nature of bonding in classical coordination and organometallic complexes of actinide elements has attracted interest since their discovery. Several workers emphasized,through various experimental techniques (31),some covalency in the metal-ligand bonding it being ambiguous,however,whether it were involving 5f or 6d metal AOs. PE spectroscopy proved useful in this area (32) and the reasons of its potentiality reside mainly on the remarkable increase of 5f relative cross-section under He-II radiation. As a consequence,within the framework of Gelius' model,growth in relative intensity of PE bands associated with MOs having some 5f contribution should be observed. However limitations to a PE spectroscopic approach in discriminating between 5f and 6d contri-

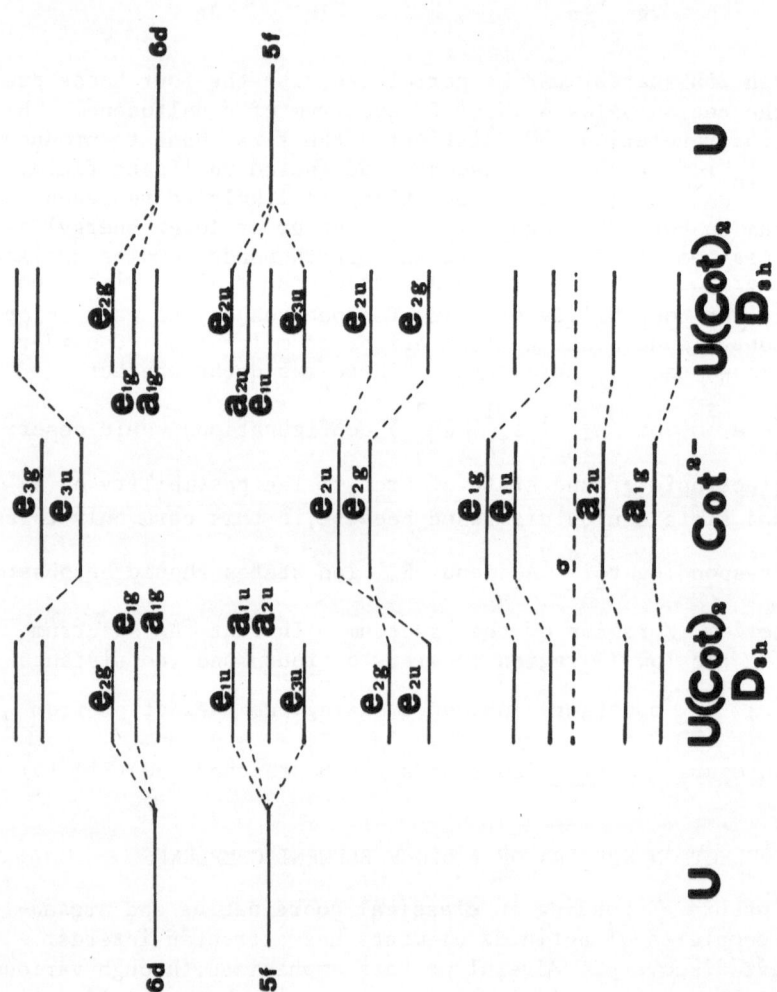

Fig.16. One-electron MO schemes of uranocene. The 6d–π (right side) and 5f–π
(left side) metal–ligand interactions are considered.

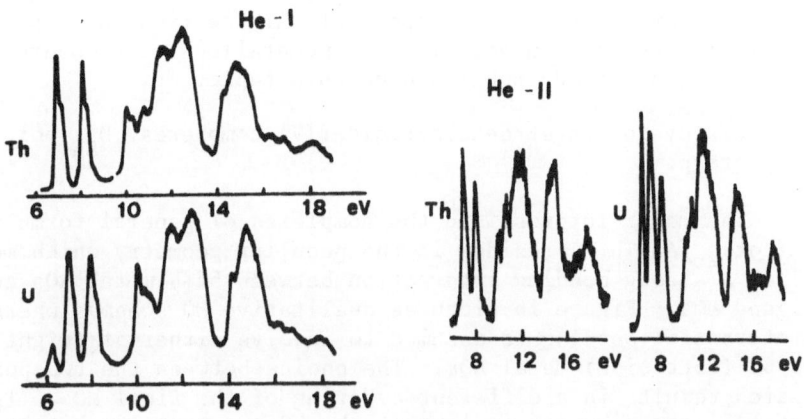

Fig.17. The He-I and He-II PE spectra of uranocene and thoracene.

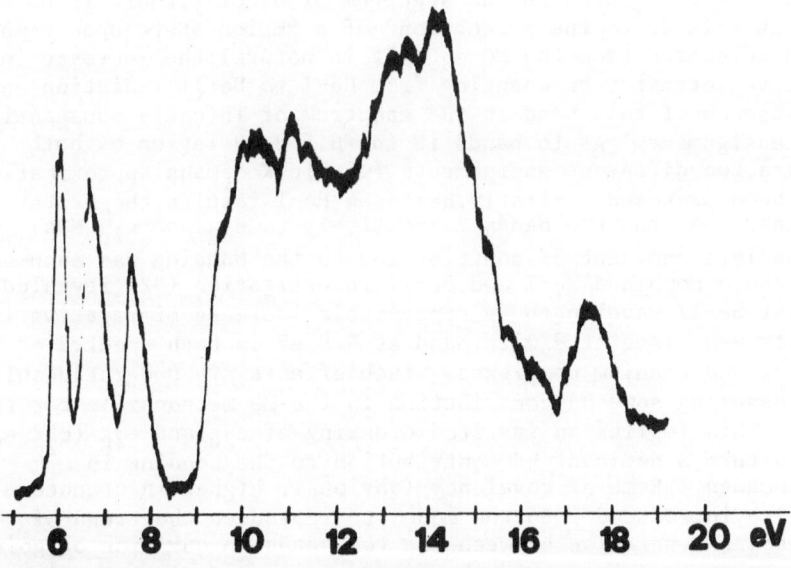

Fig.18. The He-II PE spectrum of the ring substituted trimethyl-silyl uranocene.

butions depend both upon the few theoretical and experimental
information on the'frequency dependence' of 6d cross-section and
upon the fact that the Gelius model of PE cross-section does not
allow for two-centre effects (33) which could be important even
at He-II wavelength. The purpose of this section is to survey
critically PE data on actinide organometallics. It covers almost
all the data already published on this topic.

5.2 Bis(cyclooctatetraene)actinide(IV) complexes. D_{8h} point group.

The major interest in the complexes of general formula
$An(cot)_2$ (An=Th,U) resides in the peculiar geometry which may
allow a strong bonding interaction between $5f_{\pm 2}$ metal AOs and e_{2u}
ligand MOs. Figure 16 pictures qualitative MO schemes where
preferential bonding is assumed to involve either 5f (right side)
or 6d (left side) metal AOs. The choice between the two possibilities results in a different ordering of the final MOs. If the
metal 5f contribution is largest the MO e_{2g} must lie higher in
energy than the e_{2u} shell. The reverse is true if the 6d contribution is assumed to be dominant. In both cases the e_{3u} MO,main-
ly 5f in nature,remains the upper filled one:symmetry restriction
precludes any back-bonding interaction involving metal 6d AOs.
PE spectra of $U(cot)_2$ and $Th(cot)_2$ (Fig.17) consist, respectively,
of three and two well-defined bands in the low IE region (<9 eV).
As to the first band in the spectrum of $U(cot)_2$,there is no doubt
that it relates to the production of a ^2E ion state upon removal
of one electron from the MO e_{3u} (5f in nature);the increase in
relative intensity on changing from He-I to He-II radiation and
the absence of this band in the spectrum of $Th(cot)_2$ substantiate
this assignment. As to bands in the 6.5-8 eV region of both
spectra,two different assignments (within Koopmans'approximation)
have been proposed. Firstly,based on He-I results,the writer
assigned (34) the two bands respectively to e_{2g} and e_{2u} MOs.
Obviously,a dominant 5f contribution to the bonding was assumed.
Later on,a combined He-I and He-II investigation (32) revealed
that,at He-II wavelength an appreciable increase of relative intensity was associated with band at 6.8 eV in both spectra of
thorium and uranium complexes. Such effects are understandable
only assuming some 6d contribution in the MO responsible for that
band. This implies an inverted ordering of e_{2u} and e_{2g} ($e_{2g} < e_{2u}$)
and,in turn,a dominant 6d contribution to the bonding in
actinocenes. Some 5f covalency (obviously higher in uranocene)
was also introduced into the model to reproduce the trend of
increasing separation between the two bands in question on going
from thorium to uranium complex. PE spectra of ring substituted
trimethylsilyl-cyclooctatetraene complex of uranium $U(cotSi)_2$(35)
(Fig.18) also provide indication comparable to that of unsubstituted $U(cot)_2$. Green's model seems to be substantiated also for
ring-substituted cot derivatives. In reality models involving

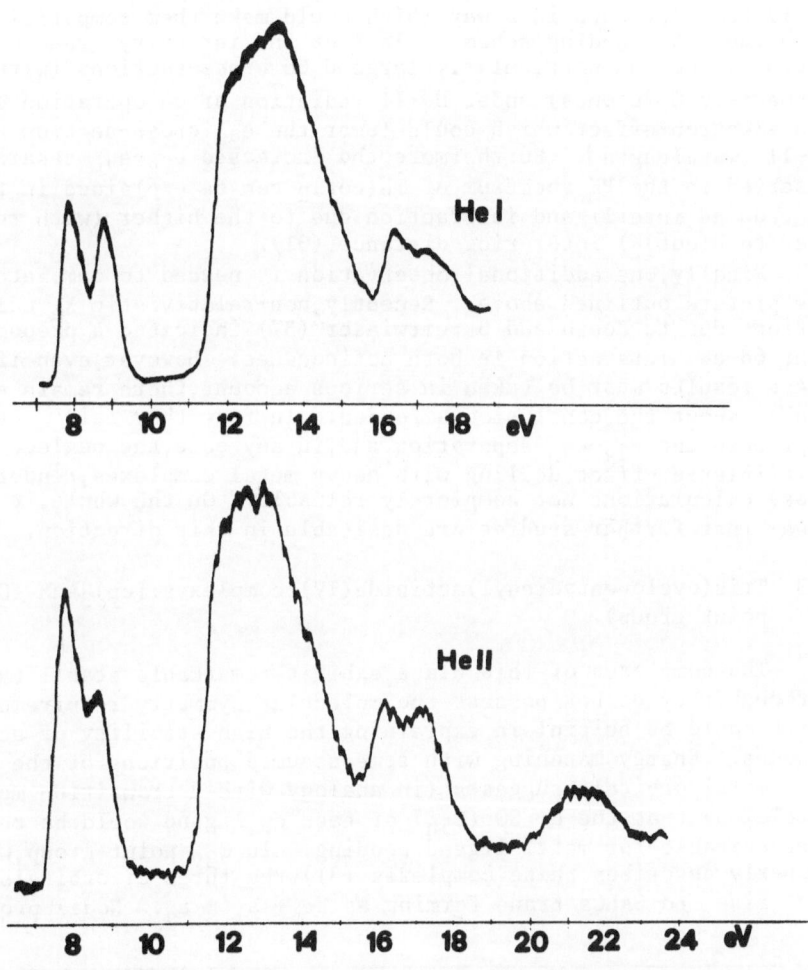

Fig.19. The He-I and He-II PE spectra of $(C_5H_5)_2Mg$. Note the reduced intensity of the second band (e_{1u} MO) under the He-II excitation.

either 5f or 6d interactions suffer from some ambiguities but, also,can be supported by some experimental observations. Consider,for instance,Green's model. Its standpoint favouring the 6d interaction resides in the peculiar intensity trend on passing from He-I to He-II radiation and in the increased e_{2u}-e_{2g} separation observed in the spectrum of Th(cot)$_2$. This latter fact has been connected with the reduced (with respect to U(cot)$_2$) $5f_{\pm2}$-e_{2u} interaction in Th(cot)$_2$. Both of the above arguments,however, could be questioned in a way which could make them compatible with the '5f' bonding scheme. In fact the intensity trend could depend either on particularly large U 6d cross-sections (with respect to C 2p ones) under He-II radiation or on operation of two-electron effect which could lower the e_{2u} cross-section at He-II wavelength†. Furthermore the increased e_{2g}-e_{2u} separation observed in the PE spectrum of Th(cot)$_2$ can be explained in terms of reduced interligand interaction due to the higher (with respect to U(cot)$_2$) inter-ring distance (31).

Finally,one additional observation is needed to complete the picture outlined above. Recently,non-relativistic X_α calculations due to Roesh and Streitwieser (37) indicated a predominant 6d-e_{2g} interaction in both actinocenes. However,even if these results must be taken in serious account,there remain some doubts about the conclusion afforded. In fact they fail to reproduce the e_{2g}-e_{2u} separation and,in any case,the neglect of relativistic effect,dealing with heavy metal complexes,render these calculations not completely reliable. On the whole,it is clear that further studies are desirable in this direction.

5.3   Tris(cyclopentadienyl)actinide(IV) complexes:(cp)$_3$AnX (C$_{3v}$ point group).

The complexes of this class exhibit remarkable stability although they do not possess the molecular symmetry requirements which could be helpful in explaining the high stability of actinocenes. Energy matching with some assumed positions of the pure metal orbitals suggests (in analogy with d transition metal complexes) that the e_1 SO (D$_{5h}$) of each cp ligand would be the more suitable for metal-ligand bonding. In C$_{3v}$ point group,which properly describes these complexes (31),the three e_1 orbitals give rise to SALCs trans forming as 2e + a_1 + a_2. Nodal pro-

---

†

A similar decrease of He-II intensity has been observed(36) in connection with the second band in PE spectrum of Mgcp$_2$(Fig.19). This band has been attributed to the e_{1u} (cp) MO which has characteristics similar to those of the e_{2u} MO in U(cot)$_2$. Based on the given interpretation (25),it is conceivable that such an effect depend only on the characteristics of e_{1u} MO.

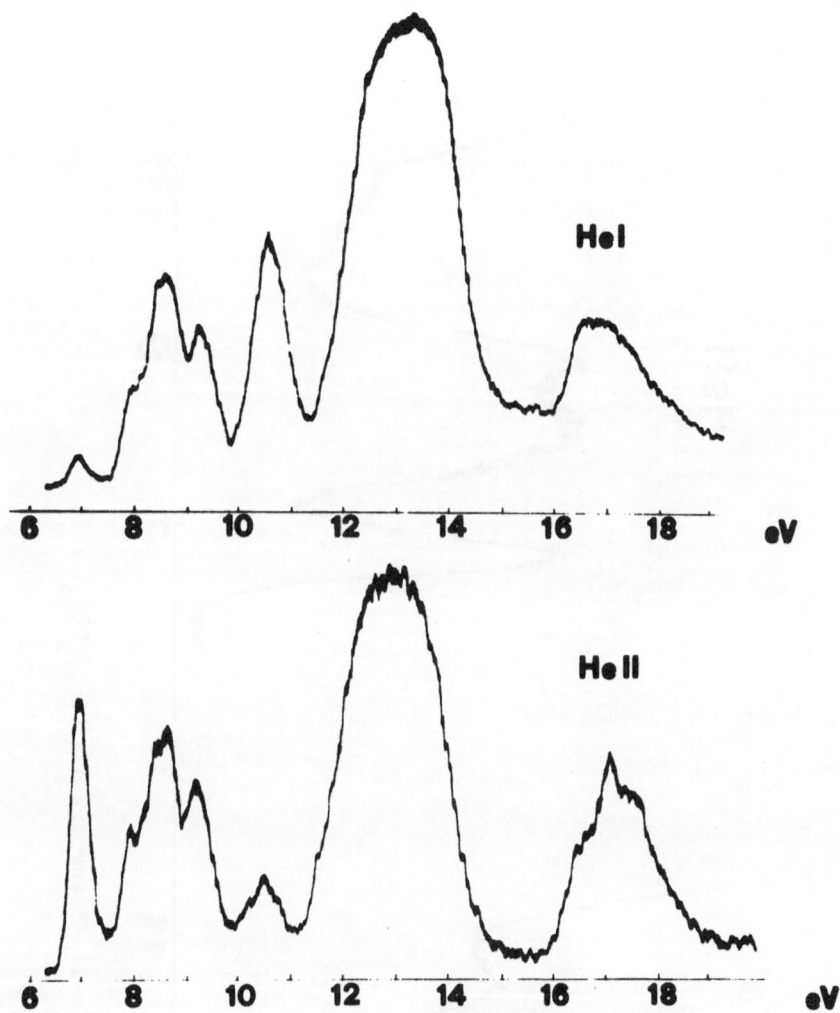

Fig.20. The He-I and He-II PE spectra of $(C_5H_5)_3UCl$.

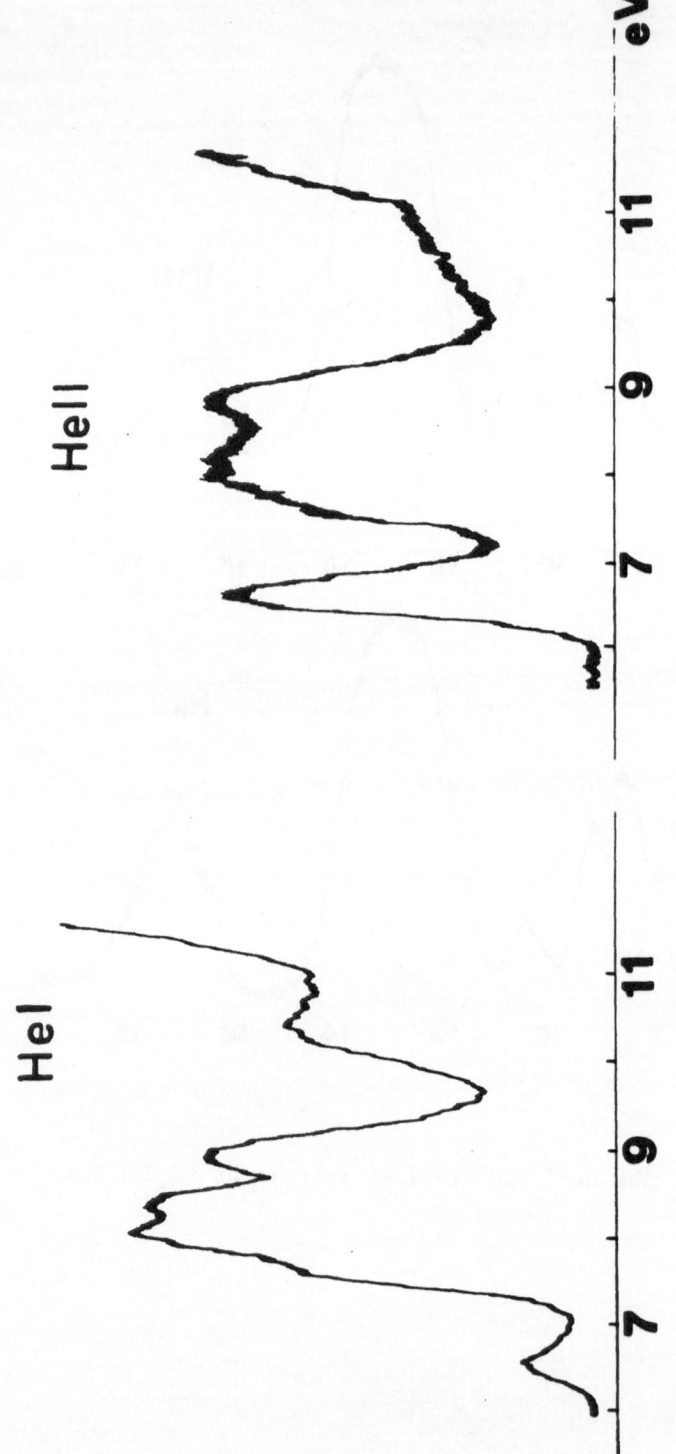

Fig.21. The He-I and He-II PE spectra of $(C_5H_5)_3UBH_4$.

perties of the above SALCs suggest the energy ordering :

$$a_2 > e(1) \simeq e(2) > a_1$$

In the same point group, the 6d and 5f metal orbital transform as:

$$6d \rightarrow 2e + a_1 \; : \; 5f \rightarrow 2e + 2a_1 + a_2$$

It is evident that all the 5f SOs are allowed, by symmetry, to interact with the cp set. If the 6d SOs are considered, the only interaction precluded is that involving the a_2 (cp) MO. In both cases it is difficult to gauge qualitatively the extent of metal-ligand interaction on overlap grounds. The spectral pattern of (cp)$_3$AnX complexes already studied (38,39) is quite similar. The spectra, in the case of uranium complexes. consist generally of a low intensity band in the 6.3-7.2 eV IE region (band a in figure)
    A series of three bands (labelled b,c and d) generally follows in the region up 9.5 eV. Furthermore a band, the characteristics of which (IE and intensity) depend on the nature of the substituent X, is present in the region up 11 eV.
    There is no doubt in assigning the first band a in the spectra of uranium complexes to $5f^{-1}$ ionization. The dramatic increase in intensity associated with this band at He-II wavelength agrees well with the well known 'frequency dependence' of the 5f cross-sections. Similarly, no ambiguities can exist in the assignment of the band in the 9.5-10.5 eV region. It belongs to ionizations of MOs localized on the ligand X (halogen lone-pair if X=Cl or Br, B-H bonds if X=BH$_4$). The shift of this band to lower IE on passing from Cl to Br and its complete absence when X=BH$_4$ clearly argue for this assignment (38). In addition, in the He-II PE spectra of (cp)$_3$UCl (Fig.20) and (cp)$_3$ThCl this band exhibits a very pronounced relative intensity fall-off (39) and it is well known that MOs Cl 3p based possess very small cross-sections at the He-II wavelength (40).
    There are somewhat more difficulties with a detailed assignment of the remaining group of bands below 9.5 eV which, irrispective of the order, certainly relate to 2e+a$_1$+a$_2$ (cp) ionizations. However, the intensity ratios generally observed in the spectra of this class of complexes (see for example Fig.20 and 21) would indicate an ordering of MOs such that a non-degenerate MO (a$_1$ or a$_2$) would correspond to first band b, a twofold degenerate MO(e) to the third band and, finally, the remaining MOs to the second band. Thus, the main point in assignment is to decide whether a$_1$ or a$_2$ MO lies higher in energy. The variations in relative intensities on switching to He-II radiation can be very helpful in this respect. For instance, in the He-II spectrum of (cp)$_3$UBH$_4$ (Fig.21), the intensity of band b appears noticeably reduced with respect to those of both bands c and d. One way to explain this intensity trend is to assume some interligand interaction involving B-H (bridging) bonds (e+a$_1$ in C$_{3v}$ point group) and MOs localized on

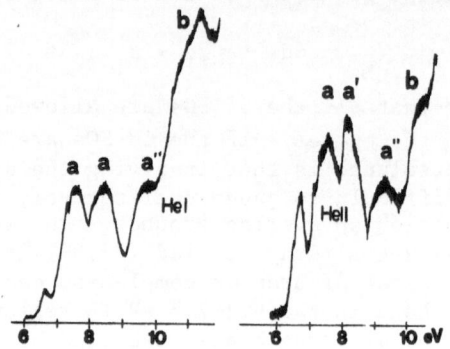

Fig.22. The He-I and He-II PE spectra of $(C_9H_7)_3UCl$.

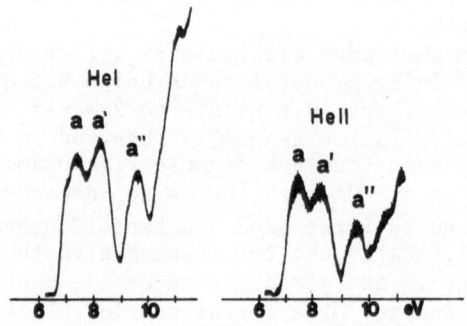

Fig.23. The He-I and He-II PE spectra of $(C_9H_7)_3ThCH_3$.

the $(cp)_3$ framework. In particular the interaction involving the
two MOs of symmetry a_1 seems the most suited on overlap grounds.
Such an interaction destabilizes the MO a_1,based on the cp rings,
which,as a consequence,will result the topmost one among the
$(cp)_3$ set. Within this scheme it becomes possible both to assign
the band b to ionization of the $a_1(cp)$ MO and to connect the aforementioned intensity trend at He-II wavelength to the contribution
of H 1s AOs (which have small He-II cross-sections) to the same
orbital. Obviously this assignment leaves the ionization of a_2
(cp) MO lying over the band c. The same procedure can be used to
interpret PE spectra of other cp derivatives (X=Cl or Br) of
thorium and uranium. The sole variation will be to assume that
the interaction destabilizing the a_1 (cp) MO will involve some
halogen based orbitals of a_1 symmetry. Furthermore,among the
bands which have been associated with MOs mainly based on the cp
framework,the band d generally shows the more pronounced increase
of intensity under the He-II radiation. This observation,without
excluding operation due to involvment of metal 6d AOs in the
bonding,substantiates a significant 5f participation to the MO
responsible for this band. This fact,consequently,underscores
some specific role played by uranium 5f AOs in the metal-ligand
bonding.

5.4   Tris(indenyl)actinide(IV) complexes: $(ind)_3AnX$ (C_{3v} point
      group).

      PE spectra of indenyl complexes (Fig.22,23),in accordance
with the more complicated electronic structure of indenyl anion,
appear more complex than those of corresponding cyclopentadienyl
systems. The electronic structure of the indenyl anion (C_{2v}
local symmetry) can be described simply assuming that this species originates from a cyclopentadienyl anion interacting with a
cis-butadiene moiety. The assumed (41) ordering of the upper
filled π MOs:

$$1a_2 > 2a_2 > 1b_1$$

is at variance to that indicated by simple Hückel theory but
results from indication of the PE spectra hereafter discussed.
In the C_{3v} point group ,which describes properly most of
$(ind)_3AnX$ systems,the indenyl π set transforms as follows:

$$3 \cdot a_2 \rightarrow e + a_2 \quad ; \quad 3 \cdot b_1 \rightarrow e + a_1$$

Simple comparison with the PE spectra of naphthalene(7)(which is
iso-π-electronic with indenyl anion)suggests that the bands corresponding to the above SALCs must be expected in the 7-10 eV
region. In this region,PE spectra of indenyl complexes consist
of a series of bands which can be arranged in three distinct
groups (labelled a,a' and a" in the Figures). In addition,in the
region up to 10.5 eV other bands are present;one (labelled b)

around 10.3 eV and, only in the spectra of uranium complexes, another of low intensity (in He-I spectra) at the onset of each spectrum. The assignment of these two latter bands does not pose serious problems (41). The band b, whose IE and relative He-I and He-II intensities srongly depend on the nature of the σ-bonded substituent X, is certainly correlated with ionization from MOs localized on the substituent itself (chlorine or oxygen lone-pairs). The band lying at the onset of uranium spectra is the analogous of similar bands observed at the onset of other PE spectra of uranium complexes (vide supra) and relates to $5f^{-1}$ ionization.

Turning now to bands a, a' and a" left unassigned, they can be confidently assigned using criteria based on IE shifts and on variation of relative intensities on passing from He-I to He-II radiation (42). Thus, the reduced intensity of bands a" in the spectra of indenyl complexes where X=Cl, Br or CH_3 and the dependence of the IEs connected with these bands from the nature of the ligand X, suggest that band a" relates to MOs, localized on the indenyl rings, which can interact with some orbital localized on the ligand X (41). The band a', which does not appreciably move by varying either the central metal atom or the σ-bonded group X, is likely related to the $e+a_2$ ($2a_2$) MOs mainly localized on the six-membered ring of the indenyl anion. Finally, the remaining band a correlates with both the $e+a_2$ ($1a_2$) MOs. The dependence of both IE and band shape on the nature of central metal atom clearly argues for a relevant involvment of these latter MOs in the metal-ligand bonding. There   again, evidence of some role played by 5f actinide AOs in this bonding. In particular, the growth in relative intensity of band a under He-II radiation can be only explained by some 5f contribution in the corresponding MOs. In this respect, it is interesting to note that, in the indenyl complexes, the relative increase of intensities of 5f bands at the He-II wavelength is the lowest (with respect to that observed in spectra of other actinide organometallics) as yet observed. Based on Gelius model, this fact would indicate that, among the actinide organometallic complexes already examined, the indenyl derivatives possess the highest degree of covalency.

5.5 Final states related to ionization of 5f subshells.

In connection with the band present at the onset of PE spectra of organometallic (vide supra) and classical coordination complexes of uranium(IV) (43) thus far measured, it has been pointed out that this band relates to ionization of uranium 5f AOs. Actually, on ionizing one electron from the U(IV) $^3H_4$ ground term , two multiplet levels $^2F_{5/2}$ and $^2F_{7/2}$ are expected. In a purely atomic model their separation is $(3+1/2)\zeta$ where ζ is the one-electron spin-orbit coupling constant ($\zeta \approx 0.2$ eV).

The ligand field can split each of these levels but it has been shown that the field generally acts as a small perturbation (43). The only evidence of such splitting has been found in connection with first band present in PE spectrum of $(C_5H_4CH_3)_3UBH_4$ (38). The relative probabilities of reaching the two final states are:

$$\frac{P_{J=5/2}}{P_{J=7/2}} = \frac{1.714}{0.286}$$

in the L-S coupling scheme (29). The probability of reaching the J=7/2 final state is reduced further in an intermediate coupling scheme and vanishes in the J-J limit (43). It transpires that the intensities of PE bands associated with $^2F_{7/2}$ final state must be expected to be extremely low. They are likely to be obscured by the onset of more intense PE bands which generally follow the f band.

References.

1.   M.H.Palmer,D.Leaver,J.D.Nisbet and R.W.Miller,J.Mol.Struc.,
      42,85 (1977).

2.   I.Fragalà and R.D.Fischer, work in progress.

3.   A.F.Orchard,Faraday Discuss.Chem.Soc.,54,252 (1973).

4.   A.Kasdan and W.C.Lineberger,Phys.Rev.A,10,1658 (1974).

5.   M.S.Banna and D.A.Shirley,J.Electron Spectrosc.,8,23 (1976).

6.   J.H.D.Eland, "Photoelectron Spectroscopy",Butterworths,
      London 1974 .

7.   D.W.Turner,C.Baker,A.D.Baker and C.R.Brundle,"Molecular
      Photoelectron Spectroscopy",Wiley-Interscience 1974.

8.   -J.A.Connor,L.M.R.Derrick,M.B.Hall,I.H.Hillier,M.F.Guest,B.R.
      Higginson and D.R.Lloyd,Mol.Phys.,28,1193 (1974) .

      -M.F.Guest,I.H.Hillier and D.R.Lloyd,Mol.Phys.,29,113(1975).

      -M.B.Hall,I.H.Hillier,J.A.Connor,M.F.Guest and D.R.Lloyd,
      Mol.Phys.,30,838(1975) .

      -S.Evans,M.F.Guest,I.H.Hillier and A.F.Orchard,J.Chem.Soc.
      Faraday II,70,417(1974) .

      -See also Refs. 16 and 20.

9.   S.Evans,J.C.Green,P.J.Joachim and A.F.Orchard,J.Chem.Soc.
      Faraday II,68,905(1972) .

10.   B.R.Higginson,D.R.Lloyd,P.Borroughs,D.M.Gibson and A.F.Orchard
      J.Chem.Soc.Faraday II,69,1659(1973) .

11.   M.Gower,L.A.P.Kane-Maguire,J.P.Maier and D.A.Sweigart,
      J.Chem.Soc.Dalton,316(1977) .

12.   J.C.Green,M.L.H.Green,P.J.Joachim,A.F.Orchard andD.W.Turner,
      Phil.Trans.Roy.Soc.A,268,111(1970) .

13.   R.G.Egdell,I.Fragalà and A.F.Orchard, to be published.

14.   U.Gelius, "Electron Spectroscopy" D.A.Shirley Ed.
      North-Holland, Amsterdam 1972.

15.   D.R.Lloyd and N.Lynaugh, ibid.

16. M.M.Rohmer,J.Demuynch and A.Veillard,Theoret.Chim.Acta, 36,93(1974) .

17. C.Batich,J.Amer.Chem.Soc.,98,7585(1976) .

18. K.H.Johnson and U.Wahlgren,Int.J.Quantum Chem.,Symposium 6, 243(1972) .

19. E.J.Baerends and P.Ros,Mol.Phys.,30,1735(1975) .

20. I.H.Hillier,M.F.Guest,B.R.Higginson and D.R.Lloyd,Mol.Phys., 27,215(1974) .

21. J.B.Johnson and W.G.Klemperer,J.Amer.Chem.Soc.,99,7132(1977).

22. I.Fragalà, to be published.

23. I.Fragalà and J.Thomas, to be published.

24. G.Condorelli,I.Fragalà,G.Centineo and E.Tondello,J.Organomet. Chem.,87,311(1975) .

25. S.Evans,M.L.H.Green,B.Jewitt,A.F.Orchard and C.F.Pygall, J.Chem.Soc.Faraday II,68,1847(1972) .

26. R.G.Egdell,I.Fragalà and A.F.Orchard,J.Electron Spectrosc., in press .

27. A.W.Potts and M.L.Lyus,J.Electron Spectrosc.,13,327(1978) .

28. M.M.Coutiere,J.Demuynch and A.Veillard,Theor.Chim.Acta, 27,281(1972) .

29. P.A.Cox,Struct.Bonding(Berlin),24,59(1975) .

30. S.Evans,M.L.H.Green,B.Jewitt,G.H.King and A.F.Orchard, J.Chem.Soc.Faraday II,70,356(1974) .

31. E.C.Baker,G.W.Halstead and K.N.Raymond,Struct.Bonding(Berlin), 25,23(1976) .

32. J.P.Clark and J.C.Green, J.Chem.Soc.Dalton,505(1977).

33. A.Schweig and W.Thiel,J.Chem.Phys.,60,951(1974) .

34. I.Fragalà,G.Condorelli,P.Zanella and E.Tondello,J.Organomet. Chem.,122,357(1976) .

35. I.Fragalà and R.D.Fischer, to be published .

36. I.Fragalà, unpublished data.

37. N.Rosch and A.Streitwieser Jr.,J.Organomet.Chem.145,195(1978).

38. I.Fragalà,E.Ciliberto,R.D.Fischer,G.R.Sienel and P.Zanella,
     J.Organomet.Chem.,120,C9(1976) .

39. I.Fragalà and R.D.Fischer, to be published.

40. R.G.Egdell,A.F.Orchard,D.R.Lloyd and N.V.Richardson,
     J.Electron Spectrosc.,12,415(1977) .

41. I.Fragalà,G.Condorelli and J.Goffart, VIII ICOMC Kyoto 1977
     paper C4

42. I.Fragalà and J.Goffart, to be published.

43. I.Fragalà,G.Condorelli,E.Tondello and A.Cassol,Inorg.Chem.,
     in press.

VIBRATIONAL SPECTROSCOPY OF SOME ORGANOMETALLIC COMPOUNDS OF
ACTINIDES.

Jean Goffart[*]

Laboratory of Radiochemistry.
University of Liège Sart Tilman, B-4000 Liège (Belgium).

INTRODUCTION.

Although vibrational spectroscopy ranks as a secondary
method for precise structural studies[1], it is still very useful
because of its convenience and its wide applicability. In reality,
there are practically no limitations to the utilization of infrared spectroscopy and only a few limitations to Raman spectroscopy;
experimental difficulties are caused by the fluorescence and the
color of some samples. Both methods can be applied to liquids,
solutions, gases, solids (crystals or powder) and independently
of the magnetic characteristics of the actinide element present
in the compound. For radiochemists, it is very easy to avoid
contamination during the preparation of the samples when radioactive materials are handled and the majority of the measurements
can be made in a small box with optical windows without risk of
contamination of the spectrometer.

Besides its possibilities for structural analysis, vibrational spectroscopy can also provide information on the type of
bonding involved in a complex[2-6].

With the facilities offered by the last generation of infrared and Raman spectrometers with cover a wide frequency range
from 10000 to 15 cm^{-1}, the recording of vibrational spectra has
become almost a routine task. The majority of publications describing primarily the preparation of a new organometallic compound
present at least typical features of the infrared spectrum. In

[*] Chargé de Recherches à l'I.I.S.N. (Brussels).

T. J. Marks and R. D. Fischer (eds.), Organometallics of the f-Elements, 467–496.
All Rights Reserved. Copyright © 1979 by D. Reidel Publishing Company, Dordrecht, Holland.

fact, the amount of information to be obtained from infrared and Raman spectra is considerable but there are usually severe limitations to an exhaustive interpretation of the experimental data.

THEORETICAL CONSIDERATIONS.

As we shall see below, infrared absorption and Raman scattering are complementary techniques : therefore in almost all attempts at structural determination or for the assignments of absorption bands, it is advisable to employ both techniques to obtain the most information.

Infrared.

If a substance is placed in a beam of infrared radiation, absorption of the incident radiation occurs at frequencies corresponding to those of the vibrational modes of the molecule. The position of these absorptions gives therefore the characteristic vibrational frequencies of the studied sample. However a vibration will be observed in the infrared spectrum only if the molecular electric dipole moment changes during the motion under study. To bring this condition into a more formal way, we can write

$$(\mu_x)_{ij} = (\delta\mu_x/\delta Q_k)_0 \int \psi_v^i \, Q_k \, \psi_v^j \, d\tau \neq 0 \qquad (1)$$

where $(\mu_x)_{ij}$ is the transition moment, $(\delta\mu_x/\delta Q_k)_0$ is the magnitude of the x-component of the dipole derivative moment with respect to the appropriate normal coordinate Q_k, the integral term gives the probability of a given transition where i and j represent two vibrational states ; $d\tau$ is a volume element in configurational space, and integration takes place over the whole of this space. Two corresponding integrals of the same type exist for μ_y and μ_z along the other Cartesian coordinate axes. Only one of these three integrals needs to be non-zero to obtain an infrared band. A detailed treatment of infrared absorption can be found in Wilson, Decius and Cross[7].

Raman.

In Raman spectroscopy, the molecules of the studied compound are generally irradiated with light from the visible region of the spectrum. The electromagnetic field associated with the incident radiation induces an oscillating dipole in the compound under investigation, the nuclei being attracted towards the negative pole and the electrons towards the positive pole. Energy is therefore taken up which can be re-irradiated with the same frequency of the incident light, or with a different frequency. The difference between these two frequencies, corresponding to the Raman scattered light, is characteristic of vibrational motion. In addition since the incident light is commonly plane-polarised light (laser), we can observe that certain frequencies give polarised, while others give depolarised scattered light.

As in the case of infrared vibrations, the probability of a transition occurring in Raman spectra is dependent on the integral

$$\int \psi_0 \ \alpha_{ij} \ \psi_j \ d\tau \tag{2}$$

where $\hat{\alpha}_{ij}$ stands for the electric polarisability operator.

The induced dipole moment P is proportional to the strength of the electric field E and to the polarisability α of the molecule under investigation

$$P = \alpha E \tag{3}$$

This polarisability α is a proportionality constant which can be visualised qualitatively as a measure of the ease with which the molecular electron distribution can be distorted. The situation is complicated by the fact that in most molecules, the direction of polarisation does not coincide with the direction of applied field, because the direction of chemical bonds also affects the direction of polarisation and the equation (3) is in general more complicated[7].

An other important feature of the theory of Raman scattering is the possibility to determine whether or not a given Raman line will be polarised, like the incident light, or depolarised. It appears that only transitions involving totally symmetric vibrational modes will give rise to polarised lines[7]. This rule can be of considerable importance for vibrational assignment.

ANALYSIS OF VIBRATIONAL SPECTRA.

A vibrational spectrum represents a series of infrared bands or Raman shifts each characterized by a certain wave number. An important factor must always be taken into consideration : in spite of the utilization of glove boxes, filled with purified argon, helium or nitrogen, and/or Schlenk vessels, the preparation of halide discs or mulls of organometallic compounds for infrared study must be done as rapidly as possible to avoid oxidation and/or hydration of these sensitive compounds. Grinding or mulling of the organometallic compounds does not always lead to suitable microparticles. Thus the resulting partial

anisotropy and the random size of the particles often preclude
an application of the theory of diffraction by isotropic sphe-
res[8]. In such instances, the band contours, the position of
band maxima and the band intensities can change slightly from
one spectrum to another with the same compound. In consequence,
it is useful to compare the spectra of a compound in different
conditions.

The interpretation of infrared or Raman spectra must take
into account three main factors :
  1.- The masses of the vibrating atoms.
  2.- The nature of the bonding between the atoms.
  3.- The symmetry of the molecule being studied.

1. The nature of the vibrating atoms.

If we consider a mass m_1 fixed on a spring with a force
constant k_1, a small displacement of the mass will give rise to
an oscillation with a frequency

$$\nu_1 = 1/2 \, \pi \sqrt{k_1/m_1} \tag{4}$$

If two systems of the same nature are weakly coupled each
will continue vibrating with its own frequency, but in the event
that the frequency of the first system is approximately equal to
that of the second system, energy will flow from one system
to the other. Experiment shows that it is possible to bring the
masses into motion in either of two described ways. In the first
of these vibrations, the two masses will be moving in-phase ;
in the latter, they will be moving out-of-phase. In terms of the
underlying mathematical terminology, these non-equivalent vibrational modes are called "normal modes".

One and only one frequency is associated with each normal
mode of vibration. If many independently vibrating pairs of the
same nature are present in one molecule, only one frequency will
be observed

$$\nu = 1/2 \, \pi \sqrt{k/\mu} \tag{5}$$

where $\mu = (m_1 + m_2)/m_1 m_2$

Table I illustrates the importance of masses on the position of the infrared absorption bands.

Table I.   Characteristic frequencies of some oscillators.

Bond	Reduced mass	Approximative wave number (cm^{-1})
C - H$_{aliph.}$	0,92	2900
C - C	6,0	1000
C = C	6,0	1650
C ≡ C	6,0	2100
C - O	6,9	1200
C = O	6,9	1720
C ≡ O	6,9	2150
C - Cl	9,0	700

2. The nature of bonding.

Generally a change of the mass imposes by the greatest effect on the value of the frequency, especially for light atoms (Table I). The second largest effect is certainly due to modifications of the bonding between the vibrating atoms. The difference between the absorption frequencies of single, double and triple bonds for the carbon-carbon and carbon-oxygen vibration, respectively, is most characteristic in this context (Table I). The approximate constancy of the vibrational frequency of a given functional group, when the molecular environment changes, has very interesting consequences in the interpretation of an unknown spectrum. Mechanical effects produced by the new environment can slightly modify the form of the vibration of a functional group resulting from variation in the coupling between motions of adjacent atoms. Electrical interactions lead to the same result. Such effects frequently arise when electron densities are changed as by conjugation or induction.

3. The symmetry of the molecule.

Qualitative as well as quantitative treatments of the vibrational states of molecules begin with an analysis of the symmetry properties of the studied compound. The theorems underlying such analyses and interpretations are derived from group theory. It is not our intent to present here this well-developed theory as it is the subject of classical texts[7,9,10]. From the combined infrared and Raman data, it is often possible to determine the symmetry of the molecule. Vibrational modes can be classified into bond-stretching and angle-deformation vibrations. Group theory helps determine the number and symmetry of the different vibrations. If an exhaustive use of group theory is made in case of a sufficiently symmetrical molecule, it is possible to arrive at a fairly accurate assignment of the total spectrum.

THE ACTINIDE CYCLOPENTADIENYL COMPOUNDS.

Although the first cyclopentadienyl-metal compound, KC_5H_5, was prepared at the beginning of this century[11], the interest in organometallic chemistry did not begin until the synthesis of bis-cyclopentadienyl iron (ferrocene) in the early fifties. In 1956, Reynolds and Wilkinson[12] reported the synthesis of the first cyclopentadienyl compound of a 5f element, tris-cyclopenta-dienyl uranium chloride, $(C_5H_5)_3UCl$. Three detailed discussions of the vibrational spectra of cyclopentadienyl compounds have been published by Fritz[13] and Maslowsky[3,14].

Cyclopentadiene.

The vibrational spectrum of cyclopentadiene has been studied by several authors[15,16] both by Raman and infrared spectroscopy. Some assignments are sometimes doubtful because it is difficult to keep cyclopentadiene for a long period in the monomeric form. Cyclopentadiene, C_5H_6, has a planar structure with the C_{2v} point group and should give rise to 27 (3 x 11 - 6) normal vibrations. Table II presents the assignment to the fundamental vibrations of cyclopentadiene.

Table II. Fundamental vibrations of cyclopentadiene[15].

	Wave number (cm^{-1})	Assignment		Wave number (cm^{-1})	Assignment
A_1	3091	CH stretching	B_1	3105	CH stretching
	3075	CH stretching		3045	CH stretching
	2886	CH_2 stretching		1580	C=C stretching
	1500	C=C stretching		1292	CH bending + ring stretching
	1441	CH bending		1239	CH bending + ring stretching
	1378	CH_2 scissoring		1090	CH_2 wagging
	1365	CH bending		959	ring stretching + CH bending
	1106	ring		805	ring bending
	994	ring			
	915	ring	B_2	2900	CH_2 stretching
A_2	(1135)	CH_2 twisting		925	CH bending
	920	CH bending		891	CH_2 rocking
	700	CH bending		664	CH bending
	515	ring bending		350	ring bending

The A_2 vibrations are infrared inactive.

The cyclopentadienyl anion.

The aromatic anion, $C_5H_5^-$, is easily obtained by reduction of the monomeric cyclopentadiene molecule. The cyclopentadienyl ligand containing ten atoms is expected to have 24 normal modes of vibration. Their irreducible representations for a D_{5h} symmetry is :

$$2A_1' + 1A_2' + 1A_2'' + 3E_1' + 1E_1'' + 4E_2' + 2E_2''$$

The character table for the D_{5h} point group permits us to predict that twelve modes are Raman active ($2A_1' + 1E_1'' + 4E_2'$), seven are infrared active ($1A_2'' + 3E_1'$) and the last five ($1A_2' + 2E_2''$) modes are totally inactive. The vibrational spectrum of an ionic compound like KC_5H_5 is expected to be simpler than the spectrum of cyclopentadiene. Experiment confirms this prediction (table III).

Table III. Vibrational spectra of KC_5H_5 in solution[13].

Wave number (cm^{-1})	Irreducible representation	Activity
3096	E_2'	Raman
3043	A_1'	Raman
3039	E_1'	IR
1455	E_1'	IR
1447	E_2'	Raman
1020	E_2'	Raman
1003	E_1'	IR
983	A_1'	Raman
710	A_2''	IR
625	E_1''	Raman
565	E_2'	Raman

From this example, it becomes clear that vibrational spectroscopy may be a powerful method for distinguishing between different possibilities of structure. Let us see what happens with more complicated organometallic compounds.

The cyclopentadienyl compounds.

The spectra of π-bonded actinide compounds are expected to be rather different from those obtained with purely ionic com-

pounds. It might appear that the spectra of these actinide compounds would be more complicated than those of the ionic compounds
because :

1. The number of normal modes of vibration is greater, resulting from the greater number of atoms, and often from less
   symmetric structures.
2. If an appreciable amount of coupling takes place between
   the organic ligands; supplementary vibrations would appear
   at different frequencies. Fortunately, it has been shown
   experimentally that the spectra originating from π-compounds
   containing more than one C_5 ring can be discussed, at least
   above ca. 500 cm^{-1}, in terms of the spectra expected for
   one of the rings. This simplicity of spectra indicates that
   in the majority of the compounds inter-ligand coupling is
   very weak[3,13,14,21].
3. At low frequencies, absorption should occur from the ring-
   to-actinide skeleton vibrations. Table IV presents some
   data for $An(\eta^5-C_5H_5)_3$ and $An(\eta^5-C_5H_5)_4$ systems.

Table IV. Infrared spectra of tris-cyclopentadienyl- and tetra-
          cyclopentadienyl-actinide compounds.

$(\eta^5-C_5H_5)_3Th$ (17) (cm^{-1})	$(\eta^5-C_5H_5)_3U$ (18) (cm^{-1})	$(\eta^5-C_5H_5)_3Pu$ (19) (cm^{-1})	$(\eta^5-C_5H_5)_3Am$ (18) (cm^{-1})	Type of vibration
3072	3095	3098	3078	ν(C-H)
1435	1441	1442	1448	ν(C-C)
1012	1014	1009	1007	δ(C-H)($\parallel$)
	780	793	795	δ(C-H)($\perp$)
785	773	779	768	

$(\eta^5-C_5H_5)_4Th$ (19) (cm^{-1})	$(\eta^5-C_5H_5)_4Pa$ (18) (cm^{-1})	$(\eta^5-C_5H_5)_4U$ (20) (cm^{-1})	$(\eta^5-C_5H_5)_4Np$ (18) (cm^{-1})	Type of vibration
3068	3078	3077	3077	ν(C-H)
1441	1445	1449	1447	ν(C-C)
1066	1070	1070	1070	δ(C-C)
1008	1008	1010	1008	δ(C-H)($\parallel$)
805	811		810	
788	784	789		δ(C-H)($\perp$)
778		782	780	

The C_5 ring in the fictitious $An(\eta^5-C_5H_5)$-model has 24 modes
of vibration. They are classified as $3A_2 + 4E_1 + 6E_2$.

Eleven ($3A_1$ + $4E_1$) are both infrared and Raman active, the six (E_2) are Raman active, and the A_2 mode is totally inactive (Table V).

Table V. Normal vibrations for "(η^5-C_5H_5)An" in C_{5v} symmetry.

Approximative type of vibration	Irreducible representation	Activity
C-H stretching C-C stretching C-H bending ($\perp$)	$3A_1$	IR + Raman
C-H bending ($\parallel$)	$1A_2$	Inactive
C-H stretching C-C stretching C-H bending ($\parallel$) C-H bending ($\perp$)	$4E_1$	IR + Raman
C-H stretching C-C stretching 2 C-H bending ($\parallel$) C-H bending ($\perp$) C-C-C bending ($\perp$)	$6E_2$	Raman

In C_{5v} symmetry, there would be three C-H stretching vibrations, but one, E_2 mode, is infrared inactive. The two other modes, A_1 and E_1, would be observed at two different frequencies, normally between 3100 and 3000 cm^{-1} for an aromatic molecule[5]. Experimentally only one band is noted in the different spectra (Table IV). The same behaviour is observed for Fe(η^5-C_5H_5)$_2$ and Ni(η^5-C_5H_5)$_2$, these two molecules having sandwich geometries with π-bonds. Cotton and Reynold[22] have suggested that the A_1 vibration might give rise to a very weak absorption, too weak in fact to be normally observed : only the E_1 mode is really infrared active. The absence of bands between 2800 and 3000 cm^{-1} is in marked contrast to a possible olefinic C-H stretching band expected for a monohaptocyclopentadienyl system. For the C-C stretching vibrations, one expects two infrared bands (A_1 + E_1). The E_1 mode is observed around 1400 cm^{-1} [21].

The symmetric ring breathing mode, A_1, has been assigned to the band around 1100 cm^{-1}. Controversy has arisen concerning the assignment of this band, but the most recent investigations[3,14] confirm the original assignment. It is a medium-strong band, characteristic of almost all organometallic compounds with a covalent bond between a cyclopentadienyl ring and a centered cation. This band is not present in the ionic compounds or in compounds in which the metal is η^1-bonded[3,14]. The C-H out-of-plane deformations ($A_1 + E_1$) give rise to a broad band combined with some very intense absorptions. The exact positions are very sensitive to the nature of the bond and the cation. The last band (ca. 1000 cm^{-1}) has been assigned to the C-H in-plane wagging vibration. These assignments have been confirmed by 23-study of the bis(pentamethylcyclopentadienyl) metal compounds [25]. The majority of cyclopentadienyl complexes of the actinides gives rise to such characteristic bands (Table IV), typical for one $C_5H_5^-$ moeity with C_{5v} symmetry[17,18,26-38].

Adopting the method of local symmetry[13], we may consider the vibrational spectra at low frequency as due to "inner" vibrations. Regarding the cyclopentadienyl ligands as rigid discs and supposing that the "inner" symmetry is C_{3v}, we see that the skeleton vibrations give rise to the following irreducible representation :

$$\Gamma_{vib.} = 3A_1 + 2A_2 + 5E$$

The A_1 and E normal vibrations are both infrared and Raman active ; the A_2 modes are totally inactive. The method of internal coordinates[7] allows us to assign the skeletal vibrations (Table VI).

Table VI. Normal skeletal vibrations for $An(\eta^5-C_5H_5)_3$ and $An(\eta^5-C_9H_7)_3$ active in vibrational spectroscopy (for C_{3v} symmetry).

Approximative type of metal-to-ligand vibrations	Irreducible representation
Symmetric stretching	A_1
Symmetric bending	A_1
Assymetric stretching	E
Assymetric bending	E
Assymetric torsion	E
Ring tilting ($\perp$) in phase	A_1
Assymetric tilting ($\parallel$) out-of-phase	E
Assymetric tilting ($\perp$) out-of-phase	E

Aleksanyan et al.[39] have pointed out the difficulties of interpretation of the spectra in this low-frequency region. Only a very few data have been published (Table VII). Three important bands appear in all the spectra. The ca. 230 cm^{-1} line noted in Raman spectra of tris(cyclopentadienyl) lanthanides is polarised and can be assigned to the symmetric metal-ring stretching vibration[39].

Table VII. Vibrational spectra of some organometallic compounds at low frequency (cm^{-1}).

Raman (38) $(\eta^5-C_5H_5)_3Th$	Raman (40) $(\eta^5-C_5H_5)_3Gd$	Raman (40) $(\eta^5-C_5H_5)_3Tb$	Raman (40) $(\eta^5-C_5H_5)_3Ho$	IR (41) $(\eta^5-C_9H_7)_3U$
254	230	230	232	239
234	220	221	218	214
157	132	135	139	90

Nevertheless the Raman lines and the infrared bands are very broad (up to 30-40 cm^{-1}) and two or more bands can be combined. Some vibrational data have been published recently for thorium and uranium compounds (Table VIII). For $U(\eta^5-C_5H_5)_3$ Hal and $Th(\eta^5-C_5H_5)_3$ Hal (Hal = F, Cl, Br and I), one observes a strong infrared band between 230 and 250 cm^{-1} [42].

Table VIII. Vibrational spectra of some cyclopentadienyl complexes.

Complex	$\nu(U-Cl)$	$\nu(An-C_5H_5)$	References
$U(\eta^5-C_5H_5)Cl_3 (dma)_2$	237	264	40
$U(\eta^5-C_5H_5)Cl_3 (dmpva)_2$	232	260	40
$U(\eta^5-C_5H_5)Br (dmpva)_2$		252	40
$U(\eta^5-C_5H_5)Cl_3 (PPh_3O)_2$	235	260	40
$U(\eta^5-C_5H_5)Br_3 (PPh_3O)_2$		255	40
$U(\eta^5-C_5H_5)Cl_3 (dppoe)_2$	232	260	40
$U(\eta^5-C_5H_5)Br_3 (dppoe)_2$		249	40
$U(\eta^5-C_5H_5)Br_2 (dppoe)_2$		255	40
$U(\eta^5-C_5H_5)_3Cl$	268	241	40-42
$U(\eta^5-C_5H_5)_3Br$		245	40-42
$U(\eta^5-C_5H_5)Cl_3 (THF)_2$	240	268	40
$U(\eta^5-C_5H_5)Br_3 (THF)_2$		266	40
$Th(\eta^5-C_5H_5)_3 (n-C_4H_9)$		256*	27
$Th(\eta^5-C_5H_5)_3 (C_3H_5)$		260*	27
$Th(\eta^5-C_5H_5)_3 R$		259*	27

dma : $CH_3.CO.N(CH_3)_2$	THF : tetrahydrofuran
dmpva : $(CH_3)_3CCON(CH_3)_2$	R : 2-cis-2-butenyl
dppoe : $Ph_2(O) P (CH_2)_2 P(O) Ph_2$	* : Raman

For some complexes, one supplementary infrared band is noted at 455 cm^{-1} for $(\eta^5-C_5H_5)_3UF$[43], at 432 cm^{-1} for $(\eta^5-C_5H_5)_3UF$. $Yb(\eta^5-C_5H_5)_3$[44] and at 423 cm^{-1} for $(\eta^5-C_5H_5)_3UFU(\eta^5-C_5H_5)_3$[44]. This band has been assigned to the metal-fluorine-metal bridge[18].

THE ACTINIDE INDENYL COMPOUNDS.

If we consider indene, C_9H_8, as a derivative of cyclopentadiene, one can expect that the indenyl anion will form complexes with most metals and, especially, with actinide elements. The first indenyl actinide compounds were prepared by Laubereau and coworkers[45], they prepared tris-indenyl uranium- and tris-indenyl thorium chlorides. Several types of bonding are possible between the actinide cation and the C_5 ring of the indenyl anion. Some of them are shown in figure 1.

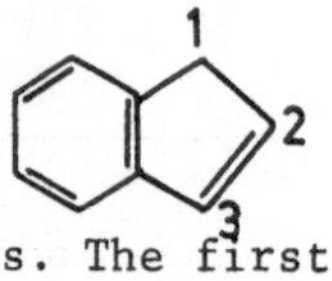

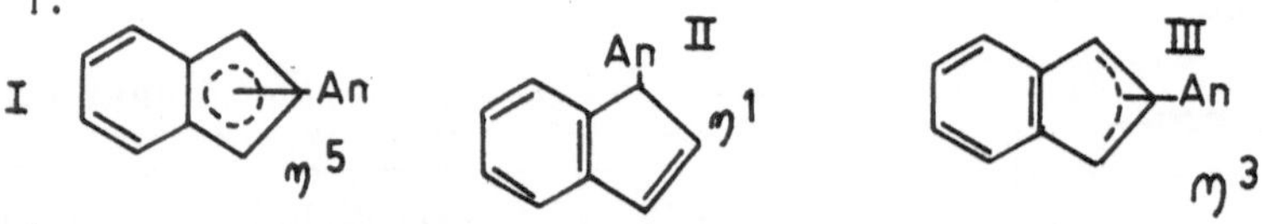

Figure 1.

Ionic bonding represents the complete transfer of one electron from the metal atom to the C_5 ring. The indenyl derivative of potassium is an example of compounds possessing this type of bonding[41]. The C_5 ring of the indenyl might also be bonded uniformly to the metal atom through a covalent bond as shown in structure I. This type of bonding could involve the overlap of the π-electron cloud of the C_5 ring with partially filled orbital of the metal atom. Bis-indenyl iron, bis-indenyl cobalt ... have this type of bonding[46]. The metal atom might also be bonded to only one carbon of the C_5 ring (structure II). This type of bonding is referred to as sigma or η^1-bonding and has been proposed for some mercury compounds[47,48]. It is also possible to form a η^3-allylic bond between the metal atom and three carbons as shown in structure III. In addition we cannot fully exclude the possibility of bonding between the C_6 ring of indenyl and the cation. In general, vibrational spectroscopy provides a straightforward method of distinguishing among these types of bonding.

The indenyl anion.

The indenyl anion is easily obtained by reduction of freshly distilled indene by powdered alkali metal. This reduction leads to important modifications of the vibrational spectrum of the indene molecule. For indenyl compounds, $C_9H_7^-$ being planar in most known complexes, a model of C_{2v} symmetry can be assumed. Our planar model has 42 (3 x 16 - 6) modes of normal vibrations.

They are classified into $15A_1$, $6A_2$, $7B_1$ and $14B_2$. All normal vibrations belonging to the A_1, B_1 and B_2 irreducible representations are both infrared and Raman active. The A_2 representations are only Raman active. For the assignment of the infrared bands of the indenyl compounds, it is very useful to prepare deuterated derivatives. The method used for the deuteration of indene gives :

1.- The best region to investigate is certainly that of the C-H stretching vibrations ($4A_1$ and $3B_2$). In MeC_9H_7 (Me = Li, Na or K), the C-H vibrations are observed between 3100 and 3000 cm^{-1} (H/D ratio = 1,34) (Table IX). No ν_{C-H} aliphatic stretching band is observed in the 2800-3000 cm^{-1} region. The C-D band at 2190 cm^{-1} of deuterated indene, which corresponds to the C_1-H aliphatic vibration, disappears almost completely in $K(1,3-C_9H_5D_2)$, showing the aromatisation of the C_5 ring.

2.- In the 2800-1700 cm^{-1} region, many weak or very weak bands are noted, chiefly between 2000 and 1700 cm^{-1}. They have been assigned to combination bands[14].

Table IX. Isotopic shifts for different metal-ring systems[26].

Ratio of normal vibrational modes	$\dfrac{Fe(C_5H_5)_2}{Fe(C_5D_5)_2}$	$\dfrac{C_6H_6}{C_6D_6}$	$\dfrac{Cr(C_6H_6)_2}{Cr(C_6D_6)_2}$	$\dfrac{(C_5H_5)NiNO}{(C_5D_5)NiNO}$
C-H stretching	1,31-1,34	1,30-1,34	1,34	1,33-1,36
C-C stretching	1,09-1,10	1,05	1,12	1,05-1,10
C-H bending ($\parallel$)	1,20-1,28	1,30	1,25	1,30
C-H bending ($\perp$)	1,22-1,30	1,20-1,28	1,19-1,23	1,24-1,33
Metal-ring tilting	1,01-1,09	–	1,03	1,05
Metal-ring stretching	1,04	–	1,08	1,08

3.- The aromatic character of the C_5 ring of indenyl anion is also demonstrated by the absence of the C=C stretching vibration bands at 1650-1600 cm^{-1}. This band is generally rather weak in the infrared spectrum but should give rise to a line of strong Raman intensity. Such a line is, however, not experimentaly observed.

4.- The infrared bands in the vicinity of 1460, 1390 and 1320 cm^{-1} in $Me(C_9H_7)$ correspond to C-C vibrations of both the C_5 and C_6 ring and belong to $6A_1$ and $4B_2$ representations. In 1,3-deuterated indenyl potassium salt, the 1460 and 1320 cm^{-1} bands are practically unaffected whereas the intensity of the 1390 cm^{-1} band is considerably reduced. The shifted band of the deuterated compound is expected around 1300 cm^{-1} (table IX) but seems masked by another band.

5.- The 1050-700 cm^{-1} region presents very intense bands which can be assigned to C-H and C-C-C bending modes ($5A_1 + 7B_1 + 7B_2$). A very intense band appears at 528 cm^{-1} in 1,3-deuterated indenyl potassium salt. The shift ratio (1,35-1,40) indicates that this band can be assigned to the out-of-plane C_1-D and C_3-D vibrations (table IX). The corresponding C-H vibration takes place around 750 cm^{-1}.

6.- A strong band around 440 cm^{-1}, sometimes a doublet, is noted in all the spectra of indenyl compounds and in indene. This band disappears in tetrahydroindenyl derivatives. It seems probable that this absorption belongs to the out-of-plane normal vibration of the C_6 ring.

7.- The band around 390 cm^{-1} in indenyl alkali metal salts is shifted to 350 cm^{-1} in $K(1,3-C_9H_5D_2)$. It can be assigned to C-C deformation vibration of the C_5 ring, because it is also present as a doublet (392-382 cm^{-1}) in the indene and is shifted to 355 cm^{-1} and 338 cm^{-1} in deuterated indene (H/D = 1.1).

The indenyl compounds.

Following the method of local symmetry[13], it is possible to consider the spectrum of any indenyl organometallic compound as the superposition of the spectrum of one C_9H_7 ligand of C_{2v} point symmetry and of a particular molecular skeleton. If we consider a compound such as $U(\eta^5-C_9H_7)_3$[49], we may suppose that the skeleton symmetry is C_{3v}. On figure 2, the spectra of KC_9H_7 and $U(C_9H_7)_3$ are compared.

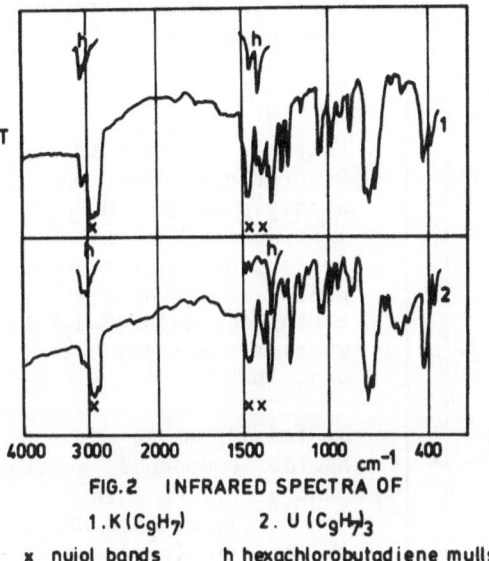

FIG.2   INFRARED SPECTRA OF

1. K (C₉H₇)   2. U (C₉H₇)₃

x nujol bands   h hexachlorobutadiene mulls

We can see that in the region 4000-300 cm^{-1} the two spectra are nearly identical, except around 550 cm^{-1} where an additional broad band appears in the spectrum of tris-indenyl uranium. Note, however, that the very strong band observed at 748 cm^{-1} in KC_9H_7 spectrum is shifted to 766 cm^{-1} in $U(\eta^5 C_9H_7)_3$. The analysis of the $K(1,3-C_9H_5D_2)$ spectrum has shown that this band can be assigned to a C-H deformation of the C_5 ring. In table X we have reported the position of the corresponding band for well known η^5-indenyl complexes.

Table X. Position of δ(C-H) in η^5-indenyl complexes[45] (cm^{-1}).

Complex	Frequency
$Ti(\eta^5-C_9H_7)_2Cl_2$	860
$Zr(\eta^5-C_9H_7)_2Cl_2$	840
$Fe(\eta^5-C_9H_7)_2Cl$	808
$Ru(\eta^5-C_9H_7)_2Cl$	808

Thus this vibration appears in the case of indenyl actinide compounds intermediate between the ones of the predominantly ionic compound KC_9H_7 and those of genuine η^5-indenyl complexes. A higher degree of ionic character could be deduced for the actinide-to-ligand bond when compared with the known η^5-indenyl complexes[50].

The general features of the $C_9H_7^-$ spectrum are unaffected by the coordination of some Lewis bases as tetrahydrofuran or cyclohexylisonitrile to the actinide ion (figure 3). Naturally some additional bands occur, especially in the 2800-3000 cm^{-1} region. The assignment of the vibration appearing below 500 cm^{-1} has not been intensively discussed for organo-actinide compounds. The same assumption can be made for indenyl compounds as for cyclopentadienyl compounds (Table VI). In the case of $U(\eta^5-C_5H_5)_3$, we expect eight infrared bands in the low frequency region, but only three bands are observed (Table VII).

Reliable assignments are under these conditions hardly possible. More information is available for organometallic complexes of tetravalent actinides, $An(\eta^5-C_9H_7)_3X$ where X = halide, CH_3, OCH_3, BH_4 ...[51].

In tris(indenyl) actinide halide compounds, a strong band at 277 cm^{-1} in $(\eta^5-C_9H_7)_3ThCl$ and at 267 cm^{-1} in $(\eta^5-1.C_2H_5-C_9H_6)_3ThCl$ is shifted to 175 cm^{-1} in $(\eta^5-C_9H_7)_3ThBr$ and disappears in tris-indenyl methyl or methoxy actinide. It certainly belongs to the Th-halide vibration. Raman spectra[51]

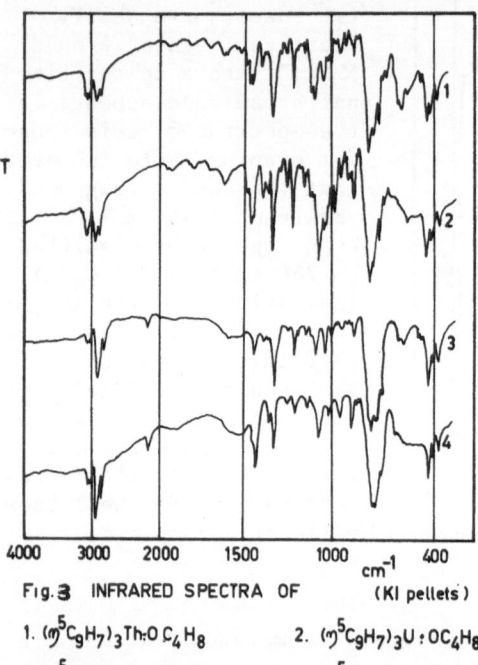

Fig.3   INFRARED SPECTRA OF   (KI pellets)

1. $(\eta^5C_9H_7)_3Th\!:\!O\,C_4H_8$ 2. $(\eta^5C_9H_7)_3U\!:\!OC_4H_8$

3. $(\eta^5C_9H_7)_3Th\!\cdot\!CNC_6H_{11}$ 4. $(\eta^5C_9H_7)_3U\!:\!CNC_6H_{11}$

present a weak line at the same frequencies. For the uranium counter-parts, $(\eta^5-C_9H_7)_3UCl$ and $(\eta^5-1.C_2H_5-C_9H_6)_3UCl$, this band is noted respectively at 268 and 267 cm^{-1}. The U-Br vibration is observed at 183 cm^{-1}. Two further infrared bands are present around 220 and 110 cm^{-1} in all spectra of indenylactinide compounds. Also, the corresponding Raman lines are always observed. These bands and lines are probably due to metal-to-ring vibrations (Table VIII).

1. The isonitrile adducts.

It is well established[52] that the isonitrile ligand shows a strong donor and a minor acceptor. From the position of the $\nu C \equiv N$ band in the vibrational spectrum, we can estimate the strength of the metal-to-ligand bond. Figure 4 presents the variation of the $\nu C \equiv N$ frequency for some well-known organometallic compounds. In every spectrum, the frequency is shifted towards higher wave numbers relative to the free ligand band (2136 cm^{-1}). Fischer and Fischer[55] have assumed that the carbon-to-metal bond has pure σ-donor character. To explain the position of the $C \equiv N$

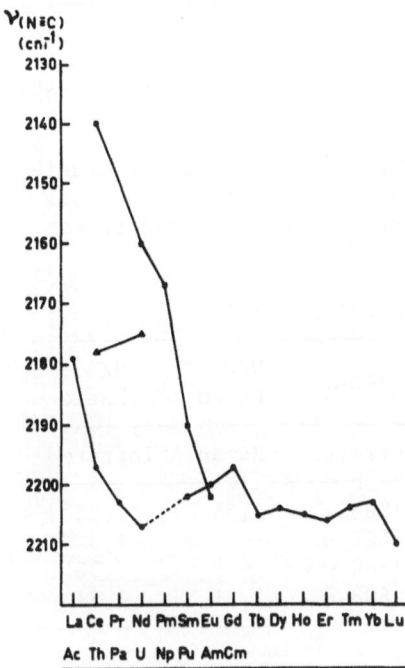

Fig. 4 INFRARED FREQUENCY OF C≡N- IN

□ $(\eta^5-C_5H_5)_3An\cdot C\equiv NC_6H_{11}$
(Ref. 53, 54).

△ $(\eta^5-C_9H_7)_3An\cdot C\equiv NC_6H_{11}$

● $(\eta^5-C_5H_5)_3Ln\cdot C\equiv NC_6H_{11}$
(Ref. 18, 36, 54, 55).

band in indenyl compounds, we must remember that this position depends on the electron density of the actinide element. It has been assumed[52] that a positive charge adjacent to the carbon atom of the isonitrile compounds enhances the importance of one of the hybrids : $R : N^+ ::: \bar{C} :$, increasing the NC bond order and, therefore, raising the N≡ C stretching frequency. As the shifted frequencies are observed at 2175 cm⁻¹ for $(\eta^5-C_9H_7)_3U\cdot CNC_6H_{11}$ and at 2178 cm⁻¹ for $(\eta^5-C_9H_7)_3\overline{3}$ Th·CNC₆H₁₁, there is a surprising difference between these values and those reported for cyclopentadienyl-actinide counterparts[53,54]. Another experimental observation must be pointed out : $(\eta^5-C_5H_5)_3U\cdot THF$ and $(\eta^5-C_5H_5)_3Th\cdot THF$ lose their tetrahydrofuran at 100°C in vacuo, while $(\eta^5-C_9H_7)_3U\cdot THF$ and $(\eta^5-C_9H_7)_3Th\cdot THF$ seem at least stable to 220°C (10^{-4} torr) attesting a relative strengthening of the An-THF bond in the presence of indenyl ligands.

2. The tetrahydroborate compounds.

Tetrahydroborate, BH_4^-, can adopt several types of bonding in organometallic compounds, R_3AnBH_4 or $R_2An(BH_4)_2$:

1.- Ionic bonding involves the transfer of one electron from the metal to the BH_4 group. Several investigations on alkali metal tetrahydroborates have confirmed the tetrahedral structure of the tetrahydroborate in these compounds and assignments of infrared and Raman spectra have accordingly been success for assuming T_d symmetry[35, 56].

2.- As shown by structural studies, the BH_4^- ion can bind to a metal via three-center (Me-H-B) bonds. Geometries having one, two or three hydrogen bridge bonds have been observed[56]. However we must point out that essentially bidentate and tridentate coordination configurations have been identified for covalent compounds.

3.- The tetrahydroborate ligand is also known to link metal atoms in a polymer structure[35,56].

Vibrational spectroscopy provides a good method for studies of these types of bonding. For the ensuing discussion we shall consider the characteristic vibrations of the complexed tetrahydroborate as given in Table XI.

Table XI. BH_4^- and BD_4^- vibration frequencies (cm^{-1}).

$(\eta^5-C_9H_7)_3ThBH_4$		$(\eta^5-C_9H_7)_3ThBH_4$		H/D Ratio	H/D Ratio
Raman	Infrared	Raman	Infrared	Raman	Infrared
2489 vw	2485 m	1851 w	1863 m	1,34	1,33
2224 vw	2225 w	–	1662 w	–	1,34
–	2155 w	1568 vw	1560 w	–	1,38
1180 vvw	1177 s	906 vw	898 m	1,31	1,31

1.- From the number of infrared bands (Table XI), it is clear that the local symmetry of the BH_4^- group is lower than T_d since for a purely ionic compound only two infrared bands (T_2) are expected.

2.- On the other hand, $(\eta^5-C_9H_7)_3ThBH_4$ exhibits an infrared band at 2485 cm^{-1} which is shifted to 1863 cm^{-1} for $(\eta^5-C_9H_7)_3ThBD_4$ both in the solid state and in benzene solution[59]. The same behaviour is observed in the Raman spectra. This band can be attributed to a boron-terminal hydrogen vibration.

$$(\eta^5-C_9H_7)_3Th \quad \begin{array}{c} H \\ H - B - H \\ H \end{array}$$

bridge terminal

These two features, along with the mass spectra and cryoscopic measurements[59] in benzene, permit us to discard the hypothesis of a purely ionic bond between thorium and the BH_4^- group as well as to discount the existence of polymeric species. Marks and coworkers[56,61] have assumed that for a monodentate structure, terminal boron-hydrogen bonds must give rise to bands above 2300 cm^{-1}. The hydrogen bridge, in a monodentate structure, is associated with a vibrational frequency in the 2100-1600 cm^{-1} region, the exact position depending of the molecular environment[35,60].

In the region between 1150 cm^{-1} and 1000 cm^{-1}, monodentate tetrahydroborates are expected to exhibit two bands (A_1 and E in
C_{3v} symmetry) belonging to the BH_3 deformation vibrations. It is
evident that the absorption spectra of tris(indenyl) tetrahydroborate as well as its cyclopentadienyl counterparts[56-58,61] do
not correspond to this latter scheme.

The last possibilities of bonding are bidentate and tridentate coordination configurations as represented above. The spectra (figure 5) appear to be in good agreement with a tridentate
structure as observed for (η^5-C_5H_5)$_3$AnBH$_4$[35,56,60] or (η^5-C_5H_5)$_2$
An(BH$_4$)$_2$[58]. The infrared spectra in benzene or THF solution also
agree with this hypothesis[56,59,61].

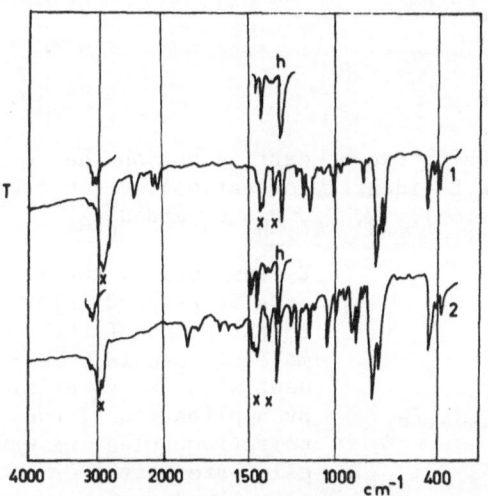

Fig.5   INFRARED   SPECTRA   OF

1. (η^5C$_9$H$_7$)$_3$Th BH$_4$ 2. (η^5C$_9$H$_7$)$_3$Th BD$_4$

x : nujol bands   h : hexachlorobutadiene mulls

The A_1 stretching vibration corresponding to
boron-hydrogen termi-
nal stretching vibra-
tions is both Raman and
infrared active. The
doublet 2225-2155 cm^{-1}
is shifted to 1662-1560
cm^{-1}, but in the Raman
spectra the correspon-
ding lines are too weak
to be seen even with
the strongest amplifi-
cation. This doublet
corresponds to B-H$_{bridge}$
vibrations.

The band most characte-
ristic of bidentate
ligation lies certainly
around 1400 cm^{-1} [56], and
has been assigned to
the bridge expansion
mode. This band is not
observed in the infra-
red spectrum of (η^5-C_9H_7)$_3$ThBH$_4$ but one might argue that it could
be masked by the strong 1337 cm^{-1} band. In view of the probable
H/D ratio (table XI) this hypothetical band should be shifted to
circa 1000 cm^{-1} in the deuterated compound. In this region, however, no additional band is observed.

The strong 1177 cm^{-1} infrared band which is shifted to 898
cm^{-1} upon deuteration belongs to the trihapto bridge
deformation.

Vibrational spectroscopy allows us to present a reasonably
unambiguous distinction between the different possibilities of

BH_4^- -interactions with actinides.

It seems from our present study that a tridentate structure prevails for this series of compounds both in solution and in the solid state.

THE ACTINIDE CYCLOOCTATETRAENYL COMPOUNDS.

The study of the planar cyclooctatetraenyl dianion, $C_8H_8^{2-}$, and the symmetric bis(cyclooctatetraenyl) compounds, $An(\eta^8-C_8H_8)_2$, is especially interesting because the symmetry of these compounds, the point group D_{8h} being one of the highest symmetries known in coordination chemistry. The corresponding D_{8h} character table can be found in the literature[62,63].

The cyclooctatetraenyl dianion.

The symmetry operations which can be carried out on the cyclooctatetraenyl dianion are E (identity operation), C_8, C_4, C_2, C_2', C_2'', i (operation of inversion), S_8, S_4, σ_h, σ_v and σ_d.

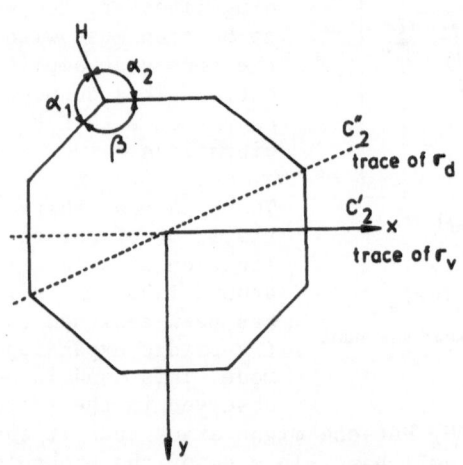

FIG. 6 : CONVENTIONS FOR $C_8H_8^{2-}$

By convention, the z axis is perpendicular to the plane of the molecule and is coincident with the C_8 axis. By application of the selection rules, we can calculate that four normal modes of vibration are infrared active (1 A_{2u} + 3 E_{1u}), seven are Raman active (2 A_{1g} + 1 E_{1g} + 4 E_{2g}) and fifteen are totally inactive. No transition can be active in both infrared and Raman spectroscopy because a center of symmetry exists in the cyclooctatetraenyl dianion (mutual exclusion rule). Table XII presents the results obtained by application of the method of internal coordinates. This method allows us to assign respective normal vibrations to the different irreducible representation of $C_8H_8^{2-}$.

As predicted by the theory, the infrared spectrum of $C_8H_8^{2-}$ is very simple[13,64] with four normal vibrations : C-H stretching (E_{1u}, 2994 cm^{-1}), C-C stretching (E_{1u}, 1431 cm^{-1}), C-H bending ($\parallel$)(E_{1u}, 880 cm^{-1}) and C-H bending ($\perp$)(A_{2u}, 684 cm^{-1}).

Table XII. Symmetry of normal vibration modes of $C_8H_8^{2-}$ [65].

D_{8h}	Total vibr.	Stret. vibr.	Deform. vibr.	ν_{C-H}	ν_{C-C}	δ_{C-H} ($\parallel$)	δ_{C-H} ($\perp$)	δ_{CCC} ($\perp$)
A_{1g}	2	2		1	1	1(a)		
A_{2g}	1		1			1		
B_{1g}	2	1	1			1	1	
B_{2g}	2	1	1	1		1		
E_{1g}	1		1				1(b)	1(b)
E_{2g}	4	2	2	1	1	2		
E_{3g}	2	2					1	1
A_{1u}								1(a)
A_{2u}	1	1					1	
B_{1u}	2	2					1	1
B_{2u}								
E_{1u}	3	2	1	1	1	2(b)		
E_{2u}	2		2				1	1
E_{3u}	4	2	2	1	1	2		

$\parallel$ and $\perp$ define the direction of the nuclear displacements with respect to the C_8H_8 ring plane.
(a) This vibrational mode is redundant : the 2 A_{1g} are stretching modes.
(b) 1 E_{1g} and 1 E_{1u} are redundant.

The cyclooctatetraenyl compounds.

The bis(cyclooctatetraenyl) complexes of the actinides with D_{8h} geometry have 58 modes of vibration which belong to a reducible representation giving rise to the following decomposition :

$$\Gamma_{vibr.} = 4A_{1g} + 1A_{2g} + 2B_{1g} + 4B_{2g} + 5E_{1g} + 6E_{2g} + 6E_{3g}$$
$$+ 2A_{1u} + 4A_{2u} + 4B_{1u} + 2B_{2u} + 6E_{1u} + 6E_{2u} + 6E_{3u}$$

Fifteen of the normal modes ($4A_{1g}$ + $5E_{1g}$ + $6E_{2g}$) are active in Raman spectroscopy, ten ($4A_{2u}$ + $6E_{1u}$) are infrared active and the thirty three remaining fundamentals are totally inactive.

It is possible to assign respective normal vibrations to the different irreducible representations of this type of molecule[65]. Table XIII presents the results.

Table XIII. Infrared and Raman active normal vibrations for D_{8h} symmetry[65].

Irreducible representation	Approximate type of vibration
4 A_{2u} (i.r.active)	C-H stretching C-C stretching C-H bending ($\perp$) asymmetric ring metal ring stretching
6 E_{1u} (i.r.active)	C-H stretching C-C stretching C-H bending ($\perp$) or CCC bending ($\perp$)[†] C-H bending ($\parallel$) ring metal ring bending asymmetric ring metal tilting
4 A_{1g} (Raman active)	C-H stretching C-C stretching C-H bending ($\perp$) symmetric ring metal ring stretching
5 E_{1g} (Raman active)	C-H stretching C-C stretching C-H bending ($\parallel$) C-H bending ($\perp$)[†] or CCC bending ($\perp$) symmetric ring metal ring tilting
6 E_{2g} (Raman active)	C-H stretching C-C stretching C-H bending ($\parallel$) C-H bending ($\parallel$) C-H bending ($\perp$) CCC bending ($\perp$)

[†] Our qualitative analysis cannot distinguish between these two possibilities. For an unambiguous interpretation, it would be necessary to use the GF matrix method.

The infrared spectra of $An(\eta^8-C_8H_8)_2$ systems present two absorption bands in the 3100-2900 cm^{-1}. They have been assigned to C-H stretching vibrations (in-phase and out-of-phase). These two modes of vibration are also observed for C-C stretching motions (Table XIV).

The C-H bending ($\parallel$) vibration is observed at the same frequency for $C_8H_8^{2-}$ and for $An(\eta^8-C_8H_8)_2$ while the C-H bending($\perp$) vibration is greatly influenced by coordination : it undergoes a shift of about 60 cm^{-1} to higher frequencies (table XIV). It seems reasonable to interpret this observation in the following way : the out-of-plane movement might become more difficult when the metal orbitals interact with the orbitals of the $C_8H_8^{2-}$ anion, the in-plane vibration being normally less sensitive to the coordination. The infrared spectra of $An(\eta^8-C_8H_8)_2$ are similar to those of $K \left[Ln(\eta^8-C_8H_8)_2 \right]^{71,72}$.

At low frequencies, three skeletal vibrations, illustrated in figure 7 are infrared active, $A_{2u} + 2E_{1u}$.

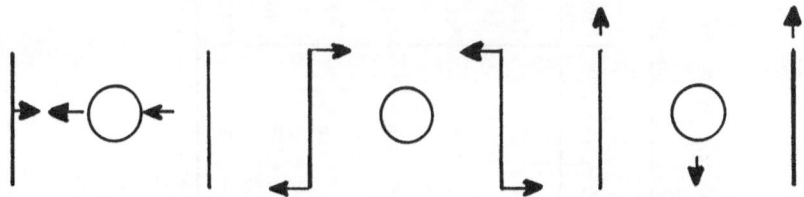

Fig. 7. IR active skeletal vibrations for $An(\eta^8-C_8H_8)_2$.

The asymmetric ring-metal-ring vibrations (E_{1u}) are always observed at the highest frequency in different sandwich compounds. It is possible to assign the 695 cm^{-1} band to such a vibration[67,70] in spite of its very high frequency compared to some other sandwich complexes[67]. The asymmetric ring-metal-ring stretching vibration is observed at 240-250 cm^{-1}. The remaining fundamental band (E_{1u}), ring-metal-ring bending, is very difficult to observe even at very low frequencies.

The Raman studies give information about the relative strengths of the metal-carbon bonds in the actinide compared to the lanthanide complexes[71,72]. The symmetrical ring-metal-ring stretch has A_{1g} symmetry (Table XIII). Since the metal does not move in this normal mode, the reduced mass of the metal is not important when we compare $\nu_{metal-carbon}$ in different homologous complexes (Table XV).

TABLE XIV INFRARED ASSIGNMENT OF An(C$_8$H$_8$)$_2$ FOR A D$_{8h}$ GEOMETRY

(hexachlorobutadiene or nujol mulls)   (frequencies: CM^{-1})

Th(η^{8}C$_8$H$_8$)$_2$ (66-68)	Pa(η^{8}C$_8$H$_8$)$_2$ (66,69)	U(η^{8}C$_8$H$_8$)$_2$ (67)	(70,71)	Np(η^{8}C$_8$H$_8$)$_2$ (70)	Irreducible representation	Assignment
3005 2920	3010 2930	3015 2920	— —	— —	A$_{2u}$ E$_{1u}$	C–H stretching
1430 1315	1450 1310	1470 1320	1470	1470	A$_{2u}$ E$_{1u}$	C–C stretching
895	895	900	890	890	E$_{1u}$	C–H bending (//)
775 742	775 745	777 746	780 740	740	A$_{2u}$ E$_{1u}$	C–H bending (⊥)• C–H bending (⊥) or C–C–C bending (⊥)
695	695	698	690	690	E$_{1u}$	asymmetric ring–metal–ring tilting
250	—	240	—	—	A$_{2u}$	asymmetric ring–metal–ring stretching

—— no recorded

Table XV. Raman spectra of some homologous complexes (cm^{-1}).

$Th(\eta^8-C_8H_8)_2$ (67,68)	$U(\eta^8-C_8H_8)_2$ (72)	$K[Ce(\eta^8-C_8H_8)_2]$ (72)	Mode
225 s	215 s	200 s	A_{1g}
242 s			E_{1g}
391 m	380 w	370 m	Ring distortion
755 s	750 m	750 m	Symmetry ring breathing

The A_{1g} mode of $An(\eta^8-C_8H_8)_2$ occurs at higher frequency than the corresponding mode for $K[Ce(\eta^8-C_8H_8)_2]$, the higher constant force implies stronger bonding.

The other assignments have been made by comparison with Raman assignments for $Fe(\eta^5-C_5H_5)_2$[73], $Cr(\eta^6-C_6H_6)_2^+$[74], $C_7H_7^+$[75] and $K[Ln(\eta^8-C_8H_8)_2]$[72].

THE ACTINIDE ALLYL COMPOUNDS.

Among the other organometallic compounds of the actinides, the allyl derivatives attract great interest from the point of view of vibrational spectroscopy. The allyl anion, $C_3H_5^-$, can form two types of bonds:

η^3 allyl compound η^1 allyl compound

Shobatake and Nakamoto[76] have studied the vibrational spectrum of allyl anion in η^3 configuration. Eighteen vibrations (3x8-6) originate in the allyl group. Thirteen assignments are given in Table XVI. The remaining five vibrations are the carbon-hydrogen stretching modes which appear as for other organometallic compounds between 3100 and 2900 cm^{-1}.

Comparison between η^3 and η^1 allyl compounds shows several differences in the vibrational spectra. The intensities of the C-H stretching bands as well as the wave numbers of the C-H deformation bands are different, but the most characteristic differences are observed in the range from 1650 to 1450 cm^{-1}. A monohaptoallyl geometry can be derived from the observation of an infrared band or a Raman line between 1575 and 1650 cm^{-1} corresponding to the $\nu(C=C)$[77]. Another band corresponding to the

Table XVI. Infrared and Raman spectra of allyl anion ion a η^3 configuration[76]. (Approximative frequencies in cm^{-1}).

Infrared	Raman	Assignment	Infrared	Raman	Assignment
1490	1491	ν (CCC)	942	953	$\delta_w(CH_2)$
1460	1462	δ (CH$_2$)	915	919	δ (CH) $(\perp)$
1382	1388	δ (CH$_2^2$)	765	773	$\delta_r(CH_2)^a$
1229	1229	$\delta_r(CH_2^2)$	765	764	$\delta_t(CH_2^2)^a$
1195	1198	δ^r(CH)	508	507	δ (CCC)
1022	1022	ν (CCC)			
998	1005	$\delta_t(CH_2)$			
964	971	$\delta_w(CH_2^2)$			

a. : overlapping bands corresponding to the rocking (δ_r) and the twisting (δ_t) of CH$_2$; δ_w : wagging.

ν(C-C) must be found at lower frequency, but is too weak to be detected.

In complexes with η^3-bonded allylic ligands, the corresponding band (asymmetric C-C stretching band) is normally found in the 1450-1550 cm^{-1} region. No infrared band can be observed in the range from 1650 to 1600 cm^{-1} but it is always better to record the Raman spectrum, the Raman line being much stronger than the corresponding infrared band.

The vibrational spectra of the known allyl compounds are in reasonable agreement with the general rule mentioned above. Table XVII presents some examples of η^1 and η^3 allyl compounds.

Table XVII. Vibrational spectra of some metal allyl compounds[77].

η^3 coordination		η^1 coordination	
Compound	ν(C-C) (cm^{-1})	Compound	ν(C=C) (cm^{-1})
$(\eta^5\text{-}C_5H_5)_2Ta(2\text{-}CH_3\text{-}C_3H_4)$	1440	$Mg(C_3H_5)_2$	1575
$(\eta^5\text{-}C_5H_5)_2Ti(2\text{-}CH_3\text{-}C_3H_4)$	1480	$(\eta^5\text{-}C_5H_5)_3U(2\text{-}CH_3\text{-}C_3H_4)$	1580
$(\eta^5\text{-}C_5H_5)_2Ti(C_3H_5)$	1509	$(\eta^5\text{-}C_5H_5)_2V(C_3H_5)$	1588
$Zr(C_3H_5)_4$	1515	$(\eta^5\text{-}C_5H_5)_2Zr(C_3H_5)_2$	1605
$(\eta^5\text{-}C_5H_5)_2Nb(C_8H_9)$	1540	$(\eta^5\text{-}C_5H_5)Ti(C_8H_9)$	1620

Table XVII (continued).

η^3 coordination		η^1 coordination	
Compound	$\nu(C \dot{=} C)$ (cm^{-1})	Compound	$\nu(C=C)$ (cm^{-1})
$U(C_3H_5)_4$	1550–1500	$(\eta^5\text{-}C_5H_5)_3U(C_3H_5)$	1630
		$(\eta^5\text{-}C_5H_5)_3Th(C_3H_5)$	1650

Nevertheless great care must be taken in the interpretation of the vibrational spectra of compounds containing both allyl and aromatic groups. The presence of an aromatic ring can give rise to a pair of infrared bands at 1600–1500 cm^{-1}. They are sometimes sharp but they may vary considerably in intensity, sometimes even being missing. Depending on the nature of the bonding, the bands may split and confusion can occur with $\nu(C-C)$ of some other groups such as η^1 allyl group.

For compounds with rather complicated structure such as $[(C_5H_5)_3U]_2 (C_8H_8)$[77] or $[(\eta^5\text{-}C_5H_5)_2 (\eta^5 : \eta^1\text{-}C_5H_4)Th]_2$[27,78] the results obtained from different techniques in addition to vibrational spectroscopy must necessarily be compared before proposing a geometry.

Nevertheless, infrared and Raman spectroscopy will continue to be important in elucidating the structure of complex actinide organometallic compounds.

ACKNOWLEDGMENTS.

My sincerest thanks are due to Prof. J. Fuger, Prof. R.D. Fischer and Prof. T. J. Marks who have contributed to this chapter with helpful suggestions and constructive criticism. It is a real pleasure to acknowledge the valuable help of Prof. L.R. Morss in revising the English manuscript.

REFERENCES.

1. I.R. Beattie, Chem. Soc. Rev. 4, 107–153 (1975).
2. G. Davidson, Organometal. Chem. Rev. Sect. A 8, 303–350 (1972).
3. E. Maslowsky, Jr., "Vibrational Spectra of Organometallic Compounds" John Wiley and Sons, New York (1975).
4. K. Nakamoto, "Characterization of Organometallic Compounds". M. Tsutsui Ed, Interscience New York, Part I,chapter 3 (1969).

5. K. Nakamoto, "Infrared Spectra of Inorganic and Coordination
    Compounds" John Wiley and Sons, New York (1970).
6. N.N. Greenwood, "Spectroscopic properties of Inorganic and
    Organometallic Compounds". Chem. Soc. London vol. 9
    (1976).
7. E.B. Wilson, J.C. Decieus and P.C. Cross, "Molecular Vibra-
    tions" Mc. Graw-Hill Book Company Ed. (1955).
8. M. Born and E. Wolf, "Principles of Optics" Pergamon Press,
    London (1959).
9. D.S. Schonland, "Molecular Symmetry" D. Van Nostrand and
    Company, London (1965).
10. F.A. Cotton, "Chemical Applications of Group Theory" Inter-
    science, New York (1963).
11. J. Thiele, Chem. Ber. 34, 68-71 (1901).
12. L.T. Reynolds and G. Wilkinson, J. Inorg. Nucl. Chem. 2,
    246-253 (1956).
13. H.P. Fritz, Adv. Organomet. Chem. 1, 239-316 (1964).
14. E. Maslowsky, Thesis, Illinois Institute of Technology (1969).
15. E. Gallinella, B. Fortunato and P. Mirone, J. Mol. Spectrosc.
    24, 345-362 (1967).
16. H.P. Fritz and L. Schäfer, Spectrochim. Acta 21, 211-212
    (1965).
17. B. Kanellakopulos, E. Dornberger and P. Baumgärtner, Inorg.
    Nucl. Chem. Letters 10, 155-160 (1974).
18. B. Kanellakopulos, Habilitationsschrift, University of Hei-
    delberg (1972).
19. E.O. Fischer and A. Treiber, Z. Naturforsch. 17b, 276-277
    (1962).
20. E.O. Fischer and Y. Hristidu, Idem 17b, 275-276 (1962).
21. E.R. Lippincott and R.D. Nelson, Spectrochim. Acta 10, 307-
    329 (1958).
22. F.A. Cotton and L.T. Reynolds, J. Am. Chem. Soc. 80, 269-
    273 (1958).
23. J.E. Bercaw, R.H. Marvich, L.G. Bell and H.H. Brintzinger,
    Idem 94, 1219-1238 (1972).
24. J.M. Manriquez, D.R. MacAlister, R.D. Sanner and J.E. Bercaw,
    Idem 98, 6733-6735 (1976).
25. J.M. Manriquez, P.J. Fagan and T.J. Marks, Idem 100, 3939-
    3941 (1978).
26. L. Hocks, unpublished results.
27. T.J. Marks and W.A. Wachter, J. Am. Chem. Soc. 98, 703-710
    (1976).
28. M. Tsutsui, N. Ely and A.E. Gebala, Inorg. Chem. 14, 78-81
    (1975).
29. T.J. Marks, A.M. Seyam and W.A. Wachter, Inorg. Synth. 16,
    147-151 (1976).
30. G.Brandi, M. Brunelli, G. Lugli and A. Mazzei, Inorg. Chim.
    Acta 7, 319-322 (1973).
31. A.E. Gebala and M. Tsutsui, Chem. Letters, 775-776 (1972).

32. Idem, J. Am. Chem. Soc. 95, 91-93 (1973).
33. T.J. Marks and A.M. Seyam, Idem 94, 6545-6546 (1972).
34. G.L. Ter Haar and M. Dubeck, Inorg. Chem. 3, 1648-1650 (1964).
35. T.J. Marks, J. Organometal. Chem. 158, 325-343 (1978) and
        references cited therein.
36. B. Kanellakopulos and K.W. Bagnall ; MTP Intern. Review of
        Science Series one 7, 299 (1972).
37. V.T. Aleksanyan and B.V. Loksin, J. Organometal. Chem. 131,
        113-120 (1977).
38. D.G. Kalina, T.J. Marks and W.A. Wachter, J. Am. Chem. Soc.
        99, 3877-3879 (1977).
39. V.T. Aleksanyan, G.K. Borisov, I.A. Gaburzova and G.G.
        Devyatykh, J. Organometal. Chem. 131, 251-255 (1977).
40. K.W. Bagnall, J. Edwards and A.G. Tempest, J. Chem. Soc.
        Dalton Trans. 295-298 (1978).
41. J. Goffart, unpublished results.
42. B. Kanellakopulos, private communication.
43. R.D. Fischer, R. von Ammon and B. Kanellakopulos, J. Organo-
        metal. Chem. 25, 123-137 (1970).
44. F. Baumgärtner, E.O. Fischer, H. Billich, E. Dornberger, B.
        Kanellakopulos, W. Roth and L. Stieglitz, Idem 22,
        C17-18 (1970).
45. P.G. Laubereau, L. Ganguly, J.H. Burns, B.M. Benjamin, J.L.
        Atwood and J. Selbin, Inorg. Chem. 10, 2274-2280
        (1971).
46. H.P. Fritz and C.G. Kreiter, J. Organometal. Chem. 4, 198-
        201 (1965).
47. E. Samuel and M. Bigorgne Idem 19, 9-15 (1969).
48. Idem 30, 235-242 (1971).
49. A recent crystallographic study has confirmed this hypothe-
        sis:
    J. Piret-Meunier and J. Goffart, to be published.
50. B. Samuel and R. Setton, J. Organometal. Chem. 4, 156-158
        (1965).
51. J. Goffart, J. Fuger, B. Gilbert, L. Hocks and G. Duyckaerts,
        Inorg. Nucl. Chem. Letters 11, 569-583 (1975).
    J. Goffart, B. Gilbert and G. Duyckaerts, Idem 13, 189-196
        (1977).
    J. Goffart and G. Duyckaerts, idem 14, 15-20 (1978).
52. F.A. Cotton and F. Zingales, J. Am. Chem. Soc. 83, 351-355
        (1961).
53. B. Kanellakopulos and C.M. Aderhold, XXIVth IUPAC Congress,
        Hamburg, 2-8 sept. 1973.
54. C.M. Aderhold, Dissertation, Universität Heidelberg (1974).
55. R.D. Fischer and H. Fischer, J. Organometal. Chem. 4, 412-
        414 (1965).
56. T.J. Marks and J.R. Kolb, Chem. Rev. 77, 263-293 (1977).
57. M.L. Anderson and L.R. Crisler, J. Organometal. Chem. 17,
        345-348 (1969).

58. P. Zanella, G. Depaoli, G. Bombieri, G. Zanotti and R. Rossi,
        Idem 142, C21-24 (1977).
59. J. Goffart, G. Michel, B.P. Gilbert and G. Duyckaerts, Inorg.
        Nucl. Chem. Letters 14, 393-403 (1978).
60. Y. Matsui and R.C. Taylor, J. Am. Chem. Soc. 90, 1363-1364
        (1968).
61. T.J. Marks, W.J. Kennely, J.R. Kolb and L.A. Shimp, Inorg.
        Chem. 11, 2540-2546 (1972).
62. L. Hocks, Thèse annexe, Université de Liège (1974).
63. A. Streitwieser, Jr., U. Müller-Westerhoff, G. Sonnichsen,
        F. Mares, D.G. Morrell, K.O. Hodgson and C.A. Harmon,
        J. Am. Chem. Soc. 95, 8644-8649 (1973).
64. H.P. Fritz and H. Keller, Z. Naturforsch. 16b, 231-234 (1961).
65. L. Hocks, J. Goffart, G. Duyckaerts and P. Teyssie, Spectro-
        chim. Acta 30A, 907-914 (1974).
66. J. Goffart and L. Hocks, unpublished results.
67. J. Goffart and L. Hocks, to be published.
68. J. Goffart, J. Fuger, B.P. Gilbert, B. Kanellakopulos and G.
        Duyckaerts, Inorg. Nucl. Chem. Letters 8, 403-412
        (1972).
69. J. Goffart, J. Fuger, D. Brown and G. Duyckaerts, Idem 10,
        413-419 (1974).
70. D.G. Karraker, J.A. Stone, E.R. Jones, Jr. and N. Edelstein,
        J. Am. Chem. Soc. 92, 4841-4845 (1970).
71. F. Mares, K.O. Hodgson and A. Streitwieser,Jr., J. Organo-
        metal. Chem. 24, C68-70 (1970).
72. K.O. Hodgson, F. Mares, D.F. Starks and A. Streitwieser, Jr.,
        J. Am. Chem. Soc. 95, 8650-8658 (1973).
73. T.V. Long, Jr. and F.R. Huege, Chem. Commun. 1239-1241 (1968).
74. H.P. Fritz, W. Lüttke, H. Stammreich and R. Forneris, Spectro-
        chim. Acta 17, 1068-1091 (1961).
75. R.D. Nelson, W.G. Fateley and E.R. Lippincott, J. Am. Chem. Soc
        Soc. 78, 4870-4872 (1956).
76. K. Shobatake and K. Nakamoto, Idem 92, 3339-3342 (1970).
77. G.R. Sienel, A.W. Spiegl and R.D. Fischer, J. Organometal.
        Chem. 160, 67-73 (1978) and references cited therein.
78. E.C. Baker, K.N. Raymond, T.J. Marks and W.A. Wachter,
        J. Am. Chem. Soc. 96, 7586-7588 (1974).

APPENDIX

Introductory Lectures

1. "Introduction to the f-Elements", R.D. Fischer.

2. "Introduction to Organometallic Chemistry", T.J. Marks.

Contributed Seminars

1. "Magnetic Properties of Trivalent Organometallic Uranium Compounds", Reinhardt Klenze, Clemens M. Aderhold, and Basil Kanellakopulos, Kernforschungszentrum Karlsruhe, Institut für Heisse Chemie and Fachbereich Physikalische Chemie der Universität Heidelberg, Federal Republic of Germany.

2. "Electrochemical Studies on Uranocene", Jared A. Butcher, Jr., James Q. Chambers, and Richard M. Pagni, Department of Chemistry, University of Tennessee, Knoxville, Tennessee 37916, USA.

3. "Organoscandium Complexes Containing Large Organic Rings", A. Westerhof, B.J. Roesink and H.J. de Liefde Meijer, Laboratorium voor Anorganische Scheikunde, Rijksuniversiteit, Nijenborgh 16, 9747 AG Groningen, The Netherlands.

4. "Bis(Pentamethylcyclopentadienyl)Actinide Alkyls: Facile Activation of Carbon Monoxide and Formation of Oxygen-Bonded Migratory Insertion Products", Juan M. Manriquez, Paul J. Fagan, and Tobin J. Marks, Department of Chemistry, Northwestern University, Evanston, IL 60201, USA; Cynthia Secaur Day and Victor W. Day, Department of Chemistry, University of Nebraska, Lincoln, NE 68588, USA.

5. "The Consistent Interpretation of the Spectroscopic and Magnetic Properties of Octahedral $5f^1$ Halide Compounds of Protactinium(IV), Uranium(V) and Neptunium(VI)", K. Eichberger and F. Lux, Institut für Radiochemie der Technischen Universität München, D-8046 Garching, Federal Republic of Germany.

6. "Synthesis and Spectroscopy of Indenyl Uranium(IV) Alkyls and Dialkyls", Afif M. Seyam and Ghaida' Ala' Eddein, Department of Chemistry, University of Jordan, Amman, Jordan.

7. "New Synthetic Routes in Organoactinide Chemistry", J. C. Green, Inorganic Chemistry Laboratory, South Parks Road, Oxford, Great Britain.

8. "Recent Advances in Uranium(III) Chemistry", David C. Moody, Los Alamos Scientific Laboratory, University of California, Los Alamos, New Mexico 87545, USA.

9. "Transferred Hyperfine Interaction Between the Rare-Earth Ions and Neighbouring Nuclei of Diamagnetic Ions in Simple Compounds Studied By NMR.", Rolf Nevald, Laboratory of Electrophysics, The Technical University of Denmark, DK-2800 Denmark.

10. "An Isotope Effect in ^{1}H-NMR and Magnetic Susceptibility of Organometallic Uranium Compounds", R. v. Ammon, C. M. Aderhold and B. Kanellakopulos, Kernforschungszentrum Karlsruhe, Federal Republic of Germany.

11. "Uranium (V) and Uranium (VI) Alkoxides: Preparation, Properties, and Photochemistry", Steven S. Miller, Tobin J. Marks, and Eric Weitz, Department of Chemistry, Northwestern University, Evanston, IL 60201, USA.

12. "Nonaqueous Reductive Lanthanide Chemistry", W. J. Evans, S. C. Engerer, A. C. Neville and A. L. Wayda, Department of Chemistry, University of Chicago, Chicago, IL 60637, USA.

13. "Studies on Uranium(V) and Uranium(VI) Alkoxides: Syntheses, ^{13}C NMR, and Magnetic Properties", P. Gary Eller and P. J. Vergamini, Los Alamos Scientific Laboratory, University of Claifornia, Los Alamos, New Mexico, 87545, USA.

14. "Reexamination of the Electronic Structure of U(NCS)$_8$(N(C$_2$H$_5$)$_4$)$_4$", Edgar Soulie and Hubert Marquet-Ellis, Division de Chimie, Service de Chimie Physique, Centre D'Etudes Nucleaires De Saclay, B. P. N$^{\circ}$ 2 - 91190 - Gif Sur Yvette, France.

15. "Anionic Tetra π -Allyl Complexes of Lanthanide Elements.", S. Poggio, M. Brunelli, U. Pedretti, G. Lugli, Snamprogetti-Diris-S. Donato Milanese, Milan, Italy.

16. "Photoionization of f Electrons With UV Radiation", Russell G. Egdell, Inorganic Chemistry Department, South Parks Road, Oxford, OX1 3QR, Great Britain.

17. "Magnetic Investigations on Highly Symmetric Octathiocyanato Complexes of Tetravalent Actinides", Arno H. Stollenwerk, Reinhardt Klenze, and Basil Kanellakopulos, Kernforschungszentrum Karlsruhe, Institut für Heisse Chemie and Fachbereich Physikalische Chemie der Universität Heidelberg, Federal Republic of Gemany.

18. "Some NMR Spectroscopic Results on Substituted Uranocenes and Thorocenes A. Streitwieser, Jr., Department of Chemistry, University of California, Berkeley, California 94770, USA.

19. "Non-Aqueous Chemistry of Uranium Pentafluoride", P. Gary Eller and G. W. Halstead, Los Alamos Scientific Laboratory, University of California, Los Alamos, New Mexico 87545 USA.

20. "Preparation and Structure of Novel Uranium-Organyls, $(\eta^5\text{-}C_5H_5)_3U(IV)XL$, with Trigonal-Bipyramidal Coordination", E. Klähne, J. Kopf and R. D. Fischer, Institut für Anorganische und Angewandte Chemie der Universität Hamburg, Martin-Luther-King-Platz 6, D-2000 Hamburg, 13, Federal Republic of Germany.

21. "Structural Aspects of Organoactininide Complexes", Victor W. Day, Sara H. Vollmer and Cynthia S. Day, Department of Chemistry, University of Nebraska, Lincoln, Nebraska 68588, USA; Tobin J. Marks, Richard D. Ernst, William J. Kennelly, Juan M. Manriquez and Paul J. Fagan, Department of Chemistry, Northwestern University, Evanston, IL 60201, USA; and Josef Takats and Anita L. Arduini, Department of Chemistry, University of Alberta, Edmonton, Alberta, Canada.

INDEX

absorption spectra of uranocenes 167
actinide
  alkyls, steric crowding and thermal stability 242,243
  catalysts 389
  complexes, lability of 221,222
  cyanides 239
  ions, ligand field splitting in 221
  tetraalkyls 242
  tetrabenzyls 241
activation barrier for hydride exchange 347
  for rotations 356
addition, oxidative 133
adducts with metal carbonyl compounds 90
adducts with metal nitrosyl compounds 90
adiabatic transitions 433
α-elimination 101,102
AIR$_3$ 385

alkylation reactions catalyzed by uranium complexes 390
alkyl lanthanide anions 241,242
alkyl lanthanide iodides 241
alkyls, actinide 113
  bis(indenyl) 126
  bis(pentamethylcyclopentadienyl) 117
  carbon monoxide, insertion of 138
  homoleptic 128
  hydrogenolysis 136
  metal 262,263
  mono(pentamethylcyclopentadienyl) 127
  protonolysis 136
  thermal stability 132
  thermolysis 131
  tris(cyclopentadienyl) 113
  tris(indenyl) 117
alkyl-μ-chloro bridges 208